A FIRST COURSE IN DYNAMICS

with a Panorama of Recent Developments

The theory of dynamical systems is a major mathematical discipline closely intertwined with all main areas of mathematics. It has greatly stimulated research in many sciences and given rise to the vast new area variously called applied dynamics, nonlinear science, or chaos theory. This introduction for senior undergraduate and beginning graduate students of mathematics, physics, and engineering combines mathematical rigor with copious examples of important applications. It covers the central topological and probabilistic notions in dynamics ranging from Newtonian mechanics to coding theory. Readers need not be familiar with manifolds or measure theory; the only prerequisite is a basic undergraduate analysis course.

The authors begin by describing the wide array of scientific and mathematical questions that dynamics can address. They then use a progression of examples to present the concepts and tools for describing asymptotic behavior in dynamical systems, gradually increasing the level of complexity. In the final chapters the panorama introduces modern developments and applications of dynamics, discussing logistic maps, hyperbolic dynamics, strange attractors, twist maps, closed geodesics, and applications to number theory.

Boris Hasselblatt is Professor of Mathematics at Tufts University.

Anatole Katok is Raymond N. Shibley Professor of Mathematics and Director of the Center for Dynamical Systems at The Pennsylvania State University.

A FIRST COURSE IN DYNAMICS

with a Panorama of Recent Developments

BORIS HASSELBLATT
Tufts University

ANATOLE KATOK
The Pennsylvania State University

CAMBRIDGE
UNIVERSITY PRESS

CAMBRIDGE UNIVERSITY PRESS
Cambridge, New York, Melbourne, Madrid, Cape Town, Singapore,
São Paulo, Delhi, Dubai, Tokyo

Cambridge University Press
32 Avenue of the Americas, New York, NY 10013-2473, USA

www.cambridge.org
Information on this title: www.cambridge.org/9780521587501

First published 2003

A catalog record for this publication is available from the British Library

Library of Congress Cataloging in Publication data

Hasselblatt, Boris.
    A first course in dynamics : with a panorama of recent developments /
Boris Hasselblatt, Anatole Katok.
        p.   cm.
    Includes bibliographical references and index.
    ISBN 0-521-58304-7 – ISBN 0-521-58750-6 (pb.)
        1. Differentiable dynamical systems.   I. Katok, A. B.   II. Title.
QA614.8 .H38   2002
514′.74–dc21   2002019246

ISBN 978-0-521-58304-6 Hardback
ISBN 978-0-521-58750-1 Paperback

Transferred to digital printing 2010

Contents

Preface

This book provides a self-contained introductory course on dynamical systems for advanced undergraduate students as well as a selection of recent developments in dynamical systems that serve to illustrate applications and refinements of the ideas from this course. The parts differ fundamentally in pedagogical approach but are closely interrelated. Either part can stand on its own; the course is complete without the panorama, and the panorama does not require this specific course as background. Scientists and engineers may use this book by picking and choosing, from both the panorama and the course text. Errata and other pertinent information can be found by visiting the first author's web page.

Introduction. The book begins with an introduction to pique interest in dynamics and to present samples of what scientific and mathematical problems dynamics can address. It adds motivation to the course but it is not a required part of it.

The Course. The undergraduate course assumes only knowledge about linear maps and eigenvalues, multivariable differential calculus, and Riemann integration with proofs. Some background is developed in Chapter 9 and the Appendix. Occasionally somewhat more involved portions of the text are set off with this different font to emphasize that the course will remain self-contained when these are omitted. These portions do not assume any more prior knowledge. Dynamics provides the concepts and tools to describe and understand complex long-term behavior in systems that evolve in time. The course accordingly develops these ideas in a gradual progression toward ever greater complexity, with proofs. Both topological and statistical points of view are developed. We know of no other text that makes both accessible at the undergraduate level.

Panorama. The panorama of dynamical systems assumes slightly stronger mathematical background in some places, but this is balanced by a more relaxed standard of proof that serves to outline and explain further developments carefully without carrying all of them out. It provides applications of the ideas in the course and connects them to topics of current interest, including ample references in the text.

Further Reading. The most natural continuation of the course presented here and of some subjects in the panorama is our book, *Introduction to the Modern Theory of Dynamical Systems* (Cambridge University Press, 1995), which also provides some reading to complement this course. We offer reading suggestions at the end of the book.

Acknowledgments. Many of the figures were produced by Boris Katok, Serge Ferleger, Roland Gunesch, Ilie Ugarcovici, and Alistair Windsor. Figure 4.4.3 was kindly provided by Sebastian van Strien, Figure 5.2.1 is due to Daniel Keesing, and Figure 13.3.2 was created by Mattias Lindkvist. The exposition of this book benefited from the Mathematics Advanced Study Semesters at The Pennsylvania State University, where early drafts were tested and many exercises developed in the fall of 1996. Thanks are also due to the Center for Dynamical Systems at The Pennsylvania State University for substantial financial support for almost all the time of collaboration. It has been a privilege and distinct pleasure to work with our editor at Cambridge University Press, Lauren Cowles. She combined patience and prodding ideally, and she procured very helpful evaluation of the text that determined the course of our work in the last year.

Finally, most thanks are due to Kathleen Hasselblatt and Svetlana Katok for their support and unlimited patience.

CHAPTER 1

Introduction

This chapter is a prelude to this book. It first describes in general terms what the discipline of dynamical systems is about. The following sections contain a large number of examples. Some of the problems treated later in the book appear here for the first time.

1.1 DYNAMICS

What is a dynamical system? It is dynamical, something happens, something changes over time. How do things change in nature? Galileo Galilei and Isaac Newton were key players in a revolution whose central tenet is *Nature obeys unchanging laws that mathematics can describe*. Things behave and evolve in a way determined by fixed rules. The prehistory of dynamics as we know it is the development of the laws of mechanics, the pursuit of exact science, and the full development of classical and celestial mechanics. The Newtonian revolution lies in the fact that the principles of nature can be expressed in terms of mathematics, and physical events can be predicted and designed with mathematical certainty. After mechanics, electricity, magnetism, and thermodynamics, other natural sciences followed suit, and in the social sciences quantitative deterministic descriptions also have taken a hold.

1.1.1 Determinism Versus Predictability

The key word is determinism: Nature obeys unchanging laws. The regularity of celestial motions has been the primary example of order in nature forever:

> God said, let there be lights in the firmament of the heavens to divide the day from the night and let them be for signs and for seasons and for days and years.

The successes of classical and especially celestial mechanics in the eighteenth and nineteenth centuries were seemingly unlimited, and Pierre Simon de Laplace felt justified in saying (in the opening passage he added to his 1812 *Philosophical Essay on Probabilities*):

> We ought then to consider the present state of the universe as the effects of its previous state and as the cause of that which is to follow. An intelligence that, at a given instant, could comprehend all the forces by which nature is animated and the respective situation of the beings that make it up, if moreover it were vast enough to submit these data to analysis, would encompass in the same formula the movements of the greatest bodies of the universe and those of the lightest atoms. For such an intelligence nothing would be uncertain, and the future, like the past, would be open to its eyes.[1]

The enthusiasm in this 1812 overture is understandable, and this forceful description of determinism is a good anchor for an understanding of one of the basic aspects of dynamical systems. Moreover, the titanic life's work of Laplace in celestial mechanics earned him the right to make such bold pronouncements. There are some problems with this statement, however, and a central mission of dynamical systems and of this book is to explore the relation between determinism and predictability, which Laplace's statement misses. The history of the modern theory of dynamical systems begins with Henri Jules Poincaré in the late nineteenth century. Almost 100 years after Laplace he wrote a summary rejoinder:

> If we could know exactly the laws of nature and the situation of the universe at the initial instant, we should be able to predict exactly the situation of this same universe at a subsequent instant. But even when the natural laws should have no further secret for us, we could know the initial situation only *approximately*. If that permits us to foresee the subsequent situation *with the same degree of approximation*, this is all we require, we say the phenomenon has been predicted, that it is ruled by laws. But this is not always the case; it may happen that slight differences in the initial conditions produce very great differences in the final phenomena; a slight error in the former would make an enormous error in the latter. Prediction becomes impossible and we have the fortuitous phenomenon.[2]

His insights led to the point of view that underlies the study of dynamics as it is practiced now and as we present it in this book: The study of long-term asymptotic behavior, and especially that of its qualitative aspects, requires direct methods that do not rely on prior explicit calculation of solutions. And in addition to the qualitative (geometric) study of a dynamical system, probabilistic phenomena play a role.

A major motivation for the study of dynamical systems is their pervasive importance in dealing with the world around us. Many systems evolve continuously in time, such as those in mechanics, but there are also systems that naturally evolve in discrete steps. We presently describe models of, for example, butterfly populations, that are clocked by natural cycles. Butterflies live in the summer, and

---

[1] Pierre Simon marquis de Laplace, *Philosophical Essay on Probabilities*, translated from the fifth French edition of 1925 by Andrew I. Dale, Springer-Verlag, New York, 1995, p. 2.

[2] Henri Jules Poincaré, Science et méthode, Section IV.II., Flammarion 1908; see *The Foundations of Science; Science and Hypothesis, The Value of science, Science and Method*, translated by George Bruce Halsted, The Science Press, Lancaster, PA, 1946, pp. 397f; *The Value of Science: Essential Writings of Henri Poincaré*, edited by Stephen Jay Gould, Modern Library, 2001.

we discuss laws describing how next summer's population size is determined by that of this summer. There are also ways of studying a continuous-time system by making it look like a discrete-time system. For example, one might check on the moon's position precisely every 24 hours. Or one could keep track of where it rises any given day. Therefore we allow dynamical systems to evolve in discrete steps, where the same rule is applied repeatedly to the result of the previous step.

This is important for another reason. Such stepwise processes do not only occur in the world around us, but also in our minds. This happens whenever we go through repeated steps on our way to the elusive perfect solution. Applied to such procedures, dynamics provides insights and methods that are useful in analysis. We show in this book that important facts in analysis are consequences of dynamical facts, even of some rather simple ones: The Contraction Principle (Proposition 2.2.8, Proposition 2.2.10, Proposition 2.6.10) gives the Inverse-Function Theorem 9.2.2 and the Implicit-Function Theorem 9.2.3. The power of dynamics in situations of this kind has to do with the fact that various problems can be approached with an iterative procedure of successive approximation by improved guesses at an answer. Dynamics naturally provides the means to understand where such a procedure leads.

1.1.2 Dynamics in Analysis

Whenever you use a systematic procedure to improve a guess at a solution you are likely to have found a way of using dynamics to solve your problem exactly. To begin to appreciate the power of this approach it is important to understand that the iterative processes dynamics can handle are not at all required to operate on numbers only. They may manipulate quite complex classes of objects: numbers, points in Euclidean space, curves, functions, sequences, mappings, and so on. The possibilities are endless, and dynamics can handle them all. We use iteration schemes on functions in Section 9.4, mappings in Section 9.2.1 and sequences in Section 9.5. The beauty of these applications lies in the elegance, power, and simplicity of the solutions and insights they provide.

1.1.3 Dynamics in Mathematics

The preceding list touches only on a portion of the utility of dynamical systems in understanding mathematical structures. There are others, where insights into certain patterns in some branches of mathematics are most easily obtained by perceiving that underlying the structure in question is something of a dynamical nature that can readily be analyzed or, sometimes, has been analyzed already. This is a range of applications of dynamical ideas that is exciting because it often involves phenomena of a rich subtlety and variety. Here the beauty of applying dynamical systems lies in the variety of behaviors, the surprising discovery of order in bewildering complexity, and in the coherence between different areas of mathematics that one may discover. A little later in this introductory chapter we give some simple examples of such situations.

■ EXERCISES

In these exercises you are asked to use a calculator to play with some simple iterative procedures. These are not random samples, and we return to several of these in due course. In each exercise you are given a function f as well as a number x_0. The assignment is to consider the sequence defined recursively by the given initial value and the rule $x_{n+1} = f(x_n)$. Compute enough terms to describe what happens in the long run. If the sequence converges, note the limit and endeavor to determine a closed expression for it. Note the number of steps you needed to compute to see the pattern or to get a good approximation of the limit.

■ **Exercise 1.1.1** $f(x) = \sqrt{2 + x}$, $x_0 = 1$.

■ **Exercise 1.1.2** $f(x) = \sin x$, $x_0 = 1$. Use the *degree* setting on your calculator – this means that (in radians) we actually compute $f(x) = \sin(\pi x/180)$.

■ **Exercise 1.1.3** $f(x) = \sin x$, $x_0 = 1$. Use the *radian* setting here and forever after.

■ **Exercise 1.1.4** $f(x) = \cos x$, $x_0 = 1$.

■ **Exercise 1.1.5**

$$f(x) = \frac{x \sin x + \cos x}{1 + \sin x}, \quad x_0 = 3/4.$$

■ **Exercise 1.1.6** $f(x) = \{10x\} = 10x - \lfloor 10x \rfloor$ (fractional part), $x_0 = \sqrt{1/2}$.

■ **Exercise 1.1.7** $f(x) = \{2x\}$, $x_0 = \sqrt{1/2}$.

■ **Exercise 1.1.8**

$$f(x) = \frac{5 + x^2}{2x}, \quad x_0 = 2.$$

■ **Exercise 1.1.9** $f(x) = x - \tan x$, $x_0 = 1$.

■ **Exercise 1.1.10** $f(x) = kx(1 - x)$, $x_0 = 1/2$, $k = 1/2, 1, 2, 3.1, 3.5, 3.83, 3.99, 4$.

■ **Exercise 1.1.11** $f(x) = x + e^{-x}$, $x_0 = 1$.

1.2 DYNAMICS IN NATURE

1.2.1 Antipodal Rabbits

Rabbits are not indigenous to Australia, but 24 wild European rabbits were introduced by one Thomas Austin near Geelong in Southern Victoria around 1860, with unfortunate consequences. Within a decade they were rampant across Victoria, and within 20 years millions had devastated the land, and a prize of £25,000 was advertized for a solution. By 1910 their descendants had spread across most of the continent. The ecological impact is deep and widespread and has been called a national tragedy. The annual cost to agriculture is estimated at AU$600 million. The unchecked growth of their population makes an interesting example of a dynamical system.

In modeling the development of this population we make a few choices. Its large size suggests to count it in millions, and when the number of rabbits is expressed as x million then x is not necessarily an integer. After all, the initial value is 0.000024 million rabbits. Therefore we measure the population by a real number x. As for time, in a mild climate rabbits – famously – reproduce continuously. (This is different for butterflies, say, whose existence and reproduction are strictly seasonal; see Section 1.2.9.) Therefore we are best served by taking the time variable to be a real number as well, t, say. Thus we are looking for ways of describing the number of rabbits as a function $x(t)$ of time.

To understand the dependence on time we look at what rabbits do: They eat and reproduce. Australia is large, so they can eat all they want, and during any given time period Δt a fixed percentage of the (female) population will give birth and a (smaller) percentage will die of old age (there are no natural enemies). Therefore the increment $x(t + \Delta t) - x(t)$ is proportional to $x(t)\Delta t$ (via the difference of birth and death rates). Taking a limit as $\Delta t \to 0$ we find that

$$(1.2.1) \qquad \frac{dx}{dt} = kx,$$

where k represents the (fixed) relative growth rate of the population. Alternatively, we sometimes write $\dot{x} = kx$, where the dot denotes differentiation with respect to t. By now you should recognize this model from your calculus class.

It is the unchanging environment (and biology) that gives rise to this unchanging evolution law and makes this a dynamical system of the kind we study. The differential equation (1.2.1), which relates x and its rate of change, is easy to solve: Separate variables (all x on the left, all t on the right) to get $(1/x)dx = k\,dt$ and integrate this with respect to t using substitution:

$$\log|x| = \int \frac{1}{x}\,dx = \int k\,dt = kt + C,$$

where log is the *natural* logarithm. Therefore, $|x(t)| = e^C e^{kt}$ with $e^C = |x(0)|$ and we find that

$$(1.2.2) \qquad x(t) = x(0)e^{kt}.$$

■ **Exercise 1.2.1** Justify the disappearance of the absolute value signs above.

■ **Exercise 1.2.2** If $x(0) = 3$ and $x(4) = 6$, find $x(2)$, $x(6)$, and $x(8)$.

1.2.2 The Leaning Rabbits of Pisa

In the year 1202, Leonardo of Pisa considered a more moderate question regarding rabbits, which we explore in Example 2.2.9 and Section 3.1.9. The main differences to the large-scale Australian model above are that the size of his urban yard limited him to small numbers of rabbits and that with such a small number the population growth does not happen continuously, but in relatively substantial discrete steps. Here is the problem as he posed it:[3]

How many pairs of rabbits can be bred from one pair in one year?

---

[3] Leonardo of Pisa: Liber abaci (1202), published in *Scritti di Leonardo Pisano*, Rome, B. Boncompagni, 1857; see p. 3 of Dirk J. Struik, *A Source Book in Mathematics 1200–1800*, Princeton, NJ, Princeton University Press, 1986.

A man has one pair of rabbits at a certain place entirely surrounded by a wall. We wish to know how many pairs can be bred from it in one year, if the nature of these rabbits is such that they breed every month one other pair and begin to breed in the second month after their birth. Let the first pair breed a pair in the first month, then duplicate it and there will be 2 pairs in a month. From these pairs one, namely the first, breeds a pair in the second month, and thus there are 3 pairs in the second month. From these in one month two will become pregnant, so that in the third month 2 pairs of rabbits will be born. Thus there are 5 pairs in this month. From these in the same month 3 will be pregnant, so that in the fourth month there will be 8 pairs ... [We have done this] by combining the first number with the second, hence 1 and 2, and the second with the third, and the third with the fourth ...

In other words, he came up with a sequence of numbers (of pairs of rabbits) governed by the recursion $b_{n+1} = b_n + b_{n-1}$ and chose starting values $b_0 = b_1 = 1$, so the sequence goes $1, 1, 2, 3, 5, 8, 13, \ldots$. Does this look familiar? (Hint: As the son of Bonaccio, Leonardo of Pisa was known as filius Bonacci or "son of good nature"; Fibonacci for short.) Here is a question that can be answered easily with a little bit of dynamics: How does his model compare with the continuous exponential-growth model above?

According to exponential growth one should expect that once the terms get large we always have $b_{n+1} \approx ab_n$ for some constant a independent of n. If we pretend that we have actual equality, then the recursion formula gives

$$a^2 b_n = ab_{n+1} = b_{n+2} = b_{n+1} + b_n = (a+1)b_n,$$

so we must have $a^2 = a + 1$. The quadratic formula then gives us the value of the growth constant a.

■ **Exercise 1.2.3** Calculate a.

Note, however, that we have only shown that *if* the growth is eventually exponential, then the growth constant is this a, not that the growth is eventually exponential. (If we *assume* the recursion $b_{n+1} = 1$ leads to exponential growth, we could come up with a growth parameter if we are quick enough to do it before getting a contradiction.) Dynamics provides us with tools that enable us to verify this property easily in various different ways (Example 2.2.9 and Section 3.1.9). In Proposition 3.1.11 we even convert this recursively defined sequence into closed form.

The value of this asymptotic ratio was known to Johannes Kepler. It is the golden mean or the divine proportion. In his 1619 book *Harmonices Mundi* he wrote (on page 273):

there is the ratio which is never fully expressed in numbers and cannot be demonstrated by numbers in any other way, except by a long series of numbers gradually approaching it: this ratio is called *divine*, when it is perfect, and it rules in various ways throughout the dodecahedral wedding. Accordingly, the following consonances begin to shadow forth that ratio: 1:2 and 2:3 and 3:5 and 5:8. For it exists most imperfectly in 1:2, more perfectly in 5:8, and still more perfectly if we add 5 and 8 to make 13 and take 8 as the numerator ....[4]

---

[4] Johannes Kepler, *Epitome of Copernican Astronomy & Harmonies of the World*, Amherst, NY, Prometheus Books, 1995.

We note in Example 15.2.5 that these Fibonacci ratios are the optimal rational approximations of the golden mean.

■ **Exercise 1.2.4** Express $1 + 1 + 2 + 3 + \cdots + b_n$ in terms of b_{n+2}.

1.2.3 Fine Dining

Once upon a time lobsters were so abundant in New England waters that they were poor man's food. It even happened that prisoners in Maine rioted to demand to be fed something other than lobsters for a change. Nowadays the haul is less abundant and lobsters have become associated with fine dining. One (optimistic?) model for the declining yields stipulates that the catch in any given year should turn out to be the average of the previous two years' catches.

Using again a_n for the number of lobsters caught in the year n, we can express this model by a simple recursion relation:

$$(1.2.3) \qquad a_{n+1} = a_{n-1}/2 + a_n/2.$$

As initial values one can take the Maine harvests of 1996 and 1997, which were 16,435 and 20,871 (metric) tons, respectively. This recursion is similar to the one for the Fibonacci numbers, but in this case no exponential growth is to be expected. One can see from the recursion that all future yields should be between the two initial data. Indeed, 1997 was a record year. In Proposition 3.1.13 we find a way of giving explicit formulas for future yields, that is, we give the yield in an arbitrary year n in a closed form as a function of n.

This situation as well as the Fibonacci rabbit problem are examples where time is measured in discrete steps. There are many other examples where this is natural. Such a scenario from population biology is discussed in Section 1.2.9. Other biological examples arise in genetics (gene frequency) or epidemiology. Social scientists use discrete-time models as well (commodity prices, rate of spread of a rumor, theories of learning that model the amount of information retained for a given time).

1.2.4 Turning Over a New Leaf

The word phyllotaxis comes from the words phyllo=leaf and taxis=order or arrangement. It refers to the way leaves are arranged on twigs, or other plant components on the next larger one. The seeds of a sunflower and of a pine cone are further examples. A beautiful description is given by Harold Scott Macdonald Coxeter in his *Introduction to Geometry*. That regular patterns often occur is familiar from sunflowers and pineapples.

In some species of trees the leaves on twigs are also arranged in regular patterns. The pattern varies by species. The simplest pattern is that of leaves alternating on opposite sides of the twig. It is called (1, 2)-phyllotaxis: Successive leaves are separated by a half-turn around the twig. The leaves of elms exhibit this pattern, as do hazel leaves.[5] Adjacent leaves may also have a (2/3) turn between them, which would be referred to as (2, 3)-phyllotaxis. Such is the case with beeches. Oak trees

---

[5] On which the first author of this book should be an expert!

show a $(3, 5)$-pattern, poplars a $(5, 8)$, and willows, $(8, 13)$-phyllotaxis. Of course, the pattern may not always be attained to full precision, and in some plants there are transitions between different patterns as they grow.

The diamond-shaped seeds of a sunflower are packed densely and regularly. One may perceive a spiral pattern in their arrangement, and, in fact, there are always two such patterns in opposite directions. The numbers of spirals in the two patterns are successive Fibonacci numbers. The seeds of a fir cone exhibit spirals as well, but on a cone rather than flat ones. These come in two families, whose numbers are again successive Fibonacci numbers.

Pineapples, too, exhibit spiral patterns, and, because their surface is composed of approximately hexagonal pieces, there are three possible directions in which one can perceive spirals. Accordingly, one may find 5, 8, and 13 spirals: 5 sloping up gently to the right, say, 8 sloping up to the left, and 13 sloping quite steeply right.

The observation and enjoyment of these beautiful patterns is not new. They were noticed systematically in the nineteenth century. But an explanation for why there are such patterns did not emerge particularly soon. In fact, the case is not entirely closed yet.

Here is a model that leads to an explanation of how phyllotaxis occurs. The basic growth process of this type consists of buds (primordia) of leaves or seeds growing out of a center and then moving away from it according to three rules proposed in 1868 by the self-taught botanist Wilhelm Friedrich Benedikt Hofmeister, while he was professor and director of the botanical garden in Heidelberg:

(1) New buds form at regular intervals, far from the old ones.
(2) Buds move radially from the center.
(3) The growth rate decreases as one moves outward.

A physical experiment designed to mimic these three *Hofmeister rules* produces spiral patterns of this Fibonacci type, so from these rules one should be able to infer that spiral patterns must occur. This has been done recently with methods of the kind that this book describes.[6]

Here is a description of how dynamics may help. To implement the Hofmeister rules we model the situation by a family of $N+1$ concentric circles of radius r^k $(k = 0, \ldots, N)$, where r stands for growth rate, and we put a bud on each circle. The angle (with respect to the origin) between one bud and the next is θ_k. Possible patterns are now parametrized by angles $(\theta_0, \ldots, \theta_N)$. This means that the "space of plants" is a *torus*; see Section 2.6.4. When a new bud appears on the unit circle, all other buds move outward one circle. The angle of the new bud depends on all previous angles, so we get a map sending old angles θ_k to new angles Θ_k by

$$\Theta_0 = f(\theta_0, \ldots, \theta_N), \quad \Theta_1 = \theta_0, \ldots, \Theta_N = \theta_{N-1}.$$

Now f has to be designed to reflect the first Hofmeister rule. One way to do this is to define a natural potential energy to reflect "repulsion" between buds and choosing

---

[6] Pau Atela, Christophe Golé, and Scott Hotton: A dynamical system for plant pattern formation: A rigorous analysis, *Journal of Nonlinear Science* **12** (2002), no. 6, pp. 641–676.

$f(\theta_0, \ldots, \theta_N)$ to be the minimum. A natural potential is

$$W(\Theta) = \sum_{k=0}^{N} U(\|r^k e^{i\theta_k} - e^{i\Theta}\|),$$

where $U(x) = 1/x^s$ for some $s > 0$. A simpler potential that gives the same qualitative behavior is $W(\Theta) = \max_{0 \leq k \leq N} U(\|r^k e^{i\theta_k} - e^{i\Theta}\|)$. With either choice one can show that regular spirals (that is, $\theta_0 = \cdots = \theta_N$) are attracting fixed points (Section 2.2.7) of this map. This means that spirals will appear naturally. A result of the analysis is furthermore that the Fibonacci numbers also must appear.

1.2.5 Variations on Exponential Growth

In the example of a rabbit population of Section 1.2.1 it is natural to expect a positive growth parameter k in the equation $\dot{x} = kx$. This coefficient, however, is the difference between rates of reproduction and death. For the people of some western societies, the reproduction rate has declined so much as to be lower than the death rate. The same model still applies, but with $k < 0$ the solution $x(t) = x(0)e^{kt}$ describes an exponentially shrinking population.

The same differential equation $\dot{x} = kx$ comes up in numerous simple models because it is the simplest differential equation in one variable.

Radioactive decay is a popular example: It is an experimental fact that of a particular radioactive substance a specific percentage will decay in a fixed time period. As before, this gives $\dot{x} = kx$ with $k < 0$. In this setting the constant k is often specified by the *half-life*, which is the time T such that $x(t + T) = x(t)/2$. Depending on the substance, this time period may be minute fractions of a second to thousands of years. This is important in regard to the disposal of radioactive waste, which often has a long half-life, or radioactive contamination. Biology laboratories use radioactive phosphorus as a marker, which has a half-life of a moderate number of days. A spill on the laboratory bench is usually covered with plexiglas for some two weeks, after which the radiation has sufficiently diminished. On the other hand, a positive effect of radioactive decay is the possibility of radioisotope dating, which can be used to assess the age of organic or geologic samples. Unlike in population biology, the exponential decay model of radioactivity needs no refinements to account for real data. It is an exact law of nature.

■ **Exercise 1.2.5** Express the half-life in terms of k, and vice versa.

The importance of the simple differential equation $\dot{x} = kx$ goes far beyond the collection of models in which it appears, however many of these there may be. It also comes up in the study of more complicated differential equations as an approximation that can illuminate some of the behavior in the more complicated setting. This approach of *linearization* is of great importance in dynamical systems.

1.2.6 The Doomsday Model

We now return to the problem of population growth. Actual population data show that the world population has grown with increasing rapidity. Therefore we should consider a modification of the basic model that takes into account the progress of civilization. Suppose that with the growth of the population the growing number of researchers manages to progressively decrease the death rate and increase fertility as well. Assuming, boldly, that these improvements make the relative rate of increase in population a small positive power x^ϵ of the present size x (rather than being constant k), we find that

$$\frac{dx}{dt} = x^{1+\epsilon}.$$

As before, this is easy to solve by separating variables:

$$t + C = \int x^{-1-\epsilon}\, dx = -x^{-\epsilon}/\epsilon$$

with $C = -x(0)^{-\epsilon}/\epsilon$, so $x(t) = (x(0)^{-\epsilon} - \epsilon t)^{-1/\epsilon}$, which becomes infinite for $t = 1/(\epsilon x(0)^\epsilon)$. Population explosion indeed!

As far as biology is concerned, this suggests refining our model. Clearly, our assumptions on the increasing growth rate were too generous (ultimately, resources are limited). As an example in differential equations this is instructive, however: There are reasonable-looking differential equations that have divergent solutions.

1.2.7 Predators

The reason rabbits have not over taken over the European continent is that there have always been predators around to kill rabbits. This has interesting effects on the population dynamics, because the populations of predators and their prey interact: A small number of rabbits decreases the predator population by starvation, which tends to increase the rabbit population. Thus one expects a stable equilibrium – or possibly oscillations.

Many models of interacting populations of predator and prey were proposed independently by Alfred Lotka and Vito Volterra. A simple one is the *Lotka–Volterra equation*:

$$\frac{dx}{dt} = a_1 x + c_1 xy$$

$$\frac{dy}{dt} = a_2 x + c_2 xy,$$

where $a_1, c_2 > 0$ and $a_2, c_1 < 0$, that is, x is the prey population, which would grow on its own ($a_1 > 0$) but is diminished by the predator ($c_1 < 0$), while y is the predator, which would starve if alone ($a_2 < 0$) and grows by feeding on its prey ($c_2 > 0$). Naturally, we take x and y positive. This model assumes that there is no delay between causes and effects due to the time of gestation or egg incubation. This is reasonable when the time scale of interest is not too short. Furthermore, choosing time continuously is most appropriate when generations overlap substantially. Populations with nonoverlapping generations will be treated shortly.

There is an equilibrium of species at $(a_2/c_2, a_1/c_1)$. Any other initial set of populations turns out to result in oscillations of the numbers of predator and prey. To see this, use the chain rule to verify that

$$E(x, y) := x^{-a_2} e^{-c_2 x} y^{a_1} e^{c_1 y}$$

is constant along orbits, that is, $(d/dt)E(x(t), y(t)) = 0$. This means that the solutions of the Lotka–Volterra equation must lie on the curves $E(x, y) = \text{const.}$ These curves are closed.

1.2.8 Horror Vacui

The Lotka–Volterra equation invites a brief digression to a physical system that shows a different kind of oscillatatory behavior. Its nonlinear oscillations have generated much interest, and the system has been important for some developments in dynamics.

The Dutch engineer Balthasar van der Pol at the Science Laboratory of the Philips Light Bulb Factory in Eindhoven modeled a vacuum tube circuit by the differential equation

$$\frac{d^2x}{dt^2} + \epsilon(x^2 - 1)\frac{dx}{dt} + x = 0,$$

which can be rewritten using $y = dx/dt$ as

$$\frac{dx}{dt} = y$$
$$\frac{dy}{dt} = \epsilon(1 - x^2)y - x.$$

If $\epsilon = 1$, the origin is a repeller (Definition 2.3.6). However, solutions do not grow indefinitely, because there is a periodic solution that circles around the origin. Indeed, for $\epsilon = 0$ there are only such solutions, and for $\epsilon = 1$ one of these circles persists in deformed shape, and all other solutions approach it ever more closely as $t \to +\infty$. The numerically computed picture in Figure 1.2.1 shows this clearly. The curve is called a *limit cycle*.

As an aside we mention that there is also the potential for horrifying complexity in a vacuum tube circuit. In 1927, van der Pol and J. van der Mark reported on experiments with a "relaxation oscillator" circuit built from a capacitor and a neon lamp (this is the nonlinear element) and a periodic driving voltage. (A driving voltage corresponds to putting a periodic term on the right-hand side of the van der Pol equation above.) They were interested in the fact that, in contrast to a linear oscillator (such as a violin string), which exhibits multiples of a base frequency, these oscillations were at "submultiples" of the basic frequency, that is, half that frequency, a third, and so on down to 1/40th, as the driving voltage increased. They

Figure 1.2.1. The van der Pol equation.

obtained these frequencies by listening "with a telephone coupled loosely in some way to the system" and reported that

> Often an irregular noise is heard in the telephone receivers before the frequency jumps to the next lower value. However, this is a subsidiary phenomenon, the main effect being the regular frequency demultiplication.

This irregular noise was one of the first experimental encounters with what was to become known as chaos, but the time was not ripe yet.[7]

1.2.9 The Other Butterfly Effect[8]

Population dynamics is naturally done in discrete-time steps when generations do not overlap. This was imposed somewhat artificially in the problem posed by Leonardo of Pisa (Section 1.2.2). For many populations this happens naturally, especially insects in temperate zones, including many crop and orchard pests. A pleasant example is a butterfly colony in an isolated location with a fairly constant seasonal cycle (unchanging rules and no external influence). There is no overlap at all between the current generation (this summer) and the next (next summer). We would like to know how the size of the population varies from summer to summer. There may be plenty of environmental factors that affect the population, but by assuming unchanging rules we ensure that next summer's population depends only on this summer's population, and this dependence is the same every year. That means that the only parameter in this model that varies at all is the population itself. Therefore, up to choosing some fixed constants, the evolution law will specify the population size next summer as a function of this summer's population only. The specific evolution law will result from modeling this situation according to our understanding of the biological processes involved.

1. Exponential growth. For instance, it is plausible that a larger population is likely to lay more eggs and produce a yet larger population next year, proportional, in fact, to the present population. Denoting the present population by x, we then find that next year's population is $f(x) = kx$ for some positive constant k, which is the average number of offspring per butterfly. If we denote the population in year i by x_i, we therefore find that $x_{i+1} = f(x_i) = kx_i$ and in particular that $x_1 = kx_0, x_2 = kx_1 = k^2 x_0$, and so on, that is, $x_i = k^i x_0$; the population grows exponentially. This looks much like the exponential–growth problem as we analyzed it in continuous time.

2. Competition. A problem familiar from public debate is sustainability, and the exponential growth model leads to large populations relatively rapidly. It is more realistic to take into account that a large population will run into problems with limited food supplies. This will, by way of malnutrition or starvation, reduce the

---

[7] B. van der Pol, J. van der Mark, Frequency demultiplication, *Nature* **120** (1927), 363–364.

[8] This is a reference to the statement of Edward Lorenz (see Section 13.3) that a butterfly may flutter by in Rio and thereby cause a typhoon in Tokyo a week later. Or maybe to butterfly ballots in the 2000 Florida election?

number of butterflies available for egg-laying when the time comes. A relatively small number of butterflies next year is the result.

The simplest rule that incorporates such more sensible qualitative properties is given by the formula $f(x) = k(1 - \alpha x)x$, where x is the present number of butterflies. This rule is the simplest because we have only adduced a linear correction to the growth rate k. In this correction α represents the rate at which fertility is reduced through competition. Alternatively, one can say that $1/\alpha$ is the maximal possible number of butterflies; that is, if there are $1/\alpha$ butterflies this year, then they will eat up all available food before getting a chance to lay their eggs; hence they will starve and there will be no butterflies next year. Thus, if again x_i denotes the butterfly population in the year i, starting with $i = 0$, then the evolution is given by $x_{i+1} = kx_i(1 - \alpha x_i) =: f(x_i)$. This is a deterministic mathematical model in which every future state (size of the butterfly colony) can be computed from this year's state. One drawback is that populations larger than $1/\alpha$ appear to give negative populations the next year, which could be avoided with a model such as $x_{i+1} = x_i e^{k(1-x_i)}$. But tractability makes the simpler model more popular, and it played a significant role in disseminating to scientists the important insight that simple models can have complicated long-term behaviors.[9]

One feature reminiscent of the exponential-growth model is that, for populations much smaller than the limit population, growth is indeed essentially exponential: If $\alpha x \ll 1$, then $1 - \alpha x \approx 1$ and thus $x_{i+1} \approx kx_i$; hence $x_n \approx k^n x_0$ – but only so long as the population stays small. This makes intuitive sense: The population is too small to suffer from competition for food, as a large population would.

Note that we made a slip in the previous paragraph: The sequence $x_n \approx k^n x_0$ grows exponentially *if* $k > 1$. If this is not the case, then the butterfly colony becomes extinct. An interesting interplay between reproduction rates and the carrying capacity influences the possibilities here.

3. Change of variable. To simplify the analysis of this system it is convenient to make a simple change of variable that eliminates the parameter α. We describe it with some care here, because changing variables is an important tool in dynamics.

Write the evolution law as $x' = kx(1 - \alpha x)$, where x is the population in one year and x' the population in the next year. If we rescale our units by writing $y = \alpha x$, then we must set

$$y' = \alpha x' = \alpha kx(1 - \alpha x) = ky(1 - y).$$

In other words, we now iterate the map $g(y) = ky(1 - y)$. The relationship between the maps f and g is given by $g(y) = h^{-1}(f(h(y)))$, where $h(y) = y/\alpha = x$. This can be read as "go from new variable to old, apply the old map, and then go to the new variable again."

[9] As its title shows, getting this message across was the aim of an influential article by Robert M. May, Simple Mathematical Models with Very Complicated Dynamics, *Nature* **261** (1976), 459–467. This article also established the quadratic model as the one to be studied. A good impression of the effects on various branches of biology is given by James Gleick, *Chaos, Making a New Science*, Viking Press, New York, 1987, pp. 78ff.

The effect of this change of variable is to normalize the competition factor α to 1. Since we never chose specific units to begin with, let's rename the variables and maps back to x and f.

4. The logistic equation. We have arrived at a model of this system that is represented by iterations of

$$f(x) = kx(1 - x).$$

This map f is called the *logistic map* (or logistic family, because there is a parameter), and the equation $x' = kx(1 - x)$ is called the logistic equation. The term logistic comes from the French *logistique*, which in turn derived from *logement*, the lodgment of soldiers. We also refer to this family of maps as the *quadratic family*. It was introduced in 1845 by the Belgian sociologist and mathematician Verhulst.[10]

From the brief discussion before the preceding subsection it appears that the case $k \leq 1$ results in inevitable extinction. This is indeed the case. For $k < 1$, this is clear because $kx(1 - x) < kx$, and for $k = 1$ it is not hard to verify either, although the population decay is not exponential in this case. By contrast, large values of k should be good for achieving a large population. Or maybe not. The problem is that too large a population will be succeeded by a less numerous generation. One would hope that the population settles to an agreeable size in due time, at which there is a balance between fertility and competition.

■ **Exercise 1.2.6** Prove that the case $k = 1$ results in extinction.

Note that, unlike in the simpler exponential growth model, we now refrained from writing down an explicit formula for x_n in terms of x_0. This formula is given by polynomials of order 2^n. Even if one were to manage to write them down for a reasonable n, the formulas would not be informative. We will, in due course, be able to say quite a bit about the behavior of this model. At the moment it makes sense to explore it a little to see what kind of behavior occurs. Whether the initial size of the population matters, we have not seen yet. But changing the parameter k certainly is likely to make a difference, or so one would hope, because it would be a sad model indeed that predicts certain extinction all the time. The reasonable range for k is from 0 to 4. [For $k > 4$, it predicts that a population size of $1/2$ is followed two years later by a negative population, which makes little biological sense. This suggests that a slightly more sophisticated (nonlinear) correction rule would be a good idea.]

5. Experiments. Increasing k should produce the possibility of a stable population, that is, to allow the species to avoid extinction. So let's start working out the model for some $k > 1$. A simpleminded choice would be $k = 2$, halfway between 0 and 4.

■ **Exercise 1.2.7** Starting with $x = 0.01$, iterate $2x(1 - x)$ until you discern a clear pattern.

---

[10] Pierre-François Verhulst, Récherches mathématiques sur la loi d'accroissement de la population, *Nouvelles Mémoires de l'Academie Royale des Sciences et Belles-Lettres de Bruxelles* **18** (1845), 1–38.

Starting from a small population, one obtains steady growth and eventually the population levels off at 1/2. This is precisely the behavior one should expect from a decent model. Note that steady states satisfy $x = 2x(1 - x)$, of which 0 and 1/2 are the only solutions.

■ **Exercise 1.2.8** Starting with $x = 0.01$ iterate $1.9x(1 - x)$ and $2.1x(1 - x)$ until you discern a clear pattern.

If k is a little less than 2, the phenomenon is rather the same, for k a little bigger it also goes that way, except for slightly overshooting the steady-state population.

■ **Exercise 1.2.9** Starting with $x = 0.01$, iterate $3x(1 - x)$ and $2.9x(1 - x)$ until you discern a clear pattern.

For $k = 3$, the ultimate behavior is about the same, but the way the population settles down is a little different. There are fairly substantial oscillations of too large and too small population that die out slowly, whereas for k near 2 there was only a hint of this behavior, and it died down fast. Nevertheless, an ultimate steady state still prevails.

■ **Exercise 1.2.10** Starting with $x = 0.01$, iterate $3.1x(1 - x)$ until you discern a clear pattern.

For $k = 3.1$, there are oscillations of too large and too small as before. They do get a little smaller, but this time they do not die down all the way. With a simple program one can iterate this for quite a while and see that no steady state is attained.

■ **Exercise 1.2.11** Starting with $x = 0.66$, iterate $3.1x(1 - x)$ until you discern a clear pattern.

In the previous experiment, there is the possibility that the oscillations die down so slowly that the numerics fail to notice. Therefore, as a control, we start the same iteration at the average of the two values. This should settle down if our diagnosis is correct. But it does not. We see oscillations that grow until their size is as it was before.

These oscillations are stable! This is our first population model that displays persistent behavior that is not monotonic. No matter at which size you start, the species with fertility 3.1 is just a little too fertile for its own good and keeps running into overpopulation every other year. Not by much, but forever.

Judging from the previous increments of k there seems only about $k = 4$ left, but to be safe let's first try something closer to 3 first. At least it is interesting to see whether these oscillations get bigger with increasing k. They should. And how big?

■ **Exercise 1.2.12** Starting with $x = 0.66$, iterate $3.45x(1 - x)$ and $3.5x(1 - x)$ until you discern a clear pattern.

The behavior is becoming more complicated around $k = 3.45$. Instead of the simple oscillation between two values, there is now a secondary dance around each of these values. The oscillations now involve four population sizes: "Big, small, big, Small" repeated in a 4-cycle. The period of oscillation has doubled.

■ **Exercise 1.2.13** Experiment in a manner as before with parameters slightly larger than 3.5.

A good numerical experimenter will see some pattern here for a while: After a rather slight parameter increase the period doubles again; there are now eight population sizes through which the model cycles relentlessly. A much more minute increment brings us to period 16, and it keeps getting more complicated by powers of two. This cascade of period doublings is complementary to what one sees in a linear oscillator such as a violin string or the column of air in wind instruments or organ pipes: There it is the frequency that has higher harmonics of double, triple, and quadruple the base frequency. Here the frequency is halved successively to give *subharmonics*, an inherently nonlinear phenomenon.

Does this period doubling continue until $k = 4$?

■ **Exercise 1.2.14** Starting with $x = .5$, iterate $3.83x(1 - x)$ until you discern a clear pattern.

When we look into $k = 3.83$ we find something rather different: There is a periodic pattern again, which we seem to have gotten used to. But the period is 3, not a power of 2. So this pattern appeared in an entirely different way. And we don't see the powers of 2, so these must have run their course somewhat earlier.

■ **Exercise 1.2.15** Try $k = 3.828$.

No obvious pattern here.

■ **Exercise 1.2.16** Try $k = 4$.

There is not much tranquility here either.

6. Outlook. In trying out a few parameter values in the simplest possible nonlinear population model we have encountered behavior that differs widely for different parameter values. Where the behavior is somewhat straightforward we do not have the means to explain how it evolves to such patterns: Why do periods double for a while? Where did the period-3 oscillation come from? And at the end, and in experiments with countless other values of the parameter you may choose to try, we see behavior we cannot even describe effectively for lack of words. At this stage there is little more we can say than that in those cases the numbers are all over the place.

We return to this model later (Section 2.5, Section 7.1.2, Section 7.4.3 and Chapter 11) to explain some of the basic mechanisms that cause these diverse behaviors in the quadratic family $f_k(x) = kx(1 - x)$. We do not provide an exhaustive analysis that covers all parameter values, but the dynamics of these maps is quite well understood. In this book we develop important concepts that are needed to describe the complex types of behavior one can see in this situation, and in many other important ones.

Already this purely numerical exploration carries several lessons. The first one is that *simple systems can exhibit complex long-term behavior*. Again, we arrived at this example from the linear one by making the most benign change possible. And immediately we ran into behavior so complex as to defy description. Therefore such complex behavior is likely to be rather more common than one would have thought.

The other lesson is that it is worth learning about ways of understanding, describing, and explaining such rich and complicated behavior. Indeed, the important insights we introduce in this book are centered on the study of systems where explicit computation is not feasible or useful. We see that even in the absence of perfectly calculated results for all time one can make precise and useful qualitative and quantitative statements about such dynamical systems. Part of the work is to develop concepts adequate for a description of phenomena of such complexity as we have begun to glimpse in this example. Our study of this particular example begins in Section 2.5, where we study the simple behaviors that occur for small parameter values. In Section 7.1.2 and Section 7.4.3 we look at large parameter values. For these the asymptotic behavior is most chaotic. In Chapter 11 we present some of the ideas used in understanding the intermediate parameter regime, where the transitions to maximal complexity occur.

As an interesting footnote we mention that the analogous population with continuous time (which is quite reasonable for other species) has none of this complexity (see Section 2.4.2).

1.2.10 A Flash of Inspiration

As another example of dynamics in nature we can take the flashing of fireflies. Possibly the earliest report of a remarkable phenomenon is from Sir Francis Drake's 1577 expedition:

> [o]ur general... sailed to a certaine little island to the southwards of Celebes,... throughly growen with wood of a large and high growth.... Among these trees night by night, through the whole land, did shew themselves an infinite swarme of fiery wormes flying in the ayre, whose bodies beeing no bigger than our common English flies, make such a shew of light, as if every twigge or tree had been a burning candle.[11]

A clearer description of what is so remarkable about these fireflies was given by Engelbert Kämpfer, a doctor from eastern Westphalia who made a 10-year voyage through Russia, Persia, southeast Asia, and Japan. On July 6, 1690, he traveled down the Chao Phraya (Meinam) River from Bangkok and observed:

> The glowworms (Cicindelae) represent another shew, which settle on some trees, like a fiery cloud, with this surprising circumstance, that a whole swarm of these

---

[11] Richard Hakluyt (pronounced Hack-loot), *A Selection of the Principal Voyages, Traffiques and Discoveries of the English Nation*, edited by Laurence Irving, Knopf, New York, 1926.

insects, having taken possession of one tree, and spread themselves over its branches, sometimes hide their light all at once, and a moment after make it appear again with the utmost regularity and exactness, as if they were in perpetual systole and diastole.[12]

So, in some locations large numbers of the right species of flashing fireflies in a bush or a collection of bushes synchronize, turning their arboreal home into a remarkable christmas-tree-like display. Or do they? This is such a striking phenomenon that for a long time reports of it had a status not entirely unlike that of tales of dragons and sea monsters. As late as 1938 they were not universally accepted among biologists. Only with increased affordability and speed of travel could doubters see it for themselves.[13] Once there was some belief that this really happens, it took many decades to develop an understanding of how this is possible. Early on it was supposed that some subtle and undetected external periodic influence caused this uniform behavior, but it is the fact that these fireflies naturally flash at close to the same rate combined with a tendency to harmonize with the neighbors that causes an entire colony to wind up in perfect synchrony.

An analogous situation much closer to home is the study of circadian rhythms, where periodic changes in our body (the sleep cycle) synchronize with the external cues of day and night. In the absence of clocks and other cues to the time of day, the human wake–sleep cycle reverts to its natural period, which is for most people slightly longer than 24 hours. Those external cues affect the system of neurons and hormones that make up our complicated internal oscillator and gently bring it up to speed. In this case, the rate at which the adjustment happens is fairly quick. Even the worst jet lag usually passes within a few days, that is, a few cycles.

These systems are instances of coupled oscillators, which also appear in numerous other guises. The earth–moon system can be viewed as such a system when one looks for an explanation why we always see the same side of the moon, that is, why the moon's rotation and revolution are synchronized. Here simple tidal friction is the coupling that has over eons forced the moon's rotation into lockstep with its revolution and will eventually synchronize the earth's rotation as well, so a day will be a month long – or a month a day long, making the moon a geostationary satellite. It is amusing to think that at some intermediate time the longer days may match up with our internal clocks, as if human evolution is slightly ahead of its time on this count.

---

[12] Engelbert Kämpfer, *The history of Japan*, edited by J. G. Scheuchzer, Scheuchzer, London, 1727. The translation is not too good. The German original apparently remained unpublished for centuries: "Einen zweiten sehr angenehmen Anblik geben die Lichtmücken (cicindelae), welche einige Bäume am Ufer mit einer Menge, wie eine brennende Wolke, beziehn. Es war mir besonders hiebei merkwürdig, daß die ganze Schaar dieser Vögel, so viel sich ihrer auf einem Baume verbunden, und durch alle Aeste desselben verbreitet haben, alle zugleich und in einem Augenblik ihr Licht verbergen und wieder von sich geben, und dies mit einer solchen Harmonie, als wenn der Baum selbst in einer beständigen Systole und Diastole begriffen wäre." (Geschichte und Beschreibung von Japan (1677–79). Internet Edition by Wolfgang Michel. In: Engelbert-Kaempfer-Forum, Kyushu University, 1999.).

[13] An account of this sea change is given by John Buck, Synchronous rhythmic flashing of fireflies, *Quarterly Review of Biology* **13**, no. 3 (September 1938), 301–314; II, *Quarterly Review of Biology* **63**, no. 3 (September 1988), 265–289. The articles include the quotes given here and many more reports of flashing fireflies from various continents.

We will look at systems made up of two simple oscillators in Section 4.4.5, where relatively simple considerations suggest that this kind of synchronization is somewhat typical.[14]

∎ EXERCISES

∎ **Exercise 1.2.17** In 1900, the global human population numbered 1.65 million, and in 1950 it was 2.52 billion. Use the exponential growth model (Equation 1.2.2) to predict the population in 1990, and to predict when the population will reach 6 billion. (The actual 1990 population was some 5.3 billion, and around July 1999 it reached 6 billion. Thus the growth of the world population is accelerating.)

∎ **Exercise 1.2.18** Denote by a_n the number of sequences of 0's and 1's of length n that do not have two consecutive 0's. Show that $a_{n+1} = a_n + a_{n-1}$. (Note that this is the same recursion as for the Fibonacci numbers, and that $a_1 = 2$ and $a_2 = 3$.)

∎ **Exercise 1.2.19** Show that any two successive Fibonacci numbers are relatively prime.

∎ **Exercise 1.2.20** Determine $\lim_{n\to\infty} a_n$ in (1.2.3) if $a_0 = 1$ and $a_1 = 0$.

1.3 DYNAMICS IN MATHEMATICS

In this section we collect a few examples of a range of mathematical activity where knowledge of dynamical systems provides novel insights.

1.3.1 Heroic Efforts with Babylonian Roots

Sometime before 250 A.D., in his textbook *Metrica*, Heron of Alexandria (often latinized to Hero of Alexandria) computed the area of a triangle with sides 7, 8, and 9 by first deriving the formula $\text{area}^2 = s(s - a)(s - b)(s - c)$, where a, b, c are the sides and $2s = a + b + c$. To compute the resulting square root of $12 \cdot 5 \cdot 4 \cdot 3 = 720$ he took the following approach, which may have been known to the Babylonians 2000 years before:

> Since [$z =$]720 has not its side rational [that is, 720 is not a perfect square], we can obtain its side within a very small difference as follows. Since the next succeeding square number is 729, which has [$x =$]27 for its side, divide 720 by 27. This gives [$y =$]$26\frac{2}{3}$. Add 27 to this, making $53\frac{2}{3}$, and take half of this or [$x' = \frac{1}{2}(x + y) =$]$26\frac{1}{2}\frac{1}{3}$. The side of 720 will therefore be very nearly $26\frac{1}{2}\frac{1}{3}$. . . If we desire to make the difference still smaller. . . we shall take [$x' = \frac{1}{2}(x + y) = 26\frac{1}{2}\frac{1}{3} = 26\frac{5}{6}$ instead of $x = 27$] and by proceeding in the same way we shall find that the resulting difference is much less. . .[15]

Heron used that, in order to find the square root of z, it suffices to find a square with area z; its sides have length $\sqrt{z}$. A geometric description of his procedure is

---

[14] We omit a full treatment of coupled linear oscillators. The subject of fireflies is treated by Renato Mirollo and Steven Strogatz, Synchronization of Pulse-Coupled Biological Oscillators, *SIAM Journal of Applied Mathematics* **50** no. 6 (1990), 1645–1662.

[15] Thomas L. Heath, *History of Greek Mathematics: From Aristarchus to Diophantus*, Dover, 1981, p. 324. This sequence of approximations also occurs in Babylonian texts; as related by Bartels van der Waerden: *Science awakening*, Oxford University Press, Oxford, 1961, p. 45, who gives a geometric interpretation on pp. 121ff. Some variant was known to Archimedes.

that as a first approximation of the desired square we take a rectangle with sides x and y, where x is an educated guess at the desired answer and $xy = z$. (If z is not as large as in Heron's example, one can simply take $x = 1$, $y = z$.) The procedure of producing from a rectangle of correct area another rectangle of the same area whose sides differ by less is to replace the sides x and y by taking one side to have the average length $(x + y)/2$ (arithmetic mean) and the other side to be such as to get the same area as before: $2xy/(x + y)$ (this is called the harmonic mean of x and y). The procedure can be written simply as repeated application of the function

$$(1.3.1) \qquad\qquad f(x, y) = \left(\frac{x + y}{2}, \frac{2xy}{x + y} \right)$$

of two variables starting with $(x_0, y_0) = (z, 1)$ [or $(x_0, y_0) = (27, 26\frac{2}{3})$] in Heron's example). Archimedes appears to have used a variant of this. One nice thing about this procedure is that the pairs of numbers obtained at every step lie on either side of the true answer (because $xy = z$ at every step), so one has explicit control over the accuracy. Even before starting the procedure Heron's initial guess bracketed the answer between $26\frac{2}{3}$ and 27.

■ **Exercise 1.3.1** To approximate $\sqrt{4}$, calculate the numbers (x_i, y_i) for $0 \le i \le 4$ using this method, starting with $(1, 4)$, and give their distance to 2.

■ **Exercise 1.3.2** Carry Heron's approximation of $\sqrt{720}$ one step further and use a calculator to determine the accuracy of that approximation.

■ **Exercise 1.3.3** Starting with initial guess 1, how many steps of this procedure are needed get a better approximation of $\sqrt{720}$ than Heron's initial guess of 27?

What happens after a few steps of this procedure is that the numbers x_n and y_n that one obtains are almost equal and therefore close to $\sqrt{z}$. With Heron's intelligent initial guess his first approximation was good enough ($26\frac{5}{6}$ is within .002% of $\sqrt{720}$), and he never seems to have carried out the repeated approximations he proposed. It is a remarkable method not only because it works, but because it works so quickly. But why does it work? And why does it work so quickly? And exactly how quickly does it work? These are questions we can answer with ease after our start in dynamical systems (Section 2.2.8).

1.3.2 The Search for Roots

Many problems asking for a specific numerical solution can be easily and profitably rephrased as looking for a solution of $f(x) = 0$ for some appropriate function f. We describe two well-known methods for addressing this question for functions of one variable.

1. Binary Search. There is a situation where we can be sure that a solution exists: The Intermediate-Value Theorem from calculus tells us that if $f : [a, b] \to \mathbb{R}$ is continuous and $f(a) < 0 < f(b)$ [or $f(b) < 0 < f(a)$, so we could say $f(a)f(b) < 0$], then there is some $c \in (a, b)$ such that $f(c) = 0$.

■ **Exercise 1.3.4** Show that this statement of the Intermediate-Value Theorem is equivalent to the standard formulation.

Knowing that a solution exists is, however, not quite the same as knowing the solution or at least having a fairly good idea where it is. Here is a simple reliable method for getting to a root.

Given that $f(a) < 0 < f(b)$, consider the midpoint $z = (a + b)/2$.

CASE 0: If $f(z) = 0$, we have found the root. Otherwise, there are two cases.

CASE 1: If $f(z) > 0$, replace the interval $[a, b]$ by the interval $[a, z]$, which is half as long and contains a root by the Intermediate-Value Theorem because $f(a) < 0 < f(z)$. Repeat the procedure on this interval.

CASE 2: If $f(z) < 0$, replace $[a, b]$ by $[z, b]$, which is also half as long, and apply the procedure here.

This binary search produces a sequence of nested intervals, cutting the length in half at every step. Each interval contains a root, so we obtain ever-better approximations and the limit of the right (or left) endpoints is a solution.

Note that this procedure is iterative, but it does not define a dynamical system. Not one that operates on numbers anyway. One could view it as a dynamical system operating on intervals on whose endpoints f does not have the same sign.

■ **Exercise 1.3.5** Carry out three steps of this procedure for $f(x) = x - \cos x$ on $[0, 1]$. Conclude with an approximate solution and its accuracy.

This method is reliable: It gives ever-better approximations to the solution at a guaranteed rate, and this rate is respectable and the error can be calculated. For example, nine steps give an error less than $(b - a)/1000$.

2. The Newton Method. The Newton Method (or Newton–Raphson Method) was devised as a solution of the same problem of finding zeros of functions. It is more flamboyant than the binary search: It is ingenious and can work rapidly, but it is not always reliable.

For this method we need to assume that the function f, whose zero we are to find, is differentiable, and, of course, that there is a zero someplace. One begins by making an educated guess x_0 at the solution. How to make this guess is up to the user and depends on the problem. A reasonable graph might help, or maybe the situation is such that the binary search can be applied. In the latter case a few steps give an excellent initial guess.

Newton's method endeavors to give you a vastly improved guess. If the function is linear, then your initial guess combined with the slope of the graph immediately gives the exact solution. Being differentiable, the function f is well approximated by tangent lines. Therefore the initial guess x_0 and the equation of the tangent line to the graph of f tell us the x-intercept of the tangent line. This is the improved guess. In terms of formulas the calculation amounts to

$$x_1 = F(x_0) := x_0 - \frac{f(x_0)}{f'(x_0)}.$$

■ **Exercise 1.3.6** Verify that this formula encodes the geometric description above.

■ **Exercise 1.3.7** Apply this method four times to $x^2 - 4 = 0$ with initial guess 1. Compare with Exercise 1.3.1. (Look also at Exercise 1.3.18 and Exercise 1.3.19.)

This simple procedure can be applied repeatedly by iterating F. It gives a sequence of hopefully ever-better guesses. In Section 2.2.8 we give a simple criterion to ensure that the method will succeed.

■ **Exercise 1.3.8** Several of the exercises in Section 1.1 are examples of Newton's method at work. Find the ones that are and give the equation whose solution they find.

Since this method defines a dynamical system, it has been studied as such. This is in large part because some initial choices provide situations where the asymptotic behavior is complicated. Especially when this is done with complex numbers, one can produce beautiful pictures by numerical calculations. An important development was an adaptation of this method to work on points in function spaces usually called the Kolmogorov–Arnol'd–Moser or KAM method, which provided a tool for one of the furthest advances in studying whether our solar system is stable. This is an outstanding example where knowledge about simple asymptotics of a dynamical system in an auxiliary space gives insight into another dynamical system.

1.3.3 Closed Geodesics

If an airplane pilot were to tie down the wheel[16] and had a lot of fuel, the plane would go around the earth all the way along a great circle, returning precisely to the starting point, and repeat. One could try the same with a vehicle on the surface, but some more attention would be required because of intervening mountains, oceans, rainforests, and such. The idealized model of this kind of activity is that of a particle moving freely on the surface of a sphere. Because there are no external forces (and no friction, we assume), such a particle moves at constant speed with no change of direction. It is quite clear that the particle always returns to the starting point periodically. So there are infinitely many ways of traveling (freely) in a periodic fashion.

What if your sphere is not as round and shiny as the perfect round sphere? It may be slightly dented, or maybe even badly deformed. One could adorn it with a mushroom-like appendage or even make it look like a barbell. Only, no tearing or glueing of the surface is allowed. And no crimping. A smooth but not ball-shaped "sphere." Now a freely moving particle has no obvious reason to automatically return home. Almost any way of deforming the sphere produces many nonperiodic motions. Here is a hard question: Are there still infinitely many ways, on a given deformed sphere, of moving freely and periodically?

---

[16] This means that the plane flies horizontally and straight, and the proper technical term would be "yoke" instead of "wheel".

One beautiful aspect of free particle motion is that the path of motion is always the shortest connection between any two points on it that are not too far apart. (Obviously, a closed path is not the shortest curve from a point to itself.) This is familiar for the round sphere when these paths are great circles, but it is universally true, and such paths are called *geodesics*. Therefore, the above question can also be asked in terms of geodesics: On any sphere, no matter how deformed, are there always infinitely many closed geodesics?

This is a question from geometry, and it was posed long ago. It was solved (not so long ago) by dynamicists using the theory of dynamical systems. We explain how geodesics are related to dynamics in Section 6.2.8 and outline an approach to this question in Section 14.5.

1.3.4 First Digits of the Powers of 2

As an illustration of the power of dynamics to discern patterns even of a subtle and intricate nature, consider the innocuous sequence of powers of 2. Here are the first 50 terms of this sequence:

2	2048	2097152	2147483648	2199023255552
4	4096	4194304	4294967296	4398046511104
8	8192	8388608	8589934592	8796093022208
16	16384	16777216	17179869184	17592186044416
32	32768	33554432	34359738368	35184372088832
64	65536	67108864	68719476736	70368744177664
128	131072	134217728	137438953472	140737488355328
256	262144	268435456	274877906944	281474976710656
512	524288	536870912	549755813888	562949953421312
1024	1048576	1073741824	1099511627776	1125899906842624.

This list looks rather complicated beyond the trivial pattern that these numbers grow. There are some interesting features to be observed, however. For example, the last digits repeat periodically: 2, 4, 8, 6. That this must be so is quite obvious: The last digit of the next power is determined by the last digit of the previous one; so once a single repetition appears, it is bound to reproduce the pattern. (Furthermore, the last digit is always even and never 0.)

A similar argument shows that the last two digits jointly must also eventually start repeating periodically: By the previous observation there are at most 40 possibilities for the last two digits, and since the last two digits of the next power are determined by those of the previous one, it is sufficient to have one repetition to establish a periodic pattern. Looking at our sequence we see that, indeed, the last two digits form the following periodic sequence with period 20 beginning from the second term: 04 08 16 32 64 28 56 12 24 48 96 92 84 68 36 72 44 88 76 52.

Note that this sequence has a few interesting patterns. Adding its first and eleventh terms gives 100, as does adding the second and twelfth, the third and thirteenth, and so on. One way of developing this sequence is to start from 04 and apply the following rule repeatedly: If the current number is under 50, double it; otherwise, double the difference to 100. The simpler 2,4,8,6 above exhibits analogous patterns.

Now look at the sequence of the *first* digits. Reading off the same list:

2	2048	2097152	2147483648	2199023255552
4	4096	4194304	4294967296	4398046511104
8	8192	8388608	8589934592	8796093022208
16	16384	16777216	17179869184	17592186044416
32	32768	33554432	34359738368	35184372088832
64	65536	67108864	68719476736	70368744177664
128	131072	134217728	137438953472	140737488355328
256	262144	268435456	274877906944	281474976710656
512	524288	536870912	549755813888	562949953421312
1024	1048576	1073741824	1099511627776	1125899906842624

one finds the first digits of the 50 entries to be

$$2481361251$$
$$2481361251$$
$$2481361251$$
$$2481361251$$
$$2481371251.$$

This is tantalizingly close to being periodic, but a small change creeps in at the end, so no truly periodic pattern appears – and there is no reason to expect any. (If you calculate further entries in this sequence, this behavior continues; little changes keep appearing here and there.)

Since this sequence is not as regular as the previous one, a statistical approach might be helpful. Look at the *frequency* of each digit – how often does a particular digit figure in this list? We have:

digit : 1 2 3 4 5 6 7 8 9
number of times : 15 10 5 5 5 4 1 5 0.

These frequencies look somewhat uneven. In particular, seven and nine seem to be disfavored. Seven appears for the first and only time at the 46th place in our sequence, and nine appears for the first time as the first digit of 2^{53}. Calculation of the first 100 entries gives slightly less lopsided frequencies, but they seem to be smaller for larger digits.

Thus, all nine digits appear as the first digit of some power of 2. We would like to know more, however. Does every digit appear infinitely many times? If yes, do they appear with any regularity? Which of the digits appear most often?

In order to discuss them we need to formulate these questions precisely. To that end we count for each digit d and every natural number n the number $F_d(n)$ of those powers 2^m, $m = 1, \ldots, n$ that begin with d. Thus, we just listed the 10 values of $F_d(50)$. The frequency with which d appears as the first digit among the first n powers of 2 is $F_d(n)/n$. Thus, one of our questions is whether each of these quantities has a limit as n goes to infinity and how these limits, if they exist, depend on d. Once these questions have been answered, one can also ask them about powers of 3 and compare the limit frequencies.

In Proposition 4.2.7 we obtain existence of these limits and give a formula for them which in particular implies that all limit frequencies are positive and that they decrease as d increases. Thus, contrary to the evidence from the first 50 powers (but in agreement with what one sees among the first 100), seven eventually appears more often than eight. The relationship between these limits for powers of 3 versus powers of 2 is also striking.

1.3.5 Last Digits of Polynomials

In the previous example we had immediate success in studying patterns of last digits and noted that some dose of dynamics provides the tools for understanding the behavior of the first digits. Let us look at another problem of integer sequences where similar questions can be asked about last digits.

Instead of an exponential sequence consider the sequence $x_n = n^2$ for $n \in \mathbb{N}_0$. The last digits come out to be 01496569410 and repeat periodically thereafter.

■ **Exercise 1.3.9** Prove that these digits repeat periodically.

■ **Exercise 1.3.10** Explain why this sequence is *palindromic*, that is, unchanged when reversed.

This is about as simple as it was earlier, so let's try $x_n = n^2 p/q$ instead, for some $p, q \in \mathbb{N}$. Unless $q = 1$, these won't all be integers, so we make explicit that we are looking at the digit before the decimal point. You may want to experiment a little, but it is easy to see directly that we still get a periodic pattern, with period at most $10q$. The reason is that

$$a_{n+10q} - a_n = (n + 10q)^2\, p/q - n^2 p/q = 10(2np + 10\, pq)$$

is an integer multiple of 10, so the digit before the decimal point (as well as all the ones after) is the same for a_{n+10q} and a_n.

■ **Exercise 1.3.11** Prove that the initial $10q$ results form a palindromic string.

This was interesting, but not subtle. It is natural to replace p/q by an irrational number, because that should cause an "infinite period," as it were, that is, no periodicity at all.

So, consider $x_n = n^2\sqrt{2}$. The sequence of last digits (before the decimal point) begins with the following 100 terms: 47764935641602207257751690074812184811 0737998503554058008492320613431613320591107257752701195034 3171.

There are no obvious reasons for periodicity, nor is any such pattern apparent. Certainly all digits make an appearance. However, the questions we asked about first digits of powers of 2 are also appropriate here: Do all digits appear infinitely often? Do they appear with well-defined relative frequencies? Relative frequencies are defined as before: Let $P_n(d)$ be the number of times the last digit is d in the set $\{n^2\sqrt{2}\}_{i=0}^{n-1}$ and consider $P_n(d)/n$ for large n. Among the first 100 values we get the frequencies

i:	0	1	2	3	4	5	6	7	8	9
$P_{100}(i)/100$:	0.14	0.15	0.09	0.10	0.09	0.11	0.06	0.13	0.06	0.07.

This list does not suggest any answer to this question, and the same list for larger n might not either.

Dynamics is able to address these questions as well as many similar ones completely and rigorously. In this particular example it turns out that all relative frequencies converge to $1/10$. Thus we have an example of *uniform distribution*, which is one of the central paradigms in dynamics as well as in nature. We outline a solution of the problem of distribution of last digits in Section 15.1.

1.3.6 Cellular Automata

A game of sorts called the game of life was popular in the 1980s. It is intended to model a simple population of somethings that live in fixed locations. Each of the "organisms" is at a point of a fixed lattice, the points in the plane with integer coordinates, say, and can have several states of health. In the simplest version such organisms might have only the two states "present" and "not there" (or 1 and 0). But one may also take a model with a larger number of possible states, including, for example, "sickly" or "happy." The rule of the game is that the population changes in discrete time steps in a particular way. Each organism checks the states of some of its neighbors (up to some distance) and, depending on all these, changes its own state accordingly. For example, the rule might say that if all immediate neighbors are present, the organism dies (overpopulation). Maybe the same happens if there are no neighbors at all (too lonely or exposed). This game was popular because from relatively simple rules one could find (or design) intriguing patterns, and because computers, even early ones, could easily go through many generations in a short time.

If the number of cells is finite, then from our perspective of asymptotic long-term behavior there is not too much to say about the system. It has only finitely many states, so at some point some state must be attained for a second time. Because the rules are unchanged, the pattern thereafter cycles again through the same sequence of states since the last time, and again and again. No matter how interesting the patterns may be that emerge, or how long the cycle, this is a complete qualitative description of the long-term behavior.

When there are infinitely many cells, however, there is no reason for this kind of cycling through the same patterns, and there may be all kinds of long-term behaviors.

Systems of this kind are called cellular automata. Since the rules are so clearly described, one can easily make mathematics out of them. To keep the notation simple we look not at the integer points in the plane, but only those on the line. Accordingly, a state of the system is a sequence, each entry of which has one of finitely many values (states). If the states are numbered $0, \ldots, N-1$, then we can denote the space of these sequences by Ω_N. All organisms have the same rule for their development. It is given by a function $f: \{0, \ldots, N-1\}^{2n+1} \to \{0, \ldots, N-1\}$, that is, a function that maps $2n+1$-characters-long strings of states $(0, \ldots, N-1)$ to a state. The input consists of the states of all neighbors up to distance n in either direction, and the output is the future state of the individual. Therefore, each step of the evolution of the whole system is given by a map $\Phi: \Omega_N \to \Omega_N$ such that $(\Phi(\omega))_i = f(\omega_{i-n}, \ldots, \omega_{i+n})$. By way of example, take $N = n = 1$ and $f(x_{-1}, x_0, x_1) = x_1$. This means that every individual just chooses to follow its right neighbor's lead (today's x_1 is tomorrow's x_0). You might call this example "the wave," because whatever pattern you begin with, it will relentlessly march leftward.

This is a general description of cellular automata, whose interest goes well beyond the game of life. The same mathematical concept admits a rather different interpretation. If one thinks of each of these sequences as a stream of data, then the map Φ transforms these data – it is a code. This particular class of codes is known as *sliding block codes*, and this kind is suitable for real-time streaming data encoding or decoding. For us, it is a transformation on a nice space that can be repeated, a dynamical system. The general class of dynamical systems whose states are given by sequences (or arrays) is called *symbolic dynamics*, and some of our most useful models are of this kind. "The wave" is actually our favorite, and we call it the (left) shift. As a class, sliding block codes play an important role, although under a different name (conjugacies).

Symbolic dynamics is introduced in Section 7.3.4 and studied in Section 7.3.7. It provides a rich supply of examples that are simple to describe but produce a variety of complicated dynamical phenomena.

▓ EXERCISES

▓ **Exercise 1.3.12** Prove that in the binary search for a root the sequences of left and right endpoints both converge and that they have the same limit.

▓ **Exercise 1.3.13** In the binary search for a root assume $a = 0$, $b = 1$ and that the procedure never terminates. Keep track of the choices at each step by noting a 0 whenever Case 1 occurs and noting a 1 whenever Case 2 occurs. Prove that the string of 0's and 1's thus obtained gives the binary representation of the solution found by the algorithm.

▓ **Exercise 1.3.14** In the preceding exercise assume that the search terminates. How does the finite string of 0's and 1's relate to the binary representation of the root?

▓ **Exercise 1.3.15** Solve $\cos x = x$ with the Newton Method and the initial guess $x_0 = 3/4$.

▓ **Exercise 1.3.16** Approximate $\sqrt{5}$ to the best possible accuracy of your calculator by the Newton Method with initial guess 2.

▓ **Exercise 1.3.17** Use the Newton Method to solve $\sin x = 0$ with initial guess 1 and note the pattern in the size of the absolute error.

▓ **Exercise 1.3.18** Try to solve $\sqrt[3]{x} = 0$ with the Newton Method, *not* taking 0 as initial guess.

▓ **Exercise 1.3.19** For the Greek method of arithmetic/harmonic mean, express the successive arithmetic means as the iterates of some function, that is, write down a recursive formula for the first components alone.

▓ **Exercise 1.3.20** Finding the root of a number z can be done in various ways. Compare the Greek method of arithmetic/harmonic mean with the Newton Method, taking 1 as the initial guess.

■ **Exercise 1.3.21** Find the lowest power of 2 such that, among all the powers up to this one, the first digit is 7 more often than 8.

■ **Exercise 1.3.22** Consider the sequence $(a_n)_{n\in\mathbb{N}}$ defined by the last two digits of powers of 2. Prove that $a_n + a_{n+10} = 100$ for every $n \geq 2$.

■ **Exercise 1.3.23** Prove that the sequence defined by the last three digits of powers of 2 (starting with 008) is periodic with period 100.

■ **Exercise 1.3.24** Consider the sequence $(a_n)_{n\in\mathbb{N}}$ defined by the last three digits of powers of 2. Prove that $a_n + a_{n+50} = 1000$ for every $n \geq 3$.

PART 1

A COURSE IN DYNAMICS: FROM SIMPLE TO COMPLICATED BEHAVIOR

Dynamics provides the concepts and tools to describe and understand complicated long-term behavior in systems that evolve in time. An excellent way to acquire and appreciate these concepts and tools is in a gradual progression from simple to complex behavior during which examples, concepts, and tools all evolve toward greater sophistication.

Accordingly, the course begins with the most simple dynamical behavior (a universal steady state), which turns out to be a fundamental tool as well. (This latter theme is developed in detail in Chapter 9.) Chapter 2 gently progresses to also include systems with several steady states. Chapter 3 studies linear maps as slightly more complicated yet quite tractable systems that double as a central tool later.

Complexity first appears in Chapter 4 (and moves to higher dimension in Chapter 5). We encounter orbits whose long-term behavior involves close but not exact returns to the initial state (recurrence), as well as distribution throughout the space of states. However, this potential complexity is tempered by the regularity of individual asymptotics and uniformity of behavior across orbits. This is also borne out by statistical analysis, which is introduced here. Mechanical systems (Chapter 6) provide natural examples where some complexity (recurrence) is forced while the intrinsic structure often limits complexity to the level of the previous two chapters.

The highest level of complexity is reached in Chapter 7, where complicated recurrence is highly irregular for individual orbits and the asymptotic behavior of orbits is thoroughly heterogeneous and inextricably intertwined. Concepts appropriate for this level of complexity are developed throughout this chapter and the next, including fundamental aspects of statistical behavior.

Although it is part of the Panorama, Chapter 9 is a natural continuation of the course, and it is written to the same standard of proof as the course.

CHAPTER 2

Systems with Stable Asymptotic Behavior

This chapter prepares the ground for much of this book in several ways. On one hand, it provides the simplest examples of dynamical behavior, with the first hints as to how more complicated behavior can arise. On the other hand, it provides some important tools and concepts that we will need frequently. There are two kinds of dynamical systems we present here as "simple". There are linear maps, whose simplicity lies in the possibility of breaking them down into components that one can study separately. Contracting maps are simple because everything moves toward a single point. We introduce linear maps briefly here and concentrate on a preview of their utility for studying nonlinear dynamical systems. Linear maps are studied systematically in Chapter 3. We present the facts about contracting maps that will be used throughout this course. Applications pervade this book and are featured prominently in Chapter 9.

2.1 LINEAR MAPS AND LINEARIZATION

2.1.1 Scalar Linear Maps

The primitive discrete-time population model $x_{i+1} = f(x_i) = kx_i$ (with $k > 0$) introduced in Section 1.2.9.1 has simple dynamics: Starting with any $x_0 \neq 0$, the sequence $(x_i)_{i \in \mathbb{N}}$ diverges if $k > 1$ and goes to 0 if $k < 1$. Part of the simplicity is that the asymptotic behavior is independent of the initial condition; scaling x_0 by a factor a scales all x_i by the same factor. Furthermore, the allowed asymptotic behaviors are quite simple.

So long as $k \neq 1$, this changes little if we replace $f(x) = kx$ by $g(x) := kx + b$. Indeed, changing variables to $y = x - (b/1 - k)$ leads to the recursion $y_{i+1} = ky_i$. Therefore we have by now fully described the dynamical possibilities for linear maps. (Except for $f(x) = kx$ when $k = \pm 1$.)

Linear maps in higher dimension admit more complex behavior of individual orbits. However, one important aspect of simplicity remains. Although not all orbits have the same long-term behavior, knowledge of the dynamics of a small number of orbits is sufficient to develop the dynamics of all of the others. This is studied systematically in Chapter 3.

2.1.2 Linearization

Some important dynamical systems studied in this book are rather directly derived from a linear one, but most interesting dynamical systems are not linear. Yet, knowledge of linear dynamical systems can be useful for studying nonlinear ones. The key ingredient is differentiability, that is, the existence of a good linear approximation near any given point.

Differentiability and linear approximation are discussed in Section 2.2.4.1 and Section A.2. The central feature of differentiability is that it guarantees a good linear approximation of a map near any given point. A simple example is the approximation of $f(x) := \sqrt{x}$ near $x = 16$ by $L(x) = f(16) + f'(16)(x - 16) = 4 + \frac{1}{8}(x - 16)$, in particular, $\sqrt{17} \approx 4\frac{1}{8}$. This corresponds to the first step in the Newton method for approximating $\sqrt{17}$ (see Section 1.3.2.2).

Such linear approximation can sometimes be useful for dynamics when the orbits of a nonlinear map stay near enough to the reference point for the linear approximation to be relevant. There are examples of this throughout the book. For now we give a special case of Proposition 2.2.17:

Proposition 2.1.1 *Suppose F is a differentiable map of the line and $F(b) = b$. If all orbits of the linearization of F at b are asymptotic to b, then all orbits of F that start near enough to b are asymptotic to b as well.*

The quintessential linearization result in analysis is the Mean Value Theorem A.2.3 (see also Lemma 2.2.12), and it is already used numerous times in the next section (for example, Proposition 2.2.3 and Section 2.2.4.4). Linearization also plays an important role in highly complicated dynamical systems (see for example, Chapter 7 and Chapter 10.

▥ EXERCISES

▥ **Exercise 2.1.1** Show that for $k \neq 1$ the change of variable $y = x - (b/1 - k)$ transforms the recursion $x_{i+1} = f(x_i) = kx_i + b$ to the recursion $y_{i+1} = ky_i$.

▥ **Exercise 2.1.2** Describe what asymptotic behaviors appear for the maps $f(x) = kx$ when $k = \pm 1$.

▥ **Exercise 2.1.3** Describe what asymptotic behaviors appear for the maps $f(x) = kx + b$ when $k = \pm 1$ and $b \neq 0$.

2.2 CONTRACTIONS IN EUCLIDEAN SPACE

Traditionally, scientists and engineers have had a preference for dynamical systems that have stable asymptotic behavior, ideally settling into a steady state, maybe after a short period of "transient behavior." Simple real-life examples abound. A desk lamp, when turned on, settles into the "on" state of constant light intensity instantly (after a very short heating period for the filament). Unless it is broken, it does not blink or flicker erratically. Likewise escalators are preferred in a steady state of constant speed. Radios, when first turned on, have complicated transient behavior for a remote fraction of a second but then settle into a steady state of reception. Our tour of dynamical systems begins with those that display such simple behavior.

Corresponding to the above continuous-time real-world examples, the simplest imaginable kind of asymptotic behavior of a discrete-time dynamical system is represented by the convergence of iterates of any given point to a particular point, a steady state. There is an important general class of situations where this kind of behavior can be established, namely, for contracting maps. These are presented here not only because their simple dynamics provides an ideal starting point, but also because we will use contractions as a tool in numerous problems in analysis and differential equations, as well as for studying dynamical systems with more complicated behavior. These applications appear throughout this book, and Chapter 9 concentrates on such applications.

We now define contractions and clarify the usage of the words "map" and "function."

2.2.1 Definitions

When we use the word "map" we usually mean that the domain and range lie in the same space, and even more often that the range is in the domain – we iterate maps and in this way they generate a dynamical system. Fibonacci's rabbits, Maine lobsters, phyllotaxis, butterflies, and methods for finding roots all provided examples of such dynamical systems. Time is discrete, and the laws of nature (or of an algorithm) have been distilled into a rule that determines present data from prior data, and the next "state" of the system from the present state. All of this is achieved by applying one map that encodes these laws. So, discrete-time dynamical systems are maps of a space to itself. Maps are almost always continuous.

"Functions," on the other hand, have numerical values even if they are defined on a rather different space, and they are not iterated. Still, we will sometimes use the conventional word "function" to denote a map of the real line or a subset of it into itself. There is a third possibility of transformations used for a change of variables. These are called *coordinate changes* or *conjugacies* (and maybe sometimes also maps). One map that we always have at our disposal is the identity, which we denote by Id. It is defined by $\mathrm{Id}(x) = x$.

Now we define contracting maps with respect to the Euclidean distance $d(x, y) := \sqrt{\sum_{i=1}^{n} (x_i - y_i)^2}$.

Definition 2.2.1 A map f of a subset X of Euclidean space is said to be *Lipschitz-continuous* with Lipschitz constant λ, or λ-Lipschitz if

(2.2.1) $$d(f(x), f(y)) \leq \lambda d(x, y)$$

for any $x, y \in X$. f is said to be a *contraction* or a λ-*contraction* if $\lambda < 1$. If a map f is Lipschitz-continuous, then we define $\mathrm{Lip}(f) := \sup_{x \neq y} d(f(x), f(y))/d(x, y)$.

Example 2.2.2 The function $f(x) = \sqrt{x}$ defines a contraction on $[1, \infty)$. To prove this, we show that for $x \geq 1$ and $t \geq 0$ we have $\sqrt{x + t} \leq \sqrt{x} + (1/2)t$ (why is this enough?). This is most easily seen by squaring:

$$\left(\sqrt{x} + \frac{t}{2}\right)^2 = x + xt + \frac{t^2}{4} \geq x + xt \geq x + t.$$

2.2.2 The Case of One Variable

We now give an easy way of checking the contraction condition that uses the derivative.

Proposition 2.2.3 *Let I be an interval and $f : I \to \mathbb{R}$ a differentiable function with $|f'(x)| \leq \lambda$ for all $x \in I$. Then f is λ-Lipschitz.*

Proof By the Mean Value Theorem A.2.3, for any two points $x, y \in I$ there exists a point c between x and y such that

$$d(f(x), f(y)) = |f(x) - f(y)| = |f'(c)(x - y)| = |f'(c)|d(x, y) \leq \lambda d(x, y). \quad \square$$

Note that we need no information about f' at the endpoints of I.

Example 2.2.4 This criterion makes it easier to check that $f(x) = \sqrt{x}$ defines a contraction on $I = [1, \infty)$ because $f'(x) = 1/2\sqrt{x} \leq 1/2$ for $x \geq 1$.

Let us point out that by Problem 2.2.14 the weaker condition $|f'(x)| < 1$ does not suffice to obtain (2.2.1). However, sometimes it does:

Proposition 2.2.5 *Let I be a closed bounded interval and $f : I \to I$ a continuously differentiable function with $|f'(x)| < 1$ for all $x \in I$. Then f is a contraction.*

Proof The maximum λ of $|f'(x)|$ is attained at some point x_0 because f' is continuous. It is less than 1 because $|f'(x_0)| < 1$. $\square$

The difference is that the real line is not closed and bounded (see also Problem 2.2.13 for a related fact).

In calculus, a favorite example of a recursively defined sequence is of the form $a_{n+1} = f(a_n)$, with a_0 given and f a function with $|f'| \leq \lambda < 1$. This is a simple dynamical system given by the map f. For each initial value a_0 a sequence is uniquely defined by $a_{n+1} = f(a_n)$. If f is invertible, then this sequence is defined for all $n \in \mathbb{Z}$.

Definition 2.2.6 For a map f and a point x, the sequence $x, f(x), f(f(x)), \ldots,$ $f^n(x), \ldots$ (if f is not invertible) or the sequence $\ldots f^{-1}(x), x, f(x), \ldots$ is called the *orbit* of x under f. A *fixed point* is a point such that $f(x) = x$. The set of fixed points is denoted by $\mathrm{Fix}(f)$. A *periodic point* is a point x such that $f^n(x) = x$ for some $n \in \mathbb{N}$, that is, a point in $\mathrm{Fix}(f^n)$. Such an n is said to be *a* period of x. The smallest such n is called the *prime period* of x.

Example 2.2.7 If $f(x) = -x^3$ on $\mathbb{R}$, then 0 is the only fixed point and ± 1 is a periodic orbit, that is, 1 and -1 are periodic points with prime period 2.

The reason the calculus examples of such sequences always converge is the following important fact:

Proposition 2.2.8 (Contraction Principle) *Let $I \subset \mathbb{R}$ be a closed interval, possibly infinite on one or both sides, and $f: I \to I$ a λ-contraction. Then f has a unique fixed point x_0 and $|f^n(x) - x_0| \le \lambda^n |x - x_0|$ for every $x \in \mathbb{R}$, that is, every orbit of f converges to x_0 exponentially.*

Proof By iterating $|f(x) - f(y)| \le \lambda |x - y|$, one sees that

$$(2.2.2) \qquad\qquad |f^n(x) - f^n(y)| \le \lambda^n |x - y|$$

for $x, y \in \mathbb{R}$ and $n \in \mathbb{N}$; so for $x \in I$ and $m \ge n$ we can use the triangle inequality to show

$$
|f^m(x) - f^n(x)| \le \sum_{k=0}^{m-n-1} |f^{n+k+1}(x) - f^{n+k}(x)|
$$

$$(2.2.3)$$

$$
\le \sum_{k=0}^{m-n-1} \lambda^{n+k} |f(x) - x| \le \frac{\lambda^n}{1 - \lambda} |f(x) - x|.
$$

Here we used the familiar fact

$$
(1 - \lambda) \sum_{k=l}^{n-1} \lambda^k = \lambda^l + \lambda^{l+1} + \cdots + \lambda^{n-1} - \lambda^{l+1} + \lambda^{l+2} + \cdots + \lambda^n = (\lambda^l - \lambda^n)
$$

about partial sums of geometric series. Since the right-hand side of (2.2.3) becomes arbitrarily small as n gets large, (2.2.3) shows that $(f^n(x))_{n \in \mathbb{N}}$ is a Cauchy sequence. Thus for any $x \in I$ the limit of $f^n(x)$ as $n \to \infty$ exists because Cauchy sequences converge. The limit is in I because I is closed. By (2.2.2), this limit is the same for all x. We denote this limit by x_0 and show that x_0 is a fixed point for f. If $x \in I$ and $n \in \mathbb{N}$, then

$$
|x_0 - f(x_0)| \le |x_0 - f^n(x)| + |f^n(x) - f^{n+1}(x)| + |f^{n+1}(x) - f(x_0)|
$$

$$(2.2.4)$$

$$
\le (1 + \lambda)|x_0 - f^n(x)| + \lambda^n |x - f(x)|.
$$

Since $|x_0 - f^n(x)| \to 0$ and $\lambda^n \to 0$ as $n \to \infty$, we have $f(x_0) = x_0$.

That $|f^n(x) - x_0| \le \lambda^n |x - x_0|$ for every $x \in \mathbb{R}$ follows from (2.2.2) with $y = x_0$. $\square$

Example 2.2.9 In contemplating his rabbits, Leonardo of Pisa, also known as Fibonacci, came up with a model according to which the number of rabbit pairs in the nth month is given by the number b_n, defined by the recursive relation $b_0 = 1$, $b_1 = 2$, $b_n = b_{n-1} + b_{n-2}$ for $n \ge 2$ (Section 1.2.2). Expecting that the growth of these numbers should be exponential, we would like to see how fast these numbers grow by finding the limit of $a_n := b_{n+1}/b_n$ as $n \to \infty$. To that end we use the Contraction Principle. Since

$$
a_{n+1} = \frac{b_{n+2}}{b_{n+1}} = \frac{b_{n+1} + b_n}{b_{n+1}} = \frac{1}{b_{n+1}/b_n} + 1 = \frac{1}{a_n} + 1,
$$

$(a_n)_{n=1}^{\infty}$ is the orbit of 1 under iteration of the map $g(x) := (1/x) + 1$. Since $g(1) = 2$, we are in fact considering the orbit of 2 under g. Now $g'(x) = -x^{-2}$. This tells us that g is not a contraction on $(0, \infty)$. Therefore we need to find a suitable (closed) interval where this is the case and that is mapped inside itself.

Since $g' < 0$, g is decreasing on $(0, \infty)$. This implies that $g([3/2, 2]) \subset [3/2, 2]$ because $3/2 < g(3/2) = 5/3 < 2$ and $g(2) = 3/2$. Furthermore, $|g'(x)| = 1/x^2 \le 4/9 < 1$ on $[3/2, 2]$, so g is a contraction on $[3/2, 2]$. By the Contraction Principle, the orbit of 2 and hence that of 1 is asymptotic to the unique fixed point x of g in $[3/2, 2]$. Thus $\lim_{n \to \infty} b_{n+1}/b_n = \lim_{n \to \infty} a_n$ exists. To find the limit we solve the equation $x = g(x) = 1 + 1/x = (x + 1)/x$, which is equivalent to $x^2 - x - 1 = 0$. There is only one positive solution: $x = (1 + \sqrt{5})/2$. (This solves Exercise 1.2.3.) Another way of obtaining this ratio and an explicit formula for the Fibonacci numbers is given in Section 3.1.9.

2.2.3 The Case of Several Variables

We now show that the Contraction Principle holds in higher dimension as well, and we use the same proof, replacing absolute values by the Euclidean distance.

Proposition 2.2.10 (Contraction Principle) *Let* $X \subset \mathbb{R}^n$ *be closed and* $f : X \to X$ *a* λ-*contraction. Then* f *has a unique fixed point* x_0 *and* $d(f^n(x), x_0) = \lambda^n d(x, x_0)$ *for every* $x \in X$.

Proof Iterating $d(f(x), f(y)) \le \lambda d(x, y)$ shows

$$(2.2.5) \qquad\qquad d(f^n(x), f^n(y)) \le \lambda^n d(x, y)$$

for $x, y \in X$ and $n \in \mathbb{N}$. Thus $(f^n(x))_{n \in \mathbb{N}}$ is a Cauchy sequence because

$$d(f^m(x), f^n(x)) \le \sum_{k=0}^{m-n-1} d(f^{n+k+1}(x), f^{n+k}(x))$$

$$(2.2.6)$$

$$\le \sum_{k=0}^{m-n-1} \lambda^{n+k} d(f(x), x) \le \frac{\lambda^n}{1 - \lambda} d(f(x), x)$$

for $m \ge n$, and $\lambda^n \to 0$ as $n \to \infty$. Thus $\lim_{n \to \infty} f^n(x)$ exists (because Cauchy sequences in $\mathbb{R}^n$ converge) and is in X because X is closed (see Figure 2.2.1). By (2.2.5) it is the same for all x. Denote this limit by x_0. Then

$$d(x_0, f(x_0)) \le d(x_0, f^n(x)) + d(f^n(x), f^{n+1}(x)) + d(f^{n+1}(x), f(x_0))$$

$$(2.2.7)$$

$$\le (1 + \lambda) d(x_0, f^n(x)) + \lambda^n d(x, f(x))$$

for $x \in X$ and $n \in \mathbb{N}$. Now $f(x_0) = x_0$ because $d(x_0, f^n(x)) \xrightarrow[n \to \infty]{} 0$. $\square$

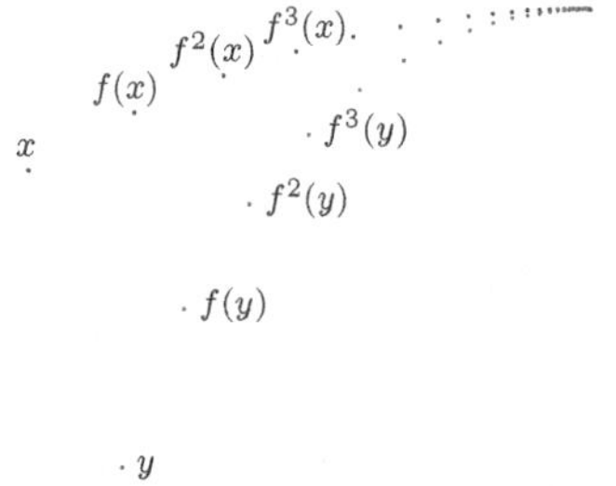

Figure 2.2.1. Convergence of iterates.

Taking the limit in (2.2.6) as $m \to \infty$ we obtain $d(f^n(x), x_0) \leq (\lambda^n/1 - \lambda)$ $d(f(x), x)$. This means that, after n iterations, we can say with certainty that the fixed point is in the $(\lambda^n/1 - \lambda)d(f(x), x)$-ball around $f^n(x)$. In other words, if we make numerical computations, then we can make a rigorous conclusion about where the fixed point must be (after accounting for roundoff errors).

Definition 2.2.11 We say that two sequences $(x_n)_{n \in \mathbb{N}}$ and $(y_n)_{n \in \mathbb{N}}$ of points in $\mathbb{R}^n$ *converge exponentially* (or *with exponential speed*) *to each other* if $d(x_n, y_n) < cd^n$ for some $c > 0$, $d < 1$. In particular, if one of the sequences is constant, that is, $y_n = y$, we say that x_n *converges exponentially* to y.

2.2.4 The Derivative Test

We now show, similarly to the case of one variable, that the contraction property can be verified using the derivative.

To that end we recall some pertinent tools from the calculus of several variables, namely, the differential and the Mean Value Theorem.

1. The Differential. Let $f : \mathbb{R}^n \to \mathbb{R}^m$ be a map with continuous partial derivatives. Then at each point one can define the derivative or differential of $f = (f_1, \dots, f_m)$ as the linear map defined by the matrix of partial derivatives

$$Df := \begin{pmatrix} \dfrac{\partial f_1}{\partial x_1} & \dfrac{\partial f_1}{\partial x_2} & \cdots & \dfrac{\partial f_1}{\partial x_n} \\[2mm] \dfrac{\partial f_2}{\partial x_1} & \dfrac{\partial f_2}{\partial x_2} & \cdots & \dfrac{\partial f_2}{\partial x_n} \\[2mm] \vdots & \vdots & \ddots & \vdots \\[2mm] \dfrac{\partial f_m}{\partial x_1} & \dfrac{\partial f_m}{\partial x_2} & \cdots & \dfrac{\partial f_m}{\partial x_n} \end{pmatrix}.$$

We say that the map is regular at x_0 if this map is invertible. We define the norm (see Definition A.1.29) of the differential by the norm of the matrix Df. In linear algebra the norm of a matrix A is defined by looking at its action as a linear map:

$$(2.2.8) \qquad \|A\| := \max_{v \neq 0} \frac{\|A(v)\|}{\|v\|} = \max_{\|v\|=1} \|A(v)\|.$$

Geometrically, this is easy to visualize by considering the second of these expressions: Consider the unit sphere $\{v \in \mathbb{R}^n \mid \|v\| = 1\}$ and notice that the second maximum is just the size of the largest vectors in the image of this unit sphere. The image of the unit sphere under a linear map is an ellipsoidal figure, and in a picture the largest vector is easy to find. Calculating this norm in particular cases may not always be easy, but there are easy ways of finding upper bounds (see Exercise 2.2.9 and Lemma 3.3.2).

2. The Mean Value Theorem

Lemma 2.2.12 *If* $g : [a, b] \to \mathbb{R}^m$ *is continuous and differentiable on* (a, b), *then there exists* $t \in [a, b]$ *such that*

$$\|g(b) - g(a)\| \leq \left\| \frac{d}{dt} g(t) \right\| (b - a).$$

Proof Let $v = g(b) - g(a)$, $\varphi(t) = \langle v, g(t) \rangle$. By the Mean Value Theorem A.2.3 for one variable there exists a $t \in (a, b)$ such that $\varphi(b) - \varphi(a) = \varphi'(t)(b - a)$, and so

$$(b-a)\|v\|\left\|\frac{d}{dt}g(t)\right\| \geq (b-a)\left\langle v, \frac{d}{dt}g(t) \right\rangle = \frac{d}{dt}\varphi(t)(b-a) = \varphi(b) - \varphi(a)$$

$$= \langle v, g(b) \rangle - \langle v, g(a) \rangle = \langle v, v \rangle = \|v\|^2.$$

Divide by $\|v\|$ to finish the proof. $\square$

3. Convexity. A further notion we need is that of a convex set.

Definition 2.2.13 A *convex set* in $\mathbb{R}^n$ is set C such that for all $a, b \in C$ the line segment with endpoints a, b is entirely contained in C. It is said to be *strictly convex* if for any points a, b in the closure of C the segment from a to b is contained in C, except possibly for one or both endpoints (see Figure 2.2.2).

For example, the disk $\{(x, y) \in \mathbb{R}^2 \mid x^2 + y^2 < 1\}$ is strictly convex. The open upper half-plane $\{(x, y) \in \mathbb{R}^2 \mid y > 0\}$ is convex but not strictly convex. A kidney shape $\{(r, \theta) \mid 0 \leq r \leq 1 + (1/2)\sin\theta\}$ (in polar coordinates) is not convex. Neither is the *annulus* $\{(x, y) \in \mathbb{R}^2 \mid 1 < x^2 + y^2 < 2\}$.

4. The Derivative Test. We can now give two versions of a derivative test for contractions in several variables.

Theorem 2.2.14 *If $C \subset \mathbb{R}^n$ is convex and open and $f : C \to \mathbb{R}^m$ is differentiable with $\|Df(x)\| \leq M$ for all $x \in C$, then $\|f(x) - f(y)\| \leq M\|x - y\|$ for $x, y \in C$.*

Proof The line segment connecting x and y is given by $c(t) = x + t(y - x)$ for $t \in [0, 1]$, and it is contained in C by convexity. Let $g(t) := f(c(t))$. Then by the chain rule

$$\left\|\frac{d}{dt}g(t)\right\| = \left\|Df(c(t))\frac{d}{dt}c(t)\right\| = \|Df(c(t))(y - x)\| \leq M\|y - x\|.$$

By Lemma 2.2.12, this implies $\|f(y) - f(x)\| = \|g(1) - g(0)\| \leq M\|y - x\|$. $\square$

Corollary 2.2.15 *If $C \subset \mathbb{R}^n$ is a convex open set, $f : C \to C$ a map with continuous partial derivatives, and $\|Df\| \leq \lambda < 1$ at every point $x \in \mathbb{R}^n$, then f is a λ-contraction.*

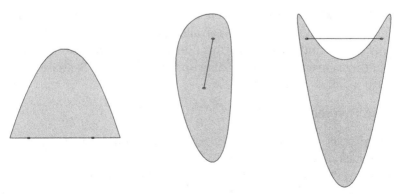

Figure 2.2.2. A convex, strictly convex, and nonconvex set.

The role of convexity in Theorem 2.2.14, and hence in Corollary 2.2.15, is elucidated in Problem 2.2.12. In particular, it is not sufficient to assume that any two points of C can be connected by a curve, however nice. It is really necessary to use a single line segment.

The preceding corollary does not quite seem to be geared toward applying the Contraction Principle because an open set may not contain the limits of Cauchy sequences in it. Therefore we give a result that holds for the closure of such a set. It is proved exactly like Theorem 2.2.14.

Theorem 2.2.16 *If $C \subset \mathbb{R}^n$ is an open strictly convex set, $\bar{C}$ its closure, $f : \bar{C} \to \mathbb{R}^n$ differentiable on C and continuous on $\bar{C}$ with $\|Df\| \leq \lambda < 1$ on C, then f has a unique fixed point $x_0 \in \bar{C}$ and $d(f^n(x), x_0) \leq \lambda^n d(x, x_0)$ for every $x \in \bar{C}$.*

Proof For $x, y \in \bar{C}$ we parameterize the line segment connecting x and y by $c(t) = x + t(y - x)$ for $t \in [0, 1]$ and let $g(t) := f(c(t))$. Then $c((0, 1))$ is contained in C by strict convexity and

$$\left\| \frac{d}{dt} g(t) \right\| = \left\| Df(c(t)) \frac{d}{dt} c(t) \right\| = \|Df(c(t))(y - x)\| \leq \lambda \|y - x\|.$$

By Lemma 2.2.12 this implies $\|f(y) - f(x)\| \leq \lambda \|y - x\|$. Thus f is a λ-contraction and has a unique fixed point x_0. Furthermore, $d(f^n(x), x_0) = \lambda^n d(x, x_0)$ for every $x \in \bar{C}$. $\square$

2.2.5 Local Contractions

Now we discuss maps that are not contracting on their entire domain but on a part of it. A prime example of a map that contracts only locally is given by the following:

Proposition 2.2.17 *Let f be a continuously differentiable map with a fixed point x_0 where $\|Df_{x_0}\| < 1$. Then there is a closed neighborhood U of x_0 such that $f(U) \subset U$ and f is a contraction on U.*

Definition 2.2.18 By a closed neighborhood of x we mean the closure of an open set containing x.

Proof Since Df is continuous, there is a small closed ball $U = \overline{B(x_0, \eta)}$ around x_0 on which $\|Df_x\| \leq \lambda < 1$ (Exercise 2.2.11). If $x, y \in U$, then $d(f(x), f(y)) \leq \lambda d(x, y)$ by Corollary 2.2.15; so f is contraction on U. Furthermore, taking $y = x_0$ shows that if $x \in U$, then $d(f(x), x_0) = d(f(x), f(x_0)) \leq \lambda d(x, x_0) \leq \lambda \eta < \eta$ and hence $f(x) \in U$. $\square$

Unfortunately the definition of $\|Df\|$ is inconvenient for calculations. However, one can avoid having to use it if one is willing to either adjust the metric or pass to an iterate. We carry this out in the next chapter.

Proposition 2.2.19 *Let f be a continuously differentiable map with a fixed point x_0 such that all eigenvalues of Df_{x_0} have absolute value less than 1. Then there is a closed neighborhood U of x_0 such that $f(U) \subset U$ and f is a contraction on U with respect to an adapted norm.*

Proof In the next chapter (Proposition 3.3.3) we will show that the assumption on the eigenvalues implies that one can choose a norm that we denote by $\| \cdot \|'$ for which $\|Df\|' < 1$. Now Proposition 2.2.17 applies. In other words, a sufficiently small closed "ball" around x_0 with respect to the norm $\| \cdot \|'$ can be chosen as the set U. This ball is in fact an ellipsoid in $\mathbb{R}^n$. □

This particular situation is interesting because of some robustness under perturbation.

2.2.6 Perturbations
We now study what happens to the fixed point when one perturbs a contraction.

Proposition 2.2.20 *Let f be a continuously differentiable map with a fixed point x_0 where $\|Df_{x_0}\| < 1$, and let U be a closed neighborhood of x_0 such that $f(U) \subset U$. Then any map g sufficiently close to f is a contraction on U.*

Specifically, if $\epsilon > 0$, then there is a $\delta > 0$ such that any map g with $\|g(x) - f(x)\| \leq \delta$ and $\|Dg(x) - Df(x)\| \leq \delta$ on U maps U into U and is a contraction on U with its unique fixed point y_0 in $B(x_0, \epsilon)$.

Proof Since the linear map Df_x depends continuously on the point x, there is a small closed ball $U = B(x_0, \eta)$ around x_0 on which $\|Df_x\| \leq \lambda < 1$ (Exercise 2.2.11). Assume $\eta, \epsilon < 1$ and take $\delta = \epsilon \eta (1 - \lambda)/2$. Then

$$\|Dg\| \leq \|Dg - Df\| + \|Df\| \leq \delta + \lambda \leq \lambda + (1 - \lambda)/2 = (1 + \lambda)/2 =: \mu < 1$$

on U, so g is a contraction on U by Corollary 2.2.15. If $x \in U$, then $d(x, x_0) \leq \eta$ and

$$(2.2.9) \quad d(g(x), x_0) \leq d(g(x), g(x_0)) + d(g(x_0), f(x_0)) + d(f(x_0), x_0)$$

$$\leq \mu d(x, x_0) + \delta + 0 \leq \mu \eta + \delta \leq \eta(1 + \lambda)/2 + \eta(1 - \lambda)/2 = \eta,$$

so $g(x) \in U$ also, that is, $g(U) \subset U$. Finally, since $g^n(x_0) \to y_0$, we have

$$d(x_0, y_0) \leq \sum_{n=0}^{\infty} d(g^n(x_0), g^{n+1}(x_0)) \leq d(g(x_0), x_0) \sum_{n=0}^{\infty} \mu^n \leq \frac{\delta}{1 - \mu} = \frac{\epsilon \eta (1 - \lambda)}{1 - \lambda},$$

which is less than ϵ. □

The previous result in particular tells us that the fixed point of a contraction depends continuously on the contraction. This part can be proved without differentiability:

Proposition 2.2.21 *If $f \colon \mathbb{R} \times (a, b) \to \mathbb{R}$ is continuous and $f_y := f(\cdot, y)$ satisfies $|f_y(x_1) - f_y(x_2)| \leq \lambda |x_1 - x_2|$ for all $x_1, x_2 \in \mathbb{R}$ and all $y \in (a, b)$, then the fixed point $g(y)$ of f_y depends continuously on y (see Figure 2.2.3).*

Proof Since

$$|x - g(y)| \leq \sum_{i=0}^{\infty} |f_y^i(x), f_y^{i+1}(x)| \leq \frac{1}{1 - \lambda} |x - f_y(x)|,$$

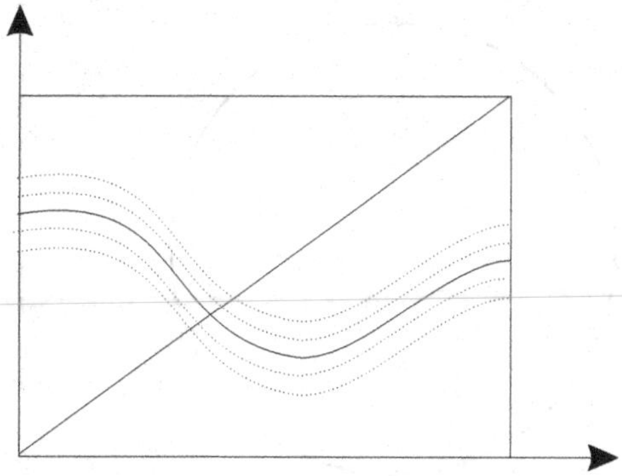

Figure 2.2.3. Continuous dependence of the fixed point.

we take $x = g(y') = f_{y'}(g(y'))$ to get

$$|g(y') - g(y)| \leq \frac{1}{1-\lambda} |f_{y'}(g(y')) - f_y(g(y'))|. \qquad \square$$

This also works in greater generality (Proposition 2.6.14), and an even stronger result in this direction is given by Theorem 9.2.4.

2.2.7 Attracting Fixed Points

At this point we have encountered two kinds of *stability*: Given a contraction, each individual orbit exhibits stable behavior in that every nearby orbit (actually, every orbit) has precisely the same asymptotics. Put differently, a little perturbation of the initial point has no effect on the asymptotic behavior. This constitutes the stability of orbits. On the other hand, Proposition 2.2.20 and Proposition 2.2.21 tell us that contractions are stable as a system; that is, when we perturb the contracting map itself, then the qualitative behavior of all orbits remains the same, and the fixed point changes only slightly.

This is a good time to make precise what we mean by a stable fixed point. As we said, we want every nearby orbit to be asymptotic to it. However, this is not all we want, as Figure 2.2.4 shows, where we have a semistable fixed point. Such a map can be given, for example, as $f(x) = x + (1/4) \sin^2 x$ if the circle is represented as $\mathbb{R}/\mathbb{Z}$ (see Section 2.6.2). We need to make sure that no nearby points ever stray far. But, as the example

$$f(x) = \begin{cases} -2x & x \leq 0 \\ -x/4 & x > 0 \end{cases}$$

(or Figure 3.1.3) shows, we must allow points to go a little further for a while.

Definition 2.2.22 A fixed point p is said to be *Poisson stable* if, for every $\epsilon > 0$, there is a $\delta > 0$ such that if a point is within δ of p then its positive semiorbit is within ϵ of p. The point p is said to be *asymptotically stable* or an *attracting fixed point* if it is Poisson stable and there is an $a > 0$ such that every point within a of p is asymptotic to p.

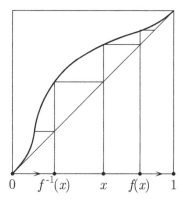

Figure 2.2.4. Not an attracting fixed point.

2.2.8 The Newton Method

A refined application of linear approximation to an otherwise difficult problem is the Newton method for finding roots of equations (see Figure 2.2.5), which we saw in Section 1.3.2.2. Doing this exactly is often difficult or impossible, and the roots are rarely expressible in closed form. The Newton method can work well to find a root in little computational time given a reasonable initial guess. To see how, consider a function f on the real line and suppose that we have a reasonable guess x_0 for a root. Unless the graph intersects the x-axis at x_0, that is, $f(x_0) = 0$, we need to improve our guess. To that end we take the tangent line and see at which point x_1 it intersects the x-axis by setting $f(x_0) + f'(x_0)(x_1 - x_0) = 0$. Thus the improved guess is

$$x_1 = x_0 - \frac{f(x_0)}{f'(x_0)}.$$

Example 2.2.23 When we start from $x_0 = 4$ for the function $x^2 - 17$, this improved guess is

$$x_1 = x_0 - \frac{x_0^2 - 17}{2x_0} = \frac{x_0}{2} + \frac{17}{2x_0} = \frac{33}{8}.$$

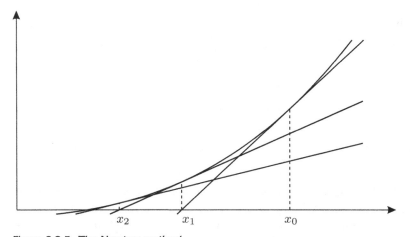

Figure 2.2.5. The Newton method.

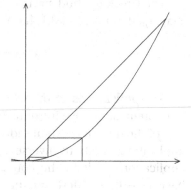

Figure 2.2.6. A superattracting fixed point.

One further step gives

$$x_2 = \frac{33}{16} + \frac{17 \cdot 8}{2 \cdot 33} = \frac{33^2 + 17 \cdot 64}{16 \cdot 33}.$$

Iteratively one can improve the guess to $x_3, \ldots$ using the same formula. With a good initial guess few steps usually give a rather accurate solution. (Indeed, x_2 is already off by less than 10^{-6}.) It is easy to see why: We are applying the map $F(x) := x - (f(x)/f'(x))$ repeatedly, and the desired point has the following property:

Definition 2.2.24 A fixed point x of a differentiable map F is said to be *superattracting* if $F'(x) = 0$ (see Figure 2.2.6).

Proposition 2.2.25 *If* $|f'(x)| > \delta$ *and* $|f''(x)| < M$ *on a neighborhood of the root* r, *then* r *is a superattracting fixed point of* $F(x) := x - f(x)/f'(x)$.

Proof $F(r) = r$ and $F'(x) = f(x)f''(x)/(f'(x))^2$. □

Remark 2.2.26 A small first derivative might cause the intersection of the tangent line with the x-axis to go quite far from x_0. The hypothesis $|f''(x)| < M$ holds whenever f'' is continuous.

At a superattracting fixed point we have quadratically (that is, superexponentially) converging iterates, as in the case of the fixed point of the quadratic map f_2 in Section 2.5. In other words, the error is approximately squared in every iteration.

This argument only works if the initial guess is fairly good. With an unfortunate initial choice the iterates under F can behave rather erratically. In other words, F has an attracting fixed point but may otherwise have quite complicated dynamics.

The special case of extracting roots by the Newton method had an ancient precursor.

Proposition 2.2.27 *Approximating* $\sqrt{z}$ *by the Newton method with initial guess* 1 *is the same as using the first components of the Greek root extraction method in* (1.3.1).

Proof With initial guess 1 the Newton method gives the recursion

$$x_0 = 1, \qquad x_{n+1} = x_n - \frac{x_n^2 - z}{2x_n} = \frac{1}{2}\left(x_n + \frac{z}{x_n}\right).$$

The Greek method starts with $(x_0, y_0) = (1, z)$, and the recursion $(x_{n+1}, y_{n+1}) = f(x_n, y_n)$ defined by (1.3.1) has the property that $y_n = z/x_n$. Therefore we have

$$x_{n+1} = \frac{x_n + y_n}{2} = \frac{1}{2}\left(x_n + \frac{z}{x_n}\right). \qquad \square$$

2.2.9 Applications of the Contraction Principle

The foremost tool we introduced in this chapter is the Contraction Principle. It is one of the most important individual facts in analysis and dynamical systems. Its applications are both diverse and fundamental. We do not only have numerous applications of it in the course of our development of dynamics, but several basic facts that underlie the theory are consequences of the Contraction Principle. Chapter 9 is devoted to such applications. While it is located in the Panorama, it is closely connected to the present material and maintains the same standard of rigor. It provides the Inverse- and Implicit-Function Theorems, which are fundamental to analysis (Theorem 9.2.2 and Theorem 9.2.3). As we mentioned, the fixed point of a contraction persists under perturbations, and Chapter 9 gives the most general condition on a fixed point for such persistence (Proposition 9.3.1). Also, the theory concerning existence and uniqueness of differential equations, on which in a manner of speaking half of dynamics is based, is derived there from the Contraction Principle (Theorem 9.4.3). A result central to dynamical systems of the type discussed in Chapter 7 is the Stable Manifold Theorem (Theorem 9.5.2; see the comments at the end of Section 10.1). It also depends crucially on the Contraction Principle.

▓ EXERCISES

▓ **Exercise 2.2.1** Show that entering any number on a calculator and repeatedly pressing the sin button gives a sequence that goes to zero. Prove that convergence is not exponential if we use the radian setting and exponential if we use the degree setting. In the latter case, find out how many iterates are needed to obtain a number less than 10^{-10} times the initial input.

▓ **Exercise 2.2.2** If one enters a number greater than 1 on a calculator and repeatedly hits the square root key, the resulting numbers settle down eventually. Prove that this always happens and determine the limit. If the calculator keeps k binary digits internally, roughly how long does it take for the sequence to settle down to this limit of accuracy?

▓ **Exercise 2.2.3** Do the previous exercise for initial values in $(0, 1]$.

▓ **Exercise 2.2.4** Show that x^2 defines a λ-contraction on $[-\lambda/2, \lambda/2]$.

▓ **Exercise 2.2.5** This is a variation on Fibonacci's problem of rabbit populations, taking mortality into account.

A population of polar lemmings evolves according to the following rules. There are equal numbers of males and females. Each lemming lives for two years and dies in the third winter of its life. Each summer, each female lemming produces an offspring of four. In the first summer there is one pair of one-year old lemmings. Let x_n be the total number of lemmings during the nth year. Use the Contraction Principle to show that x_{n+1}/x_n converges to a limit $\omega > 1$. Calculate ω.

■ **Exercise 2.2.6** Let x be a fixed of point of a map f on the real line such that $|f'(x)| = 1$ and $f''(x) \neq 0$. Show that arbitrarily close to x there is a point y such that the iterates of y do not converge to x.

■ **Exercise 2.2.7** Which of the following are convex? $\{(x, y) \in \mathbb{R}^2 \mid xy > 1\}$, $\{(x, y) \in \mathbb{R}^2 \mid xy < 1\}$, $\{(x, y) \in \mathbb{R}^2 \mid x + y > 1\}$, $\{(x, y) \in \mathbb{R}^2 \mid x > y^2\}$.

■ **Exercise 2.2.8** Prove that the norm of a matrix defined in (2.2.8) is a norm in the sense of Definition A.1.29.

■ **Exercise 2.2.9** Show that $\|A\| \leq \sqrt{\sum_{i,j} a_{ij}^2}$ for any $n \times n$ matrix $A = (a_{ij})_{1 \leq i, j \leq n}$.

■ **Exercise 2.2.10** Show that $\|A\| \geq |\det A|^{1/n}$ for any $n \times n$ matrix $A = (a_{ij})_{i, j=1,\dots,n}$.

■ **Exercise 2.2.11** Prove that the norm of a matrix is a continuous function of its coefficients.

■ **PROBLEMS FOR FURTHER STUDY**

■ **Problem 2.2.12** Construct an example of an open connected subset U of the plane $\mathbb{R}^2$ and a continuously differentiable map $f: U \to U$ such that $\|Df_x\| < \lambda < 1$ for all $x \in U$ but f is not a contraction. (Such a set cannot be convex.)

■ **Problem 2.2.13** Suppose that I is a closed bounded interval and $f: I \to I$ is such that $d(f(x), f(y)) < d(x, y)$ for any $x \neq y$ (this is weaker than the assumption of the Contraction Principle). Prove that f has a unique fixed point $x_0 \in I$ and $\lim_{n \to \infty} f^n(x) = x_0$ for any $x \in I$.

■ **Problem 2.2.14** Show that the assertion of the previous exercise is not valid for $I = \mathbb{R}$ by constructing a map $f: \mathbb{R} \to \mathbb{R}$ such that $d(f(x), f(y)) < d(x, y)$ for $x \neq y$, f has no fixed point, and $d(f^n(x), f^n(y))$ does not converge to zero for some x, y.

2.3 NONDECREASING MAPS OF AN INTERVAL AND BIFURCATIONS

As a next step we look at maps that may have several fixed points. Although the behavior of any such map is hardly more complicated than that of a contraction, it may, unlike that of contractions, change radically under perturbations.

2.3.1 Nondecreasing Interval Maps
We now study the situation where the dynamics is similar to that of a contraction, but there is no guarantee of exponentially fast convergence to a fixed point. This situation is instructive because it demonstrates an important method in low-dimensional dynamics, the systematic use of the Intermediate-Value Theorem.

Definition 2.3.1 If $I \subset \mathbb{R}$ is an interval, then $f: I \to \mathbb{R}$ is said to be *increasing* if $x > y \implies f(x) > f(y)$ and *decreasing* if $x > y \implies f(x) < f(y)$. We say that f is *nondecreasing* if $x \geq y \implies f(x) \geq f(y)$ and *nonincreasing* if $x \geq y \implies f(x) \leq f(y)$.

The simple example situation, and a useful building block in the theory of nondecreasing maps, is the following observation.

Lemma 2.3.2 *If $I = [\alpha, \beta] \subset \mathbb{R}$ is a closed bounded interval and $f: I \to I$ a nondecreasing map without fixed points in (α, β), then one endpoint of I is fixed and all orbits converge to it, except for the other endpoint if it is fixed as well. If f is invertible, then both endpoints are fixed and all orbits of points in (α, β) are positively asymptotic to one endpoint and negatively asymptotic to the other.*

Remark 2.3.3 (Cobweb Pictures/Graphical Computing) For any particular example of such a function this is quite obvious when one draws a picture. For the picture it helps to use the device of "cobweb pictures" or "graphical computing." It goes like this: To determine the orbit of a point x, find x on the horizontal axis and draw a vertical line segment that connects it to the graph. This locates the point $(x, f(x))$ on the graph. The x-coordinate of the next point of the orbit is $f(x)$, and therefore a horizontal line from $(x, f(x))$ to the diagonal [that is, to the point $(f(x), f(x))$] gives the new x-coordinate. Now draw again a vertical line to the graph followed by a horizontal line to the diagonal and repeat. Figure 2.3.1 shows how easy this is even for a slightly less simple function than in this lemma.

The idea behind this lemma is that, since the graph is not allowed to intersect the diagonal over (α, β) (no fixed points), it is either entirely above or below it. If it is above, then $f(x) > x$; so every orbit is an increasing sequence, and points move inexorably to the right. Since every orbit is bounded above, it must converge, and then one has to see that the limit is always β. If f is invertible, then points should move in the opposite direction (towards α) when one iterates the inverse. Drawing pictures helps to figure out or understand a proof, but it is not quite a substitute. Here is the proof.

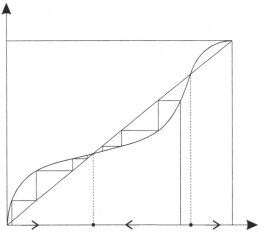

Figure 2.3.1. Dynamics of an increasing interval map.

Proof $f(\alpha) \geq \alpha$ and $f(\beta) \leq \beta$ since $f(I) \subset I$, so $(f - \text{Id})(\alpha) \geq 0$ and $(f - \text{Id})(\beta) \leq 0$, where Id is the identity. On the other hand, by the Intermediate-Value Theorem, the continuous function $f - \text{Id}$ cannot change sign on I because it is never zero on (α, β) by assumption. Thus we must have either $f(\alpha) = \alpha$ (if $f(x) < x$ on (α, β)) or $f(\beta) = \beta$ (if $f(x) > x$ on (α, β), or both. To be specific, suppose $f(x) > x$ on (α, β) and hence that β is a fixed point. Then for any $x \in (\alpha, \beta)$ the sequence $x_n := f^n(x)$ is increasing and bounded above by β; hence it is convergent to some $x_0 \in (\alpha, \beta]$. But then by continuity

$$(2.3.1) \qquad f(x_0) = f\left(\lim_{n\to\infty} x_n \right) = \lim_{n\to\infty} f(x_n) = \lim_{n\to\infty} x_{n+1} = x_0,$$

so $x_0 \in \text{Fix}(f) \cap (\alpha, \beta] = \{\beta\}$. In case $f(y) < y$ on (α, β), we would similarly conclude that $f^n(x) \to \alpha$ for all $x \in (\alpha, \beta)$ as $n \to \infty$.

If f is invertible, then $z := f(y) > y$ implies $f^{-1}(z) = y < f(y) = z$; so if $f(x) > x$ on (α, β) and hence $f^n(x) \to \beta$ as $n \to \infty$, then $f^{-1}(x) < x$ on (α, β) and $f^{-n}(x) \to \alpha$ as $n \to \infty$ by the above arguments, so every $x \in (\alpha, \beta)$ is positively and negatively asymptotic to opposite ends of $[\alpha, \beta]$. $\square$

This behavior suggests the following useful terminology:

Definition 2.3.4 If $f\colon X \to X$ is an invertible map and $x \in X$ a point such that $\lim_{n\to\infty} f^{-n}(x) = a$ and $\lim_{n\to\infty} f^n(x) = b$, then x is said to be *heteroclinic* to a and b. If $a = b$, then x is said to be a *homoclinic* point of a.

At this point we have no examples of homoclinic points. There are none in the case of nondecreasing interval maps because all orbits are monotonic. Figure 2.2.4 shows how homoclinic points can arise on the circle.

After the situation of Lemma 2.3.2, the next simple type of asymptotic behavior is the convergence of every orbit to a fixed point, but with the possibility that different orbits converge to different fixed points. This occurs in the case of increasing functions of a real variable viewed as maps.

Proposition 2.3.5 *If $I \subset \mathbb{R}$ is a closed bounded interval and $f\colon I \to I$ is a non-decreasing continuous map, then all $x \in I$ are either fixed or asymptotic to a fixed point of f. If f is increasing (hence invertible), then all $x \in I$ are either fixed or heteroclinic to adjacent fixed points.*

Proof Figure 2.3.1 shows how to prove this. The direction of motion is indicated by the sign of $f - \text{Id}$: If $(f - \text{Id})(x) < 0$, then $f(x) - x < 0$; so $f(x) < x$ and x moves left. Conversely when $(f - \text{Id})(x) > 0$.

We first show that there are fixed points. Write $I = [a, b]$ and consider the continuous map $f - \text{Id}\colon I \to \mathbb{R}$ given by $f(x) - x$. Since $f(I) \subset I$ we have $f(a) \geq a$ and $f(b) \leq b$; hence $(f - \text{Id})(a) \geq 0$ and $(f - \text{Id})(b) \leq 0$, so $(f - \text{Id})(x) = 0$ for some $x \in I$ by the Intermediate-Value Theorem. But then $f(x) = x$ and x is a fixed point.

The set $\text{Fix}(f)$ of fixed points of f is closed because it is the set of zeros of the continuous function $f - \text{Id}$. If $\text{Fix}(f) = I$, then each point is fixed and we are done.

Otherwise, $I \smallsetminus \mathrm{Fix}(f)$ is a nonempty open set and can be written as a disjoint union of open intervals. Consider an interval (α, β) from this collection. Either $\alpha, \beta \in \mathrm{Fix}(f)$ or one of α, β is an endpoint of I, so $f(\alpha) \geq \alpha$ and $f(\beta) \leq \beta$ regardless. If $y \in [\alpha, \beta]$, then $\alpha \leq f(\alpha) \leq f(y) \leq f(\beta) \leq \beta$ because f is nondecreasing. This shows that $f([\alpha, \beta]) \subset [\alpha, \beta]$. Thus all orbits in $[\alpha, \beta]$ are asymptotic to a fixed point by Lemma 2.3.2. If f is increasing, hence invertible, then Lemma 2.3.2 shows that all points of (α, β) are heteroclinic to the endpoints. of $[\alpha, \beta]$. $\square$

In Figure 2.3.1 the left one of the marked fixed points is an attracting fixed point (Definition 2.2.22); the other one is of the opposite kind. This other kind is as common as attracting fixed points and deserves a name also.

Definition 2.3.6 A fixed point x is said to be a *repelling fixed point* (or a *repeller*) if for every $\epsilon > 0$ and every y within ϵ of x there is an $n \in \mathbb{N}$ such that the positive semiorbit of $f^n(y)$ has no points within ϵ of x.

2.3.2 Bifurcations

For contracting maps the individual stability of orbits coexisted with the stability of the behavior of the whole system under perturbation. This is not necessarily the case in the present context, where the dynamics can have different qualitative features. Here, there are several fixed points, and orbits may be attracted to any given one, or not. And the number of fixed points is not prescribed either. This makes it interesting to look into the way the qualitative picture can change as one changes the map. That is, one can look for the transitions between different behaviors in families of nondecreasing maps depending on a parameter and find the values of the parameter where such changes occur.

These transitions are called *bifurcations*. An example is given by the family of maps illustrated in Figure 2.3.2. Here we first have two fixed points (left picture), one attracting and one repelling. At one specific parameter value, the *bifurcation parameter*, the two merge into one (middle picture), which immediately disappears (right picture) for larger values of the parameter. Of course, outside the pictured portion there must be another fixed point because we saw in the last proof that there always must be at least one by the Intermediate-Value Theorem. This is sometimes called a *saddle-node bifurcation*. This terminology comes from differential equations, where the corresponding bifurcation in the more visually distinctive

Figure 2.3.2. Basic (tangent) bifurcation.

two-dimensional situation consists of the merging of two equililibria, a *saddle* (Figure 3.1.6) and a *node* (Figure 3.1.2), which cancel each other and disappear.

We initially have one new stable and unstable fixed point each, and the single fixed point present for the bifurcation parameter is semistable (as in Figure 2.2.4); that is, points approach it from one side and leave from the other. For higher values of the parameter this fixed point is gone, but for moderate amounts of time this has little effect, because it takes an orbit a long time to pass by the "bottleneck" between the diagonal and the graph (lots of zigzags in a cobweb diagram).

■ EXERCISES

■ **Exercise 2.3.1** Prove that the set of zeros of a continuous function is closed.

■ **Exercise 2.3.2** Let $f: [0, 1] \to [0, 1]$ be a *nonincreasing* continuous map. What are the possible periods for periodic points for such a map?

■ **Exercise 2.3.3** In the proof of Proposition 2.2.8, show that one can prove that the limit is a fixed point using an argument like (2.3.1) (no exponential estimates) in place of (2.2.4).

■ **Exercise 2.3.4** Prove that for any closed set $E \subset \mathbb{R}$ there is a continuous strictly increasing map $f: \mathbb{R} \to \mathbb{R}$ such that $\mathrm{Fix}(f) = E$.

■ PROBLEMS FOR FURTHER STUDY

■ **Problem 2.3.5** Prove that for a closed set $E \subset \mathbb{R}$ there is a continously differentiable strictly increasing map $f: \mathbb{R} \to \mathbb{R}$ such that $\mathrm{Fix}(f) = E$.

■ **Problem 2.3.6** Looking at Figure 2.2.4, we remarked that for nondecreasing interval maps a fixed point to which all other points are positively asymptotic must be an attractor. Prove that this is true even if the map is not monotone.

2.4 DIFFERENTIAL EQUATIONS

In this section we develop the simple dynamics of maps hitherto observed in the context of differential equations. First we show the counterpart to nondecreasing interval maps, differential equations on the line. We include a specific discussion of a continuous-time logistic growth model. Both parts contain instances of attracting fixed points, the obvious analog to those for maps. However, there is a second analog to attracting fixed points for maps, and we introduce it in Section 2.4.3.

2.4.1 Differential Equations on the Line

We now prove a result analogous to Proposition 2.3.5 in the continuous-time situation. Consider the first-order differential equation $\dot{x} = f(x)$, where we assume that f is Lipschitz-continuous (see Definition 2.2.1). Consider the set of zeros of f, which are the constant solutions (equilibria).

Remark 2.4.1 The set of zeros is a closed set because f is continuous. Therefore its complement is open and can be written as a disjoint union of ("complementary") open intervals.

If we consider one of these intervals at a time, then the following result gives a full description of what happens, in perfect analogy to Lemma 2.3.2.

Lemma 2.4.2 *Consider a Lipschitz-continuous function f and suppose $f \neq 0$ on (a, b) and $f(a) = f(b) = 0$. Then, for any initial condition $x_0 \in I$, the corresponding solution of $\dot{x} = f(x)$ is monotone. It is increasing (and asymptotic to b) if $f > 0$ on I, decreasing (and asymptotic to a) otherwise.*

Proof Suppose $f(x_0) > 0$ to be definite (the other case works the same way). It is easy to see that the solution increases *so long as it is in* (a, b). The point is to show that it can't leave that interval.

Since $\dot{x}(0) = f(x(0)) = f(x_0) > 0$, the solution initially increases. If it ever becomes decreasing, then we must have a maximum $x(t_0) = c$ at that time, which implies that $f(c) = 0$ and therefore $c = b$. We need to check that this never happens, that is, $x(t) \neq b$ for all time. For this there are two ways, the honest one and the easy one. We begin with the honest one.

We can write the solution of a differential equation $\dot{x} = f(x)$ as $x(t) = x(0) + \int_0^t f(x(s)) \, ds$ [because differentiating both sides gives $\dot{x} = f(x)$ by the Fundamental Theorem of Calculus]. For our problem write $dx/dt = f(x)$ and by the Inverse-Function Theorem $dt/dx = 1/f(x)$, so in integral form $t(x) = \int_{x_0}^x (1/f(s)) \, ds$. Since f is Lipschitz-continuous, we have $f(s) = f(s) - f(b) \leq C(b - s)$ for some constant C. Therefore,

$$t(x) = \int_{x_0}^x \frac{1}{f(s)} \, ds \geq \int_{x_0}^x \frac{1}{C(b - s)} \, ds.$$

If $x = b$, then this integral diverges, that is, $t(x) = \infty$. This shows that $x(t) < b$ for all (finite) t and furthermore that $x(t) \to b$ as $t \to \infty$.

The easy way to do the last portion is to use existence and uniqueness of solutions to differential equations, which we only get to in Theorem 9.4.3. Because $\bar{x}(t) = b$ for all $t \in \mathbb{R}$ is a solution that takes the value b at some time (any time), any solution that ever reaches b is of the form $\bar{x}(t - t_0) = b$. Since we did not start from b, our solution is not this one, so it never reaches b. $\square$

Therefore also in this situation every solution has simple asymptotics: As $t \to \infty$, or $t \to -\infty$ the solution either converges to a fixed point or diverges to infinity.

By Lemma 2.4.2 it is now easy to piece together a full qualitative picture of the behavior of any differential equation in one variable. On the real line mark the points where $f = 0$. These are the fixed points. On each complementary interval determine the sign of f. If $f > 0$ on that interval, draw an arrow to the right, otherwise an arrow to the left. Figure 2.4.1 gives an example.

The type of fixed points can be classified by considering f':

Proposition 2.4.3 *When $f(x_0) = 0$ and $f'(x_0) < 0$, then x_0 is an attracting fixed point of $\dot{x} = f(x)$: Every nearby orbit is positively asymptotic to x_0. Likewise, fixed points x_0 with $f'(x_0) > 0$ are repelling: Nearby points move away from x_0.*

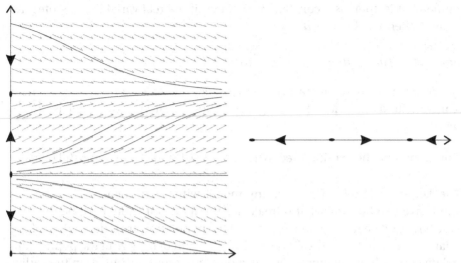

Figure 2.4.1. Solutions and phase portrait of a differential equation.

Proof If $f'(x_0) < 0$, then $f(x) < 0$ for $x > x_0$ nearby, so such points move toward x_0, and vice versa for $x < x_0$. Thus every nearby orbit is positively asymptotic to x_0. □

Therefore, the *phase portrait* of $\dot{x} = f(x)$ is immediately evident whenever all zeros of f are known and are nondegenerate (nonzero derivative). Mark all of these points and draw arrows toward those with a negative derivative and away from those with a positive derivative. The word "phase portrait" derives from physics; see Section 6.2.

Remark 2.4.4 The simple ingredient that replaces the monotonicity assumption in discrete time is that different solutions of a differential equation cannot cross. This means that order is preserved, as it is by increasing maps. We formulate this more crisply in Section 2.4.2.5.

2.4.2 The Logistic Differential Equation

In Chapter 1 we saw that some problems in population biology are naturally modeled by using discrete time; our example of a butterfly colony involved discrete-time steps of one year. Starting with Section 1.2.1, however, and from Section 1.2.5 to Section 1.2.7 we saw situations where it is appropriate to use continuous-time models for populations, notably when one tries to understand the growth of a population in a setting that changes slowly compared to the reproductive cycle of the population. Insects give typical examples again, because many species have short reproductive cycles of about a day, which makes it reasonable to study the population growth during a single summer with a continuous-time model.

1. Exponential Population Growth. The simplest model of this nature is one involving exponential growth (see Section 1.2.1): Suppose that at any given time the rate of births and deaths is a constant percentage of the total population at that time, that is, there is a constant k such that if the real variable x denotes the population then $\dot{x} = kx$ or $(d/dt)x = kx$.

Lemma 2.4.5 *The solution of $\dot{x} = kx$ is $x(t) = x(0)e^{kt}$.*

Proof Write $\dot{x}/x = k$ and integrate over t to get $\log |x| = kt + C$ or $|x| = e^{kt+C}$, which we can rewrite as $x = Ae^{kt}$. Inserting $t = 0$ shows that $A = x(0)$, that is, $x(t) = x(0)e^{kt}$. $\square$

The argument above is incorrect when $x(0) = 0$, but the statement still holds.

2. The Logistic Model. Of course, any model with such unchecked and rapid growth is likely to be somewhat unrealistic, and indeed it should not be the case that the birth and death rates as a fraction of the population are independent of the population itself, that is, that k should be independent of the size of the population. For relatively small populations this is a good approximation, but, as in the earlier butterfly example, for larger populations the limited amount of food and possibly other resources play a role. Thus there should be a saturation population that does not grow any more; and if the population were to start out at a higher number, it should shrink to the saturation level. Thus, in a manner of speaking, k should be a function of x that is zero at a (positive) saturation value L of x (no growth) and negative for larger values of x (shrinking population). If we take a linear function $k = a(L - x)$ with $a > 0$ to do this, then we get the differential equation

$$(2.4.1) \qquad \frac{d}{dt}x = ax(L - x).$$

The qualitative behavior of the solutions is easy to develop using Proposition 2.4.3. However, this situation is so simple that we can even solve the differential equation explicitly.

Lemma 2.4.6 *The solution of $\dot{x} = ax(L - x)$ is*

$$x(t) = \frac{Lx(0)}{x(0) + (L - x(0))e^{-Lat}}.$$

Proof We separate variables, that is, bring all x's to one side:

$$a = \frac{dx/dt}{x(L - x)}.$$

Integrating over t gives

$$at + C = \int \frac{1}{x(L - x)} \frac{dx}{dt} dt = \int \frac{1}{x(L - x)} dx = \int \frac{1}{Lx} dx + \int \frac{1}{L(L - x)} dx$$

using partial fractions. Thus

$$at + C = \frac{\log |x|}{L} - \frac{\log |L - x|}{L} = \frac{1}{L} \log \left| \frac{x}{L - x} \right|.$$

Taking $t = 0$ shows that

$$CL = \log \left| \frac{x(0)}{L - x(0)} \right|, \quad \text{hence } e^{-CL} = \left| \frac{L}{x(0)} - 1 \right|.$$

Changing sign and exponentiating gives

$$e^{-Lat}\left|\frac{L-x(0)}{x(0)}\right| = e^{-L(at+C)} = \left|\frac{L}{x}-1\right| = \left|\frac{L-x(t)}{x(t)}\right| = \left|\frac{L}{x(t)}-1\right|.$$

The quantities in absolute value signs turn out to always agree in sign, so we can drop the absolute values. This gives

$$x(t) = \frac{Lx(0)}{x(0) + (L-x(0))e^{-Lat}}. \quad \square$$

3. Asymptotic Behavior. We develop the asymptotic behavior of the solutions to this differential equation. For $x(0) = L$ we get the expected constant solution $x(t) = L$. When $t \to +\infty$, the exponential term goes to zero; hence $x(t) \to L$ for any positive initial condition. If $x(0) < L$, then as $t \to -\infty$ the exponential term diverges and $x(t) \to 0$. For $x(0) > L$ or $x(0) < 0$ (the latter is biologically meaningless) the denominator is zero (the solution has a singularity) for

$$t = \frac{\log(1 - [L/x(0)])}{La},$$

which is negative for $x(0) > L$ and positive for $x(0) < 0$.

Therefore the asymptotic behavior for positive time is simple: If the initial population is zero, then it remains zero forever. If the initial population is positive but below saturation (that is, less than L), then the population increases and in the long run creeps up to the saturation population. The increase is most rapid when the population is $L/2$ because $x(L - x)$ is maximal at $L/2$. Initial populations larger than L shrink to L asymptotically. The qualitative behavior is reflected in Figure 2.4.1.

In the language of dynamical systems we have found that the fixed point L is stable: All nearby solutions tend to it in positive time. The opposite is true for the constant solution zero. Any nearby solution diverges from it, either toward the equilibrium solution L or (for biologically meaningless negative solutions) to a negative singularity.

4. Interpretation of the Blowup. The fact that solutions starting with $x(0) > L$ have a singularity for negative time is not entirely "ugly." To be specific, if $x(0) = 2L$, the singularity occurs for $t = -\log^2 /La$. This means that no matter how large the initial condition $x(0)$ is, we always have $x(t) < 2L$ for $+ \geq \log 2/La$, that is, no matter how huge the initial population $x(0)$, it shrinks to reasonable size ($2L$, say) within a period of time that is bounded independently of $x(0)$: The more there are the faster they starve. This also makes clear that populations larger than L cannot arise from this model intrinsically. The excessive population must have been transplanted into the ecosystem from the outside at some recent time.

5. One-Dimensional Flow. Let us look back at these two models (exponential and logistic) in a slightly different fashion. In both cases we obtained the solutions as functions of time that depend on the initial condition, so the set of solutions is described as a family of functions of time with a parameter from the phase space. In the first example this family is given by $x_0 e^{kt}$, where x_0 is this parameter. On the other hand, we can look at this formula and decide to view x_0 as variable and t as a parameter, or as fixed for the moment. Then instead we obtain a family of functions $\phi^t(x) = xe^{kt}$, which is now parametrized by t and has x as an independent variable. That there is something interesting about this viewpoint becomes apparent when one notices that the dependence on the parameter t is not entirely arbitrary; we have $\phi^{t+s}(x) = xe^{k(t+s)} = (xe^{kt})e^{ks} = \phi^s(\phi^t(x))$. This is a simple consequence of $e^{a+b} = e^a e^b$ and may therefore look like an accident; but the same property holds in the second example as well, where one would not expect it at first glance: If we let

$$\phi^t(x) := \frac{Lx}{x + (L - x)e^{-Lat}},$$

then we also get $\phi^{t+s}(x) = \phi^s(\phi^t(x))$ (Exercise 2.4.5). This property means that if we follow a solution for time t and then further from that value for time s, then this is the same as following the solution for time $t + s$ in the first place. We present this point of view of time-t-maps in detail in Section 9.4.7 (p. 271).

Note here simply that Remark 2.4.4 tells us that in the present situation these maps are increasing in x (for fixed t).

2.4.3 Limit Cycles

We now produce the less obvious continuous-time analog of attracting fixed points for maps. We use properties of *flows*. These appear in Section 3.2, but the definitive treatment is in Section 9.4.7. While the proof here is somewhat formidable, the facts and pictures should provide good insight.

The obvious analogs are attracting fixed points for flows such as the saturation population L in the previous example. The second analog cannot be found on the line. It is an attracting periodic orbit (periodic solution) for a differential equation in the plane or in higher dimension, such as the van der Pol equation in Section 1.2.8.

We now give a simple criterion that establishes that a periodic orbit is attracting, and this criterion is a consequence of the Contraction Principle, specifically Proposition 2.2.19. We show that if ϕ^t is flow and p a periodic point with period T, then from the differential of ϕ^T at p one can infer that the orbit $\mathcal{O}(p)$ of p is attracting.

As a first step we show that the flow direction is irrelevant for the study of stability.

Lemma 2.4.7 *If p is T-periodic and not fixed for $\dot{x} = f(x)$, then 1 is an eigenvalue of $D\phi_p^T$ (see Section 2.2.4.1).*

Proof $f(p) = f(\phi^T(p)) = (d/ds)\phi^s(p)|_{s=T} = (d/ds)\phi^T \circ \phi^s(p)|_{s=0} = D\phi_p^T f(p)$. Thus, $f(p)$ is an eigenvector for $D\Phi_p^T$ with eigenvalue 1. $\square$

Therefore we ignore this eigenvalue from now on:

Definition 2.4.8 *If p is a T-periodic point and the eigenvalues of $D\phi_p^T$ are $\lambda_1, \ldots, \lambda_{n-1}, 1$ (not necessarily distinct), then $\lambda_1, \ldots, \lambda_{n-1}$ are called the eigenvalues at p.*

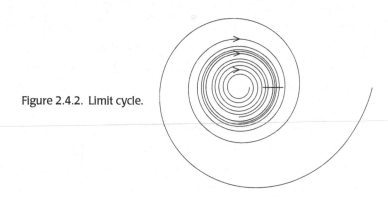

Figure 2.4.2. Limit cycle.

Remark 2.4.9 These eigenvalues depend only on the orbit: If $q = \phi^s(p)$, then $\phi^T \circ \phi^s = \phi^s \circ \phi^T$ implies $D\phi_q^T D\phi_p^s = D\phi_p^s D\phi_p^T$, that is, the linear maps $D\phi_q^T$ and $D\phi_p^T$ are conjugate via $D\phi_p^s$; hence the eigenvalues at p and q coincide.

Here is the promised criterion.

Proposition 2.4.10 *If p is a periodic point with all eigenvalues of absolute value less than 1, then the orbit $\mathcal{O}(p)$ of p is a limit cycle, that is, it has a neighborhood whose every point is positively asymptotic to $\mathcal{O}(p)$.*

Proof In order to apply Proposition 2.2.19 we construct a map that reflects the dynamics and the eigenvalue information. To that end consider the flow direction at p and pick a small piece from its orthogonal subspace. This is a little disk S containing p such that the orbit of p crosses it, as shown in Figure 2.4.2. We need to use continuity up to time $1.1 \cdot T$, say (Proposition 9.4.5) several times, which can be stated thus:

Lemma 2.4.11 *Given $\epsilon > 0$, there is a $\delta > 0$ such that any point within δ of $\mathcal{O}(p)$ will remain within ϵ of $\mathcal{O}(p)$ for time $1.1 \cdot T$.*

Taking ϵ such that S contains an ϵ-disk around p we find that, whenever $q \in S$ is sufficiently close to p, its orbit again intersects S after time less than $1.1 \cdot T$. This means that on a neighborhood of p in S there is a well-defined *return* map F_p^S. By smoothness (Proposition 9.4.6) and the Implicit-Function Theorem 9.2.3 F_p^S is smooth.

Proposition 2.4.12 *The eigenvalues at p coincide with those of DF_p^S.*

Proof If we denote the projection to S parallel to $f(p)$ by $\pi \colon \mathbb{R}^n \to S$ (see Figure 2.4.3) then the differential of $F^S(x) = \phi^{t_x}(x)\restriction_S$ as a map into S is

$$DF_p^S = \pi \left(D\phi_p^{t_p}\restriction_S + \dot{\phi}^{t_p}(p) Dt_p\restriction_S \right).$$

Applying π to $\dot{\phi}^{t_p}(p) Dt_p\restriction_S = f(\phi^{t_p}(p)) Dt_p\restriction_S = f(p) Dt_p\restriction_S$ gives zero, so $DF_p^S = \pi D\phi_p^{t_p}\restriction_S =: A$. But, on the other hand, extending a basis of S to one of $\mathbb{R}^n$ by adding $f(p)$ gives the coordinate representation $D\phi_p^{t_p} = \begin{pmatrix} A & 0 \\ * & 1 \end{pmatrix}$. $\square$

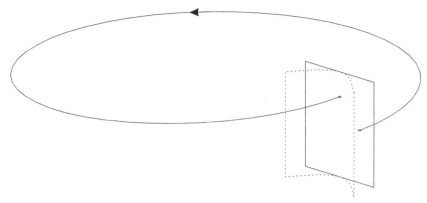

Figure 2.4.3. Projection to a section.

This means that, by Proposition 2.2.19, p is an attracting fixed point of F^S with a neighborhood $U \subset S$ of attraction. By Lemma 2.4.11, every point close enough to the orbit of p will encounter U and from then on do so at intervals less than $1.1 \cdot T$. The resulting return points converge to p. Again by Lemma 2.4.11, the entire positive semiorbit of q then converges to p. $\square$

Remark 2.4.13 Proposition 2.2.20 and Proposition 2.2.21 tell us that the fixed point of a contraction persists under perturbation. Because they are obtained from the Contraction Principle, limit cycles have the same property. If in the situation of Proposition 2.4.10 the dynamical system is perturbed slightly, then the map F_p^S on S defined above will be slightly perturbed. Since it is a contraction on $V \subset S$, the same goes for its perturbation. Accordingly, there is still a unique fixed point in S, which gives rise to a periodic point for the flow, and the last paragraph of the proof of Proposition 2.4.10 then shows that this produces a limit cycle.

▨ EXERCISES

▨ **Exercise 2.4.1** Explain the disappearance of the absolute value signs in the proof of Lemma 2.4.5.

▨ **Exercise 2.4.2** Consider the differential equation $\dot{x} = -x^k$ for $k > 1$. Denote the solution with initial condition $x_0 > 0$ by $x(t)$. Prove that there is a number $s > 0$ such that the limit $\lim_{t \to \infty} x(t)/t^s$ is finite and not zero.

▨ **Exercise 2.4.3** Give an example of a differential equation with a right-hand side for which a nonconstant solution coincides with a constant one after finite time, that is, where a nonfixed point approaches a fixed point in finite time.

▨ **Exercise 2.4.4** Give an example of a differential equation for which a solution diverges to infinity in finite time.

▨ **Exercise 2.4.5** Show that if

$$\phi^t(x) := \frac{Lx}{x + (L - x)e^{-Lat}},$$

then $\phi^{t+s}(x) = \phi^s(\phi^t(x))$.

■ **Exercise 2.4.6** Consider the differential equation $\dot{x} = f(x)$ and suppose that $f(x_0) = 0$ and $f'(x_0) < 0$. Show that for any point $x \neq x_0$ near x_0 the orbit of x converges exponentially fast to x_0.

■ **PROBLEM FOR FURTHER STUDY**

■ **Problem 2.4.7** Give an example of a differential equation on the plane with a differentiable right-hand part that has a limit cycle such that convergence of nearby orbits to it is not exponential.

2.5 QUADRATIC MAPS

After this study of the logistic differential equation it is inviting to revisit the discrete-time logistic equation that was discussed in the introduction. Although this map has complicated dynamics for large values of the "fertility" parameter, for smaller values of the parameter it is possible to make some deductions about its behavior using ideas from this chapter. The discussion of interval maps in the previous section depended on monotonicity and is therefore not directly applicable to the map $f(x) = \lambda x(1 - x)$. Nevertheless, it is sometimes possible to establish simple dynamics without monotonicity by combining local monotonicity with the concept of contraction to study the dynamics.

2.5.1 Attracting Fixed Points for Small Parameter Values

The family of maps $f_\lambda \colon [0, 1] \to [0, 1]$, $f_\lambda(x) := \lambda x(1 - x)$ for $0 \leq \lambda \leq 4$ is called the *quadratic family*. It is the most popular model in one-dimensional dynamics, both real and complex (in the latter case the maps are extended to $\mathbb{C}$, and often complex values of the parameter λ are considered). These maps are evidently not monotonic but increase on $[0, 1/2]$ and decrease on $[1/2, 0]$. On the other hand, for $\lambda < 3$ the dynamics are simple, according to the next two results. Recall that experimentation recommended in the introduction suggests this.

Proposition 2.5.1 *For $0 \leq \lambda \leq 1$ all orbits of $f_\lambda(x) = \lambda x(1 - x)$ on $[0, 1]$ are asymptotic to 0 (see Figure 2.5.1).*

Proof $f_\lambda(x) = \lambda x(1 - x) \leq x(1 - x) < x$ for $x \neq 0$ so $(f^n(x))_{n \in \mathbb{N}}$ decreases and is bounded below by zero; hence it is convergent by completeness (Section A.1.2). The limit is a fixed point by (2.3.1), and hence must be 0. (Another approach is to notice that $f_\lambda([0, 1]) \subset [0, \lambda/4] \subset [0, 1/2]$ and that f_λ is monotonic on $[0, 1/2]$, so

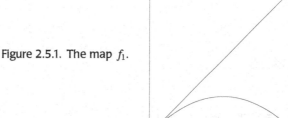

Figure 2.5.1. The map f_1.

we could use the arguments for monotone maps after the first application of f_λ.)
For $\lambda < 1$ we can use the Contraction Principle because $|f'_\lambda(x)| = \lambda|1 - 2x| \le \lambda < 1$.
This shows in addition that all orbits for $0 \le \lambda < 1$ (but not for $\lambda = 1$) are asymptotic
to the fixed point 0 with exponential speed. $\square$

For $\lambda > 1$ the situation changes slightly.

Proposition 2.5.2 *For $1 < \lambda \le 3$ all orbits of $f_\lambda(x) = \lambda x(1 - x)$ on $[0, 1]$, except for*
0 and 1, are asymptotic to the fixed point $x_\lambda = 1 - (1/\lambda)$.

Proof $f_\lambda(x) = x$ is equivalent to the quadratic equation $0 = \lambda x(1 - x) - x = \lambda x(1 - x - (1/\lambda))$ with the nonzero solution $x_\lambda = 1 - (1/\lambda)$, which is a new fixed
point in $[0, 1]$ if $\lambda > 1$. (Notice that for $\lambda < 1$ this solution is also present but negative.)

Case 1: $1 < \lambda \le 2$ In this case, $x_\lambda < 1/2$ (Figure 2.5.2) and f_λ is an increasing map
of the interval $[0, x_\lambda]$ to itself with $f_\lambda(x) = \lambda x(1 - x) > x$, and thus every point x of
$[0, x_\lambda]$ is positively asymptotic to x_λ.

Now look at points to the right of x_λ. Notice that $f_\lambda(1 - x) = f_\lambda(x)$, that is,
the function f_λ is symmetric around $x = 1/2$. Therefore, $f_\lambda([1 - x_\lambda, 1]) \subset [0, x_\lambda]$,
and every $x \in (1 - x_\lambda, 1]$ is also asymptotic to x_λ. It remains to examine points in
$(x_\lambda, 1 - x_\lambda)$. Again, since f_λ is symmetric,

$$f_\lambda([x_\lambda, 1 - x_\lambda]) \subset [x_\lambda, f(1/2)] = [x_\lambda, \lambda/4] \subset [x_\lambda, 1 - x_\lambda],$$

so f maps this interval into itself. Furthermore,

$$|f'_\lambda(x)| = \lambda|1 - 2x| \le \lambda|1 - 2x_\lambda| = \lambda\left|1 - 2\left(1 - \frac{1}{\lambda}\right)\right| = |2 - \lambda| < 1$$

for $1 < \lambda \le 3$ and $x \in [x_\lambda, 1 - x_\lambda]$, so f_λ is a contraction of $[x_\lambda, 1 - x_\lambda]$. Hence all
points of this interval are asymptotic to the only fixed point x_λ in this interval. In
conclusion, we have shown that every orbit of the map f_λ for $1 < \lambda \le 2$ (other than
0 and 1) is asymptotic to the nonzero fixed point of this map.

Case 2: $2 < \lambda < 3$. In this case (see Figure 2.5.3) we can apply a similar
argument, but it is rather more involved.

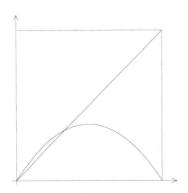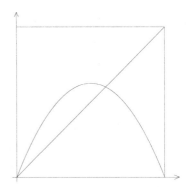

Figure 2.5.2. The maps $f_{1.5}$ and $f_{2.5}$.

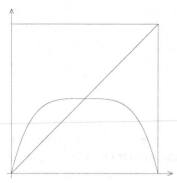 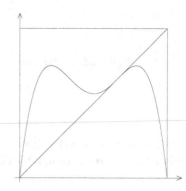

Figure 2.5.3. The maps f_2^2 and f_3^2.

The above calculation shows that the fixed point is attracting, but to show that it attracts all points is harder and we do not use the Contraction Principle. In this case the nonzero fixed point $x_\lambda = 1 - 1/\lambda$ of f_λ is to the right of $1/2$, so f_λ is no longer increasing on $[0, x_\lambda]$. First we show that, if $I := [1 - x_\lambda, f_\lambda(1/2)] = [1/\lambda, \lambda/4]$, then $f_\lambda(I) \subset I$. To that end observe that $f_\lambda(x) \le f_\lambda(1/2) = \lambda/4$ for $x \in [0, 1]$; hence for $x \in I$. On the other hand, $f(\lambda/4) = (\lambda^2/4) - (\lambda^3/16) > 1/\lambda \in I$ because the derivative of the function $q(\lambda) = \lambda^3/4 - \lambda^4/16$ is positive for $2 < \lambda < 3$, and hence $q(\lambda) > q(2) = 1$. But $f(\lambda/4)$ is the minimum of f on I because $\lambda/4$ is the point in I furthest from $1/2$. [$\lambda/4 > x_\lambda$ because $0 \le ((\lambda/2) - 1)^2/\lambda = (\lambda/4 - (1 - (1/\lambda)))$.] This shows $f(I) \subset I$.

We want to show that every orbit except those of 0 and 1 eventually enters I. For $x \in I$ there is nothing to show. For $x \in (0, 1/\lambda)$ let $x_n := f_\lambda^n(x)$ and note that $f([0, 1/\lambda]) = [0, x_\lambda]$; so if $(x_n)_{n \in \mathbb{N}}$ has no terms contained in I, then we must have $x_n \le 1/\lambda$ for all $n \in \mathbb{N}$, which implies that $x_{n+1} > x_n$ for all $n \in \mathbb{N}$ because $f(x) > x$ on $[0, 1/\lambda]$. Being increasing and bounded above, this sequence converges to a limit $x_0 \in (0, 1/\lambda]$, which is a fixed point by (2.3.1). But there is no fixed point in this interval. Finally, $f_\lambda(x) \in [0, 1/\lambda]$ for any $x \in [\lambda/4, 1]$ and hence these points also enter I by falling into the preceding case after one step.

Next we show that $f^n(x) \to x_\lambda$ as $n \to \infty$ for $x \in I$. Unfortunately, f is not a contraction on I because $f_\lambda'(\lambda/4) = \lambda(1 - 2(\lambda/4)) < -1$ for $\lambda > 1 + \sqrt{3} < 3$. To overcome this difficulty we consider f^2. Note that $f_\lambda^2(I) \subset I$, and furthermore $f_\lambda([1/\lambda, x_\lambda]) \subset [x_\lambda, \lambda/4]$, and vice versa; that is, $[1/\lambda, x_\lambda]$ is f^2-invariant. Let us locate $y(\lambda) := f_\lambda^2(1/2) = \lambda^2(4 - \lambda)/16$ in relation to $1/2$. For $\lambda = 2$ we have equality and for $\lambda = 3$ we have $y(\lambda) = 9/16 > 1/2$. Furthermore,

$$\frac{d^2}{d\lambda^2} y(\lambda) = \frac{8 - 6\lambda}{16} < 0$$

for $\lambda \ge 2$, so $y(\lambda)$ is concave down, and hence $f_\lambda^2(1/2) > 1/2$ for $2 < \lambda \le 3$. This implies, that the interval $J := [1/2, x_\lambda]$ is strictly invariant under f^2, that is, $f^2(J) \subset (1/2, x_\lambda]$. On this interval f_λ^2 is increasing, and hence all of its points are asymptotic to a fixed point of f_λ^2 by Lemma 2.3.2. The only available fixed point is x_λ. This shows that for $x \in I$ we have $f_\lambda^{2n}(x) \to x_\lambda$ or $f_\lambda^{2n+1}(x) \to x_\lambda$ as $n \to \infty$. But in the first case, for example, we have $f_\lambda^{2n+1}(x) \to f_\lambda(x_\lambda) = x_\lambda$ as well, ending the proof. This long argument becomes quite obvious when looking at the graph of

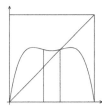

Figure 2.5.4. The map $f_{2.5}^2$.

$f_{2.5}^2$ in Figure 2.5.4. Note also that for $\lambda < 3$ the convergence to x_λ is exponential by Proposition 2.2.17, even though we did not use the Contraction Principle. $\square$

The case distinction around $\lambda = 2$ corresponds to the fact that, for this particular value of the parameter, the fixed point x_λ has derivative zero, and for larger values of λ it has negative derivative. Thus, from then on nearby points approach the fixed point by alternating around it, rather than monotonically as in the case $\lambda < 2$.

The case $\lambda = 2$ is interesting also because for f_2 the point $1/2$ is superattracting (Definition 2.2.24). Because the derivative of f_2 vanishes at $1/2$, nearby points approach $1/2$ faster than exponentially, in fact, $|f_2(x) - 1/2| = |x - 1/2|^2$, that is, the error gets squared with every step (see Figure 2.2.6). One can most conveniently see this by changing the variable x to $x' = 1/2 - x$. This *quadratic convergence* also appears in the Newton method in Section 2.2.8.

The dynamics changes substantially when $\lambda \geq 3$; this is explored in Chapter 11.

2.5.2 Stable Asymptotic Behavior

The common feature to the examples in this chapter is that all asymptotic behavior is stable; every orbit is asymptotic to a fixed point or limit cycle. In the case of contracting maps this is due to a strong assumption that forces all orbits toward the fixed point. For monotone interval maps there is the possibility of mildly more complex behavior because various fixed points are available as ultimate rest points. In this situation we saw a mechanism for increasing the dynamical complexity slightly: Bifurcations can increase the number of fixed points. Yet they do not fundamentally increase the range of behaviors. The limitations imposed by the Intermediate-Value Theorem and monotonicity prevail. Our study of the quadratic family so far has not produced much extra complexity. However, in this case this is related to the restricted range of parameters considered. The reason there is no fundamental limitation of complexity is that quadratic maps are not invertible and hence do not preserve order. This allows orbits to trade places and is geometrically reflected in the "folding" of the interval by the map. For larger parameter values this new facet produces ever more complicated dynamics.

For differential equations in the plane the complexity of the dynamics is limited in a way similar to that of nondecreasing interval maps. Any orbit that does not accumulate on a fixed point is asymptotic to a limit cycle. This *Poincaré–Bendixson Theorem* is related to the fact that, for flows in the plane, sections such as the S in the proof of Proposition 2.4.10 are intervals, and the return map F^S to them is, roughly speaking, a monotone map. (A material ingredient for this that depends also on dimension is the *Jordan Curve Theorem*, which says that, analogously to points dividing the line into two pieces, a closed curve divides the plane into two pieces.)

■ **EXERCISES**

■ **Exercise 2.5.1** Show that f_λ^2 is a contraction on $[1/2, x_\lambda]$ for $2 \le \lambda < 3$.

■ **Exercise 2.5.2** (Alternate form of the quadratic family) Given $g_\alpha(x) := \alpha - x^2$ and $h_\lambda = \lambda(x - \frac{1}{2})$, show that $h_\lambda(f_\lambda(x)) = g_\alpha(h_\lambda(x))$, where $\alpha = (\lambda^2/4) - (\lambda/2)$.

■ **Exercise 2.5.3** Suppose $f: [0, 1] \to [0, 1]$ is differentiable and $|f'(x)| \le 1$ for all x. Show that the set of fixed points of f is nonempty and connected (that is, it is a single point or an interval).

■ **Exercise 2.5.4** Suppose $f: [0, 1] \to [0, 1]$ is differentiable and $|f'(x)| \le 1$ for all x. Show that every periodic point has period 1 or 2.

2.6 METRIC SPACES

Some interesting dynamical systems do not naturally "live" in Euclidean space, and there are occasions where the study of a dynamical system benefits from considerations in an auxiliary space. Therefore we now introduce metric spaces in some generality with emphasis on three specific ones, the circle, cylinder, and torus. They are the phase spaces for many natural dynamical systems, in particular many that arise from classical mechanics (Section 5.2) and billiards. Subsequent sections pay attention to Cantor sets and "devil's" staircases. These may look pathological but appear naturally in various dynamical systems (see Proposition 4.3.19 and Section 7.4.4 for the Cantor set and the discussion near Definition 4.4.1 as well as Proposition 4.4.13 for the devil's staircase).

2.6.1 Definitions

In the arguments about contractions in Euclidean spaces we did not use the particular properties of the Euclidean distance. Indeed, we could apply the same reasoning if we were to use a different way of measuring distance on $\mathbb{R}^n$, for example, the maximum distance $d(x, y) = \max_{1 \le i \le n} |x_i - y_i|$.

Naturally, the notion of a contraction depends on the way distances are measured, so some maps that are not contractions with respect to the Euclidean distance may turn out to be contractions with respect to the maximum distance or yet another distance function on $\mathbb{R}^n$. This is useful when one is able to cleverly choose a metric that makes a given map a contraction. We come back to this reasoning in the next chapter, when we find a necessary and sufficient condition for a linear map to allow us to apply the Contraction Principle for an appropriately chosen distance function.

Even the fact that our space is a vector space or a subset of one is not important for the arguments in the proof of Proposition 2.2.10. We essentially use only the most basic properties of the Euclidean metric, such as the triangle inequality and the fact that Cauchy sequences converge. Reducing our assumptions to such basic facts brings about a fruitful generalization of the earlier situation.

Definition 2.6.1 If X is a set, then $d: X \times X \to \mathbb{R}$ is said to be a *metric* or *distance function* if

  (1) $d(x, y) = d(y, x)$ *(symmetry)*,
  (2) $d(x, y) = 0 \Leftrightarrow x = y$ *(positivity)*,

(3) $d(x, y) + d(y, z) \geq d(x, z)$ *(triangle inequality)*.

Putting $z = x$ in (3) and using (1) and (2) shows that $d(x, y) \geq 0$. If d is a metric, then (X, d) is said to be a *metric space*.

A subset of a metric space is itself a metric space by using the metric d.

We need a few basic notions related to metric spaces, which generalize familiar concepts from Euclidean space. Metric spaces are discussed further in the Appendix.

Definition 2.6.2 The set $B(x, r) := \{y \in X \mid d(x, y) < r\}$ is called the *(open) r-ball* around x. A sequence $(x_n)_{n \in \mathbb{N}}$ in X is said to *converge* to $x \in X$ if for all $\epsilon > 0$ there exists an $N \in \mathbb{N}$ such that for every $n \geq N$ we have $d(x_n, x) < \epsilon$.

Now we define a property of a metric space that distinguishes it from all those with "holes" or otherwise "missing" points.

Definition 2.6.3 A sequence $(x_i)_{i \in \mathbb{N}}$ is said to be a *Cauchy sequence* if for all $\epsilon > 0$ there exists an $N \in \mathbb{N}$ such that $d(x_i, x_j) < \epsilon$ whenever $i, j \geq \mathbb{N}$. A metric space X is said to be *complete* if every Cauchy sequence converges.

In the proofs of the Contraction Principle (Proposition 2.2.8 and Proposition 2.2.10) the fixed point is obtained as the limit of a Cauchy sequence. Therefore we assume completeness in Proposition 2.6.10 in order to be able to use the same argument.

Definition 2.6.4 Let $(X, d), (Y, d')$ be metric spaces. A map $f: X \to Y$ is said to be an *isometry* if $d'(f(x), f(y)) = d(x, y)$ for all $x, y \in X$. It is said to be *continuous* at $x \in X$ if for every $\epsilon > 0$ there exists a $\delta > 0$ such that $d(x, y) < \delta$ implies $d'(f(x), f(y)) < \epsilon$. A continuous bijection (one-to-one and onto map) with continuous inverse is said to be a *homeomorphism*. A map $f: X \to Y$ is said to be *Lipschitz-continuous* (or Lipschitz) with Lipschitz constant C, or C-Lipschitz, if $d'(f(x), f(y)) \leq Cd(x, y)$. A map is said to be a *contraction* (or, more specifically, a *λ-contraction*) if it is Lipschitz-continuous with Lipschitz constant $\lambda < 1$. We say that two metrics are *isometric* if the identity establishes an isometry between them. Two metrics are said to be *uniformly equivalent* (sometimes just equivalent) if the identity and its inverse are Lipschitz maps between the two metric spaces.

2.6.2 The Circle

The unit circle $S^1 = \{x \in \mathbb{R}^2 \mid \|x\| = 1\}$ in the plane can also be described as the set of complex numbers of modulus 1. It is the phase space for the example in Section 2.2.7 as well as for many dynamical systems that we intend to study in due time.

On the circle one can in a natural way introduce several metrics. The first choice that comes to mind is to measure the distance of two points of S^1 using the Euclidean metric of $\mathbb{R}^2$. This is in accordance with our earlier observation that a subset of a metric space is itself a metric space. Let us refer to this metric as the Euclidean metric d.

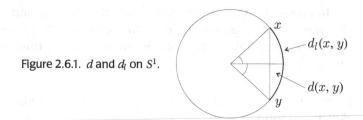

Figure 2.6.1. d and d_l on S^1.

On the other hand, one may decide that the distance between two points of S^1 should be the distance traveled when moving from one point to the other along the circle, that is, the length of the shorter arc connecting the two points. This we call the length metric d_l, because it measures lengths of arcs in order to compute distance. While these two metrics are not the same, they are not very different either.

Lemma 2.6.5 *d and d_l are uniformly equivalent.*

Proof $d(x, y) = 2\sin(d_l(x, y)/2)$, $d_l(x, y) \in [0, \pi/2]$, and $2t/\pi \le 2\sin(t/2) \le t$ for $t \in [0, \pi/2]$. Thus the identity map from the circle (S^1, d) with the Euclidean metric to the circle (S^1, d_l) with the length metric is Lipschitz-continuous with Lipschitz constant $\pi/2$ (see Figure 2.6.1). Its inverse (also the identity, but "in the other direction") is Lipschitz-continuous with Lipschitz constant 1. Therefore these two metrics are uniformly equivalent. $\square$

It will be useful later and is instructive now to see how a different construction gives rise to a metric space homeomorphic to S^1 and, in fact, isometric to (S^1, d_l) up to a constant scaling factor.

Consider the real line $\mathbb{R}$ and define the equivalence relation $\sim$ by setting $x \sim y$ if $x - y \in \mathbb{Z}$, that is, we define points to be equivalent if they differ by an integer. We define the *equivalence class* of $x \in \mathbb{R}$ by $[x] := \{y \in \mathbb{R} \mid y \sim x\}$. The equivalence class of 0 is just $\mathbb{Z}$ itself, and every equivalence class is a translate of $\mathbb{Z}$ by a member of the class, that is, $[x] = x + \mathbb{Z}$. To define a new metric space we consider the set $X = \mathbb{R}/\mathbb{Z} := \{[x] \mid x \in \mathbb{R}\}$ of all equivalence classes.

Remark 2.6.6 The notation $[\cdot]$ for equivalence classes is virtually universal, and unfortunately it looks similar to that for the integer part, $\lfloor\cdot\rfloor$, which is almost the opposite notion. $[\cdot]$ is closely related to the fractional part $\{\cdot\}$.

A metric on $\mathbb{R}/\mathbb{Z}$ is induced from that on $\mathbb{R}$:

Proposition 2.6.7 $d(x, y) := \min\{|b - a| \mid a \in x, b \in y\}$ *defines a metric on* $X = \mathbb{R}/\mathbb{Z}$.

Proof d is clearly symmetric. To check $d(x, y) = 0 \Rightarrow x = y$, note first that the metric does not change if instead we take the minimum over $a \in x$ only for a fixed $b \in y$, because the least distance from b to elements of x is the same as the least distance from any integer translate of b to elements of x. But obviously $\min\{|b - a| \mid a \in x\}$ is actually attained, and hence is only zero if $b \in x$ and hence $x = y$.

To prove the triangle inequality take $x, y, z \in \mathbb{R}/\mathbb{Z}$ and $a \in x, b \in y$ such that $d(x, y) = |b - a|$. Then for any $c \in z$ we have $d(x, z) \leq |c - a| \leq |c - b| + |b - a| = |c - b| + d(x, y)$. Taking the minimum over $c \in z$ then shows that $d(x, z) \leq d(y, z) + d(x, y)$. $\square$

Example 2.6.8 $d([\pi], [3/2]) = 7/2 - \pi = 0.5 - 0.14159265 \cdots = 0.3584073 \ldots$ and $d([0.9], [0]) = 0.1$.

Thus we obtain a metric space. To see what it looks like, note that every equivalence class has exactly one representative (that is, member) in $[0, 1)$. Therefore, as a set of points, we can naturally identify $\mathbb{R}/\mathbb{Z}$ with $[0, 1)$.

Lemma 2.6.9

(1) *If* $a, b \in [0, 1)$ *with* $|a - b| \leq 1/2$, *then* $d([a], [b]) = |a - b|$.
(2) *If* $|a - b| \geq 1/2$, *then* $d([a], [b]) = 1 - |a - b|$.

Proof

(1) $d([a], [b]) \leq |a - b|$ by definition, but the inequality cannot be strict because every integer translate of b is further from a than b itself.
(2) $d([a], [b]) = 1 - |a - b|$ because this is the smaller of $|a - (b - 1)|$ and $|a - (b + 1)|$. $\square$

For example, the distance between the classes $[1 - \epsilon]$ and $[0]$ is ϵ, if $\epsilon < 1/2$. Therefore, this construction intuitively corresponds to taking the interval $[0, 1)$ and "attaching" the open end to 0. Or, referring to the identification on the entire line $\mathbb{R}$, the construction amounts to "rolling up" the entire line onto a circle of circumference 1, so that integer translates of the same number all end up on the same point of the circle (see Figure 2.6.2). Conversely, going from the circle to the line is like rolling a bicycle wheel along and leaving periodic tire prints.

This description can be made analytically as well: To map $\mathbb{R}$ onto the unit circle in the complex plane $\mathbb{C}$ in this fashion we can employ the map given by $f(x) = e^{2\pi i x}$. Then $f(x + k) = e^{2\pi i (x+k)} = e^{2\pi i x} e^{2\pi i k} = e^{2\pi i x} = f(x)$, so this function depends only on the equivalence class $[x]$ of x under integer translations and therefore defines a map from $\mathbb{R}/\mathbb{Z}$ to S^1 by $F([x]) = f(x)$. If we use the metric $d_l/2\pi$ on S^1, then F is

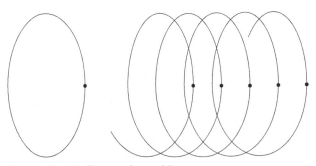

Figure 2.6.2. Rolling up the real line.

an isometry, justifying our claim of having given an alternate construction of the metric space (S^1, d_l).

Our point of view is that we regard the identification space $\mathbb{R}/\mathbb{Z}$ as *the* circle, and the unit circle in the plane as a convenient and often physically motivated and appropriate *representation* of the circle.

In dynamics and various applications of mathematics the circle, defined as $\mathbb{R}/\mathbb{Z}$, naturally arises because one may want to study periodic functions. Any collection of functions with a common period is, in fact, a collection of functions on a circle, because one can scale the independent variable to make the common period equal to 1, and the values of 1-periodic functions depend only on the equivalence class (mod $\mathbb{Z}$) of the variable. So these functions are well defined on $\mathbb{R}/\mathbb{Z}$. A particularly important class of this kind is $(e^{2k\pi ix})_{k \in \mathbb{Z}}$. In problems where periodic functions naturally arise, this may introduce a helpful geometric component.

2.6.3 The Cylinder

The cylinder is a space naturally visualized as a tube or pipe. There are several ways of defining it from more basic ingredients.

One of these is motivated by a natural parametrization of a cylinder as follows:

$$(\cos 2\pi t, \sin 2\pi t, z) \quad \text{for } t \in \mathbb{R} \quad \text{and} \quad -1 \le z \le 1.$$

Of course, taking $0 \le t \le 1$ suffices to get the whole cylinder, and by periodicity of the trigonometric functions the points $(0, z)$ and $(1, z)$ are mapped to the same point in $\mathbb{R}^3$. Therefore this parametrization can be visualized as taking a unit square and rolling it up into a tube. This is illustrated in the left half of Figure 2.6.3.

On the other hand, the previous section discussed rolling up $\mathbb{R}$ to get S^1. The present parametrization does the same, except that there is an inert variable z added in. Therefore we can also describe the cyclinder as a product of a circle (t-variable) with an interval (z-variable). This is an instance of the product construction described in Section A.1.6.

2.6.4 The Torus

The torus is the surface usually visualized as the surface of a doughnut. One can think of this surface as obtained by taking a circle in the xz-plane of $\mathbb{R}^3$ that does not intersect the z-axis and sweeping out a surface by revolving it around the z-axis, that is, by moving its center around a circle in the xy-plane. Doing this with the circle parametrized by

$$(R + r \cos 2\pi\theta, 0, \sin 2\pi\theta)$$

Figure 2.6.3. The torus.

gives the parametrization

$$((R + r\cos 2\pi\theta)\cos 2\pi\phi, (R + r\cos 2\pi\theta)\sin 2\pi\phi, \sin 2\pi\theta)$$

of the torus. Seeing that the angles θ and ϕ show up in periodic functions only, it is natural to directly think of the torus as the cartesian product of two circles in $\mathbb{R}^2$ embedded in $\mathbb{R}^4$. Once we view the torus as $\mathbb{T}^2 = S^1 \times S^1$, however, we can utilize the description of S^1 as $\mathbb{R}/\mathbb{Z}$ just given and describe $\mathbb{T}^2$ directly as $\mathbb{R}^2/\mathbb{Z}^2$ by considering equivalence classes of points $(x_1, x_2) \in \mathbb{R}^2$ under translations by integer vectors $(k_1, k_2) \in \mathbb{Z}^2$, that is, $[(x, y)] = ([x], [y])$. As before, the Euclidean metric on $\mathbb{R}^2$ induces a metric on $\mathbb{T}^2$, which is the same as the product metric $d((x_1, x_2), (y_1, y_2)) = \sqrt{(d(x_1, y_1))^2 + (d(x_2, y_2))^2}$. Continuing the rolling-up construction of the cylinder one more step (to roll the z-interval up into a circle as well) we obtain a description of $\mathbb{T}^2$ by taking the unit square $[0, 1) \times [0, 1)$ and gluing the right and left edges together to obtain a cylinder, and then gluing the top and bottom circles together as well to get the torus. One can picture the "seams" as the equator of the doughnut hole and a meridian around the ring. Likewise, we can construct and describe tori $\mathbb{T}^n$ of any dimension as n-fold products of circles or as $\mathbb{R}^n/\mathbb{Z}^n$.

As with the circle, we regard the unit square (or cube) with the identifications described above as *the* torus, whereas the "doughnut" surface is a canonical *representation* of the abstract torus as a concrete surface. This is contrary to viewing the unit square merely as a convenient domain for a parametrization of the torus.

Here, too, periodic functions are naturally related: A function $f: \mathbb{R}^2 \to \mathbb{R}$ such that $f(x + i, y + k) = f(x, y)$ whenever $i, k \in \mathbb{Z}$ is naturally defined on the torus. Analogously to a circle, a torus may give rise to helpful visualizations when such doubly periodic functions arise prominently in applications. Such cyclic coordinates often appear in mechanics, and this is why tori often play a role in that context.

We repeat that cylinders and tori are both examples of products of metric spaces as described generally in Section A.1.6.

2.6.5 Contracting and Eventually Contracting Maps

Having defined the necessary general notions, we can now show that the Contraction Principle holds in any complete metric space:

Proposition 2.6.10 (Contraction Principle)

Let X be a complete metric space. Under the action of iterates of a contraction $f: X \to X$, all points converge with exponential speed to the unique fixed point of f.

Proof As in Euclidean space, iterating $d(f(x), f(y)) \le \lambda d(x, y)$ gives

$$d(f^n(x), f^n(y)) \to 0 \qquad \text{as } n \to \infty,$$

so the asymptotic behavior of all points is the same. On the other hand, (2.2.6) shows that for any $x \in X$ the sequence $(f^n(x))_{n\in\mathbb{N}}$ is a Cauchy sequence. Thus for any $x \in X$ the limit of $f^n(x)$ as $n \to \infty$ exists if the space is complete, and by (2.2.5) this limit is the same for all x. Equation (2.2.7) shows that it is a fixed point x_0 of f. (Note that uniqueness of the fixed point does not depend on completeness.) $\square$

As in the Euclidean case, we see that $d(f^n(x), x_0) \leq (\lambda^n/1 - \lambda)d(f(x), x)$, that is, all orbits converge to x_0 exponentially fast. If x_0 is already known or an estimate in terms of initial data is not required, then one can use (2.2.5) to see that $d(f^n(x), x_0) \leq \lambda^n d(x, x_0)$ to get the same conclusion in a more straightforward way.

It is at times useful that the Contraction Principle can be applied under weaker hypotheses than the one we used. Indeed, looking at the proof one can see that it would suffice to assume the following property:

Definition 2.6.11 A map f of a metric space is said to be *eventually contracting* if there are constants $C > 0$, $\lambda \in (0, 1)$ such that

(2.6.1) $d(f^n(x), f^n(y)) \leq C\lambda^n d(x, y)$

for all $n \in \mathbb{N}$.

It is, however, not only possible to reproduce the proof of the Contraction Principle under this weakened hypothesis, but we can find a metric for which such a map becomes a contraction. Indeed, this metric is uniformly equivalent to the original one.

The change of metric that turns an eventually contracting map into a contraction has an analog for maps that are not necessarily contracting, so we prove a useful and slightly more general statement.

Proposition 2.6.12 *If* $f: X \to X$ *is a map of a metric space and there are* $C, \lambda > 0$ *such that* $d(f^n(x), f^n(y)) \leq C\lambda^n d(x, y)$ *for all* $x, y \in X$, $n \in \mathbb{N}_0$, *then for every* $\mu > \lambda$ *there exists a metric* d_μ *uniformly equivalent to* d *such that* $d_\mu(f(x), f(y)) \leq \mu d(x, y)$ *for all* $x, y \in X$.

Proof Take $n \in \mathbb{N}$ such that $C(\lambda/\mu)^n < 1$ and set

$$d_\mu(x, y) := \sum_{i=0}^{n-1} d(f^i(x), f^i(y))/\mu^i.$$

This is called an *adapted* or *Lyapunov metric* for f. The two metrics are uniformly equivalent:

$$d(x, y) \leq d_\mu(x, y) \leq \sum_{i=0}^{n-1} C(\lambda/\mu)^i d(x, y) \leq \frac{C}{1 - (\lambda/\mu)} d(x, y).$$

Note now that

$$d_\mu(f(x), f(y)) = \sum_{i=1}^{n} \frac{d(f^i(x), f^i(y))}{\mu^{i-1}} = \mu(d_\mu(x, y) + \frac{d(f^n(x), f^n(y))}{\mu^n} - d(x, y))$$

$$\leq \mu d_\mu(x, y) - (1 - C(\lambda/\mu)^n)d(x, y) \leq \mu d_\mu(x, y). \qquad \square$$

As an immediate consequence we see that eventually contracting maps can be made contracting by a change of metric because for $\lambda < 1$ as in Definition 2.6.11

we can take $\mu \in (\lambda, 1)$ in Proposition 2.6.12:

Corollary 2.6.13 *Let X be a complete metric space and $f: X \to X$ an eventually contracting map (Definition 2.6.11). Then, under the iterates of f, all points converge to the unique fixed point of f with exponential speed.*

Let us point out one of the major strengths of the notion of an eventually contracting map. As we just found, whether or not a map is a contraction can depend on the metric. This is not the case for eventually contracting maps: If a map f satisfies (2.6.1) and d' is a metric uniformly equivalent to d, specifically $md'(x, y) \le d(x, y) \le Md'(x, y)$, then

$$d'(f^n(x), f^n(y)) \le Md(f^n(x), f^n(y)) \le MC\lambda^n d(x, y) \le \frac{MC}{m}\lambda^n d'(x, y).$$

In other words, only the constant C depends on the metric, not the existence of such a constant.

Even without considering smooth maps, as we did in Proposition 2.2.20, the fixed point of a contraction depends continuously on the contraction. This is useful in applications, and therefore it is worthwhile to develop this idea further. The natural way to express continuous dependence is to consider families of contractions parametrized by a member of another metric space.

Proposition 2.6.14 *If X, Y are metric spaces, X is complete, $f: X \times Y \to X$ a continuous map such that $f_y := f(\cdot, y)$ is λ-contraction for all $y \in Y$, then the fixed point $g(y)$ of f_y depends continuously on y.*

Proof Apply

$$d(x, g(y)) \le \sum_{i=0}^{\infty} d\left(f_y^i(x), f_y^{i+1}(x)\right) \le \frac{1}{1-\lambda}d(x, f_y(x))$$

to $x = g(y') = f(g(y'), y')$ to get

$$d(g(y), g(y')) \le \frac{1}{1-\lambda}d(f(g(y'), y'), f(g(y'), y)). \qquad \square$$

▓ EXERCISES

▓ **Exercise 2.6.1** Show that an open r-ball is an open set.

▓ **Exercise 2.6.2** Show that any union (not necessarily finite or countable) of open sets is open, and that any intersection of closed sets is closed.

▓ **Exercise 2.6.3** Consider the set $\mathbb{Z}$ of integers as a metric space with the Euclidean metric $d(n, m) = |n - m|$. Describe the balls $\{n \in \mathbb{Z} \mid d(n, 0) < 1\}$ and $\{n \in \mathbb{Z} \mid d(n, 0) \le 1\}$. Which of these is open and which is closed?

▓ **Exercise 2.6.4** Describe all open sets of $\mathbb{Z}$ [with the Euclidean metric $d(n, m) = |n - m|$].

▓ **Exercise 2.6.5** Show that the interior of any set is open and that the closure of any set is closed.

■ **Exercise 2.6.6** Show that the boundary of a subset of a metric space is a closed set and that the boundary of an open set is nowhere dense. Conclude that the boundary of a boundary is nowhere dense.

■ **Exercise 2.6.7** Decide, with proof, which of the following are complete metric spaces (with the usual metric): $\mathbb{R}$, $\mathbb{Q}$, $\mathbb{Z}$, $[0, 1]$.

■ **Exercise 2.6.8** Prove that a closed subset of a complete metric space is complete.

■ **PROBLEMS FOR FURTHER STUDY**

■ **Problem 2.6.9** Suppose that X is a compact metric space (see Definition A.1.17) and $f\colon X \to X$ is such that $d(f(x), f(y)) < d(x, y)$ for any $x \neq y$. Prove that f has a unique fixed point $x_0 \in I$ and $\lim_{n\to\infty} f^n(x) = x_0$ for any $x \in I$.

■ **Problem 2.6.10** Suppose that X is a complete metric space such that the distance function is at most 1, and $f\colon X \to X$ is such that $d(f(x), f(y)) \leq d(x, y) - 1/2(d(f(x), f(y)))^2$. Prove that f has a unique fixed point $x_0 \in I$ and $\lim_{n\to\infty} f^n(x) = x_0$ for any $x \in I$.

2.7 FRACTALS

2.7.1 The Cantor Set

We next consider a space that is often seen as an oddity in an analysis course, the Cantor set. We will see, however, that sets like this arise naturally and frequently in dynamics and constitute one of the most important spaces that we encounter.

1. Geometric Definition. The *ternary Cantor set* or *middle-third Cantor set* is described as follows. Consider the unit interval $C_0 = [0, 1]$ and remove from it the open middle third $(1/3, 2/3)$ to retain two intervals of length $1/3$ whose union we denote by C_1. Apply the same prescription to these intervals, that is, remove their middle thirds. The remaining set C_2 consists of four intervals of length $1/9$ from each of which we again remove the middle third. Continuing inductively we obtain nested sets C_n consisting of 2^n intervals of length 3^{-n} [for a total length of $(2/3)^n \to 0$]. The intersection C of all of these sets is nonempty (because they are closed and bounded, and by Proposition A.1.24) and closed and bounded because all C_n are. It is called the middle-third or ternary Cantor set (see Figure 2.7.1).

2. Analytic Definition. It is useful to describe this construction analytically as follows.

Lemma 2.7.1 *C is the collection of numbers in $[0, 1]$ that can be written in ternary expansion (that is, written with respect to base 3 as opposed to base 10) without using 1 as a digit.*

Figure 2.7.1. The ternary Cantor set.

Proof The open middle third $(1/3, 2/3)$ is exactly the set of numbers that must have a 1 as the first digit after the (ternary) point, that is, that cannot be written in base 3 as $0.0\ldots$ or $0.2\ldots$. (Note that $1/3$ can be written as $0.02222\ldots$ and $2/3$ as $0.20000000\ldots$.) Correspondingly, the middle thirds of the remaining intervals are exactly those remaining numbers whose second digit after the point must be 1, and so on. $\square$

3. Properties. This set is clearly small in some sense (being the intersection of sets consisting of intervals whose lengths add up to ever smaller numbers) and certainly cannot contain any interval. Moreover, we have:

Lemma 2.7.2 *The ternary Cantor set is totally disconnected (see Definition A.1.8).*

Proof Any two points of C are in different components of some C_n. Taking a sufficiently small open neighborhood of one of these together with the interior of its complement gives two disjoint open sets whose union contains C and each contains one of the points in question. $\square$

In contrast, we have

Lemma 2.7.3 *The ternary Cantor set is uncountable.*

Proof Mapping each point $x = 0.\alpha_1\alpha_2\alpha_3\cdots = \sum_{i=1}^{\infty}(\alpha_i/3^i) \in C$ ($\alpha_i \neq 1$) to the number $f(x) := \sum_{i=1}^{\infty}(\alpha_i/2/2^i) = \sum_{i=1}^{\infty}\alpha_i 2^{-i-1} \in [0, 1]$ defines a surjective map because all binary expansions indeed occur here. The fact that the image is uncountable implies that C is uncountable. $\square$

Deformed versions of the ternary Cantor set abound in this book. Hence we attach the same name to these:

Definition 2.7.4 A set homeomorphic to the ternary Cantor set will be referred to as *a* Cantor set.

Proposition A.1.7 shows that Cantor sets can be described intrinsically.

4. Self-Similarity. There is an interesting example of a contraction on the middle-third Cantor set, namely, $f: [0, 1] \rightarrow [0, 1]$, $f(x) = x/3$. Since f is a contraction, it is also a contraction on every invariant subset, and in particular on the Cantor set. The unique fixed point is obviously 0. This property of invariance under a linear contraction is often referred to as *self-similarity* or *rescaling property*. Its meaning is quite clear and rather striking: The microscopic structure of the Cantor set is exactly the same as its global structure; it does not become any simpler at any smaller scale.

5. The Devil's Staircase. The only points whose ternary expansion is not unique are those that can be written with a terminating expansion, that is, the ternary rationals. These are exactly the countably many endpoints involved in the construction of the Cantor set. Consider the function $f(\sum_{i=1}^{\infty}\alpha_i 3^{-i}) = \sum_{i=1}^{\infty}\alpha_i 2^{-i-1}$ from the

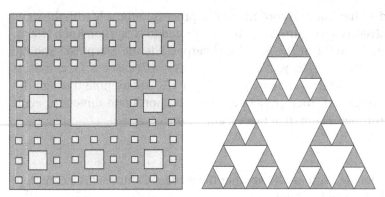

Figure 2.7.2. The square and triangular Sierpinski carpets.

proof of Lemma 2.7.3. Then $f(1/3) = f(0.02222222\ldots) = 0.011111\ldots$ (binary) $=$ 0.1 (binary) $= 1/2$. At the same time, $f(2/3) = f(0.2000000\ldots) = 0.1$ (binary) $=$ $1/2$ also. Likewise, one sees that all endpoints are identified in corresponding pairs under this map. It is also not hard to see that f is nondecreasing. Sometimes it is interesting to extend the map f to $[0, 1]$ by taking it to be constant on complementary intervals. The resulting continuous function has several exotic properties and is therefore called a "devil's staircase." As is the case with Cantor sets, it turns out that such functions arise naturally in the study of dynamical systems (see Definition 4.4.1 and Figure 4.4.1 (p. 136)).

2.7.2 Other Self-Similar Sets

Let us describe some other interesting self-similar metric spaces that are of a different form. The *Sierpinski carpet* (see Figure 2.7.2) is obtained from the unit square by removing the "middle-ninth" square $(1/3, 2/3) \times (1/3, 2/3)$, then removing from each square $(i/3, i+1/3) \times (j/3, j+1/3)$ its "middle ninth," and so on. This construction can easily be described in terms of ternary expansion in a way that immediately suggests higher-dimensional analogs (Exercise 2.7.4). Alternatively, one can start from an equilateral triangle with the bottom side horizontal, say, and divide it into four congruent equilateral triangles of which the central one has a horizontal top side. Then one deletes this central triangle and continues this construction on the remaining three triangles. The *von Koch snowflake* is obtained from an equilateral triangle by erecting on each side an equilateral triangle whose base

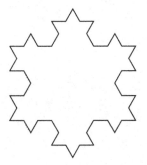
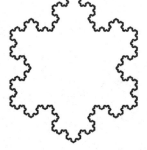

Figure 2.7.3. The Koch snowflake.

is the middle third of that side and continuing this process iteratively with the sides of the resulting polygon (see Figure 2.7.3). It is attributed to Helge von Koch (1904). A three-dimensional variant of the Sierpinski carpet S is the Sierpinski sponge or Menger curve defined by $\{(x, y, z) \in [0, 1]^3 \mid (x, y) \in S, \ (x, z) \in S \ (y, z) \in S\}$. It is obtained from the solid unit cube by punching a 1/3-square hole through the center from each direction, then punching, in each coordinate direction, eight 1/9-square holes through in the right places, and so on.

■ **EXERCISES**

■ **Exercise 2.7.1** Decide, with proof, whether the ternary Cantor set is a complete metric space (with the usual metric).

■ **Exercise 2.7.2** Show that the map $x \mapsto f(x) = 1 - (x/3)$ on $[0, 1]$ is a contraction and maps the ternary Cantor set into itself. Where is its fixed point?

■ **Exercise 2.7.3** Repeat the previous exercise with the map given by $f(x) = (x + 2)/3$.

■ **Exercise 2.7.4** Prove that the square Sierpinski carpet is the set of points in the unit square of which at least one coordinate can be represented by a ternary expansion without the digit 1.

■ **PROBLEMS FOR FURTHER STUDY**

■ **Problem 2.7.5** Prove that the ternary Cantor set C is homeomorphic to its cartesian double $C \times C$. (This appears naturally in Section 7.3.3.)

■ **Problem 2.7.6** If f is the function in the proof of Lemma 2.7.3 and $(h_1, h_2) \colon C \to C \times C$ the homeomorphism from the previous exercise, then $F(x) := (f(h_1(x)), f(h_2(x)))$ defines a surjective map $F \colon C \to [0, 1] \times [0, 1]$. Show that f (and hence F) is continuous and that F extends to a continuous map on $[0, 1]$. (The resulting map is a continuous surjective map $[0, 1] \to [0, 1] \times [0, 1]$, that is, a *space-filling curve* or *Peano curve*.)

■ **Problem 2.7.7** Show that the set C' of those $x \in [0, 1]$ that have a base-5 expansion without odd digits is a Cantor set.

■ **Problem 2.7.8** Decide whether the set of those $x \in [0, 1]$ that have a base-10 expansion without odd digits is a Cantor set.

CHAPTER 3

Linear Maps and Linear Differential Equations

In this chapter the complexity of the dynamical behavior increases slightly over that observed in the examples in Chapter 2. In particular, periodic motions are now present in both discrete and continuous time. At the same time, in most linear systems, that is, linear maps and linear differential equations, the orbit structure is easy to understand (the limited *elliptic* complexity that arises from complex eigenvalues on the unit circle is discussed in the first sections of the next two chapters). We describe it carefully here. This involves linear algebra, but it is not simply a repetition of it because we investigate the dynamical aspects of linear systems, paying attention to the asymptotic behavior under iteration. Thus, this chapter serves to augment the range of asymptotic behaviors that we are able to describe. Our development takes place first in the plane and then in Euclidean spaces of any dimension.

Aside from widening our horizons in terms of the possibilities for asymptotic behavior, understanding linear maps is useful for the study of nonlinear maps by way of linearization, which was first discussed in Section 2.1.2. This is most directly the case when one wants to understand the asymptotic behavior of orbits near a fixed point of a nonlinear dynamical system, but it can also help study the relative behavior of orbits and help describe the global orbit structure. One place where this is discussed explicitly is in Section 6.2.2.7.

3.1 LINEAR MAPS IN THE PLANE

In trying to understand linear maps as dynamical systems, that is, to understand how points move under iteration of these maps, we do not need to understand all details of the dynamics of the linear maps in question. The primary attention is to the coarse aspects of behavior, such as going to the origin, diverging, asymptotic to a line, and spiraling.

3.1.1 The Line

We begin with dimension one. Linear maps of the line are easily described: They are of the form $x \mapsto \lambda x$ and either $|\lambda| < 1$, in which case the map is contracting with 0 as the attracting fixed point, or $|\lambda| > 1$, in which case all nonzero orbits tend to infinity, or, finally, $|\lambda| = 1$, which means that the map is the identity or $x \mapsto -x$, when all orbits have period 2. Thus these maps are (eventually) contracting if and only if $|\lambda| < 1$.

3.1.2 Eigenvalues

Now consider maps $x \mapsto Ax$ in the plane from the same viewpoint of examining the asymptotic behavior under repeated application of the map. In this case there are more possibilities, and we will use some simple linear algebra to understand them. The crucial role in the analysis is played by the eigenvalues of the 2×2 matrix A representing a map with respect to a basis. For a matrix

$$A = \begin{pmatrix} a & b \\ c & d \end{pmatrix}$$

the real eigenvalues are those numbers λ for which there is a nonzero vector $\begin{pmatrix} x \\ y \end{pmatrix}$ such that

$$\begin{pmatrix} a & b \\ c & d \end{pmatrix} = \lambda \begin{pmatrix} x \\ y \end{pmatrix} \quad \text{(an eigenvector)}.$$

Geometrically this means that there is a line [the subspace spanned by $\begin{pmatrix} x \\ y \end{pmatrix}$] that is preserved by A. The dynamics on this line is then an instance of what was just discussed in Section 3.1.1, which thus becomes a building block for the planar picture. Complex eigenvalues are defined by the same formulas but allowing complex vectors w. In this situation the real vectors $w + \bar{w}$ and $i(w - \bar{w})$ span a real subspace (eigenspace).

The condition

$$\begin{pmatrix} a & b \\ c & d \end{pmatrix} \begin{pmatrix} x \\ y \end{pmatrix} = \lambda \begin{pmatrix} x \\ y \end{pmatrix}$$

is equivalent to

$$\begin{pmatrix} a - \lambda & b \\ c & d - \lambda \end{pmatrix} \begin{pmatrix} x \\ y \end{pmatrix} = 0.$$

This last equation always holds for

$$\begin{pmatrix} x \\ y \end{pmatrix} = 0,$$

and there is a nonzero solution as well if and only if

$$\begin{pmatrix} a - \lambda & b \\ c & d - \lambda \end{pmatrix}$$

is not invertible. Thus, nonzero eigenvectors exist precisely for those λ that make

$$\begin{pmatrix} a - \lambda & b \\ c & d - \lambda \end{pmatrix}$$

noninvertible, that is, for which its determinant is zero. These are the roots of the characteristic polynomial

$$(a - \lambda)(d - \lambda) - bc = \lambda^2 - (a+d)\lambda + ad - bc = \lambda^2 - (\text{tr } A)\lambda + \det A,$$

where tr $A := a + d$ and det $A := ad - bc$. From the quadratic formula

(3.1.1) $$2\lambda = -\text{tr } A \pm \sqrt{(\text{tr } A)^2 - 4 \det A}$$

we know that there are three cases: We can have two real solutions, or a single one, or two complex conjugate ones, according to whether the *discriminant* $(\text{tr } A)^2 - 4 \det A$ is positive, zero, or negative.

3.1.3 Distinct Real Eigenvalues

Let us consider the first case. When there are two different real eigenvalues λ and μ, the equations $Av = \lambda v$ and $Aw = \mu w$ can be solved for nonzero *eigenvectors* v and w. Then A preserves the *eigenspaces* $\mathbb{R}v = \{tv \mid t \in \mathbb{R}\}$ and $\mathbb{R}w = \{tw \mid t \in \mathbb{R}\}$ and thus reduces to a one-dimensional linear map on each of these lines. We would like these two lines to be the coordinate axes of a new coordinate system. This means that we want to express our matrix A with respect to the basis consisting of v and w instead of the standard unit vectors e_1 and e_2. From $Av = \lambda v$ and $Aw = \mu w$ it is clear that this representation of A is given by the matrix

$$B = \begin{pmatrix} \lambda & 0 \\ 0 & \mu \end{pmatrix}.$$

Proposition 3.1.1 *A linear map with real eigenvalues $\lambda \neq \mu$ can be diagonalized to*

$$B = \begin{pmatrix} \lambda & 0 \\ 0 & \mu \end{pmatrix}$$

by a linear coordinate change.

It is good to be aware of how this coordinate change looks in terms of matrix calculations. The required coordinate change is the one sending v to e_1 and w to e_2. Its inverse therefore has matrix C (with respect to standard coordinates) whose columns are v and w – this matrix sends e_1 to v and e_2 to w. Thus, the representation of A with respect to these coordinates is $B = C^{-1}AC$. C changes from new coordinates to old, A then gives the transformation, and C^{-1} changes coordinates back. This is analogous to the discussion in Section 1.2.9.3 (p. 13).

Example 3.1.2 Consider

$$A = \begin{pmatrix} 2 & 1 \\ 1 & 2 \end{pmatrix}.$$

Then $2\lambda = 4 \pm \sqrt{16 - 4 \cdot 3} = 4 \pm 2$ by (3.1.1), so $\lambda = 1$, $\mu = 3$. An eigenvector v for $\lambda = 1$ is found by

$$\begin{pmatrix} 1 & 1 \\ 1 & 1 \end{pmatrix} \begin{pmatrix} x \\ y \end{pmatrix} = 0,$$

for example,

$$x = -y = 1; \quad v = \begin{pmatrix} 1 \\ -1 \end{pmatrix}.$$

Likewise,

$$w = \begin{pmatrix} 1 \\ 1 \end{pmatrix}$$

is an eigenvector for the second eigenvalue $\mu = 3$, so

$$C = \begin{pmatrix} 1 & 1 \\ -1 & 1 \end{pmatrix}$$

will do and

$$B = C^{-1}AC = \begin{pmatrix} 1 & 0 \\ 0 & 3 \end{pmatrix}.$$

The particular case $|\lambda| < 1 < |\mu|$ (or vice versa) has a name.

Definition 3.1.3 A linear map of $\mathbb{R}^2$ with one eigenvalue in $(-1, 1)$ and one eigenvalue of absolute value greater than 1 is said to be *hyperbolic*.

3.1.4 Single Real Eigenvalue
In the second case of a single real eigenvalue λ the map $A - \lambda \operatorname{Id}$ is noninvertible and there are two possibilities: The first is that it is zero, in which case $A = \lambda \operatorname{Id}$. These scalings are simple but important.

Definition 3.1.4 A map $A = \lambda \operatorname{Id}$ is called a *homothety* or a *scaling*.

The other case is that in which up to a scale factor one has only one nonzero solution v to $Av = \lambda v$. Let C be an invertible matrix with v as the first column. Then the first column of $B = C^{-1}AC$ is $\binom{\lambda}{0}$ and the other diagonal entry is also λ (it is an eigenvalue). Thus

$$B = \begin{pmatrix} \lambda & s \\ 0 & \lambda \end{pmatrix}$$

for some $s \neq 0$ and

$$B^n = \begin{pmatrix} \lambda^n & ns\lambda^{n-1} \\ 0 & \lambda^n \end{pmatrix} = \lambda^n \begin{pmatrix} 1 & ns/\lambda \\ 0 & 1 \end{pmatrix}.$$

Actually, we can go further. Given any $a \neq 0$ we can take

$$C' = \begin{pmatrix} 1 & 0 \\ 0 & 1/sa \end{pmatrix}$$

to get

$$B' := C'^{-1}BC = \begin{pmatrix} 1 & 0 \\ 0 & sa \end{pmatrix}\begin{pmatrix} \lambda & s \\ 0 & \lambda \end{pmatrix}\begin{pmatrix} 1 & 0 \\ 0 & 1/sa \end{pmatrix} = \begin{pmatrix} \lambda & a \\ 0 & \lambda \end{pmatrix}.$$

Proposition 3.1.5 *Suppose $a \neq 0$ and A is any linear map of the plane with a double eigenvalue λ but only one linearly independent eigenvector. Then A is conjugate to*

$$\begin{pmatrix} \lambda & a \\ 0 & \lambda \end{pmatrix}.$$

In particular, A is conjugate to

$$\begin{pmatrix} \lambda & 1 \\ 0 & \lambda \end{pmatrix}.$$

The case $\lambda = 1$ has distictive asymptotic behavior and a special name.

Definition 3.1.6 A map conjugate to

$$\begin{pmatrix} 1 & s \\ 0 & 1 \end{pmatrix}$$

for some $s \in \mathbb{R}$ is called a *shear* or a *parabolic linear map*.

3.1.5 Complex Conjugate Eigenvalues

In the case of a complex conjugate pair of roots of the characteristic polynomial, consider the matrix A as representing a map of the two-dimensional complex space $\mathbb{C}^2$. This is the set of pairs of complex numbers with addition and multiplication by (complex) scalars defined componentwise. Any linear map of $\mathbb{R}^2$ defines a map of $\mathbb{C}^2$ by using matrix multiplication in the same way as for real vectors. (One can view $\mathbb{C}^2$ as a four-dimensional real vector space in much the same way as one can view $\mathbb{C}$ as a two-dimensional real vector space, but we do not need this here.)

On $\mathbb{C}^2$ the map $A - \lambda \operatorname{Id}$ is noninvertible and thus there are nonzero complex solutions to $Av = \lambda v$ and $Aw = \bar{\lambda}w$, which can be taken as a complex conjugate pair, $w = \bar{v}$ (because $\bar{A} = A$). Taking C with columns v and w as before would give a diagonal matrix $B = C^{-1}AC$ with complex entries. So instead we take the real vectors $v + \bar{v}$ and $-i(v - \bar{v})$ in $\mathbb{R}^2$ as columns of a matrix C (which is invertible). One can calculate that

$$B := C^{-1}AC = \rho \begin{pmatrix} \cos\theta & \sin\theta \\ -\sin\theta & \cos\theta \end{pmatrix},$$

where $\lambda = \rho e^{i\theta}$. This is the rotation by an angle θ followed by the scaling map $\rho \operatorname{Id}$.

Proposition 3.1.7 *A linear map of* $\mathbb{R}^2$ *with a complex eigenvalue* $\rho e^{i\theta}$ *is conjugate to the map*

$$\rho \begin{pmatrix} \cos\theta & \sin\theta \\ -\sin\theta & \cos\theta \end{pmatrix}.$$

Again $\rho = 1$ gives following important case:

Definition 3.1.8 A linear map of $\mathbb{R}^2$ conjugate to a rotation

$$\begin{pmatrix} \cos\theta & \sin\theta \\ -\sin\theta & \cos\theta \end{pmatrix}$$

is said to be *elliptic*.

Note that if θ/π is rational than all orbits are periodic.

Example 3.1.9 For

$$A = \begin{pmatrix} 7 & 8 \\ -4 & -1 \end{pmatrix}$$

we find

$$C = \begin{pmatrix} 1 & 1 \\ -1 & 0 \end{pmatrix} \quad \text{and} \quad B = \begin{pmatrix} 3 & 4 \\ -4 & 3 \end{pmatrix}.$$

Figure 3.1.1 shows the action of

$$\begin{pmatrix} 2 & 0 \\ 0 & 1/2 \end{pmatrix}, \begin{pmatrix} 1 & 1 \\ 0 & 1 \end{pmatrix}, \text{and} \begin{pmatrix} 1 & 1 \\ -1 & 1 \end{pmatrix} / \sqrt{2}$$

on a square. These are hyperbolic, parabolic, and elliptic, respectively.

3.1.6 Asymptotic Behavior

Let us see what the asymptotic behavior of orbits of such maps can look like. To that end it is useful to clarify the relation between changing coordinates and changing norms, because in all matrix calculations we use the Euclidean norm for the coordinate representations of the vectors involved. Consider, then, a vector v and its image Cv under an invertible map (the coordinate change). Often $\|Cv\| \neq \|v\|$. But we can simply define a new norm $\|v\|' := \|Cv\|$ (this is a norm because C is linear and invertible; see Section 2.6.1 for this "pullback" construction). In this way we see that taking

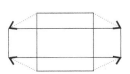 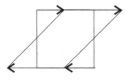 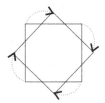

Figure 3.1.1. Hyperbolic, parabolic, and elliptic maps.

the Euclidean norm in our matrix calculations with respect to a different basis simply reflects a different choice of norm. Therefore, any conclusions obtained from such calculations that are independent of the choice of norm will give us statements about the map with respect to the Euclidean norm and independent of a choice of basis.

Proposition 3.1.10 *A linear map of* $\mathbb{R}^2$ *is eventually contracting (Definition 2.6.11) if and only if all eigenvalues are of absolute value less than one.*

Proof There are three cases to consider: distinct real eigenvalues, a double real eigenvalue, and complex eigenvalues. In all three cases the map can only be eventually contracting if the eigenvalues are of absolute value less than one. This is clear when one considers the canonical forms in each case, that is, diagonal matrices, upper triangular matrices, and a rotation with scaling. Since the condition of being eventually contracting is independent of the norm (because all norms are equivalent), the general case then follows.

We need to show that having eigenvalues of absolute value less than one is a sufficient condition. In the first case this is clear by diagonalization: The change of coordinates that diagonalizes the matrix defines a pullback norm equivalent to the standard one. But the diagonalized matrix is clearly contracting, and the property of being eventually contracting is unaffected by changing to an equivalent metric.

In the case of only one real eigenvalue, consider the matrix

$$B = \begin{pmatrix} \lambda & a \\ 0 & \lambda \end{pmatrix}$$

with $0 < 2a < 1 - |\lambda|$ (Proposition 3.1.5). Then

$$\left\| B \begin{pmatrix} x \\ y \end{pmatrix} \right\| = \left\| \begin{pmatrix} \lambda x + ay \\ \lambda y \end{pmatrix} \right\| \le |\lambda| \left\| \begin{pmatrix} x \\ y \end{pmatrix} \right\| + a \left\| \begin{pmatrix} 0 \\ y \end{pmatrix} \right\| \le (|\lambda| + a) \left\| \begin{pmatrix} x \\ y \end{pmatrix} \right\| ,$$

so B is a contraction and in particular eventually contracting. As we noted before, the same then goes for a matrix A conjugate to B, that is, any matrix with eigenvalue λ only, because conjugation amounts to changing the norm to an equivalent one. As we noted after Corollary 2.6.13, a change in norm does not affect the property of being eventually contracting.

In the case of complex eigenvalues, note that the rotation by θ does not change the Euclidean norm of any vector and the subsequent application of ρ Id reduces their norm by a factor $\rho < 1$ if the eigenvalues have absolute value $\rho < 1$. Again, every matrix with complex eigenvalues is conjugate to a rotation with scaling. $\square$

3.1.7 Structure at the Origin
We can study the way in which points approach the origin under any of these contractions in some more detail.

1. Distinct Real Eigenvalues. In the first case of two distinct real eigenvalues we note that

$$B^n \begin{pmatrix} x \\ y \end{pmatrix} = \begin{pmatrix} \lambda^n x \\ \mu^n y \end{pmatrix} .$$

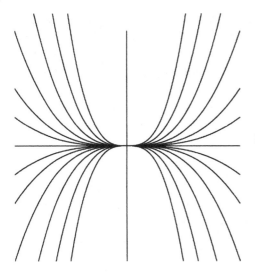

Figure 3.1.2. A node, $|y| = C|x|^{\log|\mu|/\log|\lambda|}$.

Suppose $|\mu| < |\lambda| < 1$ and rewrite

$$\begin{pmatrix} \lambda^n x \\ \mu^n y \end{pmatrix} = \lambda^n \begin{pmatrix} x \\ (\mu/\lambda)^n x \end{pmatrix}$$

to see that all orbits off the y-axis approach zero at a rate $|\lambda^n|$, the "slower" of the two. In fact, in this case all orbits of points $\begin{pmatrix} x \\ y \end{pmatrix}$ with $x \neq 0$ move along curves preserved by B, which are tangent to the x-axis at zero. Indeed, these curves are given by the equation $|y| = C|x|^{\log|\mu|/\log|\lambda|}$. For each value of C this gives four "branches," one per quadrant. Altogether we get a picture with symmetry about both axes, called a *node*. If one or both of the eigenvalues are negative, then orbits will alternate between two branches (see Figure 3.1.2).

To verify that these are indeed invariant, take $x, y, \lambda, \mu > 0$ for simplicity and note that invariance of $y = Cx^\alpha$ implies

$$\mu^n \cdot (Cx^\alpha) = \mu^n y = C(\lambda^n x)^\alpha = C\lambda^{\alpha n} x^\alpha,$$

and hence $\mu = \lambda^\alpha$, which gives $\log \mu = \alpha \log \lambda$.

Note that these curves are smooth (infititely differentiable) when $|\mu| = |\lambda|^n$ for some $n \in \mathbb{N}$ (otherwise they are only differentiable finitely many times). This coincidence is called a *resonance*.

2. Single Real Eigenvalue. In the case of one real eigenvalue λ and $A \neq \lambda \operatorname{Id}$ we conjugate A to

$$B = \begin{pmatrix} \lambda & \lambda \\ 0 & \lambda \end{pmatrix}$$

as described above to get

$$B^n = \lambda^n \begin{pmatrix} 1 & n \\ 0 & 1 \end{pmatrix}.$$

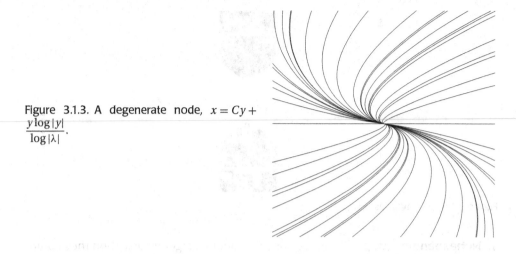

Figure 3.1.3. A degenerate node, $x = Cy + \dfrac{y \log|y|}{\log|\lambda|}$.

Here we find that the second coordinate of $B^n \left(\begin{smallmatrix} x \\ y \end{smallmatrix}\right)$ is $\lambda^n y$, which converges monotonically to zero; whereas the first coordinate $\lambda^n(x + ny)$ also converges to 0 but not necessarily monotonically, as is the case when $x = 0$. We will see shortly that there are invariant curves in this case as well, given by

$$(3.1.2) \qquad\qquad x = Cy + \frac{y \log|y|}{\log|\lambda|}.$$

This family of curves is symmetric with respect to the origin, and a fixed point of this kind is called a *degenerate node*. Here, too, a negative eigenvalue produces orbits that alternate between branches.

3. Complex Eigenvalues. In the case of complex conjugate eigenvalues $\rho e^{\pm i\theta}$, the orbits lie on spirals $r = \text{const.}\, e^{-(\theta^{-1}\log\rho)\phi}$ in polar coordinates (r, ϕ); this is called a *focus* (see Figure 3.1.4).

3.1.8 The Noncontracting Case
We now consider these maps in the noncontracting case, beginning with the cases in which the eigenvalues are real.

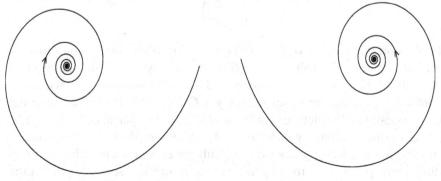

Figure 3.1.4. Foci, $r = \text{const.}\, e^{-(\theta^{-1}\log\rho)\phi}$.

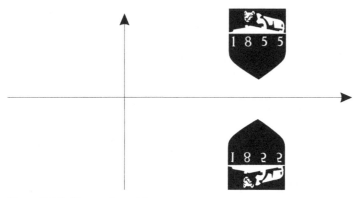

Figure 3.1.5. Eigenvalues ± 1.

1. Nonexpanding Maps. If we have two distinct real eigenvalues, then they could be 1 and -1, which corresponds to a reflection (see Figure 3.1.5), or there could be one of absolute value 1 or none with absolute value 1. If one eigenvalue λ has absolute value 1 and the other eigenvalue μ has absolute value less than 1, then the eigenspace for λ consists of fixed or period-2 points, and all other points are approaching this eigenspace along lines parallel to the eigenspace for μ. Another way of putting this is to say that in this case $\mathbb{R}^2$ decomposes into two one-dimensional subspaces and on one of them the map A is a linear contraction; on the other neither A nor A^{-1} are contracting.

2. One Expanding and One Neutral Direction. If $|\mu| > 1$, then all other points move away from this eigenspace along these lines and we get a similar decomposition of $\mathbb{R}^2$ into an expanding and a neutral subspace in this case.

3. Hyperbolic Cases. If both eigenvalues have absolute value greater than one, then all orbits diverge to infinity along the same invariant curves described for the contracting case (because these arise for the inverse map).

The most novel remaining possibility is that of $|\lambda| > 1$ and $|\mu| < 1$, called the *hyperbolic (saddle)* case. Diagonalizing as before we find that points on the x-axis diverge along the x-axis under

$$B^n = \begin{pmatrix} \lambda^n & 0 \\ 0 & \mu^n \end{pmatrix},$$

and all points off the y-axis diverge to the right or left while their y-coordinate is approaching 0, that is, all orbits approach the x-axis asymptotically. Conversely, under B^{-n} all orbits off the x-axis diverge and have the y-axis as an asymptote. Again, orbits move along invariant curves $y = Cx^{\log|\mu|/\log|\lambda|}$. Note that here the exponent is negative. This picture is called a *saddle*. In the special case of $\mu = 1/\lambda$, these curves are the standard hyperbolas $y = 1/x$, justifying the term hyperbolicity. Thus, in this case $\mathbb{R}^2$ decomposes into two subspaces, one contracting and one expanding, corresponding to the eigenspaces for μ and λ, respectively. As in the node cases, negative eigenvalues produce orbits that alternate between branches.

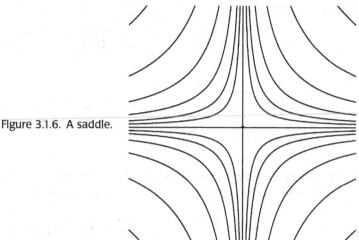

Figure 3.1.6. A saddle.

4. Complex Eigenvalues. Finally, the case, of complex eigenvalues is quite simple: Either both eigenvalues lie on the unit circle, in which case the map is conjugate to a rotation. In this case, the dynamics turns out to be rather interesting even when restricted to any (invariant) circle $r = \text{const}$. This is analyzed in the next chapters. If the eigenvalues are outside the unit circle, all orbits spiral outward. The calculations are the same as for the contracting case.

3.1.9 Fibonacci Revisited

1. A Hyperbolic Matrix. A particular example of a hyperbolic map is worth a closer look. The matrix

$$A := \begin{pmatrix} 0 & 1 \\ 1 & 1 \end{pmatrix}$$

has characteristic polynomial

$$\det \begin{pmatrix} -\lambda & 1 \\ 1 & 1 - \lambda \end{pmatrix} = \lambda^2 - \lambda - 1,$$

so the eigenvalues are $(1 \pm \sqrt{5})/2$. Since $2 < \sqrt{5} < 3$, one of these is greater than 1 and the other is in $(-1, 0)$, which makes this a hyperbolic matrix. Since the two eigenvectors are orthogonal (because A is symmetric, or by explicit verification), this implies that the orbit picture for this map looks like a rotated version of Figure 3.1.6. The eigenvector for the "expanding" eigenvalue is obtained by solving

$$0 = (A - \lambda \, \text{Id}) \begin{pmatrix} x \\ y \end{pmatrix} = \begin{pmatrix} (-1 - \sqrt{5})x/2 + y \\ x + (1 - \sqrt{5})y/2 \end{pmatrix},$$

Thus the eigenspace is the line given by $y = (1 + \sqrt{5})x/2$.

Under repeated application of a hyperbolic map all points approach the expanding subspace ever more closely, as illustrated in Figure 3.1.6. In particular, if we start with $\binom{1}{1}$ and set

$$\binom{x_n}{y_n} := A^n \binom{1}{1},$$

then $\lim_{n \to \infty} y_n/x_n = (1 + \sqrt{5})/2$.

Notice the successive values of x_n: 1, 1, 2, 3, 5, 8, 13, ... This is the Fibonacci sequence we encountered in Section 1.2.2 and Example 2.2.9. The reason is interesting in itself.

2. Fibonacci Numbers. As in Example 2.2.9, denote the Fibonacci numbers by b_n and remember that $b_{n+2} = b_{n+1} + b_n$. This implies that

(3.1.3)
$$\binom{b_{n+1}}{b_{n+2}} = \binom{b_{n+1}}{b_{n+1} + b_n} = \begin{pmatrix} 0 & 1 \\ 1 & 1 \end{pmatrix} \binom{b_n}{b_{n+1}} = A \binom{b_n}{b_{n+1}}.$$

This shows that $x_n = b_n$ and $y_n = b_{n+1}$. And the observation that $\lim_{n \to \infty} y_n/x_n = (1 + \sqrt{5})/2$ gives a new proof of the asymptotic ratio we first derived in Example 2.2.9.

Take another look at (3.1.3). It converted the original two-step recursion $b_{n+2} = b_{n+1} + b_n$ in one variable into a one-step recursion in two variables. This is an example of *reduction to first order* that is useful in differential equations. In this discrete-time situation we can use this conversion to develop an explicit (that is, nonrecursive) formula for the Fibonacci numbers.

Proposition 3.1.11 *The Fibonacci numbers $b_0 = 1$, $b_1 = 1$, ... are given by*

$$b_n = \frac{(1 + \sqrt{5})^{n+1} - (1 - \sqrt{5})^{n+1}}{2^{n+1}\sqrt{5}}.$$

Proof As in Section 3.1.7.1, we diagonalize A by taking the matrix C whose columns are the eigenvalues of A:

$$C = \frac{1}{2}\begin{pmatrix} 2 & 2 \\ 1 + \sqrt{5} & 1 - \sqrt{5} \end{pmatrix}, \qquad C^{-1} = \frac{1}{2\sqrt{5}}\begin{pmatrix} \sqrt{5} - 1 & 2 \\ \sqrt{5} + 1 & -2 \end{pmatrix},$$

$$C^{-1}AC = \frac{1}{2}\begin{pmatrix} 1 + \sqrt{5} & 0 \\ 0 & 1 - \sqrt{5} \end{pmatrix}.$$

Conversely, this gives

$$A = \frac{1}{2}\begin{pmatrix} 2 & 2 \\ 1 + \sqrt{5} & 1 - \sqrt{5} \end{pmatrix}\frac{1}{2}\begin{pmatrix} 1 + \sqrt{5} & 0 \\ 0 & 1 - \sqrt{5} \end{pmatrix}\frac{1}{2\sqrt{5}}\begin{pmatrix} \sqrt{5} - 1 & 2 \\ 1 + \sqrt{5} & -2 \end{pmatrix}$$

and

$$A^n = \frac{1}{2^{n+2}\sqrt{5}}\begin{pmatrix} 2 & 2 \\ 1 + \sqrt{5} & 1 - \sqrt{5} \end{pmatrix}\begin{pmatrix} 1 + \sqrt{5} & 0 \\ 0 & 1 - \sqrt{5} \end{pmatrix}^n\begin{pmatrix} \sqrt{5} - 1 & 2 \\ 1 + \sqrt{5} & -2 \end{pmatrix}$$

$$= \begin{pmatrix} 2(1 + \sqrt{5})^n(\sqrt{5} - 1) + 2(1 - \sqrt{5})^n(1 + \sqrt{5}) & 4(1 + \sqrt{5})^n - 4(1 - \sqrt{5})^n \\ (1 + \sqrt{5})^{n+1}(\sqrt{5} - 1) - (1 - \sqrt{5})^{n+1}(1 + \sqrt{5}) & 2(1 + \sqrt{5})^{n+1} - 2(1 - \sqrt{5})^{n+1} \end{pmatrix}.$$

Consequently,

$$\begin{pmatrix} b_n \\ b_{n+1} \end{pmatrix} = \begin{pmatrix} x_n \\ y_n \end{pmatrix} = A^n \begin{pmatrix} 1 \\ 1 \end{pmatrix} = \frac{1}{\sqrt{5}} \begin{pmatrix} \left(\frac{1+\sqrt{5}}{2}\right)^{n+1} - \left(\frac{1-\sqrt{5}}{2}\right)^{n+1} \\ \left(\frac{1+\sqrt{5}}{2}\right)^{n+2} - \left(\frac{1-\sqrt{5}}{2}\right)^{n+2} \end{pmatrix}. \qquad \square$$

Remark 3.1.12 Proposition 3.1.11 shows that b_n is the nearest integer to $(1/\sqrt{5})((1+\sqrt{5})/2)^{n+1}$ (and lies alternatingly above and below). In that sense, the Fibonacci sequence is as close to a strict exponential growth model as one can get with integers. Example 15.2.5 embeds this in a general pattern.

3. Second-Order Difference Equations. What we just did with the Fibonacci recursion is, of course, a general method. Whenever a sequence is defined using a linear second-order recursion $a_{n+1} = pa_{n-1} + qa_n$, one can convert this to a first-order vector recursion

$$\begin{pmatrix} a_n \\ a_{n+1} \end{pmatrix} = \begin{pmatrix} 0 & 1 \\ p & q \end{pmatrix} \begin{pmatrix} a_{n-1} \\ a_n \end{pmatrix}.$$

In most cases these recursions are as easy to solve as they were in the case of the Fibonacci sequence. To see this, note that the closed form of the Fibonacci numbers is a linear combination of like powers of the two eigenvalues. This is not surprising:

Proposition 3.1.13 If $\begin{pmatrix} 0 & 1 \\ p & q \end{pmatrix}$ has two distinct eigenvalues λ and μ, then every solution of the recursion $a_{n+1} = pa_{n-1} + qa_n$ is of the form $a_n = x\lambda^n + y\mu^n$.

Proof Let v and w be eigenvectors for λ and μ, respectively, and write

$$\begin{pmatrix} a_0 \\ a_1 \end{pmatrix} = \alpha v + \beta w.$$

Then

$$\begin{pmatrix} a_n \\ a_{n+1} \end{pmatrix} = \alpha \lambda^n v + \beta \mu^n w,$$

that is, $x = \alpha v_1$ and $y = \beta w_1$ are as required. $\square$

Remark 3.1.14 Note that x and y can be determined from the initial conditions directly, without finding eigenvectors.

Example 3.1.15 The recursion $a_{n+1} = a_{n-1}/2 + a_n/2$ from Section 1.2.3 corresponds to the matrix

$$\begin{pmatrix} 0 & 1 \\ 1/2 & 1/2 \end{pmatrix}$$

with eigenvalues $1, -1/2$, so the lobster harvests modeled by it are expected to be given by $a_n = x + (-1/2)^n y$ for some x, y. This corresponds to a relatively rapid stabilization with oscillations and also occurs in the logistic population model for parameters between 2 and 3 (see Case 2 of Proposition 2.5.2).

▓ EXERCISES

▓ **Exercise 3.1.1** Show that $\| \cdot \|'$ defined in Section 3.1.6 is a norm.

▓ **Exercise 3.1.2** Suppose A is a symmetric matrix and $\lambda \neq \mu$ are eigenvalues with eigenvectors v, w, respectively. Show that $v \perp w$, that is, v and w are orthogonal.

▓ **Exercise 3.1.3** Show that every 2×2 matrix with real eigenvalues and two orthogonal eigenvectors is symmetric.

▓ **Exercise 3.1.4** State and prove an analog of Proposition 3.1.13, for the case of a double eigenvalue with a single eigenvector.

▓ **Exercise 3.1.5** Suppose you have an unlimited supply of tiles with side lengths 1×2 and 2×2 that can be shifted but not rotated. In how many different ways can you tile a strip of dimensions $n \times 2$?

3.2 LINEAR DIFFERENTIAL EQUATIONS IN THE PLANE

The continuous-time companions to linear maps are linear differential equations. We begin to study these now, with a view to asymptotic behavior.

The appearance of invariant curves in the above examples is not accidental. The linear maps we described above arise from solutions of closely related differential equations, whose solutions interpolate iterates of the maps above. These are of the form

$$\begin{pmatrix} \dot{x} \\ \dot{y} \end{pmatrix} = A \begin{pmatrix} x \\ y \end{pmatrix},$$

or, more explicitly,

$$\dot{x} = a_{11}x + a_{12}y$$

$$\dot{y} = a_{21}x + a_{22}y.$$

3.2.1 Node

The case of a right-hand side with two distinct positive (real) eigenvalues is represented by the differential equation

$$\begin{pmatrix} \dot{x} \\ \dot{y} \end{pmatrix} = \begin{pmatrix} \log \lambda & 0 \\ 0 & \log \mu \end{pmatrix} \begin{pmatrix} x \\ y \end{pmatrix},$$

where $\lambda, \mu > 0$, whose solutions are given by

$$\begin{pmatrix} x(t) \\ y(t) \end{pmatrix} = \begin{pmatrix} x(0)e^{t\log\lambda} \\ y(0)e^{t\log\mu} \end{pmatrix} = \begin{pmatrix} x(0)\lambda^t \\ y(t)\mu^t \end{pmatrix} = \begin{pmatrix} \lambda^t & 0 \\ 0 & \mu^t \end{pmatrix} \begin{pmatrix} x(0) \\ y(0) \end{pmatrix}.$$

At this point it is natural to look back to Section 2.4.2.5, where we took the point of view that the set of solutions should be considered as a family of maps of the space with t as a parameter. In this example the maps are given by the matrices

$$\begin{pmatrix} \lambda^t & 0 \\ 0 & \mu^t \end{pmatrix}.$$

For $t = 1$ we get the map

$$\begin{pmatrix} \lambda & 0 \\ 0 & \mu \end{pmatrix},$$

and therefore the solution curves parametrize the invariant curves that we found before. As in the corresponding discrete-time case, this orbit picture is called a *node*.

For the special case $A = \log \lambda \, \mathrm{Id}$ and $t = 1$ we get $\lambda \, \mathrm{Id}$.

3.2.2 Degenerate Node

Corresponding to the linear map

$$\begin{pmatrix} \lambda & 1 \\ 0 & \lambda \end{pmatrix}$$

with $\lambda > 1$ consider the differential equation

$$\begin{pmatrix} \dot{x} \\ \dot{y} \end{pmatrix} = \begin{pmatrix} \log \lambda & 1 \\ 0 & \log \lambda \end{pmatrix} \begin{pmatrix} x \\ y \end{pmatrix}$$

whose solutions are given by

$$\begin{pmatrix} x(t) \\ y(t) \end{pmatrix} = \begin{pmatrix} x(0)e^{t \log \lambda} + y(0)te^{t \log \lambda} \\ y(0)e^{t \log \lambda} \end{pmatrix} = \begin{pmatrix} \lambda^t & t\lambda^t \\ 0 & \lambda^t \end{pmatrix} \begin{pmatrix} x(0) \\ y(0) \end{pmatrix} = \lambda^t \begin{pmatrix} x(0) + ty(0) \\ y(0) \end{pmatrix}.$$

Here the maps of the space with parameter t are given by the matrices

$$\begin{pmatrix} \lambda^t & t\lambda^t \\ 0 & \lambda^t \end{pmatrix},$$

and for $t = n$ this gives the action of

$$\begin{pmatrix} \lambda & 1 \\ 0 & \lambda \end{pmatrix}^n.$$

From here it is easy to obtain the invariant curves (3.1.2): Take absolute values and logarithms of the second component to solve for t and substitute this t into the first component.

3.2.3 Focus

Linear maps with eigenvalues $\rho e^{\pm i\theta}$ arise from the linear differential equation

$$\begin{pmatrix} \dot{x} \\ \dot{y} \end{pmatrix} = \begin{pmatrix} \log \rho & \theta \\ -\theta & \log \rho \end{pmatrix} \begin{pmatrix} x \\ y \end{pmatrix}$$

with solutions

$$\begin{pmatrix} x(t) \\ y(t) \end{pmatrix} = \rho^t \begin{pmatrix} x(0) \cos \theta t + y(0) \sin \theta t \\ y(0) \cos \theta t - x(0) \sin \theta t \end{pmatrix} = \rho^t \begin{pmatrix} \cos \theta t & \sin \theta t \\ -\sin \theta t & \cos \theta t \end{pmatrix} \begin{pmatrix} x(0) \\ y(0) \end{pmatrix}.$$

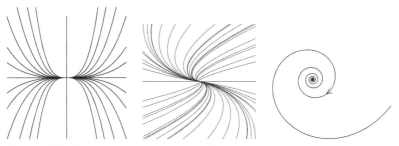

Figure 3.2.1. Node, degenerate node, and focus.

Here the solution family gives maps with matrices

$$\rho^t \begin{pmatrix} \cos\theta t & \sin\theta t \\ -\sin\theta t & \cos\theta t \end{pmatrix},$$

the expected rotation with scaling, and these solutions parametrize the invariant spirals. Following the discrete-time case, this picture is called a *focus*. The exceptional case is $|\rho| = 1$, where the solutions are pure rotations and every circle $r = \text{const.}$ is a periodic orbit and is called a *center*. Thus, unlike the discrete-time case (Section 3.1.8.4), the dynamics is completely understood at this point.

3.2.4 Saddle

The continuous-time picture of a *saddle* is obtained from

$$\begin{pmatrix} \dot{x} \\ \dot{y} \end{pmatrix} = A \begin{pmatrix} x \\ y \end{pmatrix},$$

where A has one positive and one negative eigenvalue, that is, by considering the differential equation

$$\begin{pmatrix} \dot{x} \\ \dot{y} \end{pmatrix} = \begin{pmatrix} \log\lambda & 0 \\ 0 & \log\mu \end{pmatrix} \begin{pmatrix} x \\ y \end{pmatrix},$$

whose solutions are given by

$$\begin{pmatrix} x(t) \\ y(t) \end{pmatrix} = \begin{pmatrix} x(0)e^{t\log\lambda} \\ y(0)e^{t\log\mu} \end{pmatrix} = \begin{pmatrix} x(0)\lambda^t \\ y(t)\mu^t \end{pmatrix} = \begin{pmatrix} \lambda^t & 0 \\ 0 & \mu^t \end{pmatrix} \begin{pmatrix} x(0) \\ y(0) \end{pmatrix}.$$

For $t = 1$ this is the map $\begin{pmatrix} \lambda & 0 \\ 0 & \mu \end{pmatrix}$, and therefore the solution curves that parametrize the invariant curves that we found before. As in the corresponding discrete-time case, this orbit picture is called a *saddle*.

Notice that we obtained all pictures for the case of linear maps as pictures for solutions of differential equations, except that from solutions of differential equations we never get maps with simple negative eigenvalues. This cannot be, because a solution of a differential equation that starts on an eigenspace cannot leave the eigenspace, yet it cannot pass through 0 either (by uniqueness of solutions).

3.2.5 The Matrix Exponential

The connection between linear maps on the plane and two-dimensional linear differential equations with constant coefficients can be made explicit by recalling that the solution to the differential equation $\dot{x} = ax$ with $x(0) = x_0$ is given by $x(t) = e^{at}x_0$. Analogously, the solution of $\dot{x} = Ax$ with $x \in \mathbb{R}^n$ and an $n \times n$ matrix A is

$$x(t) = e^{At}x(0), \quad \text{where} \quad e^{At} := \sum_{i=0}^{\infty} \frac{A^i t^i}{i!}.$$

Each term of the series is an $n \times n$ matrix, so the addition makes sense. The series converges absolutely because every entry of A^i is bounded in absolute value by $\|A^n\| \le \|A\|^n$. For example, if

$$A = \begin{pmatrix} \log \lambda & 0 \\ 0 & \log \mu \end{pmatrix},$$

then

$$e^{At} = \begin{pmatrix} \lambda^t & 0 \\ 0 & \mu^t \end{pmatrix}$$

because

$$\sum_{i=0}^{\infty} \begin{pmatrix} (\log \lambda)^i & 0 \\ 0 & (\log \mu)^i \end{pmatrix} \frac{t^i}{i!} = \begin{pmatrix} \sum_{i=0}^{\infty} \frac{(\log \lambda)^i t^i}{i!} & 0 \\ 0 & \sum_{i=0}^{\infty} \frac{(\log \mu)^i t^i}{i!} \end{pmatrix} = \begin{pmatrix} e^{t \log \lambda} & 0 \\ 0 & e^{t \log \mu} \end{pmatrix}.$$

It is only slightly less straightforward to check that

$$e^{\begin{pmatrix} 0 & \theta \\ -\theta & 0 \end{pmatrix} t} = \begin{pmatrix} \cos \theta t & \sin \theta t \\ -\sin \theta t & \cos \theta t \end{pmatrix} \quad \text{and} \quad e^{\begin{pmatrix} \log \rho & \theta \\ -\theta & \log \rho \end{pmatrix} t} = \rho^t \begin{pmatrix} \cos \theta t & \sin \theta t \\ -\sin \theta t & \cos \theta t \end{pmatrix}.$$

3.2.6 Periodic Coefficients

In the previous situation the (linear) time-1-map is e^A. For an initial condition $x(0)$ it gives the solution $x(1)$ at time 1. To get $x(n)$, we iterate this map n times. This works because the differential equations do not involve the time parameter, or $e^{Ai} = (e^A)^i$.

This also works when time enters as a parameter in the differential equation in a periodic way. If $\dot{x} = A(t)x$, where $A(t+1) = A(t)$ and M is such that $x(1) = Mx(0)$ for any solution $x(\cdot)$, then we find $x(2)$ from $x(0)$ by solving the differential equation $\dot{x} = A(t+1)x = A(t)x$ with initial condition $x(1)$ for time 1; hence $x(2) = M^2 x(0)$ and inductively $x(i) = M^i x(0)$. Therefore differential equations with periodic coefficients are within the scope of our methods.

Notice that this observation did not use linearity of the differential equation. The same reasoning and conclusion apply to differential equations $\dot{x} = f(x, t)$ with $f(x, t+1) = f(x, t)$ for $x \in \mathbb{R}^n$.

■ **EXERCISES**

■ **Exercise 3.2.1** For each of the following matrices A determine which, if any, of the above cases the differential equation

$$\begin{pmatrix} \dot{x} \\ \dot{y} \end{pmatrix} = A \begin{pmatrix} x \\ y \end{pmatrix}$$

belongs to, and sketch the phase portrait (including arrows) analogously to Figure 3.2.1. (Note that the preferred axes will be askew.)

a) $\begin{pmatrix} 0 & 2 \\ -1 & 3 \end{pmatrix}$ b) $\begin{pmatrix} 1 & 1 \\ 3 & -1 \end{pmatrix}$ c) $\begin{pmatrix} 3 & -2 \\ 2 & 3 \end{pmatrix}$ d) $\begin{pmatrix} 1 & -1 \\ 1 & 3 \end{pmatrix}$

■ **Exercise 3.2.2** Obtain the invariant curves (3.1.2) for $\begin{pmatrix} \lambda^t & t\lambda^t \\ 0 & \lambda^t \end{pmatrix}$.

■ **Exercise 3.2.3** Derive an equation for the invariant spirals of a focus in polar coordinates.

■ **Exercise 3.2.4** Consider the differential equations

$$\begin{pmatrix} \dot{x} \\ \dot{y} \end{pmatrix} = \begin{pmatrix} 2 & 0 \\ 0 & 6 \end{pmatrix} \begin{pmatrix} x \\ y \end{pmatrix}, \quad \begin{pmatrix} \dot{x} \\ \dot{y} \end{pmatrix} = \begin{pmatrix} \sqrt{2} & 0 \\ 0 & \sqrt{3} \end{pmatrix} \begin{pmatrix} x \\ y \end{pmatrix}, \quad \begin{pmatrix} \dot{x} \\ \dot{y} \end{pmatrix} = \begin{pmatrix} 3 & 0 \\ 0 & 10 \end{pmatrix} \begin{pmatrix} x \\ y \end{pmatrix}$$

and consider the solution curves as graphs of the form $y = \varphi(x)$. In each case determine how many times the corresponding function can be differentiated at 0.

■ **PROBLEMS FOR FURTHER STUDY**

■ **Problem 3.2.5** Suppose a linear map with a double negative eigenvalue arises from the solution of a linear differential equation. Show that it is proportional to the identity.

3.3 LINEAR MAPS AND DIFFERENTIAL EQUATIONS IN HIGHER DIMENSION

Next we study linear maps in higher-dimensional spaces. There is greater variety here than for maps of the plane because there are more possibilities for the combinations of eigenvalues – real and complex ones may coexist, and there can be multiple real or complex eigenvalues, possibly coexisting with simple or other multiple ones. Therefore it is reasonable not to attempt a classification as fine as we achieved for planar maps, but rather to divide the possible behaviors into sufficiently distinct ones according to asymptotic behavior. In fact, for most applications, this is entirely sufficient.

The first step is suggested by Proposition 3.1.10, but it takes a little more effort in this general case. We first establish the basic pertinent notions more carefully.

3.3.1 Spectral Radius

Definition 3.3.1 Let $A: \mathbb{R}^n \to \mathbb{R}^n$ be a linear map. We call the set of eigenvalues the *spectrum* of A and denote it by sp A. We denote the maximal absolute value of an eigenvalue of A by $r(A)$ and call it the *spectral radius* of A (see Figure 3.3.1).

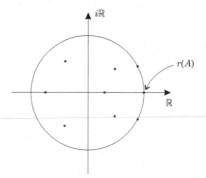

Figure 3.3.1. Eigenvalues and spectral radius.

For any choice of norm (Definition A.1.29) the spectral radius is bounded above by the norm of A as defined in (2.2.8): $r(A) \leq \|A\|$ by considering an eigenvector for the eigenvalue of largest absolute value, if this eigenvalue is real, or otherwise taking a complex eigenvector w for the (complex) eigenvalue of largest absolute value and applying A to $v := w + \bar{w}$. With respect to the Euclidean norm we have $\|A\| = r(A)$ whenever A is diagonal (or diagonalizable over the complex numbers). It is occasionally useful to have some estimates relating $\|A\|$ and the size of the entries of A:

Lemma 3.3.2 *For an $n \times n$ matrix A denote its entries by a_{ij} and define the norm $|A| := \max_{ij} |a_{ij}|$. Then $|A| \leq \|A\| \leq \sqrt{n}|A|$.*

Proof $\|Av\| = \sqrt{\sum_{i=1}^{n}(\sum_{i=1}^{n} a_{ij}v_j)^2} \leq |A|\sqrt{\sum_{i=1}^{n}(\sum_{j=1}^{n} v_j)^2} = \sqrt{n}|A|\|v\|$, and conversely $|a_{ij}| = \langle e_i, Ae_j \rangle \leq \|A\|$. $\square$

The following fact is useful for the understanding of dynamics of linear maps, even if they cannot be diagonalized.

Proposition 3.3.3 *For every $\delta > 0$ there is a norm in $\mathbb{R}^n$ such that $\|A\| < r(A) + \delta$.*

The proof uses a lemma that is analogous to Proposition 2.6.12.

Lemma 3.3.4 *Consider $\mathbb{R}^n$ with any norm $\| \cdot \|$ and a linear map $A \colon \mathbb{R}^n \to \mathbb{R}^n$. If $C, \lambda > 0$ are constants such that $\|A^n\| \leq C\lambda^n$ for all $n \in \mathbb{N}$, and if $\mu > \lambda$, then there is a norm $\| \cdot \|'$ on $\mathbb{R}^n$ with respect to which $\|A\| \leq \mu$.*

Proof If $n \in \mathbb{N}$ is such that $C(\lambda/\mu)^n < 1$, then $\|v\|' := \sum_{i=0}^{n-1} \|A^iv\|/\mu^i$ defines a norm with

$$\|Av\|' = \sum_{i=1}^{n} \|A^iv\|/\mu^{i-1} = \mu \left(\|Av\|' + \frac{\|A^nv\|}{\mu^n} - \|v\| \right)$$

$$\leq \mu\|Av\|' - \left(1 - C\frac{\lambda^n}{\mu^n}\right) \|v\| \leq \mu\|Av\|'.$$

$\square$

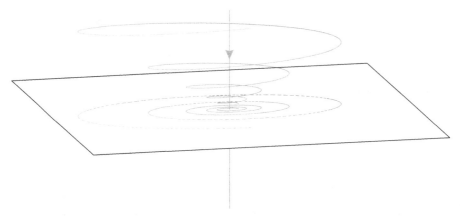

Figure 3.3.2. Attracting fixed point with real and complex eigenvalues.

Proof of Proposition 3.3.3 By the lemma it suffices to show that there is a coordinate change and a norm for which $\|A^n\| \leq C(r(A) + (\delta/2))^n$ for all $n \in \mathbb{N}$.

For each real eigenvalue λ (with multiplicity k) consider the *generalized eigenspace* or *root space* $E_\lambda := \{v \mid (A - \lambda \operatorname{Id})^k v = 0\}$. With a little linear algebra one can see that $\dim(E_\lambda) = k$. [This makes sense because the image of the unit cube under $A - t \operatorname{Id}$ has volume $\det(A - t \operatorname{Id}) \approx (t - \lambda)^k$, so there must be k directions of "collapse" as $t \to \lambda$.] Thus, these spaces generate the whole space. Alternatively, this follows from the Jordan normal form.

On E_λ the binomial formula gives

$$A^n = (\lambda \operatorname{Id} + \Delta)^n = \sum_{l=0}^{k-1} \binom{n}{l} \lambda^{n-l} \Delta^l = \lambda^n \sum_{l=0}^{k-1} \binom{n}{l} \lambda^{-l} \Delta^l.$$

The entries of $\Delta_n := \sum_{l=0}^{k-1} \binom{n}{l} \lambda^{-l} \Delta^l$ are polynomials in n, so Lemma 3.3.2 implies that $\|A^n\|/|\lambda|^n$ is bounded by a polynomial $p(n)$ in n. If $\delta > 0$, then $p(n)|\lambda|^n/(|\lambda| + \delta)^n \to 0$; so there is a $C > 0$ such that $\|A^n\|/|\lambda|^n \leq p(n) \leq C(|\lambda| + (\delta/2))^n/|\lambda|^n$ for all n.

One can analyze complex eigenvalues separately, or one can look at A as a linear map of $\mathbb{C}^n$ by allowing complex numbers as components of vectors. Then the preceding analysis applies to the root spaces for complex eigenvalues as well.

To deduce the complete result from the result for root spaces we use a norm of the desired kind on each of these root spaces. If we write vectors v as $(v_1, \ldots, v_l)$ with every v_l in a different root space with norm $\|\cdot\|_l$, then we can define the desired norm by $\|v\| := \sum_{i=1}^{l} \|v_l\|_l$. $\square$

Exercise 3.3.2 shows that any norm has the property obtained in the proof, although C depends on the norm. However, the norm in Proposition 3.3.3 is special. One could also obtain it from an argument as above without using the lemma by employing finer tools from linear algebra (Jordan normal form plus a linear coordinate change that makes the off-diagonal terms as small as desired).

Corollary 3.3.5 *If $r(A) < 1$, then A is eventually contracting. In particular, the positive iterates of every point converge to the origin with exponential speed. If, in addition, A is an invertible map, that is, if zero is not an eigenvalue for A, then negative iterates of every point go to infinity exponentially.*

The reverse result evidently applies to maps all of whose eigenvalues have absolute value greater than one.

3.3.2 Nonlinear Contractions

Our insights into the asymptotics of linear maps can sometimes be transferred to nonlinear systems. Chapter 9 uses this many times. A useful simple example arises from the fact that for an appropriate norm we can arrange for $\|A^n\| \leq C(r(A) + (\delta/2))^n$ (proof of Proposition 3.3.3, or a consequence of the statement of Proposition 3.3.3). Invoking the Mean Value Theorem 2.2.14 as in the proof of Proposition 2.2.17, we get

Lemma 3.3.6 *Let f be a continuously differentiable map with a fixed point x_0 where $r(Df_{x_0}) < 1$. Then there is a closed neighborhood U of x_0 such that $f(U) \subset U$ and f is eventually contracting on U.*

3.3.3 The Noncontracting Case

It remains to understand the mixed situations with only some eigenvalues inside or outside the unit circle. Analogously to the two-dimensional situation, it is possible to decompose $\mathbb{R}^n$ into subspaces that are contracting, expanding, or neutral, except that now all three possibilities can coexist. As in the two-dimensional case, these subspaces correspond to sets of eigenvalues inside, outside, and on the unit circle, respectively. But, similarly to the case of only one real eigenvalue in $\mathbb{R}^2$, it does not quite suffice to consider eigenspaces only. Instead, we have to consider the *generalized eigenspaces* or *root spaces* of A introduced in the preceding proof for the case of real eigenvalues (while this much generality is desirable, the reader may choose to assume diagonalizability to make the argument more transparent). For a pair of complex conjugate eigenvalues λ, $\bar{\lambda}$ we let $E_{\lambda,\bar{\lambda}}$ be the intersection of $\mathbb{R}^n$ with the sum of root spaces corresponding to E_λ and $E_{\bar{\lambda}}$ for the complexification of A (that is, the extension to the complex space $\mathbb{C}^n$). For brevity we call $E_{\lambda,\bar{\lambda}}$ a root space, too. Let

$$(3.3.1) \qquad E^- = E^-(A) = \bigoplus_{-1<\lambda<1} E_\lambda \oplus \bigoplus_{|\lambda|<1} E_{\lambda,\bar{\lambda}}$$

be the space jointly spanned by all root spaces for eigenvalues inside the unit circle and similarly

$$(3.3.2) \qquad E^+ = E^+(A) = \bigoplus_{|\lambda|>1} E_\lambda \oplus \bigoplus_{|\lambda|>1} E_{\lambda,\bar{\lambda}}.$$

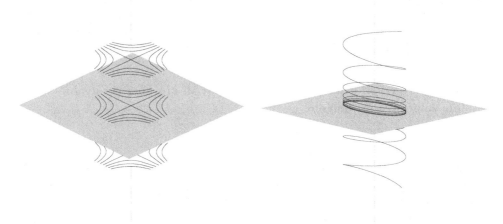

Figure 3.3.3. Stack of saddles and spirals onto a plane.

If the map A is invertible, then $E^+(A) = E^-(A^{-1})$. Finally, let

$$(3.3.3) \qquad E^0 = E^0(A) = E_1 \oplus E_{-1} \oplus \bigoplus_{|\lambda|=1} E_{\lambda,\bar{\lambda}}.$$

The spaces E^-, E^+, E^0 are obviously invariant with respect to A and $\mathbb{R}^n = E^- \oplus E^+ \oplus E^0$. Since the restriction of A to the space $E^-(A)$ is a linear map with all eigenvalues of absolute value less than one, Corollary 3.3.5 and Proposition 3.3.3 imply

Corollary 3.3.7 *The restriction* $A\!\restriction_{E^-(a)}$ *of a linear map A to the space $E^-(A)$ is eventually contracting. If A is invertible, then in addition* $A^{-1}\!\restriction_{E^+(A)}$ *is eventually contracting. Furthermore, for any $\delta > 0$ there is a norm with respect to which* $\|A\!\restriction_{E^-(a)}\| \leq r(A\!\restriction_{E^-(a)}) + \delta$ *and* $\|A^{-1}\!\restriction_{E^+(A)}\| \leq r(A^{-1}\!\restriction_{E^+(A)}) + \delta$.

To obtain this *Lyapunov norm* apply Proposition 3.3.3 on $E^-(A)$ and $E^+(a)$ separately to get norms $\|\cdot\|_-$ and $\|\cdot\|_+$, respectively, on these subspaces. Then define a norm for points $x = (x_-, x_0, x_+)$ by $\|(x_-, x_0, x_+)\| := \|x_-\|_- + \|x_0\| + \|x_+\|_+$.

Definition 3.3.8 The space $E^-(A)$ above is called the *contracting* subspace and the space $E^+(A)$, the *expanding* subspace. We say that A is *hyperbolic* if $E^0 = \{0\}$ or, equivalently, if $\mathbb{R}^n = E^+ \oplus E^-$.

▨ EXERCISES

▨ **Exercise 3.3.1** For each of the following matrices determine the spectral radius of the corresponding linear map and describe the possible long-term behaviors of orbits $(A^n v)_{n \in \mathbb{N}}$.

$$
\text{a)} \begin{pmatrix} 0 & -2 & 2 \\ 1 & 3 & -2 \\ 2 & 4 & -3 \end{pmatrix} \qquad \text{b)} \frac{1}{2} \begin{pmatrix} -1 & -1 & 0 \\ 2 & -1 & 1 \\ 0 & 1 & -1 \end{pmatrix} \qquad \text{c)} \begin{pmatrix} 2 & 0 & 1 & 0 \\ 0 & 1 & 0 & 1 \\ 0 & 0 & 2 & 1 \\ 0 & -1 & 0 & 1 \end{pmatrix}.
$$

▦ **Exercise 3.3.2** Given a linear map A show that for any norm there is a C such that $\|A^n\| \le C(r(A) + (\delta/2))^n$ for all $n \in \mathbb{N}$.

▦ **Exercise 3.3.3** Prove that $r(A) = \lim_{n \to \infty} \|A^n\|^{1/n}$.

▦ **Exercise 3.3.4** Suppose the linear map $x \mapsto Ax$ is a contraction on $\mathbb{R}^n$. Prove that $\operatorname{tr} A < n$.

▦ **Exercise 3.3.5** Suppose A is a 3×3 matrix with $\det A = 1/10$ and $\operatorname{tr} A = 2.7$. Show that A does not define a contraction.

▦ **PROBLEMS FOR FURTHER STUDY**

▦ **Problem 3.3.6** Suppose all eigenvalues of a 3×3 matrix A have absolute value one. Prove that there exists a constant C, such that for any $v \in R^3$ and $n \ne 0$, $\|A^n v\| \le C n^2 \|v\|$.

▦ **Problem 3.3.7** Generalize the statement of the previous problem to $n \times n$ matrices.

▦ **Problem 3.3.8** The norm obtained in the proof of Lemma 3.3.4 does not arise from an inner product. Show that there is an "adapted" inner product whose norm $\|x\| = \sqrt{\langle x, x \rangle}$ is an adapted norm.

CHAPTER 4

Recurrence and Equidistribution on the Circle

So far we concentrated on dynamical systems where the asymptotic behavior can be described simply: Every orbit was either fixed (sometimes periodic) or was attracted to (possibly different) fixed points as the time approached positive and negative infinity. In several situations, such as Proposition 2.3.5, we showed that no other behavior is possible.

In this chapter we study a fundamentally different type of behavior. Analysts use the rather innocuous term "quasiperiodic" to describe it and to signify that it is not much more than a generalization of periodic behavior. But from the dynamical point of view this is a starting point for the understanding of *nontrivial recurrence*, the central paradigm of the theory of dynamical systems.

We begin with a careful study of this phenomenon in the simplest possible situation, circle rotations. In the second section this already gives a remarkable array of interesting applications. The final section extends some of our insights to nonlinear circle maps.

4.1 ROTATIONS OF THE CIRCLE

The description of our first example is surprisingly simple; it is, in fact, closely related to some of the linear dynamical systems that appeared in Chapter 3, specifically Section 3.1.8.4 with $\rho = 1$: For a linear system with a pair of complex conjugate eigenvalues of absolute value 1, complex behavior may appear on the invariant circles $r = \text{const}$. We now study these rotations of a circle.

4.1.1 Circle Rotations

In Section 2.6.2 we saw two different convenient ways to represent the circle that allow us to write various formulas in a nice fashion. One can use either multiplicative notation, in which the circle is represented as the unit circle in the complex plane

$$S^1 = \{z \in \mathbb{C} \mid |z| = 1\} = \{e^{2\pi i \phi} \mid \phi \in \mathbb{R}\},$$

or additive notation, where

$$S^1 = \mathbb{R}/\mathbb{Z}$$

consists of the real numbers with integer translates identified (recall Figure 2.6.2). In multiplicative notation all algebraic operations make sense as operations over complex numbers. In additive notation we can use addition and subtraction (but not multiplication or division) just as the usual operations over real numbers, but we have to keep in mind that all equalities make sense up to an integer. It is customary to add "(mod 1)" to such equalities. Thus, the expression $a = b$ (mod 1), where a and b are real numbers, means that $a - b$ is an integer.

The logarithm map

$$e^{2\pi i\phi} \mapsto \phi$$

establishes an isomorphism between these representations. Let us measure the length of arcs on the circle by the parameter ϕ; that is, the length of the whole circle is equal to one. Let $\ell(\Delta)$ denote the length of the arc Δ measured in such a way. To similarly define a distance introduce a metric on the set $X = \mathbb{R}/\mathbb{Z} := \{[x] \mid x \in \mathbb{R}\}$ of equivalence classes by setting $d(x, y) := \min\{|b - a| \mid a \in x, b \in y\}$ as in Proposition 2.6.7.

We use the symbol R_α to denote the rotation by the angle $2\pi\alpha$. In multiplicative notation

$$R_\alpha(z) = z_0 z \text{ with } z_0 = e^{2\pi i\alpha}.$$

Not surprisingly, in additive notation we have

(4.1.1) $$R_\alpha(x) = x + \alpha \quad (\text{mod } 1).$$

The iterates of the rotation are correspondingly

$$R_\alpha^n(z) = R_{n\alpha}(z) = z_0^n z$$

in multiplicative notation and

$$R_\alpha^n(x) = x + n\alpha \quad (\text{mod } 1).$$

in additive notation.

A crucial distinction in the dynamics of rotations appears between the cases of the rotation parameter α being rational and irrational.

In the former case, write $\alpha = p/q$, where p, q are relatively prime integers. Then $R_\alpha^q(x) = x$ for all x, so R_α^q is the identity map and after q iterates the transformation simply repeats itself. Thus the total orbit of any point is a finite set and all orbits are q-periodic.

4.1.2 Density of Orbits
The case of irrational α is much more interesting. First, it is clear from the above formulas for the iterates that the orbit of every point is an infinite set. We can, however, say much more.

Proposition 4.1.1 *If $\alpha \notin \mathbb{Q}$, then every positive semiorbit of R_α is dense.*

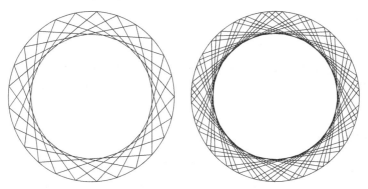

Figure 4.1.1. Periodic orbit and segment of a dense orbit.

Proof Suppose $x, z \in S^1$. To show that z is in the closure of the positive semiorbit of x, let $\epsilon > 0$. The positive semiorbit of x is infinite and no set of $k \geq \lfloor 1/\epsilon \rfloor + 1$ points has pairwise distances all exceeding ϵ. Thus there are $l, k \in \mathbb{N}$ such that $l < k \leq \lfloor 1/\epsilon \rfloor$ and $d(R_\alpha^k(x), R_\alpha^l(x)) < \epsilon$. Then $d(R_\alpha^{k-l}(x), x) < \epsilon$ because R_α^{-l} preserves distances. By the way, this latter distance is independent of x because, if $y \in S^1$, then $y = R_{y-x}(x)$ and

$$d\big(R_\alpha^{k-l}(y), y\big) = d\big(R_\alpha^{k-l}(R_{y-x}(x)), R_{y-x}(x)\big) = d\big(R_{(k-l)\alpha+y-x}(x), R_{y-x}(x)\big)$$
$$= d\big(R_{y-x}(R_\alpha^{k-l}(x)), R_{y-x}(x)\big) = d\big(R_\alpha^{k-l}(x), x\big);$$

so k and l can be chosen independently of x.

Take $\theta \in [-1/2, 1/2]$ such that $\theta = (k-l)\alpha \pmod 1$. Then $\rho := |\theta| < \epsilon$ and $R_\alpha^{k-l} = R_\theta$. Let $N = \lfloor 1/\rho \rfloor + 1$ (independently of x). Then the subset $\{R_{i\theta}(x) \mid i = 0, 1, \ldots, N\}$ of the positive semiorbit of x divides the circle into intervals of length less than $\rho < \epsilon$, so there is an $n \leq N(k-l)$ such that $d(R_\alpha^n(x), z) < \epsilon$. $\square$

Remark 4.1.2 Since the negative semiorbit of R_α is the positive semiorbit of $R_{-\alpha}$, we also proved the density of negative semiorbits.

An alternate proof of minimality shows the absence of proper invariant closed subsets by contradiction:

Alternate proof of Proposition 4.1.1 Let $A \subset S^1$ be an invariant closed set. The complement $S^1 \smallsetminus A$ is a nonempty open invariant set that consists of disjoint intervals. Let I be the longest of those intervals (or one of the longest, if there are several of the same length). Since rotation preserves the length of any interval, the iterates $R_\alpha^n(I)$ do not overlap. Otherwise, $S^1 \smallsetminus A$ would contain an interval longer than I. Since α is irrational, no iterates of I can coincide, because then an endpoint x of an iterate of I would come back to itself and we would have $x + k\alpha = x \pmod 1$ with $k\alpha = l$ an integer and $\alpha = l/k$ a rational number. Thus the intervals $R_\alpha^n(I)$ are all of equal length and all disjoint, but this is impossible because the circle has finite length and the sum of lengths of disjoint intervals cannot exceed the length of the circle. $\square$

Proposition 4.1.1 motivates the following general definitions.

Definition 4.1.3 A homeomorphism (see Definition A.1.16) $f\colon X \to X$ is said to be *topologically transitive* if there exists a point $x \in X$ such that its orbit $\mathcal{O}_f(x) := (f^n(x))_{n \in \mathbb{Z}}$ is dense in X. Equivalently, every f-invariant open invariant set is dense. A noninvertible map f is said to be topologically transitive if there exists a point $x \in X$ such that its (positive) orbit $\mathcal{O}_f^+(x) := (f^n(x))_{n \in \mathbb{N}_0}$ is dense in X.

The definitions for continuous-time systems are similar.

Definition 4.1.4 A homeomorphism $f\colon X \to X$ is said to be *minimal* if the orbit of every point $x \in X$ is dense in X or, equivalently, if f has no proper closed invariant sets. A closed invariant set is said to be minimal if it contains no proper closed invariant subsets or, equivalently, if it is the orbit closure of any of its points.

Thus Proposition 4.1.1 establishes that any rotation of the circle by an angle incommensurable with π, that is, by an irrational number of degrees (we shall call such a rotation simply an *irrational rotation*), is minimal and hence topologically transitive.

While minimality always implies topological transitivity, the converse is by no means true. Chapter 7 contains various examples that combine topological transitivity (existence of some dense orbits) with the existence of many orbits of different types, for example, infinitely many periodic (finite) orbits whose union is in turn dense.

4.1.3 Dense Orbits

It may be interesting to get a good picture of how an orbit fills the circle densely. We do this in a specific example by following the orbit of 0 under a rotation R_α, where we take

$$\alpha = \cfrac{1}{3 + \cfrac{1}{5 + \cfrac{1}{c}}}$$

for some $c > 1$. $\alpha \in \mathbb{Q}$ if and only if $c \in \mathbb{Q}$. The unusual form of α will seem more natural at the end.

Since $1/4 < \alpha < 1/3$ and hence $3\alpha < 1 < 4\alpha$, the first time the orbit returns more closely to 0 than ever before is after three steps. The first three points, α, 2α, and 3α, are evenly spaced, and since $4\alpha > 1$, 3α is closer to an integer than the previous points. The precise distance is

$$\delta := 1 - 3\alpha = 1 - \cfrac{3}{3 + \cfrac{1}{5 + \cfrac{1}{c}}} = \cfrac{\cfrac{1}{5 + \cfrac{1}{c}}}{3 + \cfrac{1}{5 + \cfrac{1}{c}}} = \cfrac{1}{16 + \cfrac{3}{c}}.$$

To find the next time of closest return we start from the fourth step, using $4\alpha = \alpha - \delta$ (mod 1). So three α-steps take us from α to $\alpha - \delta$. How many of these 3α-steps does it take to get the next closest approach? As before, it should be about α/δ, and the desired number n must satisfy $n\delta < \alpha < (n+1)\delta$. Indeed, $n = 5$ works:

$$5\delta = \frac{5}{15 + \left(1 + \dfrac{3}{c}\right)} = \frac{1}{3 + \left(\dfrac{1}{5} + \dfrac{3}{5c}\right)} < \frac{1}{3 + \dfrac{1}{5}} < \frac{1}{3 + \dfrac{1}{5 + \dfrac{1}{c}}} = \alpha,$$

and

$$6\delta = \frac{6}{16 + \dfrac{3}{c}} > \frac{6}{18} = \frac{1}{3} > \alpha.$$

These five 3α-steps evenly fill the interval $(0, \alpha)$ and simultaneously its three image intervals. When this next closest return is reached, the orbit segment is a δ-dense subset of the circle spaced evenly (except for the smaller interval of the new closest return). The next closest return after this is determined by c, and it is safe to guess that it will happen after about c steps.

If c were about a billion, this would mean that it takes about a billion 5δ-steps until the next closest return, which is some 15 billion iterations of R_α. In particular, the first 7 billion iterations are guaranteed to leave gaps of $\delta/2 > 1/35$. So large entries in this *continued fraction* form of α are not a good thing for filling the circle well. Continued fractions are discussed in greater detail in Section 15.2.

In conclusion, there is a natural sequence of ever longer time scales during each of which the orbit achieves a finer degree of density in a fairly homogeneous way. Thus, the behavior of an orbit is periodic, except for a little error δ, which produces a perturbation with much higher period – up to an even smaller error, and so on.

4.1.4 Uniform Distribution for Intervals

The preceding discussion suggests that we look into the way orbits of an irrational rotation are distributed on the circle in a quantitative fashion by finding the *frequencies* with which iterates of a point visit various parts of a circle. To be specific, fix an arc $\Delta \subset S^1$, and for $x \in S^1$ and $n \in \mathbb{N}$ let

$$F_\Delta(x, n) := \operatorname{card}\{k \in \mathbb{Z} \mid 0 \le k < n, \ R_\alpha^k(x) \in \Delta\}.$$

This function is nondecreasing in n for fixed x and Δ. Since the positive semiorbit of any point is dense, there are arbitrarily large positive iterates of x that belong to Δ. Hence

$$F_\Delta(x, n) \to \infty, \text{ as } n \to \infty.$$

The natural measure of how often these visits happen is the *relative frequency* of visits:

(4.1.2)
$$\frac{F_\Delta(x, n)}{n}.$$

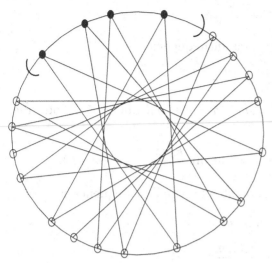

Figure 4.1.2. Frequencies.

Recall that $\ell(\Delta)$ denotes the length of the arc Δ measured by the parameter ϕ introduced at the beginning of Section 4.1.1.

The argument from the proof of Proposition 4.1.1 gives

Proposition 4.1.5 *Suppose α is irrational and consider the rotation R_α. Let Δ, Δ' be arcs such that $\ell(\Delta) < \ell(\Delta')$. Then there exists an $N_0 \in \mathbb{N}$ such that, if $x \in S^1$, $N \geq N_0$, and $n \in \mathbb{N}$, then*

$$F_{\Delta'}(x, n+N) \geq F_\Delta(x, n).$$

Proof By the density of the positive semiorbit of the left end of the arc Δ we can find an $N_0 \in \mathbb{N}$ such that $R_\alpha^{N_0}(\Delta) \subset \Delta'$. Then $R_\alpha^n(x) \in \Delta$ implies $R_\alpha^{n+N_0}(x) \in \Delta'$ and $F_{\Delta'}(x, n+N) \geq F_{\Delta'}(x, n+N_0) \geq F_\Delta(x, n)$ for $N \geq N_0$. $\square$

So far we have not specified what kinds of arcs we consider: open, closed, or half-open. There is no difference as far as limit behavior of the frequencies is concerned, since the difference between the number of visits for an open arc and its closure is at most two. So it is convenient to always take arcs closed on the left and open on the right. For such arcs we have the following *additivity* property: If the right end of Δ_1 coincides with the left end of Δ_2, then $\Delta_1 \cap \Delta_2 = \emptyset$, $\Delta_1 \cup \Delta_2$ is an arc and

$$F_{\Delta_1}(x, n) + F_{\Delta_2}(x, n) = F_{\Delta_1 \cup \Delta_2}(x, n).$$

It is also convenient to define $F_A(x, n) := \text{card}\{k \in \mathbb{Z} \mid 0 \leq k < n, \ R_\alpha^k(x) \in A\}$ for any set A that is a union of disjoint arcs. So far we do not know that the limits of relative frequencies exist. However, one can consider the upper limits:

$$\bar{f}_x(A) := \limsup_{n \to \infty} \frac{F_A(x, n)}{n}.$$

These quantities are obviously *subadditive*:

$$\bar{f}_x(A_1 \cup A_2) \leq \bar{f}_x(A_1) + \bar{f}_x(A_2).$$

In particular, if $\bigcup_{i=1}^n A_i = S^1$, then $\sum_{i=1}^n \bar{f}_x(A_i) \geq 1$. Proposition 4.1.5 implies

Corollary 4.1.6 *If $\ell(\Delta) < \ell(\Delta')$, then $\bar{f}_x(\Delta) \leq \bar{f}_x(\Delta')$.*

Similarly we introduce the lower asymptotic frequencies:

$$\underline{f}_x(A) := \liminf_{n \to \infty} \frac{F_A(x, n)}{n}.$$

Obviously, for any set A we have $F_A(x, n) = n - F_{A^c}(x, n)$, where A^c denotes the complement $S^1 \setminus A$ of A and hence

$$(4.1.3) \qquad \bar{f}_x(A) = \limsup_{n \to \infty} \frac{F_A(x, n)}{n} = 1 - \liminf_{n \to \infty} \frac{F_{A^c}(x, n)}{n} = 1 - \underline{f}_x(A^c).$$

Now we can formulate our main statement about asymptotic frequencies:

Proposition 4.1.7 *For any arc $\Delta \subset S^1$ and any $x \in S^1$*

$$f(\Delta) := \lim_{n \to \infty} \frac{F_\Delta(x, n)}{n} = \ell(\Delta),$$

and the limit is uniform in x.

Remark 4.1.8 The property of the sequence $a_n := R_\alpha^n(x)$, $n = 0, 1, 2, \ldots$ expressed by this proposition is called *uniform distribution* or *equidistribution*: The asymptotic frequency of visits is the same for arcs of equal length, regardless of where on the circle they are.

Proof First we show that the frequency of visits cannot be too high.

Lemma 4.1.9 *If $\ell(\Delta) = 1/k$, then $\bar{f}_x(\Delta) \leq 1/(k-1)$.*

Proof Consider $k-1$ disjoint arcs $\Delta_1, \Delta_2, \ldots, \Delta_{k-1}$ of length $1/(k-1)$ each. For $1 \leq i < k$, Proposition 4.1.5 gives natural numbers N_i such that, if $x \in S^1$, then

$$F_{\Delta_i}(x, n + N_i) \geq F_\Delta(x, n);$$

hence $F_{\Delta_i}(x, n + N) \geq F_\Delta(x, n)$, where $N = \max_i N_i$ and

$$(k-1) F_\Delta(x, n) \leq \sum_{i=1}^{k-1} F_{\Delta_i}(x, n + N).$$

Since N is fixed, we let $n \to \infty$ to obtain

$$(k-1) \bar{f}_x(\Delta) \leq \bar{f}_x \left(\bigcup_{i=1}^{k-1} \Delta_i \right) = 1. \quad \square$$

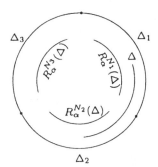

Figure 4.1.3. Upper asymptotic frequencies.

For an arc Δ and $\epsilon > 0$ find k and an arc $\Delta' \supset \Delta$ of length $l/k < \ell(\Delta) + \epsilon$. Then

$$\bar{f}_x(\Delta) < \bar{f}_x(\Delta') < \frac{l}{k-1} < (\ell(\Delta) + \epsilon)\frac{k}{k-1}$$

by Lemma 4.1.9. Letting $\epsilon \to 0$ and thus $k \to \infty$ gives $\bar{f}_x(\Delta) \le \ell(\Delta)$. Combined with (4.1.3) for $A = \Delta^c$, this also gives $\underline{f}_x(\Delta) \ge \ell(\Delta)$. This proves that the limit exists and equals $\ell(\Delta)$. $\square$

4.1.5 Uniform Distribution for Functions

Clearly, frequencies also can be defined for any set A that is a finite union of arcs. To do this in a suggestive way we call

$$\chi_A(x) := \begin{cases} 1 & \text{if } x \in A \\ 0 & \text{if } x \notin A \end{cases}$$

the *characteristic function* of A. Then we define

$$F_A(x, n) := \sum_{k=0}^{n-1} \chi_A\big(R_\alpha^k(x)\big),$$

and accordingly the relative frequency is $\sum_{k=0}^{n-1} \chi_A(R_\alpha^k(x))/n$. Since, by definition of the integral, $\ell(\Delta) = \int_{S^1} \chi_\Delta(\phi)d\phi$, Proposition 4.1.7 can be reformulated as

$$(4.1.4) \qquad \lim_{n\to\infty} \frac{1}{n} \sum_{k=0}^{n-1} \chi_A\big(R_\alpha^k(x)\big) = \int_{S^1} \chi_\Delta(\phi)d\phi.$$

1. Birkhoff Averaging. We can also consider similar expressions for functions φ other than characteristic functions.

Definition 4.1.10 The *Birkhoff averaging operator* $\mathcal{B}_n$ is the operator that associates to a function φ the function $\mathcal{B}_n(\varphi) := \sum_{k=0}^{n-1} \varphi \circ R_\alpha^k/n$ given by

$$(4.1.5) \qquad \mathcal{B}_n(\varphi)(x) = \frac{1}{n} \sum_{k=0}^{n-1} \varphi\big(R_\alpha^k(x)\big).$$

Remark 4.1.11 Some useful properties of $\mathcal{B}_n$ are

(1) $\mathcal{B}_n$ is linear: $\mathcal{B}_n(a\varphi + b\psi) = a\mathcal{B}_n(\varphi) + b\mathcal{B}(\psi)$.
(2) $\mathcal{B}_n$ is nonnegative: If $\varphi \ge 0$, then $\mathcal{B}_n(\varphi) \ge 0$. Also, $\mathcal{B}_n$ is positive (or monotone): If $\varphi > 0$, then $\mathcal{B}(\varphi) > 0$.
(3) $\mathcal{B}_n$ is nonexpanding: $\sup_{x\in S^1} \mathcal{B}_n(\varphi)(x) \le \sup_{x\in S^1} \varphi(x)$.
(4) $\mathcal{B}_n$ preserves the average: $\int_{S^1} \mathcal{B}_n(\varphi)(\phi)\,d\phi = \int_{S^1} \varphi(\phi)\,d\phi$.

This leads to the following conclusions:

Proposition 4.1.12

(1) *For any step function φ that is a linear combination of characteristic functions of arcs*, $\lim_{n\to\infty} \mathcal{B}_n(\varphi) = \int_{S^1} \varphi(\phi)\,d\phi$.

(2) *For any function φ that is a uniform limit of step functions we also have*
$\lim_{n\to\infty} \mathcal{B}_n(\varphi) = \int_{S^1} \varphi(\phi)\, d\phi$.

Proof Since the map associating to an integrable function its integral over S^1 has properties analogous to those in the remark, we can start from (4.1.4), pass to linear combinations and uniform limits, and compare results.

For the second claim fix $\epsilon > 0$, take a step function φ_ϵ with $\sup_{\phi\in S^1} |\varphi(\phi) - \varphi_\epsilon(\phi)| < \epsilon$, and apply the operators $\mathcal{B}_n$ to $\varphi = \varphi_\epsilon + (\varphi - \varphi_\epsilon)$ to get

$$(4.1.6) \quad \int_{S^1} \varphi(\phi)\, d\phi - 2\epsilon \leq \int_{S^1} \varphi(\phi) - \epsilon\, d\phi - \epsilon \leq \int_{S^1} \varphi_\epsilon(\phi)\, d\phi - \epsilon$$

$$= \lim_{n\to\infty} \mathcal{B}_n(\varphi_\epsilon) - \epsilon \leq \liminf_{n\to\infty} \mathcal{B}_n(\varphi) \leq \limsup_{n\to\infty} \mathcal{B}_n(\varphi) \leq \lim_{n\to\infty} \mathcal{B}_n(\varphi_\epsilon) + \epsilon$$

$$= \int_{S^1} \varphi_\epsilon(\phi)\, d\phi + \epsilon \leq \int_{S^1} \varphi(\phi) + \epsilon\, d\phi + \epsilon \leq \int_{S^1} \varphi(\phi)\, d\phi + 2\epsilon$$

for any $\epsilon > 0$. $\square$

Lemma 4.1.13 *Every continuous function is the uniform limit of step functions, as is every function with finitely many discontinuity points and with one-sided limits at these points (piecewise continuous functions).*

Proof Every continuous function on S^1 is uniformly continuous; that is, for every $\epsilon > 0$ one can find an $n \in \mathbb{N}$ such that, on every arc of length $1/n$, the function varies by less than ϵ. Dividing S^1 into n such arcs gives a step function that is constant on each arc and differs from the given function by less than ϵ. Essentially the same argument applies to functions with finitely many discontinuity points and one-sided limits at these points. $\square$

The last two results give:

Proposition 4.1.14 *If α is irrational and φ is continuous, then*

$$\lim_{n\to\infty} \frac{1}{n} \sum_{k=0}^{n-1} \varphi\big(R_\alpha^k(x)\big) = \int_{S^1} \varphi(\phi)\, d\phi$$

uniformly in x.

There is a more general class of functions for which the Birkhoff average converges to the integral, namely, all functions integrable in the usual (Riemann) sense.

Theorem 4.1.15 *If α is irrational and φ is Riemann integrable, then*

$$(4.1.7) \qquad\qquad \lim_{n\to\infty} \frac{1}{n} \sum_{k=0}^{n-1} \varphi\big(R_\alpha^k(x)\big) = \int_{S^1} \varphi(\phi)\, d\phi$$

uniformly in x.

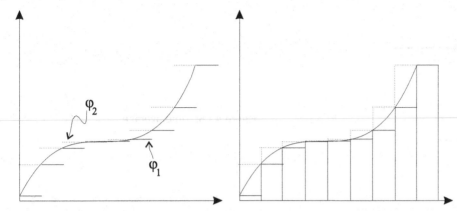

Figure 4.1.4. Approximation by step functions, Riemann sums.

Proof Pick a partition of S^1 into a finite number of arcs I_i. The corresponding lower and upper Riemann sums $\sum_i \min \varphi_{\upharpoonright I_i} l(I_i)$ and $\sum_i \max \varphi_{\upharpoonright I_i} l(I_i)$ can be interpreted as integrals of step functions φ_1 and φ_2 defined by $\varphi_1 = \min \varphi_{\upharpoonright I_i}$ on I_i and $\varphi_2 = \max \varphi_{\upharpoonright I_i}$ on I_i. By definition of Riemann integrability, the partition can be chosen such that

$$\int_{S^1} \varphi(\phi)\, d\phi - \epsilon \leq \int_{S^1} \varphi_1(\phi)\, d\phi \leq \int_{S^1} \varphi_2(\phi)\, d\phi. \leq \int_{S^1} \varphi(\phi)\, d\phi + \epsilon.$$

This implies that

$$\int_{S^1} \varphi(\phi)\, d\phi - \epsilon \leq \int_{S^1} \varphi_1(\phi)\, d\phi = \lim_{n\to\infty} \mathcal{B}_n(\varphi_1) \leq \liminf_{n\to\infty} \mathcal{B}_n(\varphi)$$

(4.1.8)

$$\leq \limsup_{n\to\infty} \mathcal{B}_n(\varphi) \leq \lim_{n\to\infty} \mathcal{B}_n(\varphi_2) = \int_{S^1} \varphi_2(\phi)\, d\phi \leq \int_{S^1} \varphi(\phi)\, d\phi + \epsilon.$$

Letting $\epsilon \to 0$ gives the result. $\square$

Remark 4.1.16 The condition of Riemann integrability is essential. To see this, take a point x_0 and define the set A as the union of the arcs of length 2^{-k+2} centered at $R_\alpha^k(x_0)$ for $k \geq 0$. Although some of these arcs overlap, A is a union of arcs the sum of whose lengths is less than $1/2$, whereas $\lim_{n\to\infty} \frac{1}{n} \sum_{k=0}^{n-1} \chi_A(R_\alpha^k(x)) = 1$. Of course, χ_A is not Riemann integrable.

2. Time Average and Space Average. The quantities on either side of (4.1.7) are averages.

Definition 4.1.17 Given a function φ, we call

$$\lim_{n\to\infty} \frac{1}{n} \sum_{k=0}^{n-1} \varphi\left(R_\alpha^k(x)\right)$$

its *time average* as sampled by following the orbit of x under the iterates of the rotation R_α. (Figure 4.1.5 illustrates this for $\varphi = x_{(0,1/2)}$.) The integral $\int_{S^1} \varphi(\phi)\, d\phi$ is called the *space average* of the function φ.

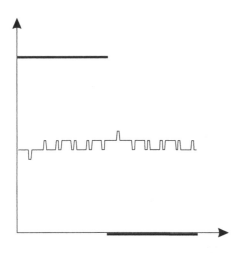

Figure 4.1.5. Time average.

These notions are both borrowed from physics, which is concerned with the measurement of observable quantities associated with a dynamical system. This means that there is a (measurable) quantity associated with the dynamical system in question that varies with the state of the dynamical system – in other words, one has a function defined on phase space whose value at a particular state of the dynamical system is displayed by the measuring device. Especially for systems that behave in unpredictable ways, it is quite natural to take a large number of successive measurements and average them. The limit of these averages is exactly the time average for the initial condition at which the measurements were begun.

The space average is more likely obtained as a result of calculations with a mathematical model of the physical system at hand. If one knows, as we do in our simple example, that space averages and time averages are supposed to coincide for the model one is testing, then the space average constitutes a prediction of the time average that is being measured, thus providing a means of verifying or falsifying the proposed model.

Returning to our situation, we note that the preceding result says that for any Riemann-integrable function the time average exists for the orbit of any point x and always coincides with the space average. This important property of irrational rotations is equivalent to uniform distribution and is referred to as *unique ergodicity*. This notion can be defined in the abstract setting of a continuous map of a compact metric space, even though there is no notion of an integral.

Definition 4.1.18 If X is a compact metric space and $f\colon X \to X$ a continuous map, then f is said to be *uniquely ergodic* if

$$\frac{1}{n} \sum_{k=0}^{n-1} \varphi(f^k(x))$$

converges to a constant uniformly (in x) for every continuous function φ.

4.1.6 The Kronecker–Weyl Method

In our arguments step functions played a special role. One can prove unique ergodicity of an irrational rotation in a much simpler, yet less elementary, way by using trigonometric polynomials to approximate continuous functions. This is possible due to the classical theorem of Weierstraß that says that continuous functions are uniform limits of trigonometric polynomials. This theorem is a close counterpart of a more familiar Weierstraß theorem that deals with the uniform approximation of a continuous function on an interval by polynomials. In this argument it is more convenient to use complex-valued functions.

Alternate proof of Proposition 4.1.14 Define the *characters* $c_m(x) := e^{2\pi imx} = \cos 2\pi mx + i \sin 2\pi mx$. If $m \neq 0$, then

$$c_m(R_\alpha(x)) = e^{2\pi im(x+\alpha)} = e^{2\pi im\alpha} e^{2\pi imx} = e^{2\pi im\alpha} c_m(x)$$

and

$$\left| \frac{1}{n} \sum_{k=0}^{n-1} c_m(R_\alpha^k(x)) \right| = \left| \frac{1}{n} \sum_{k=0}^{n-1} e^{2\pi imk\alpha} \right| = \frac{|1 - e^{2\pi imn\alpha}|}{n|1 - e^{2\pi im\alpha}|} \leq \frac{2}{n|1 - e^{2\pi im\alpha}|} \to 0$$

as $n \to \infty$, because $\sum_{k=0}^{n} x^k = (1 - x^{n+1})/(1 - x)$.

Birkhoff averaging operators are linear; so if $p(x) = \sum_{i=-l}^{l} a_i c_i(x)$ is a trigonometric polynomial, then $\lim_{n\to\infty} \mathcal{B}_n(p)(x)$ exists and is constant. It is a_0 because this constant has to be the integral of p over S^1 (the operators $\mathcal{B}_n$ do not change the integral). The same arguments as above allow us to pass to uniform limits of trigonometric polynomials, that is, all continuous functions. $\square$

This argument is more analytic and involves a much more straightforward calculation than the proof using step functions. Notice, however, that it does not give the original uniform distribution statement (Proposition 4.1.7), since characteristic functions are obviously not the uniform limit of trigonometric polynomials. To obtain uniform distribution for intervals from that for functions one can use the argument from the proof of Theorem 4.1.15 backwards: Approximate χ_A by continuous functions $\varphi_1 \leq \chi_A \leq \varphi_2$ such that $\int(\varphi_2 - \varphi_1) < \epsilon$ (Figure 4.1.6) and repeat the calculation (4.1.8).

4.1.7 Group Translations

Irrational rotations serve as the starting point for a number of fruitful generalizations. Let us discuss one of them. The circle is a compact abelian *group*, and the rotation can be represented in group terms as the group

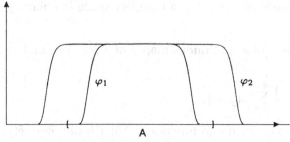

Figure 4.1.6. Approximation by continuous functions.

multiplication or translation

$$L_{g_0}: G \to G, \quad L_{g_0}g = g_0 g.$$

The orbit of the unit element $e \in G$ is the cyclic subgroup $\{g_0^n\}_{n \in \mathbb{Z}}$. Proposition 4.1.1 is closely related to the fact that the circle does not have proper infinite closed subgroups. To say that an orbit is dense requires a notion of approximation, so we define a *topological group* to be a group with a metric for which every L_g is a homeomorphism and taking inverses is continuous.

Proposition 4.1.19 *If the translation L_{g_0} on a topological group G is topologically transitive, then it is minimal.*

Proof For $g, g' \in G$ denote by $A, A' \subset G$ the closures of the orbits of g and g', respectively. Now $g_0^n g' = g_0^n g(g^{-1}g')$, so $A' = Ag^{-1}g'$ and $A' = G$ if and only if $A = G$. $\square$

■ EXERCISES

■ **Exercise 4.1.1** Prove that for the metric $d(x, y) := \min\{|b - a| \mid a \in x, b \in y\}$ on the set $X = \mathbb{R}/\mathbb{Z} := \{[x] \mid x \in \mathbb{R}\}$ every rotation is an isometry (as in Definition A.1.16).

■ **Exercise 4.1.2** Take $c = 7.1$ in Section 4.1.3 and determine the next closest return.

■ **Exercise 4.1.3** Prove the properties in Remark 4.1.11.

■ **Exercise 4.1.4** For a rotation R_α find $N \in \mathbb{N}$ in terms of α such that $F_{(0,1/2)}(x, n)/n \geq 0.45$ for all $n \geq N$ (see Section 4.1.4).

■ **Exercise 4.1.5** Suppose the motion of the sun and moon as observed from a specific place on earth are strictly periodic and that the time difference between sunrise and moonrise is never twice the same. Prove that this difference is uniformly distributed.

■ **Exercise 4.1.6** Give an example of a homeomorphism of a complete metric space that has a dense orbit but no dense semiorbit.

■ **Exercise 4.1.7** Give an example of a homeomorphism of a compact metric space that has a dense orbit but no dense semiorbit.

■ **Exercise 4.1.8** Prove that two minimal sets (Definition 4.1.4) are either disjoint or equal.

■ **Exercise 4.1.9** Prove that a contracting map of a compact space is uniquely ergodic

■ **Exercise 4.1.10** Give an example of a continuous map f of a compact metric space X such that

$$\frac{1}{n} \sum_{k=0}^{n-1} \varphi(f^k(x))$$

converges uniformly (in x) for every continuous function φ, but f is not uniquely ergodic.

■ **Exercise 4.1.11** Using enough digits of the decimal expansion of π, find the classical approximations 21/7 and 355/113 and write the result in the form described in Section 4.1.3. Find the fourth term in the continued-fraction approximation and explain how the size of this number is reflected in the quality of the approximation.

■ **PROBLEMS FOR FURTHER STUDY**

■ **Problem 4.1.12** Let G be a metrizable compact topological group. Suppose for some $g_0 \in G$ the translation L_{g_0} is topologically transitive. Prove that G is abelian.

■ **Problem 4.1.13** Show that a finite abelian group has a uniquely ergodic translation if and only if it is cyclic.

■ **Problem 4.1.14** Prove that the circle map $x \mapsto x + (1/4) \sin^2 \pi x$ shown in Figure 2.2.4 is uniquely ergodic.

■ **Problem 4.1.15** Define the following metric d_2 on the group $\mathbb{Z}$ of all integers: $d_2(m, n) = \|m - n\|_2$, where

$$\|n\|_2 = 2^{-k} \qquad \text{if } n = 2^k l \text{ with an odd number } l.$$

The completion of $\mathbb{Z}$ with respect to that metric is called the group of *2-adic* or *dyadic integers* and is usually denoted by $\mathbb{Z}_2$. It is a compact topological group. Let $\mathbb{Z}_2^+$ be the closure of the even integers with respect to the metric d_2. $\mathbb{Z}_2^+$ is a subgroup of $\mathbb{Z}_2$ of index two.

Prove that for $g_0 \in \mathbb{Z}_2$ the translation $L_{g_0} : \mathbb{Z}_2 \to \mathbb{Z}_2$ is topologically transitive if and only if $g_0 \in \mathbb{Z}_2 \setminus \mathbb{Z}_2^+$.

This is an example of a class of systems called *adding machines*. An equivalent description is given in Definition 11.3.10, and Theorem 11.3.11 shows that this dynamical system is a subsystem of the quadratic map $f_\lambda : [0, 1] \to [0, 1]$, $f_\lambda(x) := \lambda x(1 - x)$ from Section 2.5 for a particular value of λ.

4.2 SOME APPLICATIONS OF DENSITY AND UNIFORM DISTRIBUTION

There are numerous situations in which one would like to obtain information of some asymptotic nature and where the dynamics of circle rotations or toral translations, which are considered in the next chapter, plays a role, possibly behind the scenes, that makes it possible to obtain this asymptotic information from the knowledge we have acquired so far. In this section we show some such examples.

4.2.1 Distribution of Values for Periodic Functions

Let $(x_n)_{n \in \mathbb{N}}$ be a sequence of real numbers. A natural way to describe the distribution of values of such a sequence would be to consider the asymptotic frequencies with which this sequence "visits" various intervals.

Definition 4.2.1 Given a sequence $(x_n)_{n \in \mathbb{N}}$ and $a < b$, let $F_{a,b}(n)$ be the number of integers k such that $1 \le k \le n$ and $a < x_k < b$. We say that $(x_n)_{n \in \mathbb{N}}$ has an *asymptotic distribution* if for any $a, b, -\infty \le a < b \le \infty$ the limit

$$\lim_{n \to \infty} \frac{F_{a,b}(n)}{n}$$

exists. In this case the function

$$\Phi_{(x_n)_{n \in \mathbb{N}}}(t) := \lim_{n \to \infty} \frac{F_{-\infty, t}(n)}{n}$$

is called the *distribution function* of the sequence.

For sequences of the form $y_n = \varphi(x_n)$, we can give the distribution function in terms of information about φ.

Definition 4.2.2 If $A \subset \mathbb{R}$ is a finite union of disjoint intervals, then we define its *measure* $m(A)$ to be the sum of the lengths of these intervals.

A function φ on an interval is said to be *piecewise monotone* if the domain can be partitioned into finitely many intervals on which φ is monotone. In this case the preimage of every interval I is a finite union of intervals, and so we can define $m_\varphi(I) := m(\varphi^{-1}(I))$. For a piecewise monotone function we define the *distribution function* $\Psi_\varphi : \mathbb{R} \to \mathbb{R}$ of φ by $\Psi_\varphi(t) := m_\varphi((-\infty, t))$.

Note that $\Psi_\varphi(t)$ is the sum of lengths of the intervals comprising the set $\varphi^{-1}((-\infty, t))$ or, equivalently, $\Psi_\varphi(t) = \int \chi_{\varphi^{-1}((-\infty, t))}$.

Theorem 4.1.15 can be used to show that sequences obtained by calculating values of periodic functions along arithmetic progressions have asymptotic distribution and to calculate the distribution functions of such sequences:

Theorem 4.2.3 *Let φ be a function on the real line of period T such that the restriction $\varphi_T := \varphi\!\restriction_{[0,T]}$ of φ to the interval $[0, T]$ is piecewise monotone. If $\alpha \notin \mathbb{Q}$ and $t_0 \in \mathbb{R}$, then the sequence $x_n := \varphi(t_0 + n\alpha T)$ has an asymptotic distribution with distribution function Ψ_{φ_T}/T.*

Proof By introducing the new variable $s = t/T$ we may assume that $T = 1$. The periodic function φ can be viewed as a function on the circle. It will not cause any confusion to call this function φ. Then the sequence x_n coincides with the sequence of values of this function on the circle along the orbit of the irrational rotation R_α. Theorem 4.1.15 applied to the characteristic function of the set $\varphi^{-1}((a, b))$ shows that the sequence x_n has an asymptotic distribution: $\varphi(s_0 + n\alpha) \in \varphi^{-1}((a, b))$ if and only if $a < x_n < b$. This also implies the statement about the distribution function. $\square$

Example 4.2.4 Consider the sequence $(\sin n)_{n \in \mathbb{N}}$. Since sin is a 2π-periodic function and $\pi \notin \mathbb{Q}$, we can apply Theorem 4.2.3 with $\varphi(t) = \sin(t)$, $T = 2\pi$, $t_0 = 0$, $\alpha = (2\pi)^{-1}$ to establish the existence of the asymptotic distribution. In this case the distribution function Ψ can be calculated explicitly (see Figure 4.2.1):

$$(4.2.1) \qquad \Phi_{(\sin n)_{n \in \mathbb{N}}}(t) = \Psi_{\sin_{2\pi}}(t)/(2\pi) = \begin{cases} 0 & \text{for } t < -1 \\ \frac{1}{2} + \frac{1}{\pi}\arcsin(t) & \text{for } t \in [-1, 1] \\ 1 & \text{for } t > 1. \end{cases}$$

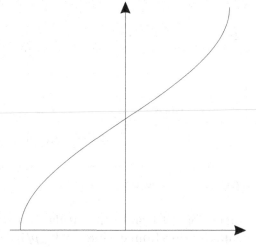

Figure 4.2.1. The distribution function.

4.2.2 Distribution of the First Digits of Powers

As an interesting arithmetic application of Theorem 4.2.3, or actually Proposition 4.1.7, we can now answer the questions posed in Section 1.3.4 about the distribution of the first digits of powers of 2 – in fact powers of any integer. Furthermore, we can answer an analogous question when an entire string of initial digits is prescribed.

If k is a power of 10, then its powers have leading digit 1, so we need not discuss this case further. We now show that otherwise any given string of digits does indeed occur as the initial string of digits of some power of k.

Proposition 4.2.5 *Let $k \in \mathbb{N}$ be a natural number other than a power of 10 and $p \in \mathbb{N}$. Then there exists an $n \in \mathbb{N}$ such that p gives the initial digits of the decimal expansion of k^n.*

Example 4.2.6 To clarify the conclusion take $k = 2$ and $p = 81$. Then $n = 13$ works because $2^{13} = 8192$.

Proof of Proposition 4.2.5 The conclusion can be rephrased by saying that there is an $l \in \mathbb{N}$ for which $k^n = 10^l p + q$, where $0 \le q < 10^l$. This, in turn, is equivalent to $10^l p \le k^n < 10^l(p+1)$ or

$$l + \lg p \le n\lg k < l + \lg(p+1),$$

where $\lg = \log_{10}$ is the logarithm to base 10. Now let $m = \lfloor \lg p \rfloor + 1$ be the number of digits of p. Then

$$0 \le \lg p - (m-1) \le n\lg k - l - (m-1) < \lg(p+1) - (m-1) \le 1,$$

which can be rewritten as

$$(4.2.2) \qquad \lg(p/10^{m-1}) \le \{n\lg k\} \le \lg((p+1)/10^{m-1}),$$

where $\{\cdot\}$ denotes the fractional part. Since $\lg k$ is irrational,[1] the sequence $(\lfloor n\lg k \rfloor)_{n=1}^{\infty}$ in $\mathbb{R}/\mathbb{Z}$ (see Section 4.1.1 or Section 2.6.2) is dense on the circle by

---

[1] If $\lg k = p/q$, then $2^p 5^p = 10^p = k^q = 2^{mq} 5^{nq}$ using prime factorization. Then $n = m$ and $k = 10^m$.

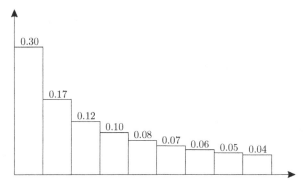

Figure 4.2.2. Distribution of first digits of 2^n.

Proposition 4.1.1 and hence $\{\{n\lg k\} \mid n \in \mathbb{N}\}$ is dense in $[0, 1)$. In particular, it contains points from the interval $[\lg(p/10^{m-1}), \lg((p+1)/10^{m-1}]$. $\square$

Our uniform distribution results give not only the existence of a given string of first digits, but also the asymptotic frequency:

Proposition 4.2.7 *For $k \in \mathbb{N}$ not a power of 10 and $p \in \mathbb{N}$, let $F_p^k(n)$ be the number of integers i between 0 and $n - 1$ such that p gives the initial digits of the decimal expansion of k^i. Then*

$$\lim_{n \to \infty} \frac{F_p^k(n)}{n} = \lg(p+1) - \lg p,$$

independently of k.

Proof We use either Proposition 4.1.7 or Theorem 4.2.3 for the function $\varphi(t) := \{t\}$ to see from the condition (4.2.2) that $\lim_{n \to \infty} (F_p^k(n)/n) = \lg((p+1)/10^{m-1}) - \lg(p/10^{m-1}) = \lg(p+1) - \lg p$. $\square$

Figure 4.2.2 shows approximate values of the asymptotic frequencies of the first digits of powers of 2 (or 3, or 7, ...), which Section 1.3.4 asked about.

4.2.3 Linear Flow on the 2-Torus

Some further applications of uniform distribution appear in conjunction with flows related to rotations and toral translations. In particular, our analysis of irrational rotations has immediate implications for the following system of differential equations on the 2-torus (we use additive notation; see Section 2.6.4)

$$(4.2.3) \qquad \frac{dx_1}{dt} = \omega_1, \qquad \frac{dx_2}{dt} = \omega_2.$$

This system of differential equations can be easily solved explicitly. The resulting flow $(T_\omega^t)_{t \in \mathbb{R}}$ has the form

$$(4.2.4) \qquad T_\omega^t(x_1, x_2) = (x_1 + \omega_1 t, x_2 + \omega_2 t) \pmod 1.$$

We present a geometric picture of this flow in Figure 4.2.3. As we already mentioned in Section 2.6.4, the torus $\mathbb{T}^2 = \mathbb{R}^2/\mathbb{Z}^2$ can be represented as the unit square

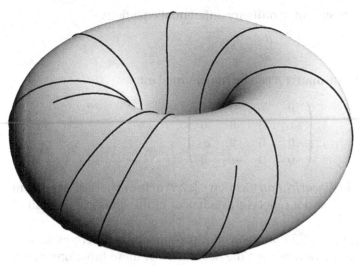

Figure 4.2.3. Linear flow on the embedded torus.

$I^2 = \{(x_1, x_2) \mid 0 \le x_1 \le 1, \ 0 \le x_2 \le 1\}$ with pairs of opposite sides identified: $(x, 0) \sim (x, 1)$ and $(0, x) \sim (1, x)$. In this representation the integral curves of the system (4.2.3) are pieces of straight lines with slope $\gamma = \omega_2/\omega_1$. The motion along the orbits is uniform with instantaneous "jumps" to the corresponding points when the orbit reaches the boundary of the square.

Proposition 4.2.8 *If γ is irrational, every orbit is dense in $\mathbb{T}^2$ and the flow is minimal in a sense analogous to that of Definition 4.1.4. If γ is rational, then every orbit is closed.*

Proof The circle $C_1 = \{x_1 = 0\}$ defines what is called a global section. This means that every positive and negative semiorbit crosses it infinitely often. Therefore we can define a *return map* to C_1 by assigning to each point of C_1 the point at which its positive semiorbit first returns to C_1.

This map is exactly the rotation R_γ, because between successive moments at which an orbit intersects C_1 the x_2-coordinate changes by exactly γ (mod 1). Thus, by Proposition 4.1.1, if γ is irrational, the closure of every orbit contains the circle C_1; and since the images of this circle under the flow $\{T_\omega^t\}$ cover the whole torus, every orbit is dense in $\mathbb{T}^2$. Thus, the flow is minimal in a sense analogous to that of Definition 4.1.4. If γ is rational, then every orbit is closed by (4.2.4). $\square$

4.2.4 Linear Differential Equations and Lissajous Figures

Flows on the 2-torus appear naturally in many problems from ordinary differential equations. Often a proper choice of coordinates brings a system of differential equations to the standard form (4.2.3). The most direct connection appears with certain systems of linear differential equations with constant coefficients. Let A be a 4×4 real matrix with two pairs of distinct, purely imaginary eigenvalues

$\pm i\alpha_1$, $\pm i\alpha_2$. Consider the system of ordinary differential equations

(4.2.5)
$$\frac{dx}{dt} = Ax.$$

By a coordinate change the matrix A can be transformed into

$$\begin{pmatrix} 0 & \alpha_1 & 0 & 0 \\ -\alpha_1 & 0 & 0 & 0 \\ 0 & 0 & 0 & \alpha_2 \\ 0 & 0 & -\alpha_2 & 0 \end{pmatrix}.$$

This system is, in fact, the *linearization* (Section 6.2.2.7) of the spherical pendulum discussed in Section 6.2.7, and it reflects the behavior of the spherical pendulum for small oscillations.

Now let $x = (x_1, x_2, x_3, x_4) \in \mathbb{R}^4$ and consider two quadratic functions $x_1^2 + x_2^2$ and $x_3^2 + x_4^2$. A direct calculation of the time derivatives of these functions shows that both of them are invariant under (4.2.5). Thus, for any two positive numbers r_1, r_2, the torus T_{r_1, r_2} determined by the equations $x_1^2 + x_2^2 = r_1^2$, $x_3^2 + x_4^2 = r_2^2$ is invariant. In the natural normalized angular coordinates φ_1, φ_2 on the torus defined by $x_1 = r_1 \cos 2\pi\varphi_1$, $x_2 = r_1 \sin 2\pi\varphi_1$, $x_3 = r_2 \cos 2\pi\varphi_2$, $x_4 = r_2 \sin 2\pi\varphi_2$ and with $\alpha_1/2\pi = \omega_1$, $\alpha_2/2\pi = \omega_2$, (4.2.5) becomes

$$\frac{d\varphi_1}{dt} = -\omega_1, \quad \frac{d\varphi_2}{dt} = -\omega_2,$$

and the solutions of (4.2.5) have the form

(4.2.6)
$$x_1(t) = r_1 \cos \omega_1 (t - t_0), \qquad x_2(t) = r_1 \sin \omega_1 (t - t_0),$$
$$x_3(t) = r_2 \cos \omega_2 (t - t_0), \qquad x_4(t) = r_2 \sin \omega_2 (t - t_0).$$

Thus if the ratio $\alpha_2/\alpha_1 = \omega_2/\omega_1$ is irrational, then on each torus T_{r_1, r_2} the flow defined by (4.2.5) is minimal.

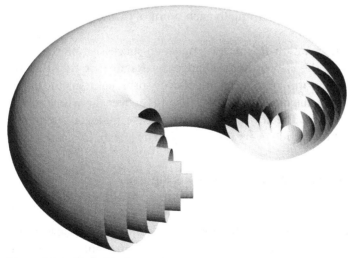

Figure 4.2.4. Nested tori.

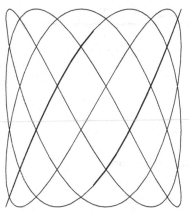
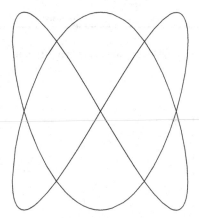

Figure 4.2.5. Lissajous figures.

Now consider the projection of solutions of (4.2.6) to the x_1, x_3-plane. The resulting curves are called *Lissajous figures*. From (4.2.6) we obtain the parametric representation of a Lissajous figure (we use coordinates x and y rather than x_1 and x_3):

$$x(t) = r_1 \cos \omega_1(t - t_0), \qquad y(t) = r_2 \cos \omega_2(t - t_0).$$

A simple physical interpretation of a Lissajous figure is as the configuration trajectory $x(t)$, $y(t)$ of a pair of *independent linear* or *harmonic oscillators* (Example 6.2.2, Section 6.2.2) given by second-order differential equations

$$\ddot{x} = \omega_1 x, \qquad \ddot{y} = \omega_2 y.$$

For commensurable frequencies ω_1, and ω_2, the Lissajous figures are families of closed curves usually self-intersecting and fairly complicated (Figure 4.2.5). But if the frequencies are incommensurable, then any Lissajous figure fills the rectangle $|x| \le r_1$, $|y| \le r_2$ densely. Moreover, by using a proper version of Theorem 4.2.3 one can prove existence and calculate the *limit density* of a Lissajous figure.

A physical simulation is easy to achieve by connecting simple oscillating circuits to the x and y inputs of an oscilloscope. With a little tuning one can reproduce various Lissajous figures easily. When one tunes away slightly from commensurate frequencies the picture twists slowly. One sees a figure such as one on the right of Figure 4.2.5, and it looks like a curve projected from a slowly rotating cylinder.[2]

4.2.5 Particles on the Interval and Billiards

There are various mechanical problems whose analysis can be reduced to that of rotations of the circle and linear flows on $\mathbb{T}^2$. This subsection analyzes a simple one carefully.

1. Particles on the Interval. Consider two point masses confined to an interval that collide elastically with the ends of the interval and with each other. Assume that the interval is the standard unit interval $[0, 1]$ and consider the case of both

---

[2] A careful description can be found in V. I. Arnold, *Mathematical Methods in Classical Mechanics*, Section 2.5. (Springer-Verlag, New York, Berlin, Heidelberg, 1978.)

Figure 4.2.6. Two particles on an interval.

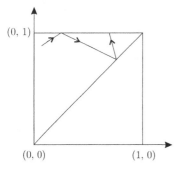

Figure 4.2.7. The configuration space.

masses equal to one. The configuration space of this system is the two-dimensional set $T = \{(x_1, x_2) \mid 0 \leq x_1 \leq x_2 \leq 1\}$, which is a right isosceles triangle, as shown in Figure 4.2.7. The motion of both masses can now be described as the motion of a point inside T.

The next few pages are dedicated to this system and contain a proof of the following

Proposition 4.2.9 *For any initial condition the system exhibits at most eight differ-ent values of the ratio of the velocities at any time. If the initial ratio is rational, then the motion is periodic; if it is irrational, then the motion eventually comes arbitrarily close to any specified configuration and has a uniform distribution property.*

The developments that lead to this result are a rich source of ideas for other questions about this system, and for other models to study. For example, it will afterwards be geometrically evident that the case of unequal masses is different in a fundamental way.

2. Billiard Flow. Between collisions both particles move at constant speed. This means that the "configuration" point in T that represents both of them has constant vertical and horizontal speed and hence traces out straight lines with constant speed. When it reaches the horizontal or vertical component of the boundary, which corresponds to a collision with one of the endpoints of the interval, the normal component of the velocity vector changes sign. At a collision with the hypothenuse the two components of the velocity change places (in an elastic collision two particles exchange momenta, that is, since the masses are equal, velocity). Thus in all three collision cases the configuration point (x_1, x_2) reflects from the boundary according to the rule "the angle of incidence is equal to the angle of reflection". In other words, the motion of the configuration point is like that of a small billiard ball inside the triangular table T. This is the reason why the continuous-time dynamical system describing this motion is usually called the *billiard flow*.

To study the billiard flow we first describe its phase space. Evidently, it consists of tangent vectors with footpoints in T. We need some conventions about the vectors at the boundary. Since the velocity changes instantaneously at a collision, we consider only outgoing velocities, so of the vectors with footpoint at the boundary, only those belong to the phase space that point inside the triangle. The corners correspond to simultaneous collisions, and we should see whether the further motion can be defined for a configuration point that hits a corner. The point $(0, 1)$, that is, the vertex at the right angle, corresponds to the simultaneous collision of two particles with the ends of the interval and presents no problem; each particle changes the sign of its velocity, so the velocity vector simply reverses itself. Thus in this case the outgoing velocity vector also points inside T. The other two vertices are more tricky. They correspond to a simultaneous collision of the particles with the same endpoint. A natural way to see how the motion can be extended is to approximate this event by a series of simple collisions and take the limit of the outcomes if it exists. It turns out that such a series always contains four collisions and the limit does exist. Rather than doing this tedious analysis directly, we ignore the two corners until we obtain the desired extension as a byproduct of geometric considerations that will give us better insight into the picture of the motion.

To summarize, the billiard flow is a continuous-time dynamical system in the phase space described above. A vector moves with speed equal to its length along its line and reflects from the sides. It is clearly sufficient to consider only vectors of a fixed, say, unit, length.

3. Unfolding. Now imagine that, when a point reaches the boundary, rather than reflecting the velocity vector we reflect the domain T and consider the motion with the same speed inside the reflected copy, as shown in Figure 4.2.8. This can be repeated when the point reaches the boundary (of the reflected copy) again and so on. Thus, instead of a complicated broken line inside T we obtain a straight-line motion that passes from one copy of T to another. Naturally, the original motion can be recovered by folding these copies back onto the original triangle. Different initial vectors lead to different sequences of reflections and produce different strings of copies of the triangle T along the unfolding trajectory. So one should try to consider all possible sequences of reflections at the same time. While for domains other than this triangle the result could be a mess with copies of the domain returning and

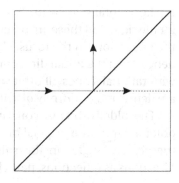

Figure 4.2.8. Unfolding the triangle.

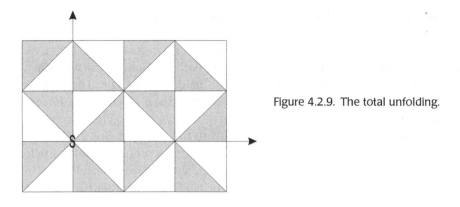

Figure 4.2.9. The total unfolding.

overlapping with the original one without coinciding with it, in this case any overlap with the interior of the original domain leads to exact matching. Thus, we obtain a *tiling* of the plane $\mathbb{R}^2$ by copies of the triangle T obtained by various compositions of reflections. Not all of those compositions are themselves reflections. For example, a reflection in the vertical side followed by a reflection in the horizontal side of the image produces a rotation by π around the vertex with the right angle. Again, careful inspection shows that the elements of the tiling come in eight different orientations (see Figure 4.2.9); all triangles in each of the eight classes are obtained from each other by *parallel translations*. As representatives of the eight classes one can take the original triangle T, its reflection in the hypotenuse, and the triangles obtained by the reflections of the unit square in the coordinate axes as well as their composition (these commute). In other words, the eight representatives exactly fill the square $S = \{(x, y) \mid \max\{|x|, |y|\} \leq 1\}$, and any other element of the tiling is obtained from one of the eight by parallel translation. The parallel translations that appear in the process are compositions of the iterates of the vertical and horizontal translations by two. We call S a *fundamental domain* for the translation group of the tiling.

4. Identification Space. Now replace the complete unfolding of a billiard flow orbit on $\mathbb{R}^2$ described above with a partial unfolding by forcing it to stay within the square S as follows. Each time the unfolded orbit reaches the boundary of S, instead of moving away to a triangle T_1 move to the unique triangle T_2 in S that is obtained from T_1 by parallel translation (see Figure 4.2.10). This creates discontinuities in the orbit when it is considered on the plane, namely, translation by $(0, -2)$ for the top horizontal side, by $(0, 2)$ for the bottom one, by $(2, 0)$ for the left, and by $(-2, 0)$ for the right. But these are precisely the identifications described in Section 4.2.3 for the linear flow on the torus, albeit of twice the size. The linear flow thus obtained depends on the initial direction of the billiard flow orbit in T. For initial directions with rational slopes all orbits are closed, and for those with irrational slopes they are dense and uniformly distributed on the torus.

The folded orbit in T consists of pieces of no more than eight different directions obtained from the original one by those compositions of reflections that produce the eight triangles in the square S. Four of these transformations act on the directions as reflections, namely, in the horizontal, vertical, and two diagonal lines;

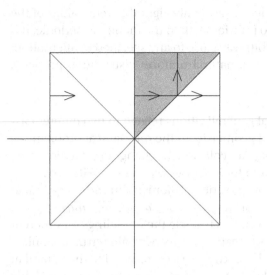

Figure 4.2.10. Partial unfolding.

the remaining ones are rotations by multiples of $\pi/2$. Thus, if the original angle with the positive horizontal direction is α, the allowable angles are $\pm\alpha + (k\pi/2)$; $k = 0, 1, 2, 3$. (See Figure 4.2.11.)

Density (corr. periodicity) of the orbits on the torus for initial directions with irrational (corr. rational) slopes translates into density (corr. periodicity) of the billiard orbits in the triangle. The same applies to the uniform distribution of the billiard orbits with initial directions with irrational slopes.

We also resolved the problem of extension of billiard orbits that hit a corner. On the torus such an extension is unambiguous, and after folding we obtain the precise prescription: An orbit that hits the origin under the angle α with the positive direction on the real axis bounces back under the angle $\pi - \alpha$, that is, it retraces itself exactly as in the case of the vertex $(0, 1)$.

Let us transfer our conclusion to the original mechanical problem. The slope of the phase vector in the billiard is the ratio of the velocities of the two particles on

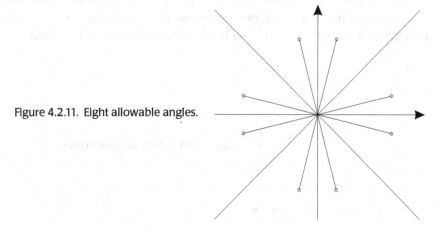

Figure 4.2.11. Eight allowable angles.

the interval. Thus, from any initial velocity ratio only eight different values of the ratio can be reached. If the initial ratio is rational, then the motion is periodic; if it is irrational, it will eventually come arbitrarily close to any specified configuration. Furthermore, in the latter case there is a natural uniform distribution property. This in particular gives Proposition 4.2.9.

5. Unequal Masses. There are natural generalizations related to two different aspects of this problem, mechanical and geometric. In the mechanical problem the masses of the particles may not be equal. At a collision the change in velocities of the two particles is inversely proportional to their masses m_1 and m_2. With respect to coordinates $q_1 = \sqrt{m_1} x_1, q_2 = \sqrt{m_2} x_2$, preservation of momentum and energy translates into the billiard law in the billiard table $\{(q_1, q_2) \mid 0 \leq q_1 \leq \sqrt{m_1/m_2} q_2 \leq \sqrt{m_1}\}$. (See Figure 4.2.12.) For the billiard flow in this triangle the unfoldings for different sequences of reflections overlap without creating any recognizable structure unless $m_1 = m_2, m_1 = 3m_2$, or $m_2 = 3m_1$. The latter two cases correspond to the billiard in the right triangle with the smallest angle $\pi/6$. Some other cases can be analyzed fairly well using similar ideas and this development. The first nontrivial case would be that of a right triangle with one angle $\pi/8$. This corresponds to a mass ratio of $3 + 4\sqrt{2}$ and is described in Section 14.4 of Katok and Hasselblatt, *Introduction to the Modern Theory of Dynamical Systems*, (New York, Cambridge University Press, 1995).

Now we turn to a generalization suggested by the geometric structure of our problem.

6. Polygonal Billiards. We now list those polygons where the approach we outlined works equally well as for the equal-masses model.

Consider billiard flows inside a polygon P with the property that the images of of P under the compositions of reflections in the sides of P tile the plane. For such polygons it is possible to unfold simultaneously all of the orbits of the billiard flow in a way similar to that for the right isosceles triangle. After that, we need to find a counterpart of the square S, that is, a nice fundamental domain of the translation group of the tiling to replace the complete unfolding into the infinite plane by a partial unfolding confined to a compact set.

The list of polygons satisfying this property is rather short. In addition to the right isosceles triangle it includes rectangles, the equilateral triangle, and the right triangle with an angle $\pi/6$. In order to find the counterpart of the square S for these

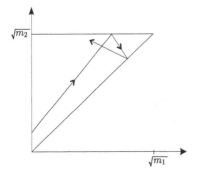

Figure 4.2.12. Two unequal masses.

Figure 4.2.13. Tilings.

cases, consider classes of domains in each tiling that differ by a parallel translation and then group one representative from each class in a nice way. The results are presented in Figure 4.2.13. There are four classes in the first case, six in the second, and twelve in the last. Natural fundamental domains are the twice dilated rectangle R in the first case and the regular hexagons in the other two (Figure 4.2.14).

We study these systems case by case. The analysis of the billiard flow in the triangle T translates to the case of a rectangle essentially verbatim, with a slight simplification. Indeed, the double rectangle R is naturally identified with the torus by identifying the pairs of opposite sides via corresponding parallel translations, and the billiard orbit in the original rectangle unfolds into a linear flow on that torus.

The other two cases can also be reduced to the linear flow on the torus because a torus may be obtained from any parallelogram by identifying pairs of opposite sides via translations. This is just a matter of a proper linear coordinate change on the plane. One can find a fundamental domain of the translation group of the two tilings in the shape of a parallelogram with an angle $\pi/3$ (Figure 4.2.15), so the billiard flow in the triangles reduces to the linear flows on the respective domains.

Remark 4.2.10 The classical example of a linear flow on the 2-torus arising from nonlinear systems of differential equations is the Kepler problem of a point mass in an inverse-square central gravitational field (Section 6.2.5). Such a point mass has three degrees of freedom, but preservation of angular momentum (Section 6.2.7) confines it to a plane, reducing the number of degrees of freedom to two. Since the equation of motion is of second order, this gives a four-dimensional phase space.

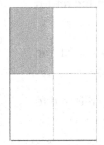 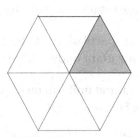 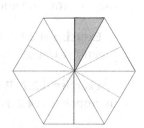

Figure 4.2.14. Fundamental domains.

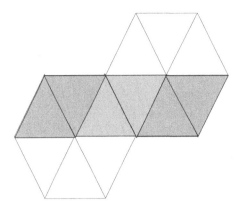

Figure 4.2.15. Parallelogram.

Energy conservation reduces this to three dimensions, and conservation of the size of angular momentum to two. These two dimensions are parametrized by a time parameter along an ellipse and a perihelion angle. This is therefore a system similar to the mathematical pendulum (Section 6.2.2) where one gets flows on circles, that is, one-dimensional invariant tori. However, the Kepler problem for several planets *without mutual interaction* gives higher-dimensional invariant tori with linear flows on them. This is the central feature of *complete integrability* in Hamiltonian dynamical systems.

▓ EXERCISES

▓ **Exercise 4.2.1** Give a detailed proof of (4.2.1).

▓ **Exercise 4.2.2** Verify directly that for any fixed number m the sum of $\lg(p+1) - \lg p$ over all p with exactly m digits is 1, as it should be according to Proposition 4.2.7.

▓ **Exercise 4.2.3** Verify the calculation needed to deduce Proposition 4.2.7 from Proposition 4.1.7 or Theorem 4.2.3.

▓ **Exercise 4.2.4** Referring to Proposition 4.2.7, determine $\lim_{n\to\infty} F_{10}^2(n)/n$ and find the asymptotic frequencies of 0 and 9, respectively, as the *second* digit of powers of 2.

▓ **Exercise 4.2.5** Referring to the proof of Proposition 4.2.8, assume $\gamma \neq 0$ and replace the section C_1 by the section $C_2 := \{x_2 = 0\}$. Prove that the resulting return map is a rotation and determine the rotation angle in terms of γ.

▓ **Exercise 4.2.6** Verify by direct calculation of the time derivatives that the functions $x_1^2 + x_2^2$ and $x_3^2 + x_4^2$ are invariant under (4.2.5).

▓ **Exercise 4.2.7** Formulate the natural uniform distribution property referred to in Proposition 4.2.9 and proved in Section 4.2.5.4.

▓ **Exercise 4.2.8** Prove that any closed proper subgroup Γ of $\mathbb{R}$ is cyclic, that is, $\Gamma = \{na\}_{n\in\mathbb{Z}}$ for some $a \in \mathbb{R}$.

■ **Exercise 4.2.9** Given an initial direction, how many slopes are there for the billiard flow in a square and in each of the two triangles, and what are they?

■ **Exercise 4.2.10** Suppose a horizontal light beam enters a circular room with mirrored walls. Describe the possibilities for which areas of the room will be best lit.

■ **Exercise 4.2.11** Prove that a complete unfolding of a regular pentagon covers every point of the plane infinitely many times.

■ **Exercise 4.2.12** Obtain the continuation of orbits in the billiard description of the 2-particle system by interpreting double collisions as limits of a series of simple collisions.

4.3 INVERTIBLE CIRCLE MAPS

The success in analyzing circle rotations is due in large part to the fact that these come from linear dynamical systems, namely, from rotations of the plane (Section 3.1). This causes the great homogeneity of the orbit structure that gives uniform density of orbits and uniform distribution. However, another ingredient, perhaps less apparent, is the simple structure of the circle itself. Analogously to the study of interval homeomorphisms (Section 2.3.1) this makes it possible to give a fairly satisfactory analysis of the orbit structure of any invertible map of the circle. One-dimensionality of the circle provides two (related) features that make a fairly detailed analysis possible: the (cyclic) ordering of its points and the Intermediate-Value Theorem. These have the effect of tying together different orbits tightly enough to make the possible orbit structures relatively easy to describe. The importance of the order structure will become particularly apparent in Proposition 4.3.11 and Proposition 4.3.15.

For noninvertible maps of an interval or of the circle the order of points may not be preserved and hence use of this first property fails, while the Intermediate-Value Theorem can still be used so long as we have continuity. Accordingly, the structural features are much more complicated while still amenable to rather extensive analysis. Chapter 11 outlines this for some interval maps.

One principle that will manifest itself in various guises throughout this section is that while, unlike the situation with rotations, the orbit structure of invertible circle maps is not always entirely homogeneous, the asymptotic behavior is in various different ways about as homogeneous, or at least coherent, as the entire orbit structure of a rotation and, in fact, ultimately turns out to look much like a rotation.

In this section a fundamental dichotomy is central: A circle homeomorphism (Definition A.1.16) may or may not have periodic points. Every orbit has the same type of asymptotic behavior, and it corresponds in a precise sense to the behavior of an orbit of a rational or an irrational rotation, respectively. The tool that leads to this conclusion is a parameter that reflects asymptotic rotation rates and is rational or not according to whether there are periodic points.

4.3.1 Lift and Degree

Recall the relation between the circle $S^1 = \mathbb{R}/\mathbb{Z}$ and the line $\mathbb{R}$ (see Section 2.6.2). There is a projection $\pi : \mathbb{R} \to S^1$, $x \mapsto [x]$, where $[x]$ is the equivalence class of x in

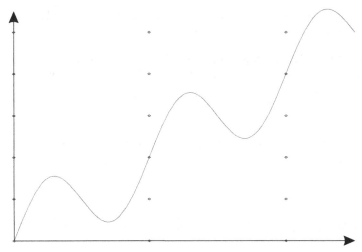

Figure 4.3.1. A lift and degree.

$\mathbb{R}/\mathbb{Z}$ as in Section 2.6.2. Here $[\cdot]$ denotes an equivalence class, whereas the integer part of a number is written $\lfloor \cdot \rfloor$. We use $\{\cdot\}$ for the fractional part.

Proposition 4.3.1 *If $f\colon S^1 \to S^1$ is continuous, then there exists a continuous $F\colon \mathbb{R} \to \mathbb{R}$, called a* lift *of f to $\mathbb{R}$, such that*

(4.3.1) $$f \circ \pi = \pi \circ F,$$

that is, $f([z]) = [F(z)]$. Such a lift is unique up to an additive integer constant, and $\deg(f) := F(x+1) - F(x)$ is an integer independent of $x \in \mathbb{R}$ and the lift F. It is called the degree *of f. If f is a homeomorphism, then $|\deg(f)| = 1$.*

Proof *Existence:* Pick a point $p \in S^1$. Then $p = [x_0]$ for some $x_0 \in \mathbb{R}$ and $f(p) = [y_0]$ for some $y_0 \in \mathbb{R}$. From these choices of x_0 and y_0 define $F\colon \mathbb{R} \to \mathbb{R}$ by requiring that $F(x_0) = y_0$, F is continuous, and $f([z]) = [F(z)]$ for all $z \in \mathbb{R}$. One can construct such an F by varying the initial point p continuously, which causes $f(p)$ to vary continuously. Then there is no ambiguity of how to vary x and y continuously, and thus $F(x) = y$ defines a continuous map.[3]

Uniqueness: Suppose $\tilde{F}$ is another lift. Then $[\tilde{F}(x)] = f([x]) = [F(x)]$ for all x, meaning $\tilde{F} - F$ is always an integer. Because it is continuous it must be constant.

Degree: $F(x+1) - F(x)$ is an integer (now evidently independent of the choice of lift) because $[F(x+1)] = f([x+1]) = f([x]) = [F(x)]$. By continuity, $F(x+1) - F(x) =: \deg(f)$ must be a constant.

Invertibility: If $\deg(f) = 0$, then $F(x+1) = F(x)$ and thus F is not monotone. Then f is noninvertible because it cannot be monotone. If $|\deg(f)| > 1$, then $|F(x+1) - F(x)| > 1$ and, by the Intermediate-Value Theorem, there exists a

---

[3] To elaborate, take $\delta > 0$ such that $d([x], [x']) \le \delta$ implies $d(f([x]), f([x'])) < 1/2$. Then define F on $[x_0 - \delta, x_0 + \delta]$ as follows: If $|x - x_0| \le \delta$, then $d(f([x]), q) < 1/2$ and there is a unique $y \in (y_0 - 1/2, y_0 + 1/2)$ such that $[y] = f([x])$. Define $F(x) = y$. Analogous steps extend the domain by another δ at a time, until F is defined on an interval of unit length. Then $f([z]) = [F(z)]$ defines F on $\mathbb{R}$.

$y \in (x, x + 1)$ with $|F(y) - F(x)| = 1$. Then $f([y]) = f([x])$ and $[y] \neq [x]$, so f is noninvertible. $\square$

Definition 4.3.2 Suppose f is invertible. If $\deg(f) = 1$, then we say that f is orientation-preserving; if $\deg(f) = -1$, then f is said to reverse orientation.

Remark 4.3.3 The function $F(x) - x \deg(f)$ is periodic because

$$F(x + 1) - (x + 1) \deg(f) = F(x) + \deg(f) - (x + 1) \deg(f) = F(x) + x \deg(f)$$

for all x. In particular, if f is an orientation-preserving homeomorphism, then $F(x) - x$ is periodic and so $F - \mathrm{Id}$ is bounded. A slightly stronger observation will come in handy soon.

Lemma 4.3.4 *If f is an orientation-preserving circle homeomorphism and F a lift, then $F(y) - y \leq F(x) - x + 1$ for all $x, y \in \mathbb{R}$.*

Proof Let $k = \lfloor y - x \rfloor$. Then

$$(4.3.2) \qquad F(y) - y = F(y) + F(x + k) - F(x + k) + (x + k) - (x + k) - y$$

$$= (F(x + k) - (x + k)) + (F(y) - F(x + k)) - (y - (x + k)).$$

Now $F(x + k) - (x + k) = F(x) - x$ and $0 \leq y - (x + k) < 1$ by choice of k, so $F(y) - F(x + k) \leq 1$. Thus the right-hand side above is at most $F(x) - x + 1 - 0$. $\square$

4.3.2 Rotation Number

The presence or absence of periodic points is determined by a single parameter called the rotation number. It also tells us which rotation to compare a circle homeomorphism to.

Proposition 4.3.5 *Let $f \colon S^1 \to S^1$ be an orientation-preserving homeomorphism and $F \colon \mathbb{R} \to \mathbb{R}$ a lift of f. Then*

$$(4.3.3) \qquad \rho(F) := \lim_{|n| \to \infty} \frac{1}{n}(F^n(x) - x)$$

exists for all $x \in \mathbb{R}$. $\rho(F)$ is independent of x and well defined up to an integer; that is, if $\tilde{F}$ is another lift of f, then $\rho(F) - \rho(\tilde{F}) = F - \tilde{F} \in \mathbb{Z}$. $\rho(F)$ is rational if and only if f has a periodic point.

The fact that the rotation number is independent of the point is the first manifestation of the coherent asymptotic behavior of orbits that we will come to expect. This proposition justifies the following terminology:

Definition 4.3.6 $\rho(f) := [\rho(F)]$ is called the *rotation number* of f.

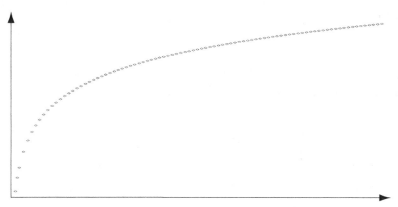

Figure 4.3.2. Subadditivity.

A sequence $(a_n)_{n \in \mathbb{N}}$ with $a_{n+m} \leq a_n + a_m$ is said to be *subadditive*. Existence of the rotation number is due to a similar property of the right-hand side of (4.3.3).

Lemma 4.3.7 *If a sequence* $(a_n)_{n \in \mathbb{N}}$ *satisfies* $a_{m+n} \leq a_n + a_{m+k} + L$ *for all* $m, n \in \mathbb{N}$ *and some k and L, then* $\lim_{n \to \infty} a_n/n \in \mathbb{R} \cup \{-\infty\}$ *exists.*

Proof $a_{m+k} \leq a_m + a_{2k} + L$ gives $a_{m+n} \leq a_m + a_n + a_{2k} + 2L = a_m + a_n + L'$, so we may take $k = 0$. Let $a := \underline{\lim}_{n \to \infty} a_n/n \in \mathbb{R} \cup \{-\infty\}$.

If $a < b < c$ and $n > 2L/(c - b)$ such that $a_n/n < b$, then for any $l \geq n$ that satisfies $l(c - b) > 2 \max_{r < n} a_r$ we can write $l = nk + r$ with $r < n$. This implies $a_l/l \leq (ka_n + a_r + kL)/l \leq a_n/n + a_r/l + (L/n) < c$, so $\overline{\lim}_{l \to \infty} a_l/l \leq c$. Since $c > a$ was arbitrary, this proves the lemma. $\square$

Proof of Proposition 4.3.5 *Independence of x:* Remark 4.3.3 gives $F(x + 1) = F(x) + 1$. If $|x - y| < 1$, then $|F(y) - F(x)| < 1$ and

$$\left| \frac{1}{n}|F^n(x) - x| - \frac{1}{n}|F^n(y) - y| \right| \leq \frac{1}{n}(|F^n(x) - F^n(y)| + |x - y|) \leq \frac{2}{n}.$$

Thus the rotation numbers of x and y coincide, if one of them exists.

Existence: Take $x \in \mathbb{R}$ and $a_n := F^n(x) - x$. Then

$$a_{m+n} = F^{m+n}(x) - x = F^m(F^n(x)) - F^n(x) + a_n \leq a_m + 1 + a_n$$

by Lemma 4.3.4 applied to f^m and F^m. Thus Lemma 4.3.7 shows that a_n/n converges, but possibly, to $-\infty$. However,

$$\frac{a_n}{n} = \frac{1}{n} \sum_{i=0}^{n-1} (F^{i+1}(x) - F^i(x)) = \frac{1}{n} \sum_{i=0}^{n-1} (F(x_i) - x_i) \geq \min F(y) - y,$$

so the limit is a real number $\rho(F)$.

Also, $\rho(F + m) = \lim_{|n| \to \infty} (1/n)(F^n(x) + nm - x) = \rho(F) + m$ for $m \in \mathbb{Z}$, that is, $\rho(F)$ is well defined (mod 1).

Periodic points: If f has a q-periodic point, then $F^q(x) = x + p$ for a lift x of it and some $p \in \mathbb{Z}$. If $m \in \mathbb{N}$, then

$$\frac{F^{mq}(x) - x}{mq} = \frac{1}{mq} \sum_{i=0}^{m-1} F^q(F^{iq}(x)) - F^{iq}(x) = \frac{mp}{mq} = \frac{p}{q};$$

so $\rho(F) = p/q$.

Conversely, for any lift F the definition of rotation number yields

$$\rho(F^m) = \lim_{n \to \infty} \frac{1}{n}((F^m)^n(x) - x) = m \lim_{n \to \infty} \frac{1}{mn}(F^{mn}(x) - x) = m\rho(F);$$

so if $\rho(f) = p/q \in \mathbb{Q}$, then $\rho(f^q) = 0$ since the rotation number is defined up to an integer. Therefore we need only show:

Claim If $\rho(f) = 0$, then f has a fixed point.

Suppose f has no fixed point and let F be a lift such that $F(0) \in [0, 1)$. Then $F(x) - x \notin \mathbb{Z}$ for all $x \in \mathbb{R}$ since $F(x) - x \in \mathbb{Z}$ would imply that $[x]$ is a fixed point for f. Therefore, $0 < F(x) - x < 1$ for all $x \in \mathbb{R}$. Since $F - \mathrm{Id}$ is continuous and periodic, it attains its minimum and maximum and therefore there exists a $\delta > 0$ such that

$$0 < \delta \le F(x) - x \le 1 - \delta < 1$$

for all $x \in \mathbb{R}$. In particular, we can take $x = F^i(0)$ and use $F^n(0) = F^n(0) - 0 = \sum_{i=0}^{n-1} F^{i+1}(0) - F^i(0)$ to get

$$n\delta \le F^n(0) \le (1 - \delta)n$$

or

$$\delta \le \frac{F^n(0)}{n} \le 1 - \delta.$$

As $n \to \infty$, this gives $\rho(F) \neq 0$, which proves the claim by contraposition. $\square$

All periodic orbits, if any, have the same period:

Proposition 4.3.8 *Let $f: S^1 \to S^1$ be an orientation-preserving homeomorphism. Then all periodic orbits have the same period.*

In fact, if $\rho(f) = [p/q]$ with $p, q \in \mathbb{Z}$ relatively prime, then the lift F of f, with $\rho(F) = p/q$ satisfies $F^q(x) = x + p$ whenever $[x]$ is a periodic point, that is, the set of periodic points of f lifts to the set of fixed points of $F^q - \mathrm{Id} - p$.

Proof If $[x]$ is a periodic point, then $F^r(x) = x + s$ for some $r, s \in \mathbb{Z}$ and

$$\frac{p}{q} = \rho(F) = \lim_{n \to \infty} \frac{F^n(x) - x}{nr} = \lim_{n \to \infty} \frac{ns}{nr} = \frac{s}{r}.$$

This means that $s = mp$ and $r = mq$ and that therefore $F^{mq}(x) = x + mp$.

Claim $F^q(x) = x + p$.

If $F^q(x) - p > x$, then monotonicity of F implies

$$F^{2q}(x) - 2p = F^q(F^q(x) - p) - p \ge F^q(x) - p > x$$

and inductively $F^{mq}(x) - mp > x$, which is impossible. Likewise, $F^q(x) - p < x$ is impossible because it implies $F^{mq}(x) - mp < x$. This proves the claim. $\square$

4.3.3 Conjugacy Invariance

The notion of topological conjugacy is central to many aspects of dynamics and will be introduced later (Definition 7.3.3). The rotation number provides the first nontrivial example of a conjugacy invariant, due to the following result:

Proposition 4.3.9 *If $f, h \colon S^1 \to S^1$ are orientation-preserving homeomorphisms, then $\rho(h^{-1}fh) = \rho(f)$.*

Proof Let F and H be lifts of f and h, respectively, that is, $\pi F = f\pi$ and $\pi H = h\pi$. Then $\pi H^{-1} = h^{-1}h\pi H^{-1} = h^{-1}\pi H H^{-1} = h^{-1}\pi$, so H^{-1} is a lift of h^{-1}. Also, $H^{-1}FH$ is a lift of $h^{-1}fh$ since $\pi H^{-1}FH = h^{-1}\pi FH = h^{-1}f\pi H = h^{-1}fh\pi$.

Suppose H is such that $H(0) \in [0, 1)$. We need to estimate

$$|H^{-1}F^n H(x) - F^n(x)| = |(H^{-1}FH)^n(x) - F^n(x)|.$$

(1) For $x \in [0, 1)$ we have $0 - 1 < H(x) - x < H(x) < H(1) < 2$, and by periodicity $|H(x) - x| < 2$ for $x \in \mathbb{R}$. Similarly, $|H^{-1}(x) - x| < 2$ for $x \in \mathbb{R}$.

(2) If $|y - x| < 2$, then $|F^n(y) - F^n(x)| < 3$ since $|[y] - [x]| \le 2$ and thus

$$-3 \le [y] - [x] - 1 = F^n([y]) - F^n([x] + 1) < F^n(y) - F^n(x)$$

$$< F^n([y] + 1) - F^n([x]) = [y] + 1 - [x] \le 3.$$

Those two estimates yield

$$|H^{-1}F^n H(x) - F^n(x)| \le |H^{-1}F^n H(x) - F^n H(x)| + |F^n H(x) - F^n(x)| < 2 + 3,$$

so $|(H^{-1}FH)^n(x) - F^n(x)|/n < 5/n$ and $\rho(H^{-1}FH) = \rho(F)$ by (4.3.3). $\square$

The behavior of the rotation number under orientation-reversing conjugacies is the subject of Exercise 4.3.6.

4.3.4 Circle Homeomorphisms with Periodic Points

The orbit structure of a circle homeomorphism can be described in a fairly complete fashion. We first do this for the case with periodic points.

The first level of description is that each periodic orbit is ordered in the same way as those of the corresponding rotation. This means that the periodic orbits of an orientation-preserving circle homeomorphism behave like those of the circle rotation with the same rotation number. So not only is there an internal coherence of the various periodic orbits as described by Proposition 4.3.8, but they also are qualitatively compatible with those of a rotation. This was, in fact, presaged by the proof of Proposition 4.3.8.

Before proving this, the "ordering" of an orbit has to be defined. It is the sequence in which one encounters the points of the orbit when moving from its initial point in the positive direction. Formally, one can define this using lifts:

Definition 4.3.10 Given $x_0, \ldots, x_{n-1} \in S^1$, take $\tilde{x}_0, \ldots, \tilde{x}_{n-1} \in [\tilde{x}_0, \tilde{x}_0 + 1) \subset \mathbb{R}$ such that $[\tilde{x}_i] = x_i$. Then the *ordering* of $(x_0, \ldots, x_{n-1})$ on S^1 is the permutation σ of $\{1, \ldots, n-1\}$ such that $\tilde{x}_0 < \tilde{x}_{\sigma(1)} < \cdots < \tilde{x}_{\sigma(n-1)} < \tilde{x}_0 + 1$.

As a warmup, we find the ordering σ of $\pi(\{0, p/q, 2p/q, \ldots, (q-1)p/q\})$ on S^1, to which we later compare that of a periodic orbit. Define $k \in \mathbb{N}$ by $0 < k < q$ and $kp \equiv 1 (\mathrm{mod}\ q)$. Then k minimizes the fractional part $\{ip/q\}$ for $0 < i < q$ and hence $k = \sigma(1)$. Inductively, $ki \equiv \sigma(i)\ (\mathrm{mod}\ q)$. This defines the ordering σ of

$$\pi\left(\left\{0, \frac{p}{q}, \frac{2p}{q}, \ldots, \frac{(q-1)p}{q}\right\}\right).$$

Therefore, we want to prove:

Proposition 4.3.11 *Let $f \colon S^1 \to S^1$ be an orientation-preserving homeomorphism. Suppose p and q are relatively prime and there is an $x \in S^1$ such that $f^q(x) = x$. Then the ordering of $\{x, f(x), f^2(x), \ldots, f^{q-1}(x)\}$ on S^1 is given by $\sigma(i) = ki\ (\mathrm{mod}\ q)$, where $kp \equiv 1\ (\mathrm{mod}\ q)$.*

Proof Fix $\tilde{x} \in \pi^{-1}([x])$ and a lift F of f such that $F^q(\tilde{x}) = \tilde{x} + p$ (Proposition 4.3.8). Then $[\tilde{x}, \tilde{x} + p]$ is partitioned (up to common endpoints) into $p \cdot q$ subintervals by $A := \pi^{-1}(\{x, f(x), f^2(x), \ldots, f^{q-1}(x)\})$, and into q subintervals $I_i = [F^i(\tilde{x}), F^{i+1}(\tilde{x})]$, $i = 0 \ldots q - 1$. Since F is a bijection between any I_i and I_{i+1} and preserves A, each I_i contains $p + 1$ points of A. Take $k, r \in \mathbb{Z}$ such that the right neighbor of $\tilde{x}$ in A is $\tilde{x}_1 = F^k(\tilde{x}) - r$. Since $\bar{F} = F^k - r$ is increasing on $\mathbb{R}$ and preserves A, the facts that $\tilde{x}_1 = \bar{F}(\tilde{x})$ is the nearest right neighbor of $\tilde{x}$ in A and that $[\tilde{x}, F(\tilde{x})]$ is divided into p subintervals by A show that $\bar{F}^p(\tilde{x}) = F(\tilde{x})$ and hence $f^{kp}(x) = f(x)$. Therefore k is the unique integer between 0 and $q - 1$ such that $kp \equiv 1\ (\mathrm{mod}\ q)$, and the ordering of the orbit $\{x, f(x), f^2(x), \ldots, f^{q-1}(x)\}$ is given by $ki \equiv \sigma(i)\ (\mathrm{mod}\ q)$. $\square$

The next proposition says that for circle homeomorphisms with rational rotation number all nonperiodic orbits are asymptotic to periodic orbits. This yields a complete classification of possible orbits with rational rotation numbers.

Proposition 4.3.12 *Let $f \colon S^1 \to S^1$ be an orientation-preserving homeomorphism with rational rotation number $\rho(f) = p/q \in \mathbb{Q}$. Then there are two possible types of nonperiodic orbits for f:*

(1) *If f has exactly one periodic orbit, then every other point is heteroclinic under f^q to two points on the periodic orbit (Definition 2.3.4). These points are different if the period is greater than one. (If the period is one, then all orbits are homoclinic to the fixed point, as shown in Figure 4.3.3.)*

(2) *If f has more than one periodic orbit, then each nonperiodic point is heteroclinic under f^q to two points on different periodic orbits.*

Proof We can identify f^q with a homeomorphism of an interval by taking a lift z of a fixed point of f^q and restricting a lift $F^q(\cdot) - p$ of f to $[z, z+1]$. Then the statement follows from Proposition 2.3.5 applied to this interval map, except for the last part of (2), that the two periodic orbits in question

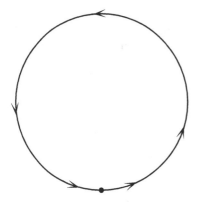
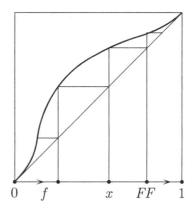

Figure 4.3.3. A semistable point.

are different. But if there is an interval $I = [a, b] \subset \mathbb{R}$ such that a and b are adjacent zeros of $F^q - \mathrm{Id} - p$ and a, b project to the same periodic orbit, then f has only one periodic orbit because, if $[a] = x \in S^1$, $[b] = f^k(x) \in S^1$, then $\bigcup_{n=0}^{q-1} f^{nk}(\pi((a, b)))$ covers the complement of $\{f^n(x)\}_{n=0}^{q-1}$ in S^1 and contains no periodic points. By invariance, $f^{nk}(\pi((a, b)))$ does not either. $\square$

Remark 4.3.13 This means that the asymptotic behavior is highly coherent for all orbits, not only periodic ones, and also coherent with the structure of the corresponding rotation.

As a particular case, if there is only one periodic orbit, then it is semistable. It "repels on one side and attracts on the other", as, for example, the point $x = 0$ under the diffeomorphism $f \colon S^1 \to S^1$ induced by the map

$$x \mapsto x + \frac{1}{4}\sin^2 \pi x \quad (\mathrm{mod}\ 1).$$

Nonperiodic points are not just individually asymptotic to periodic points, but this behavior is coherent for iterates of points under f; so for a nonperiodic point x the points $x, f(x), \ldots, f^{q-1}(x)$ are all forward asymptotic to the corresponding iterate $y, f(y), \ldots, f^{q-1}(y)$ of a periodic point, and they are moving in the same direction. This follows immediately from monotonicity (compare Lemma 2.3.2):

Lemma 4.3.14 *If $I \subset \mathbb{R}$ is an interval whose endpoints are adjacent zeros of $F^q - \mathrm{Id} - p$, then $F^q - \mathrm{Id} - p$ has the same sign on the interiors of I and $F(I)$.*

Proof If $F^q - \mathrm{Id} - p > 0$ on I, then $F^q(x) > x + p$ for all $x \in I$, and monotonicity of F implies $F^q(F(x)) = F(F^q(x)) > F(x + p) = F(x) + p$ for all $x \in I$. Therefore $F^q - \mathrm{Id} - p > 0$ on $F(I)$.

The case of $F^q - \mathrm{Id} - p < 0$. is similar. $\square$

Thus for a circle homeomorphism with a periodic point all orbits are asymptotically periodic with the same period and in a coherent way.

4.3.5 Circle Homeomorphisms Without Periodic Points

We show, analogously to Proposition 4.3.11, that the orbits of a circle home-omorphism without periodic points are ordered as those for the corresponding rotation.

Proposition 4.3.15 *Let $F : \mathbb{R} \to \mathbb{R}$ be a lift of an orientation-preserving home-omorphism $f : S^1 \to S^1$ with $\rho := \rho(F) \notin \mathbb{Q}$. Then, for $n_1, n_2, m_1, m_2 \in \mathbb{Z}$ and $x \in \mathbb{R}$,*

$$n_1 \rho + m_1 < n_2 \rho + m_2 \quad \text{if and only if} \quad F^{n_1}(x) + m_1 < F^{n_2}(x) + m_2.$$

The left of these inequalities is the special case of the one on the right when F is the rotation by ρ.

Proof We do not have equality on the right for any x because this would imply $F^{n_1}(x) - F^{n_2}(x) \in \mathbb{Z}$, and hence that $[x]$ is periodic. Thus, for given $n_1, n_2, m_1, m_2 \in \mathbb{Z}$, the continuous expression $F^{n_1}(x) + m_1 - F^{n_2}(x) - m_2$ never changes sign and the second inequality is independent of x.

Now assume $F^{n_1}(x) + m_1 < F^{n_2}(x) + m_2$ for all x. Substituting $y := F^{n_2}(x)$ shows that this is equivalent to

$$F^{n_1 - n_2}(y) - y < m_2 - m_1 \quad \text{for all } y \in \mathbb{R}.$$

In particular, for $y = 0$ we get $F^{n_1 - n_2}(0) < m_2 - m_1$, and $y = F^{n_1 - n_2}(0)$ gives

$$F^{2(n_1 - n_2)}(0) < (m_2 - m_1) + F^{n_1 - n_2}(0) < 2(m_2 - m_1).$$

Inductively, $F^{n(n_1 - n_2)}(0) < n(m_2 - m_1)$ and

$$\rho = \lim_{n \to \infty} \frac{F^{n(n_1 - n_2)}(0)}{n(n_1 - n_2)} < \lim_{n \to \infty} \frac{n(m_2 - m_1)}{n(n_1 - n_2)} = \frac{m_2 - m_1}{n_1 - n_2}$$

(with strict inequality since $\rho \notin \mathbb{Q}$). Consequently, $n_1 \rho + m_1 < n_2 \rho + m_2$. This proves "if". Reversing all inequalities proves the converse. $\square$

The preceding proposition bears some resemblance to the earlier result that *periodic* orbits are ordered like those for the corresponding rotation. It is stronger because it applies to every orbit, rather than a naturally distinguished subset.

This helps us in our study of the asymptotic behavior of orbits for homeomorphisms without periodic points.

Lemma 4.3.16 *Let $f : S^1 \to S^1$ be an orientation-preserving homeomorphism without periodic points, $m, n \in \mathbb{Z}$, $m \neq n$, $x \in S^1$, and $I \subset S^1$ a closed interval with endpoints $f^m(x)$ and $f^n(x)$. Then every semiorbit meets I.*

Remark 4.3.17 For $x \neq y \in S^1$ there are exactly two intervals in S^1 with endpoints x and y. The lemma holds for either choice. Since x is not periodic, I is not a point.

Proof Consider positive semiorbits $(f^n(y))_{n \in \mathbb{N}}$. The proof for negative semiorbits is exactly the same. To prove the lemma it suffices to show that the backward iterates of I cover S^1, that is, $S^1 \subset \bigcup_{k \in \mathbb{N}} f^{-k}(I)$.

Let $I_k := f^{-k(n-m)}(I)$ and note that these are all contiguous: If $k \in \mathbb{N}$, then I_k and I_{k-1} have a common endpoint. Consequently, if $S^1 \neq \bigcup_{k \in \mathbb{N}} I_k$, then the sequence of endpoints converges to some $z \in S^1$. But then

$$z = \lim_{k \to \infty} f^{-k(n-m)}(f^m(x)) = \lim_{k \to \infty} f^{(-k+1)(n-m)}(f^m(x))$$

$$= \lim_{k \to \infty} f^{(n-m)}\big(f^{-k(n-m)}(f^m(x))\big) = f^{(n-m)}\big(\lim_{k \to \infty} f^{-k(n-m)}(f^m(x))\big)$$

$$= f^{(n-m)}(z)$$

is periodic, contrary to the assumption. $\square$

If there are periodic points, they provide all the accumulation points of orbits. Now we see what set plays this role when the rotation number is irrational.

Definition 4.3.18 The set $\omega(x) := \bigcap_{n \in \mathbb{N}} \overline{\{f^i(x) \mid i \geq n\}}$ of accumulation points of the positive semiorbit of x is called the ω-limit set of x.

If there are periodic points, all ω-limit sets are periodic orbits. If there are no periodic points, the ω-limit sets for different orbits still look the same; in fact, they are the same.

Proposition 4.3.19 Let $f : S^1 \to S^1$ be an orientation-preserving homeomorphism of S^1 without periodic points. Then $\omega(x)$ is independent of x and $E := \omega(x)$ is perfect and either S^1 or nowhere dense (see Definition A.1.5).

By Proposition A.1.7, perfect nowhere dense sets are Cantor sets, that is, they are homeomorphic to the standard middle-third Cantor set. Therefore, this result produces Cantor sets directly from the dynamics of a circle map – at least when we give an example where this is the possibility that is actually realized.

Proof Independence of x: We need to show that $\omega(x) = \omega(y)$ for $x, y \in S^1$. Let $z \in \omega(x)$. Then there is a sequence l_n in $\mathbb{N}$ such that $f^{l_n}(x) \to z$. If $y \in S^1$, then by Lemma 4.3.16 there exist $k_m \in \mathbb{N}$ such that $f^{k_m}(y) \in I_m := [f^{l_m}(x), f^{l_{m+1}}(x)]$. But then $\lim_{m \to \infty} f^{k_m}(y) = z$ and thus $z \in \omega(y)$.

Therefore $\omega(x) \subset \omega(y)$ for all $x, y \in S^1$ and by symmetry $\omega(x) = \omega(y)$ for all $x, y \in S^1$.

$E := \omega(x)$ is either S^1 or nowhere dense: Let us first show that E is the smallest closed nonempty f-invariant set. If $\varnothing \neq A \subset S^1$ is closed and f-invariant and $x \in A$, then $\{f^k(x)\}_{k \in \mathbb{Z}} \subset A$ since A is invariant and $E = \omega(x) \subset \overline{\{f^k(x)\}_{k \in \mathbb{Z}}} \subset A$ since A is closed. Thus any closed invariant set A is either empty or contains E. In particular, $\varnothing$ and E are the only closed invariant subsets of E itself. Since E is closed, it contains its boundary, which is itself a closed set (Exercise 2.6.6). The boundary is also invariant because a boundary point is a point for any neighborhood U of which we have $U \cap E \neq \varnothing$ and $U \setminus E \neq \varnothing$, a property that persists when we apply a homeomorphism. Therefore the boundary ∂E of E is a closed invariant subset of E and as such we must have either $\partial E = \varnothing$ and hence $E = S^1$, or else $\partial E = E$, which implies that E is nowhere dense (Exercise 2.6.6).

It remains to show that E is perfect. Let $x \in E$. Since $E = \omega(x)$, there is a sequence k_n such that $\lim_{n \to \infty} f^{k_n}(x) = x$. Since there are no periodic orbits, $f^{k_n}(x) \neq x$ for all n. Consequently, x is an accumulation point of E since $f^{k_n}(x) \in E$ for all n by invariance. $\square$

4.3.6 Comparison and Classification

Both in Proposition 4.3.12 and in Proposition 4.3.19 there is a set of distinguished orbits (either periodic or in E) to which all others are asymptotic. This distinguished set corresponds most closely to the rotation with the same rotation number (for irrational rotation number this becomes clear with Theorem 4.3.20). Thus if there are periodic points, there is a remnant of the rotation that may be as small as a single periodic orbit or a finite set of them; otherwise, the corresponding remnant is at least a Cantor set. It is in this distinction that Proposition 4.3.19 shows that the orbit structure of maps without periodic points is quite different from that of maps with periodic points. If there are periodic points, all orbits are either periodic or asymptotic to a periodic orbit; otherwise, either all orbits are dense or all orbits are asymptotic to or in a Cantor set. Moreover, a further difference appears when we compare the orbit structure of a circle map with that of a rotation with the same rotation number. The vast majority of circle maps with periodic points possess nonperiodic orbits – Proposition 4.4.10 and Lemma 4.4.12 show that having nonperiodic orbits occurs over entire parameter intervals in a family of maps, whereas having all orbits periodic happens only for an instant. (Furthermore, similar arguments show that even having infinitely many periodic points is unstable, and hence rare.) Thus, the presence of nonperiodic orbits, which is a qualitative difference to a rational rotation, is the most common behavior for maps with rational rotation number.

For irrational rotation number the picture is different. The greatest qualitative similarity to an irrational rotation occurs when $E = S^1$ in Proposition 4.3.19. In this case all orbits are dense ($\omega(x) = S^1$ for all $x \in S^1$), which is the same situation as for an irrational rotation. Unlike in the case of rational rotation number, there is no indication that the alternative situation (E is a Cantor set) occurs more frequently (in fact, it never happens at all for C^2 maps). Indeed, a map with irrational rotation number ρ is either equivalent to or "contains" R_ρ, up to some distortion, according to whether its orbits are dense:

Theorem 4.3.20 (Poincaré Classification Theorem) *Let* $f: S^1 \to S^1$ *be an orientation-preserving homeomorphism with irrational rotation number* ρ. *Then there is a continuous monotone map* $h: S^1 \to S^1$ *with* $h \circ f = R_\rho \circ h$.

(1) *If* f *is transitive, then* h *is a homeomorphism.*
(2) *If* f *is not transitive, then* h *is noninvertible.*

The map h here plays the role of the changes of variable or conjugacies that we encountered in Section 1.2.9.3 and Section 3.1.3, except that it may not be invertible. Section 4.3.3 rules out the nontransitive case for smooth f.

Proof We first construct the lift of h only on the lift of a single orbit and show that it is monotone. We then extend it to the closure of that lift and, using monotonicity, "fill in" any gaps that may be left. Finally we define h as the projection.

Pick a lift $F : \mathbb{R} \to \mathbb{R}$ of f and $x \in \mathbb{R}$. Let $B := \{F^n(x) + m\}_{n,m \in \mathbb{Z}}$ be the total lift of the orbit of $[x]$. Define $H : B \to \mathbb{R}$, $F^n(x) + m \mapsto n\rho + m$, where $\rho := \rho(F)$. By Proposition 4.3.15, this map is monotone, and $H(B)$ is dense in $\mathbb{R}$ by Proposition 4.1.1. If we write $\tilde{R}_\rho : \mathbb{R} \to \mathbb{R}$, $x \mapsto x + \rho$, then $H \circ F = \tilde{R}_\rho \circ H$ on B because

$$H \circ F(F^n(x) + m) = H(F^{n+1}(x) + m) = (n+1)\rho + m$$

and

$$\tilde{R}_\rho \circ H(F^n(x) + m) = \tilde{R}_\rho(n\rho + m) = (n+1)\rho + m.$$

Lemma 4.3.21 *H has a continuous extension to the closure $\bar{B}$ of B.*

Proof If $y \in \bar{B}$, then there is a sequence $(x_n)_{n \in \mathbb{N}}$ in B such that $y = \lim_{n \to \infty} x_n$. To show that $H(y) := \lim_{n \to \infty} H(x_n)$ exists and is independent of the choice of a sequence approximating y, observe first that the left and right limits exist and are independent of the sequence since H is monotone. If the left and right limits disagree, then $\mathbb{R} \smallsetminus H(B)$ contains an interval, which contradicts the density of $H(B)$. $\square$

H can now easily be extended to $\mathbb{R}$: Since $H : \bar{B} \to \mathbb{R}$ is monotone and surjective [because H is monotone and continuous on B, $\bar{B}$ is closed, and $H(B)$ is dense in $\mathbb{R}$] there is no choice in defining H on the intervals complementary to $\bar{B}$: Set $H = \text{const.}$ on those intervals, choosing the constant equal to the values at the endpoints. This gives a map $H : \mathbb{R} \to \mathbb{R}$ such that $H \circ F = \tilde{R}_\rho \circ H$ and thus the desired map $h : S^1 \to S^1$ since for $z \in B$ we have

$$H(z + 1) = H(F^n(x) + m + 1) = n\rho + m + 1 = H(z) + 1,$$

and this property persists under continuous extension.

To decide invertibility note that in the transitive case we start from a dense orbit and so $\bar{B} = \mathbb{R}$ and h is a bijection. In the nontransitive case, H is constant on the intervals complementary to the orbit closure that we used. $\square$

Remark 4.3.22 In the transitive case of Theorem 4.3.20, when h is invertible we say that h *conjugates* f to R_ρ; in the case of noninvertible h we say that R_ρ is a *factor* of f via h. These notions are explored in Chapter 7 (Definition 7.3.3).

▨ EXERCISES

▨ **Exercise 4.3.1** For which values of a does the function $F(x) = 2x + a$ define the lift of a circle map?

▨ **Exercise 4.3.2** Referring to (4.1.1), prove that $\rho(R_\alpha) = [\alpha]$.

▨ **Exercise 4.3.3** Consider $F(x) := x + (1/2)\sin x$. Decide whether F is the lift of a circle homeomorphism.

■ **Exercise 4.3.4** Consider $F(x) := x + (1/4\pi) \sin 2\pi x$. Decide whether F is the lift of a circle homeomorphism, and, if so, decide whether that homeomorphism is orientation-preserving. If it is, determine the rotation number.

■ **Exercise 4.3.5** Let $f: S^1 \to S^1$ be a monotone (but not necessarily invertible) map of degree one, that is, its lift is a monotone function $F: \mathbb{R} \to \mathbb{R}$ such that $F(x+1) = F(x) + 1$. Prove that the assertions of Proposition 4.3.5, Proposition 4.3.8 and Proposition 4.3.9 hold for f.

■ **Exercise 4.3.6** Referring to Proposition 4.3.9, what happens with the rotation number under an orientation-reversing conjugacy?

■ **Exercise 4.3.7** Let $f: S^1 \to S^1$ be a continuous map of degree one (not necessarily monotone) and $F: \mathbb{R} \to \mathbb{R}$ its lift. Prove that

$$\rho^+(F) := \lim_{n \to \infty} \max_{x \in S^1} \frac{F^n(x) - x}{n} \quad \text{and} \quad \rho^-(F) := \lim_{n \to \infty} \min_{x \in S^1} \frac{F^n(x) - x}{n}$$

both exist.

■ **PROBLEMS FOR FURTHER STUDY**

■ **Problem 4.3.8** Under the assumptions of the previous exercise call

$$R(F) := \left\{ \rho \in \mathbb{R} \mid \exists x \in \mathbb{R} \lim_{n \to \infty} \frac{F^n(x) - x}{n} = \rho \right\}$$

the *rotation set* of F. Prove that $R(F) \neq \varnothing$.

■ **Problem 4.3.9** Prove that a circle homeomorphism with finitely many fixed points and an attracting fixed point has a repelling fixed point.

Show that there exists a circle homeomorphism with an attracting fixed point and without repelling fixed points.

4.4 CANTOR PHENOMENA

In Proposition 4.3.19, a Cantor set appears naturally for some circle homeomorphisms without periodic points. There are several other ways in which Cantor sets and related structures appear in this context. The conjugacy above is a case in point in the nontransitive case. The dependence of the rotation number on a parameter is an example with interesting physical implications.

4.4.1 Devil's Staircases
In the nontransitive case the map h in Theorem 4.3.20 is necessarily an example of the following interesting phenomenon:

Definition 4.4.1 A monotone continuous function $\phi: [0, 1] \to \mathbb{R}$ (or $\phi: [0, 1] \to S^1$) is called a *devil's staircase* if there exists a family $\{I_\alpha\}_{\alpha \in A}$ of disjoint closed subintervals of $[0,1]$ of nonzero length with dense union such that ϕ takes distinct constant values on these subintervals. (See Figure 4.4.1.)

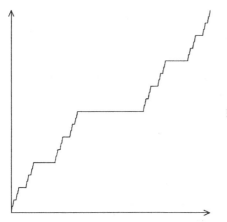

Figure 4.4.1. Devil's staircasse.

Example 4.4.2 A devil's staircase can be constructed in a fairly explicit way. For $x = 0.\alpha_1\alpha_2\alpha_3\cdots = \sum_{i=1}^{\infty} \alpha_i 3^{-i}$ ($\alpha_i \neq 1$) in the ternary Cantor set C define $f(x) := \sum_{i=1}^{\infty} \alpha_i 2^{-i-1} \in [0, 1]$ as in Lemma 2.7.3. In Section 2.7.1 we found that f is surjective and nondecreasing, and that the two endpoints of a deleted interval are mapped to the same point. It is not hard to see that f is continuous (this is used in Problem 2.7.6). We can extend f to a nondecreasing continuous map on $[0, 1]$ by defining it to be constant on each deleted interval, the constant being the common value of f at the endpoints. This is then a devil's staircase, also called a Cantor function.

The graph of this function has some self-similarity: The transformation given by $\begin{pmatrix} 1/3 & 0 \\ 0 & 1/2 \end{pmatrix}$ in the plane maps the graph to a proper subset of itself because $f(x) = f(3x)/2$ on $[0, 1/3]$.

The terminology "devil's staircase" refers to the odd situation that the graph of this function consists entirely of "steps", namely, the horizontal portions over the deleted intervals, yet there are no jumps at all; the function is continuous. Thus the tops of the steps are there, but not their "faces". In analysis this provides a quaint example with several odd properties, but we have now seen that in dynamics such functions come up rather naturally.

Let us revisit the construction of the map h above in order to understand the nontransitive case better. Since the set $\bar{B}$ from the proof of Theorem 4.3.20 projects to the closure of the orbit of $[x]$, it contains the ω-limit set $E = \omega([x])$ of $[x]$, and, by choosing $x \in \pi^{-1}(E)$, we obtain $\pi(\bar{B}) = E$, where E is the universal ω-limit set discussed previously. In the transitive case $\bar{B} = \mathbb{R}$ and $E = S^1$, but in the nontransitive case we find that if $x \in \pi^{-1}(E)$, then $\pi(\bar{B}) = E$ is a Cantor set. Consequently, in the nontransitive case h is a bijection of the identification space $E/\sim$ (identifying the two endpoints of each complementary interval) to S^1 and conjugates $f_{\lceil E/\sim}$ to $R_{\rho(f)}$. All orbits of f in E are dense in E (by the definition of E). On the other hand, the construction of $E = \omega(x)$ yields that all points outside E are attracted to E in both positive and negative time because iterates of such a point have to stay inside disjoint complementary intervals of E and the length of these goes to zero.

4.4.2 Wandering Domains

Conversely, one can think of the nontransitive map as being obtained from an irrational rotation by "blowing up" some orbits to intervals whose union then makes up the complement of E. These complementary intervals are thus permuted like the points on an orbit for an irrational rotation. All interior points in those intervals are "wandering" in the sense below since they stay within those intervals whose images are all disjoint. The next subsection has an explicit construction of such an example.

Definition 4.4.3 A point is said to be *wandering* if it has a neighborhood all of whose images and preimages are pairwise disjoint.

This behavior is the extreme opposite of recurrence, which we introduce in Definition 6.1.8.

To return to our comparison with the case of rational rotation number we note that in that case a map f is only conjugate to a rotation if all orbits are periodic with the same period and hence $f^q = \mathrm{Id}$ for some $q \in \mathbb{Z}$. Furthermore, a rational rotation can only be a factor when there are infinitely many periodic points, which, as we noted earlier, is unstable.

4.4.3 The Denjoy Example

We now give an example of a nontransitive circle diffeomorphism without periodic points. The construction starts with an irrational rotation and replaces the points of one orbit by suitably chosen intervals. The resulting map is not transitive. This example due to Arnaud Denjoy proves:

Proposition 4.4.4 *For* $\rho \in \mathbb{R} \smallsetminus \mathbb{Q}$ *there is a nontransitive* C^1 *diffeomorphism* $f \colon S^1 \to S^1$ *with* $\rho(f) = \rho$.

Proof If $l_n := (|n| + 3)^{-2}$ and $c_n := 2\left((l_{n+1}/l_n) - 1\right) \geq -1$, then

$$\sum_{n \in \mathbb{Z}} l_n < 2 \sum_{n=0}^{\infty} l_n = 2 \sum_{n=3}^{\infty} \frac{1}{n^2} < 2 \int_2^{\infty} \frac{1}{x^2}\, dx = 1.$$

To "blow up" the orbit $x_n = R_\rho^n x$ of the irrational rotation R_ρ to intervals I_n of length l_n, insert the intervals I_n into S^1 so that they are ordered in the same way as the points x_n and the space between any two such intervals I_m and I_n is

$$\left(1 - \sum_{n \in \mathbb{Z}} l_n\right) d\left(x_m, x_n\right) + \sum_{x_k \in (x_m, x_n)} l_k \,.$$

(This is the sum of the lengths of the intervals I_k inserted in between and the length of the arc of the circle between x_m and x_n, appropriately scaled to reflect the fact that the total length of $S^1 \smallsetminus \bigcup_{n \in \mathbb{Z}} I_n$ is $1 - \sum_{n \in \mathbb{Z}} l_n$.) To define a circle homeomorphism f such that $f(I_n) = I_{n+1}$ and $f\restriction_{S^1 \smallsetminus \bigcup_{n \in \mathbb{Z}} I_n}$ is semiconjugate to a rotation it suffices to specify the derivative $f'(x)$ since f is then obtained by integration.

On the interval $[a, a + l]$ define the tent function

$$h(a, l, x) := 1 - \frac{1}{l}|2(x - a) - l|.$$

Then $h(a, l, a + l/2) = 1$ and $\int_a^{a+l} h(a, l, x)\, dx = l/2$. Denote the left endpoint of I_n by a_n and let

$$f'(x) = \begin{cases} 1 & \text{for } x \in S^1 \smallsetminus \bigcup_{n \in \mathbb{Z}} I_n, \\ 1 + c_n h(a_n, l_n, x) & \text{for } x \in I_n. \end{cases}$$

The choice $c_n = 2\left((l_{n+1}/l_n) - 1\right) = 2\left(l_{n+1} - l_n\right)/l_n$ implies

$$\int_{I_n} f'(x)\, dx = \int_{I_n} (1 + c_n h(a_n, l_n, x))\, dx = l_n + \frac{l_n}{2} c_n = l_{n+1},$$

so indeed $f(I_n) = I_{n+1}$. $\square$

Close inspection of this proof reveals that the derivative of the function f has to be somewhat distorted in order to contract intervals fast enough to fit into the interstices of the universal Cantor set. A systematic careful analysis shows that no sufficiently smooth circle homeomorphism exhibits this phenomenon.

A C^2 diffeomorphism $f \colon S^1 \to S^1$ with irrational rotation number $\rho(f) \in \mathbb{R} \smallsetminus \mathbb{Q}$ is transitive and hence topologically conjugate to $R_{\rho(f)}$.

In fact, slightly weaker regularity hypotheses suffice. The most natural weakening is to assume merely that the derivative has bounded variation. A function $g \colon S^1 \to \mathbb{R}$ is said to be of *bounded variation* if its total variation $\mathrm{Var}(g) := \sup \sum_{k=1}^n |g(x_k) - g(x_k')|$ is finite. Here the sup is taken over all finite collections $\{x_k, x_k'\}_{k=1}^n$ such that x_k, x_k' are endpoints of an interval I_k and $I_k \cap I_j = \varnothing$ for $k \neq j$. Every Lipschitz function and hence every continuously differentiable function has bounded variation.

4.4.4 Dependence of the Rotation Number on a Parameter

Here we examine the dependence of the rotation number on the map as the map is varied. To begin with, it is continuous and monotone.

Proposition 4.4.5 $\rho(\cdot)$ *is continuous in the uniform topology.*

Proof If $\rho(f) = \rho$, take $p'/q', p/q \in \mathbb{Q}$ such that $p'/q' < \rho < p/q$. Pick the lift F of f for which $-1 < F^q(x) - x - p \leq 0$ for some $x \in \mathbb{R}$. Then $F^q(x) < x + p$ for all $x \in \mathbb{R}$, since otherwise $F^q(x) = x + p$ for some $x \in \mathbb{R}$ by the Intermediate-Value Theorem and $\rho = p/q$. Since the function $F^q - \mathrm{Id}$ is periodic and continuous, it attains its maximum. Thus there exists $\delta > 0$ such that $F^q(x) < x + p - \delta$ for all $x \in \mathbb{R}$. This implies that every sufficiently small perturbation $\bar{F}$ of F in the uniform topology also satisfies $\bar{F}^q(x) < x + p$ for all $x \in \mathbb{R}$ and thus $\rho(\bar{F}) < p/q$. A similar argument involving p'/q' completes the proof. $\square$

The definition of the rotation number further suggests that it is monotone: If $F_1 > F_2$, then $\rho(F_1) \geq \rho(F_2)$ follows from the definition. This leads to the following concepts of ordering on the circle and for maps of the circle:

Definition 4.4.6 Define "$<$" on S^1 by $[x] < [y] :\Leftrightarrow y - x \in (0, 1/2) \pmod 1$ and define a partial ordering "$\prec$" on the collection of orientation-preserving circle homeomorphisms by $f_0 \prec f_1 :\Leftrightarrow f_0(x) < f_1(x)$ for all $x \in S^1$.

Notice that neither of these orderings is transitive. Indeed, $[0] < [1/3] < [2/3] < [0]$ and correspondingly $R_0 \prec R_{1/3} \prec R_{2/3} \prec R_0$, where R_α is the rotation as in Section 4.1. The definition of rotation number now immediately implies:

Proposition 4.4.7 $\rho(\cdot)$ *is monotone: If $f_1 \prec f_2$, then $\rho(f_1) \leq \rho(f_2)$.*

Remark 4.4.8 In particular, if $\{f_t\}$ is a family of orientation-preserving circle homeomorphisms such that $f_t(x)$ is increasing in t for every $x \in \mathbb{R}$, then $\rho(f_t)$ is nondecreasing in t.

At irrational values the rotation number is strictly increasing:

Proposition 4.4.9 *If $f_0 \prec f_1$ and $\rho(f_0) \notin \mathbb{Q}$, then $\rho(f_0) < \rho(f_1)$.*

Proof If F_0 and F_1 are lifts with $0 < F_1(x) - F_0(x) < 1/2$ for all $x \in \mathbb{R}$, then by continuity and periodicity $F_1(x) - F_0(x) > \delta$ for some $\delta > 0$ and all $x \in \mathbb{R}$. Take $p/q \in \mathbb{Q}$ such that $p/q - \delta/q < \rho(F_0) < p/q$. Then there exists $x_0 \in \mathbb{R}$ such that $F_0^q(x_0) - x_0 > p - \delta$ [because otherwise $\rho(F_0) = \lim_{n\to\infty}(F_0^{nq}(x) - x/nq) \leq \lim_{n\to\infty}(n(p - \delta)/nq) = p/q - \delta/q$]. Since

$$F_1^q(x_0) = F_1\big(F_1^{q-1}(x_0)\big) > F_0\big(F_1^{q-1}(x_0)\big) + \delta$$

$$> F_0\big(F_0^{q-1}(x_0)\big) + \delta = F_0^q(x_0) + \delta > x_0 + p,$$

we either have $F_1^q(x) > x + p$ for all $x \in \mathbb{R}$ or $F_1^q(x_1) = x_1 + p$ for some $x_1 \in \mathbb{R}$. In either case $\rho(F_0) < p/q \leq \rho(F_1)$. $\square$

While Proposition 4.4.9 shows that having irrational rotation number is not stable, the situation is different for rational rotation number:

Proposition 4.4.10 *Let $f\colon S^1 \to S^1$ be an orientation-preserving homeomorphism with rational rotation number $\rho(f) = p/q$ and some nonperiodic points. Then all sufficiently nearby perturbations $\bar{f}$ with $\bar{f} \prec f$ or all sufficiently nearby perturbations $\bar{f}$ with $f \prec \bar{f}$ (or both) have rotation number p/q.*

The basic issue is whether the graph of $F^q - p$ has portions above and below the diagonal, in which case small perturbations either way cannot get rid of intersections with the diagonal (Figure 4.4.2). The borderline case, in which the graph lies entirely on one side, is exactly the one where the bifurcation to different dynamics occurs (see also Figure 2.3.2).

Proof Since f has nonperiodic points, $F^q - \mathrm{Id} - p$ does not vanish identically for any lift F of f. (It does have zeros by assumption.) If there exists $x \in \mathbb{R}$ with

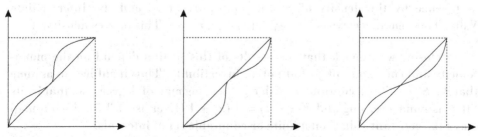

Figure 4.4.2. One-sided and two-sided stability.

$F^q(x) - x - p > 0$, then for any sufficiently small perturbation $\bar{f} \prec f$ the corresponding lift $\bar{F}$ of $\bar{f}$ is such that $\bar{F}^q(x) - x - p > 0$ and hence $\rho(\bar{f}) \geq p/q$; so $\rho(\bar{f}) = p/q$ by Proposition 4.4.7. Otherwise, the same holds for perturbations with $f \prec \bar{f}$. □

Remark 4.4.11 The proof shows that circle maps that have an attracting or repelling periodic orbit (an orbit that lifts to a point where $F^q - \text{Id} - p$ changes sign) can be perturbed (in either direction) without changing the rotation number.

On the other hand, if $F^q - \text{Id} - p$ does not change sign, for example, $F^q - \text{Id} - p \geq 0$, then any perturbation $\bar{f}$ with $f \prec \bar{f}$ has rotation number $\rho(\bar{f}) > p/q$ since $\bar{F}^q - \text{Id} - p \geq \delta > 0$. In this case the zeros of $F^q - \text{Id} - p$ project to "parabolic" or *semistable* periodic orbits. These are orbits p that attract on one side and repel on the other side; that is, there is some open neighborhood U of p such that for all x in one component of $U \smallsetminus \{p\}$ we have $\lim_{n\to\infty} d(f^n(x), f^n(p)) = 0$, and for all x in the other component $\lim_{n\to-\infty} d(f^n(x), f^n(p)) = 0$ (see Figure 4.3.3).

Here is an extreme case.

Lemma 4.4.12 *If all points of a map $f \colon S^1 \to S^1$ are periodic, then the rotation number is strictly increasing at f.*

To see that the rotation number depends on f in a nonsmooth way we reformulate these conclusions: The rotation number as a function of a parameter can (and usually will) be a devil's staircase (see Definition 4.4.1).

Proposition 4.4.13 *Suppose that $(f_t)_{t\in[0,1]}$ is a monotone continuous family of orientation-preserving circle homeomorphisms such that $\rho \colon t \mapsto \rho(f_t)$ is nonconstant and there exists a dense set $S \subset \mathbb{Q}$ such that, for each map f_t, either $\rho(f_t) \notin S$ or f_t has some nonperiodic points. Then ρ is a devil's staircase.*

Proof By Proposition 4.4.5, ρ is monotone and continuous. Together with Proposition 4.4.10, this also implies that $\rho^{-1}(S)$ is a disjoint union of closed intervals of positive length.

We need to show that $\rho^{-1}(S)$ is dense. Assume, by enlarging S if necessary, that whenever $\rho(f_t) = p/q \in \mathbb{Q} \smallsetminus S$, f_t has only periodic points. Then Proposition 4.4.9 and Lemma 4.4.12 imply that ρ is strictly monotone at points $t \in \rho^{-1}([0,1] \smallsetminus S)$. Thus for $t \in [0,1) \smallsetminus \rho^{-1}(S)$ and $\epsilon > 0$ we have $\rho(t) \neq \rho(t + \epsilon)$, and hence by the density of S, the continuity of ρ, and the Intermediate Value-Theorem there exists a $t_1 \in \rho^{-1}(S) \cap [t, t + \epsilon]$. This proves density. □

In closing we remark that the results of this section depend on the monotonicity and continuity of f, but not on invertibility. Thus it suffices to assume that $f \colon S^1 \to S^1$ is a continuous order-preserving map of degree one, that is, its lift F is nondecreasing and $F(x + 1) = F(x) + 1$ (Exercise 4.3.5). Such a map may take constant values on a finite or countable set of intervals.

4.4.5 Frequency Locking

The understanding gained in the preceding subsection about the dependence of the rotation number on a parameter also leads to insights about flows on the 2-torus, and in particular about some systems of differential equations arising in applications. The phenomenon that we are able to shed some light on now is that coupled oscillators tend to become *synchronized*, that is, their frequencies will coincide or be at least rationally related.

Where do we stand? The problems of flashing fireflies and circadian rhythms were introduced in Section 1.2.10 as situations that one might model as coupled oscillators. We can simplistically model those biological clocks as a harmonic oscillator or something close. Indeed, the harmonic oscillator is a good starting point, as we will see by linearization in Section 6.2.2.

Now, in Section 4.2.4 we found that two uncoupled harmonic oscillators produce, on a joint level set, a linear toral flow. This linear flow on $\mathbb{T}^2$ satisfies the differential equations

$$\dot{x}_1 = \omega_1$$

$$\dot{x}_2 = \omega_2.$$

To get an impression of the effects of coupling the two oscillators, modify the preceding differential equations by including "mixed" terms:

(4.4.1)
$$\dot{x}_1 = \omega_1 + c_1 \sin 2\pi (x_2 - x_1)$$

$$\dot{x}_2 = \omega_2 + c_2 \sin 2\pi (x_1 - x_2).$$

This is not exactly the same as coupling the original second-order equations in Section 4.2.4, but it is a good way to get some insight.

A small detail here is the choice of sines to produce the mixed terms. This makes sense because both variables are only defined mod 1. The constants c_1 and c_2 indicate the strength of the coupling. When they are both zero, there is no coupling and we are back to a linear flow on the 2-torus. If they are positive, the right-hand side acts to increase the slower rate of change of the two ω's and to slow down the faster one, which could plausibly lead to synchronization.

In Section 4.2.3 we learned to study flows on the 2-torus by looking at the section map of the flow of (4.4.1) for the section $x_2 = 0$, say. In the absence of the coupling terms, that is, when $c_1 = c_2 = 0$, this section map is just the rotation with rotation number ω_1/ω_2. For small values of the coupling constants the section map is therefore a perturbation of this rotation. In "most" cases this perturbation has a rational rotation number, because this is the stable situation. And whenever the rotation number is rational all asymptotic behavior is periodic (with the same period).

To explore this a little more carefully, suppose that ω_1 and ω_2 are close to each other. In fact, assume first that $\omega_1 = \omega_2 =: \omega$. In that case, $x(t) = y(t) = \omega t$ is a solution of (4.4.1). This particular solution works for all values of c_1 and c_2, so the section map always has a fixed point, and hence rotation number 0.

For $(c_1, c_2) \neq (0, 0)$, the section maps are not conjugate to rotations and therefore their rotation number persists under small perturbations by Proposition 4.4.10. In particular, when we fix c_1 and c_2, then for small values of $\omega_1 - \omega_2$ the flow of (4.4.1) has a section map with a fixed point, all of whose orbits are asymptotic to a fixed one (or fixed). This means that the corresponding solutions of (4.4.1) are all asymptotic to a periodic solution with equal frequencies; that is, experimentally, the two a priori different frequencies of the oscillators are locked into a common compromise frequency by the coupling. Thus the two oscillators synchronize as long as their natural frequencies are sufficiently close to each other.

Observe, by the way, that there should be a phase difference (that is, $x - y$) of size comparable to the difference in natural frequencies.

■ EXERCISES

■ Exercise 4.4.1 Consider the set C of monotone functions $f: [0, 1] \to [0, 1]$ with $f(0) = 0$ and $f(1) = 1$. Define a map T on C as follows. Fix $f \in C$ and denote its graph by G. Let G_1 be the image of G under the transformation given by $\left(\begin{smallmatrix} 1/3 & 0 \\ 0 & 1/2 \end{smallmatrix}\right)$ in the plane, and let $G_2 = G_1 + (2/3 \ 1/2)$. Let G' be the union of G_1, G_2 and the line segment from $(1/3 \ 1/2)$ to $(2/3 \ 1/2)$. Prove that G' is the graph of some $f' \in C$ and set $T(f) = f'$. Next prove that T is a contraction of C with respect to the norm of uniform convergence. (Since C is a closed subset of a complete space, the Contraction Principle then applies. Since the ternary Cantor function is a fixed point, this gives an effective approximation procedure. This was used to create Figure 4.4.1.)

■ Exercise 4.4.2 Prove Lemma 4.4.12.

■ PROBLEMS FOR FURTHER STUDY

The following few exercises contain a brief introduction into a more detailed explanation of frequency locking for a typical example.

■ Problem 4.4.3 (Arnold Tongues) For $a, b \in [0, 1]$, let $f_{a,b}: S^1 \to S^1$, $x \mapsto x + a + b \sin 2\pi x \pmod 1$. Show that for $p/q \in \mathbb{Q} \cap [0, 1]$ the regions $A_{p/q} := \{(a, b) \in [0, 1] \times [0, 1/2\pi] \mid \rho(f_{a,b}) = p/q\}$ are closed. These regions intersect $[0, 1] \times \{0\}$ in the point p/q. (See Figure 4.4.3.)

■ Problem 4.4.4 Show that $A_{p/q}$ intersects every line $b = $ const. in a nonempty closed interval and that this interval has nonzero length except for $b = 0$.

■ Problem 4.4.5 Are the $A_{p/q}$ connected?

■ Problem 4.4.6 Show that the union of the $A_{p/q}$ is dense in $[0, 1] \times [0, 1]$.

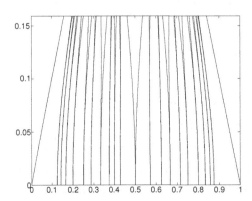

Figure 4.4.3. Arnold tongues.

CHAPTER 5

Recurrence and Equidistribution in Higher Dimension

5.1 TRANSLATIONS AND LINEAR FLOWS ON THE TORUS

5.1.1 The Torus

Consider the n-dimensional torus

$$\mathbb{T}^n = \underbrace{S^1 \times \cdots \times S^1}_{n \text{ times}} = \mathbb{R}^n/\mathbb{Z}^n = \underbrace{\mathbb{R}/\mathbb{Z} \times \cdots \times \mathbb{R}/\mathbb{Z}}_{n \text{ times}},$$

which, for $n = 2$, was first mentioned in Section 2.6.4. We already had several opportunities to represent the 2-torus in various ways. Now we discuss the n-torus in more detail. A natural **fundamental domain** for $\mathbb{R}^n/\mathbb{Z}^n$ is the unit cube:

$$I^n = \{(x_1, \ldots, x_n) \in \mathbb{R}^n \mid 0 \le x_i \le 1 \text{ for } i = 1, \ldots, n\}.$$

This means that, to represent the torus, we identify opposite faces of I^n, that is, we identify $(x_1, \ldots, x_{i-1}, 0, x_{i+1}, \ldots, x_n)$ with $(x_1, \ldots, x_{i-1}, 1, x_{i+1}, \ldots, x_n)$. (See Figure 5.1.1.) These two points represent the same element of the torus. Similar to the case of the circle, there are two convenient coordinate systems on $\mathbb{T}^n$, namely,

(1) multiplicative, where the elements of $\mathbb{T}^n$ are represented as $(z_1, \ldots, z_n)$ with $z_i \in \mathbb{C}$ and $|z_i| = 1$ for $i = 1, \ldots, n$; and

(2) additive, when they are represented by n-vectors $(x_1, \ldots, x_n)$, where each coordinate is defined mod 1.

The correspondence $(x_1, \ldots, x_n) \mapsto (e^{2\pi i x_1}, \ldots, e^{2\pi i x_n})$ establishes an isomorphism between these two representations. By the way, these coordinate systems are called multiplicative and additive, respectively, because there is a "group" structure on the torus that can be viewed as multiplication or as addition: For any two elements $x = (x_1, \ldots, x_n)$, $y = (y_1, \ldots, y_n)$ there is an element $x + y$ defined by $x + y = (x_1 + y_1, \ldots, x_n + y_n)$ (in additive notation), and this addition has negatives, just like that in $\mathbb{R}^n$. In multiplicative notation, the same structure is defined by taking products coordinatewise, and inverses are just reciprocals. In fact, the additive interpretation of this structure is just addition modulo 1 and hence "inherited" from $\mathbb{R}^n$

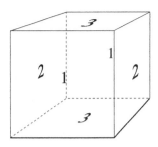

Figure 5.1.1. The 3-torus.

under the identification of vectors modulo 1. (Thus, this is a "factor" of the additive group $\mathbb{R}^n$.)

In additive notation let $\gamma = (\gamma_1, \ldots, \gamma_n) \in \mathbb{T}^n$. Consider the natural multidimensional generalization of rotations given by the translation

$$T_\gamma(x_1, \ldots, x_n) = (x_1 + \gamma_1, \ldots, x_n + \gamma_n) \quad (\text{mod } 1).$$

If all coordinates of the vector γ are rational numbers, say $\gamma_i = p_i/q_i$ with relatively prime p_i and q_i for each $i = 1, \ldots, n$, then T_γ is periodic. Its minimal period is the least common multiple of the denominators $q_1, \ldots, q_n$.

However, unlike the cases of the circle and linear flows on the 2-torus, minimality is not the only alternative to periodicity. For example, if $n = 2$ and $\gamma = (\alpha, 0)$, where α is an irrational number, then the torus $\mathbb{T}^2$ splits into a family of invariant circles $x_2 = \text{const.}$, and every orbit stays on one of these circles and fills it densely.

5.1.2 Criterion for Minimality

The right condition for minimality (see Definition 4.1.4 and Proposition 4.1.1) for the translation T_γ is a sort of mutual irrationality of the coordinates of the translation vector γ. The numbers $\gamma_1, \ldots, \gamma_n$ and 1 must be *rationally independent*.

Definition 5.1.1 A set $A \subset \mathbb{R}$ is said to be rationally independent if $x_1, \ldots, x_n \in A$ and $(k_1, \ldots, k_n) \in \mathbb{Z}^{n+1} \smallsetminus \{0\}$ imply $\sum_{i=1}^n k_i \gamma_i \neq 0$.

Rational independence of $\gamma_1, \ldots, \gamma_n$ and 1 means that $k_0 + \sum_{i=1}^n k_i \gamma_i \neq 0$ for $(k_0, k_1, \ldots, k_n) \in \mathbb{Z}^{n+1} \smallsetminus \{0\}$. Equivalently, $\sum_{i=1}^n k_i \gamma_i$ is not an integer for any collection of integers $k_1, \ldots, k_n$, except for $k_1 = k_2 = \cdots = k_n = 0$. Note that in the case of a single number this is exactly irrationality.

Proposition 5.1.2 *The translation T_γ on $\mathbb{T}^2$ is minimal if and only if the numbers $\gamma_1, \gamma_2,$ and 1 are rationally independent, that is, there are no two nonzero integers k_1, k_2 such that $k_1 \gamma_1 + k_2 \gamma_2 \in \mathbb{Z}$.*

We postpone the proof to Section 5.1.5 because the arguments are a bit long, although they are elementary. We only carry the proof out in dimension 2 to make the notation and arguments a bit easier and allow for suggestive geometric illustrations.

The reason for the requirement on the translation vector is not so hard to see, however. In Section 4.2.3 we saw that a linear flow $(T_\gamma^t)_{t \in \mathbb{R}}$ on the 2-torus is minimal if its translation vector γ has irrational slope. Therefore, a translation T_γ can only be minimal if $\gamma_1/\gamma_2 \notin \mathbb{Q}$ or $k_1\gamma_1 + k_2\gamma_2 \neq 0$ for any $(k_1, k_2) \in \mathbb{Z}^2 \setminus \{0\}$. On the other hand, this is not quite sufficient, because if γ_1 is rational, say, then the first coordinate of any orbit can only take finitely many values and the orbit cannot be dense. To rule out such problems requires the minimality condition above.

5.1.3 Linear Flows

In Section 4.2.3 we introduced linear flows on the 2-torus. On the n-torus linear flows are likewise given as a one-parameter group of translations

$$T_\omega^t(x_1, \ldots, x_n) = (x_1 + t\omega_1, \ldots, x_n + t\omega_n) \quad (\text{mod } 1).$$

Since the flow $\{T_\omega^t\}$ is minimal if for some t_0 the transformation $T_\omega^{t_0}$ is minimal, Proposition 5.1.2 allows us to establish the criterion for minimality for this case.

Proposition 5.1.3 *The flow $\{T_\omega^t\}$ is minimal if and only if the numbers $\omega_1, \ldots, \omega_n$ are rationally independent.*

Proof Since $T_\omega^t = T_{t\omega}$, minimality follows from Proposition 5.1.2 once we find $t \in \mathbb{R}$ such that $\sum_{i=1}^n tk_i\omega_i \notin \mathbb{Z}$ for any nonzero integer vector $(k_1, \ldots, k_n)$. To this end note that if $k \in \mathbb{Z}$ then $s \sum_{i=1}^n k_i\omega_i = k$ implies $s = k/\sum k_i\omega_i$ (because $\sum_{i=1}^n k_i\omega_i \neq 0$ by rational independence). Only countably many such s's arise, so any $t \in \mathbb{R} \setminus \{k/\sum k_i\omega_i \mid k_1, \ldots, k_n, k \in \mathbb{Z}, (k_1, \ldots, k_n) \neq 0\}$ is as required.

On the other hand, if $\sum_{i=1}^n k_i\omega_i = 0$ for some nonzero vector $(k_1, \ldots, k_n)$, then the function $\varphi(x) = \sin 2\pi \left(\sum_{i=1}^n k_i x_i \right)$ is continuous, nonconstant, and invariant under the flow $\{T_\omega^t\}$. Therefore the flow is not minimal, because $\varphi^{-1}([0, 1])$ is a closed invariant set (Definition 4.1.4). $\square$

5.1.4 Uniform Distribution: Elementary Proof

Similarly to Section 4.1.4 we can look at the frequencies with which an orbit of a minimal translation visits various parts of the torus. In the one-dimensional case we used arcs (intervals) as natural "windows" through which to measure the frequency of visits. A natural counterpart for the n-torus will be n-parallelepipeds, $\Delta = \Delta_1 \times \cdots \times \Delta_n$, where $\Delta_1, \ldots, \Delta_n$ are arcs. For $n = 2$ it is natural to call a parallelepiped a *rectangle*. The *volume* $\text{vol}(\Delta)$ of Δ is defined as the product of the lengths of the arcs $\Delta_1, \ldots, \Delta_n$. Thus we arrive at the following natural generalization of the notion of uniform distribution, which appeared in the remark after Proposition 4.1.7.

Definition 5.1.4 A sequence $(x_m)_{m \in \mathbb{N}}$ in T^n is said to be *uniformly distributed* if

$$\lim_{m \to \infty} \frac{\text{card}\{k \in \{1, \ldots, m\} \mid x_k \in \Delta\}}{m} = \text{vol}(\Delta).$$

for every n-parallelepiped $\Delta \subset \mathbb{T}^n$.

The proof of Proposition 4.1.7 uses only that an irrational rotation is minimal and isometric. Specifically, one needs that arcs are taken into arcs of the same length and that the circle can be decomposed into a union of arbitrarily short isometric arcs with disjoint interiors. Toral translations are isometries, and a counterpart of the decomposition property holds for the torus with arcs replaced by parallelepipeds; so it is not surprising that the proof of Proposition 4.1.7 can be adapted to this situation. We give a detailed argument in the two-dimensional case. Unlike in the case of minimality, where this assumption brings a genuine simplification, this is purely a matter of keeping notation simple. The extension of the statement "minimality implies uniform distribution" to translations in arbitrary dimension is completely straightforward.

Theorem 5.1.5 *If* $(\gamma_1, \gamma_2, 1)$ *are rationally independent, then every semiorbit of the translation* $T_{(\gamma_1, \gamma_2)}$ *is uniformly distributed.*

As in Section 4.1.4, define

$$F_\Delta(x, n) := \text{card}\left\{k \in \mathbb{Z} \mid 0 \le k < n, \ T_\gamma^k(x) \in \Delta\right\}$$

for any $x \in \mathbb{T}^2$ and any rectangle Δ. This definition extends to unions of disjoint rectangles. By Proposition 5.1.2, the translation T_γ is minimal, where $\gamma = (\gamma_1, \gamma_2)$. This allows us to extend Proposition 4.1.5:

Proposition 5.1.6 *Consider two rectangles* $\Delta = \Delta_1 \times \Delta_2$ *and* $\Delta' = \Delta'_1 \times \Delta'_2$ *such that* $\ell(\Delta_i) < \ell(\Delta'_i)$, $i = 1, 2$. *There is an* $N_0 \in \mathbb{N}$, *which depends on* Δ, Δ', *and* γ, *such that if* $x \in \mathbb{T}^2$, $N \ge N_0$, *and* $n \in \mathbb{N}$, *then* $F_{\Delta'}(x, n + N) \ge F_\Delta(x, n)$.

Proof By assumption there is a translation T_β of the rectangle Δ that lies inside Δ'. By minimality of T_γ we can find $N_0 \in \mathbb{N}$ such that the translation $T_\gamma^{N_0}\Delta$ is so close to $T_\beta\Delta$ that $T_\gamma^{N_0}\Delta \subset \Delta'$. Thus $T_\gamma^n(x) \in \Delta$ implies $T_\gamma^{n+N_0}(x) \in \Delta'$ and $F_{\Delta'}(x, n + N) \ge F_{\Delta'}(x, n + N_0) \ge F_\Delta(x, n)$ for $n \ge N_0$. $\square$

Proof of Theorem 5.1.5 Now, similarly to the one-dimensional case, take a rectangle $\Delta = \Delta_1 \times \Delta_2$, where $\ell(\Delta_1) = \ell(\Delta_2) = 1/k$. Divide the torus $\mathbb{T}^2$ into $(k-1)^2$ disjoint rectangles, each being the product of two arcs of length $1/k - 1$ (Figure 5.1.2), and

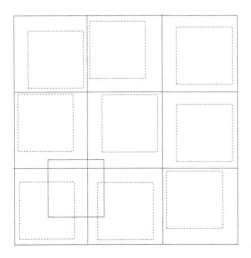

Figure 5.1.2. Torus with rectangles.

apply Proposition 5.1.6 in exactly the same way as in the proof of Lemma 4.1.9 to get

$$(5.1.1) \qquad \bar{f}(\Delta) := \limsup_{n \to \infty} \frac{F_\Delta(x, n)}{n} \leq 1/(k-1)^2.$$

Finally, let $\Delta = \Delta_1 \times \Delta_2$ be an arbitrary rectangle. Fix $\epsilon > 0$ and a rectangle $\Delta' = \Delta_1' \times \Delta_2'$ such that $\Delta_i \subset \Delta_i'$ for $i = 1, 2$; the lengths of Δ_i' are l_i/k; and $\text{vol } \Delta' < \text{vol } \Delta + \epsilon$. By using (5.1.1) and the subadditivity of $\bar{f}$ we obtain

$$\bar{f}(\Delta) \leq \bar{f}(\Delta') \leq \left(\frac{k}{k-1}\right)^2 \text{vol } \Delta' < \left(\frac{k}{k-1}\right)^2 (\text{vol } \Delta + \epsilon).$$

Since ϵ is arbitrarily small and k arbitrarily large, this implies that $\bar{f}(\Delta) \leq \text{vol } \Delta$ for any rectangle Δ and hence (by subadditivity of $\bar{f}$) for any finite union of disjoint rectangles. In particular, since $\mathbb{T}^2 \smallsetminus \Delta$ is the union of three disjoint rectangles, this implies that

$$\underline{f}(\Delta) := \liminf_{n \to \infty} \frac{F_\Delta(x, n)}{n} = 1 - \bar{f}(\mathbb{T}^2 \smallsetminus \Delta) \geq 1 - \text{vol}(\mathbb{T}^2 \smallsetminus \Delta) = \text{vol } \Delta,$$

and hence $\underline{f}(\Delta) = \bar{f}(\Delta) = \text{vol } \Delta$. $\square$

There is an obvious extension of Theorem 4.1.15 from rotations of the circle to translations of the torus:

Theorem 5.1.7 *Let $\gamma = (\gamma_1, \gamma_2)$ and φ any Riemann-integrable function on $\mathbb{T}^2$. If the numbers $1, \gamma_1, \gamma_2$ are rationally independent, then*

$$\lim_{n \to \infty} \frac{1}{n} \sum_{k=0}^{n-1} \varphi\big(T_\gamma^k(x_1, x_2)\big) = \int_{\mathbb{T}^2} \varphi(\theta_1, \theta_2) \, d\theta_1 \, d\theta_2$$

uniformly in $(x_1, x_2) \in \mathbb{T}^2$.

Proof The passage from uniform distribution for rectangles to uniform distribution for continuous and, more generally, Riemann-integrable functions goes exactly as in the one-dimensional case (Proposition 4.1.14 and Theorem 4.1.15). If Δ is a rectangle, then

$$\text{vol } \Delta = \int_{\mathbb{T}^2} \chi_\Delta(\theta_1, \theta_2) \, d\theta_1 \, d\theta_2,$$

and, by definition, a function φ is Riemann-integrable if for any $\epsilon > 0$ there exist finite linear combinations φ_1, φ_2 of characteristic functions of rectangles such that $\varphi_1 \leq \varphi \leq \varphi_2$ and

$$\int_{\mathbb{T}^2} \varphi_1(\theta_1, \theta_2) \, d\theta_1 \, d\theta_2 < \int_{\mathbb{T}^2} \varphi_2(\theta_1, \theta_2) \, d\theta_1 \, d\theta_2 + \epsilon.$$

(In particular, any continuous function or any bounded function with finitely many discontinuity points is Riemann integrable.) $\square$

5.1.5 Proof of the Minimality Criterion

Now we prove that toral translations are minimal if and only if the translation vector is "completely irrational". This condition implies that γ_1 and γ_2 as well as their ratio are irrational. However, the condition is stronger than that, as the simple example of $\gamma_2 = 1 - \gamma_1$ with any irrational γ_1 shows.

The proof is considerably more elaborate than the simple argument from the proof of Proposition 4.1.1. The main idea, however, is the same: Unless the points on an orbit are aligned in a particular fashion, they will crowd all around, and this produces minimality. The main difference with the one-dimensional case is that then a "special alignment" simply meant finiteness of the orbit and hence periodicity, while now we have to capture an intermediate case and show that it appears only if orbits lie on parallel rational lines spiraling around the torus.

Proof of Proposition 5.1.2 We use additive notation. Such a translation is minimal if and only if the orbit of 0 is dense, because if $x \in \mathbb{T}^2$, then

$$T_\gamma(x) = x + \gamma = 0 + \gamma + x = T_\gamma(0) + x \pmod 1;$$

that is, the orbit $\mathcal{O}(x)$ of x is $T_x(\mathcal{O}(0))$, and therefore it is dense if and only if $\mathcal{O}(0)$ is because T_x is a homeomorphism. (This argument is the same as that in the proof of the more general Proposition 4.1.19.)

Pick $\epsilon > 0$ and consider the set D_ϵ of all iterates $T_\gamma^m(0)$ that are in the ϵ-ball $B(0, \epsilon)$ around 0. There are two possibilities:

(1) For some $\epsilon > 0$ the set D_ϵ is linearly dependent (that is, lies on a line).
(2) For any $\epsilon > 0$ the set D_ϵ contains two linearly independent vectors.

Below we prove three corresponding lemmas.

Lemma 5.1.8 *(2) $\Rightarrow$ minimality.*

Lemma 5.1.9 *(1) $\Rightarrow$ rational dependence.*

Lemma 5.1.10 *Rational dependence $\Rightarrow$ (1).*

Minimality clearly excludes (1) and hence implies (2), so minimality is equivalent to (2). Thus minimality $\Longleftrightarrow$ (2) $\Longleftrightarrow$ *not* (1) $\Longleftrightarrow$ rational independence. $\square$

Proof of Lemma 5.1.8 This argument is similar to the proof of Proposition 4.1.1, albeit more complicated. It suffices to show that the orbit of 0 is dense. Take

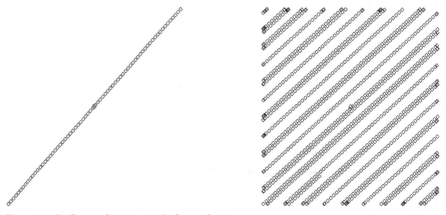

Figure 5.1.3. Dependent versus independent.

$\epsilon > 0$ and suppose $v_1, v_2 \in D_\epsilon$ are linearly independent. This means that they span a small parallelogram $\{av_1 + bv_2 \mid a, b \in [0, 1]\}$. The vertices of this parallelogram are all part of $\mathcal{O}(0)$: This is already known for 0, v_1, and v_2, and for $v_1 + v_2$ this is easy to see by representing v_1 and v_2 as $V_1 = 0 + m_1\gamma - k(m_1)$ and $V_2 = 0 + m_2\gamma - k(m_2)$ in $\mathbb{R}^2$, respectively, where $k(m_1)$ and $k(m_2)$ are those integer vectors for which $\|V_1\| < \epsilon$ and $\|V_2\| < \epsilon$. Then $V_1 + V_2 = 0 + (m_1 + m_2)\gamma - (k(m_1) + k(m_2)) = T_\gamma^{m_1+m_2}(0) \pmod{1}$ and hence $v_1 + v_2 = T_\gamma^{m_1+m_2}(0)$.

Furthermore, the orbit of 0 contains all integer linear combinations of v_1 and v_2 [because $kV_1 + lV_2 = T_\gamma^{km_1+lm_2}(0) \pmod{1}$]. Therefore, consider the tiling of the plane defined by the translates of $R := \{aV_1 + bV_2 \mid a, b \in [0, 1]\}$ by integer multiples of V_1 and V_2. This covers the plane with similar parallelograms, which have only boundary points in common, and every point of the plane is within ϵ of one of the vertices of these tiles (Figure 5.1.3). In particular, every point of $[0, 1] \times [0, 1]$ is within ϵ of some vertex, that is, every point of $\mathbb{T}^2$ is within ϵ of some point of $\mathcal{O}(0)$. According to the hypothesis of case (2), this is the case for any $\epsilon > 0$, that is, $\mathcal{O}(0)$ is dense in $\mathbb{T}^2$. $\square$

Proof of Lemma 5.1.9 If 0 is periodic, then γ_1 and γ_2 are rational and we are done.

From now on assume that the orbit of 0 is infinite. Then for any $\epsilon > 0$ it contains two points $p = T_\gamma^m(0)$ and $q = T_\gamma^n(0)$ such that $\|q - p\| < \epsilon$. Then there are points $P = m\gamma \in \mathbb{R}^2$ and $Q = n\gamma + k \in \mathbb{R}^2$ such that $\epsilon > \|P - Q\| = \|m\gamma - n\gamma - k\| = \|(m - n)\gamma - k\|$, which means that $T_\gamma^{m-n}(0) - k \in B(0, \epsilon)$ and $D_\epsilon \neq \{0\}$ for all $\epsilon > 0$.

If $\epsilon > 0$ is as in (1), then $\{0\} \neq D_{\epsilon'} \subset D_\epsilon$ is linearly dependent for all $\epsilon' < \epsilon$. Thus D_ϵ lies on a unique line L through 0 given by an equation $ax + by = 0$.

Claim $\mathcal{O}(0)$ is dense on the projection of L. (See Figure 5.1.4.)

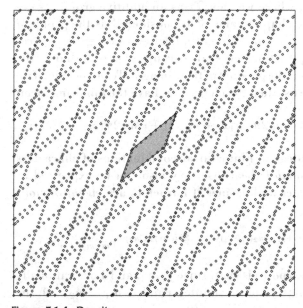

Figure 5.1.4. Density.

Since $D_{\epsilon'} \neq \{0\}$ for all $\epsilon' < \epsilon$, there are points $0 \neq p_{\epsilon'} \in D_{\epsilon'}$ and hence points $P = n\gamma - k \in L \cap B(0, \epsilon')$ (with $n \in \mathbb{Z}$, $k \in \mathbb{Z}^2$). But then $\{mP \mid m \in \mathbb{Z}\}$ is ϵ'-dense in L and projects into $\mathcal{O}(0)$.

Now a and b are rationally dependent because otherwise the slope of L is irrational, so the projection of L to $\mathbb{T}^2$ is dense and by the claim so is $\mathcal{O}(0)$. Therefore there exists $(k_1, k_2) \in \mathbb{Z}^2 \setminus \{0\}$ such that $ak_1 - bk_2 = 0$. If $a = 0$ (or $b = 0$), then $ax + by = 0 \Leftrightarrow y = 0$ (or $x = 0$). Otherwise, multiply $ax + by = 0$ by $k_1/b = k_2/a$ to get $k_2 x + k_1 y = 0$, that is, we may take $a, b \in \mathbb{Z}$. If $n\gamma - k$ lies on the line $ax + by = 0$, then $an\gamma_1 - k_1 + bn\gamma_2 - k_2 = 0$ or $an\gamma_1 + bn\gamma_2 = k_1 + k_2$, which gives rational dependence. $\square$

Proof of Lemma 5.1.10 Suppose $k_1\gamma_1 + k_2\gamma_2 = N \in \mathbb{Z}$ and divide by γ_1 to get $\gamma_2/\gamma_1 = (N - k_1)/k_2 =: s \in \mathbb{Q}$ (if $k_2 \neq 0$), that is, the iterates $(n\gamma_1, n\gamma_2)$ of 0 under repeated translation by γ lie on the line $y = sx$ with rational slope s. This projects to the torus as an orbit of the linear flow T_γ^t, which we found in Section 4.2.3 to be closed and hence not dense when $\gamma_2/\gamma_1 \in \mathbb{Q}$. Therefore the orbit of 0 under T_γ is not dense either, implying (1). (If $k_2 = 0$, then $k_1 \neq 0$ and the same argument works after exchanging x and y.) $\square$

5.1.6 Uniform Distribution: The Kronecker–Weyl Method

The Kronecker–Weyl method of proving uniform distribution starting from trigonometric polynomials, then proceeding to continuous functions, and finally to characteristic functions, described in Section 4.1.6 also works in higher dimension. Again, to simplify notation we consider the two-dimensional case, leaving the extension to arbitrary dimension to the reader.

The *characters* corresponding to those in Section 4.1.6 are defined as group "homomorphisms" of $\mathbb{T}^2$ to S^1, where we view $\mathbb{T}^2$ as an additive group (as described at the beginning of this chapter) and S^1 is considered as the group of complex numbers of absolute value one with multiplication as the group operation. A homomorphism is a map that preserves this group structure, that is, the image of the sum of two elements is the product of their images. To be specific, if we use additive notation for the torus, then the characters have the following form:

$$c_{m_1,m_2}(x_1, x_2) = e^{2\pi i (m_1 x_1 + m_2 x_2)} = \cos 2\pi (m_1 x_1 + m_2 x_2) + i \sin 2\pi (m_1 x_1 + m_2 x_2),$$

where (m_1, m_2) is any pair of integers. Finite linear combinations of characters are called *trigonometric polynomials* beacause they also can be expressed as finite linear combinations of sines and cosines. Characters are *eigenfunctions* for the translation because

$$c_{m_1,m_2}(T_\gamma(x_1, x_2)) = e^{2\pi i (m_1(x_1+\gamma_1)+m_2(x_2+\gamma_2))} = e^{2\pi i (m_1\gamma_1 + m_2\gamma_2)} c_{m_1,m_2}(x_1, x_2).$$

A crucial observation for our purposes is that, since γ_1, γ_2, and 1 are rationally independent, that is, $m_1\gamma_1 + m_2\gamma_2$ is never an integer unless $m_1 = m_2 = 0$, the *eigenvalue* $e^{2\pi i (m_1\gamma_1 + m_2\gamma_2)} \neq 1$ unless $m_1 = m_2 = 0$.

The trivial character $c_{0,0} = 1$ is not changed by averaging. For the other characters we use summation of the geometric series as in Section 4.1.6 to obtain

$$\left| \frac{1}{n} \sum_{k=0}^{n-1} c_{m_1, m_2} \left(T_\gamma^k (x_1, x_2) \right) \right| = \left| \frac{1}{n} \sum_{k=0}^{n-1} e^{2\pi i k (m_1 \gamma_1 + m_2 \gamma_2)} \right| \left| c_{m_1, m_2}(x_1, x_2) \right|$$

$$= \left| \frac{1 - e^{2\pi i n (m_1 \gamma_1 + m_2 \gamma_2)}}{n \left(1 - e^{2\pi i (m_1 \gamma_1 + m_2 \gamma_2)} \right)} \right|$$

$$\leq \frac{2}{n \left(1 - e^{2\pi i (m_1 \gamma_1 + m_2 \gamma_2)} \right)} \xrightarrow[n \to \infty]{} 0 = \int_{\mathbb{T}^2} c_{m_1, m_2}.$$

Using linearity of the integral one deduces that, for any finite linear combination φ of characters, that is, for any trigonometric polynomial, we have

(5.1.2)
$$\lim_{n \to \infty} \frac{1}{n} \sum_{k=0}^{n-1} \varphi \left(T_\gamma^k (x_1, x_2) \right) = \int_{\mathbb{T}^2} \varphi.$$

Now we can invoke a multidimensional version of the Weierstraß Approximation Theorem (a continuous function on the plane that is 1-periodic in both variables is a uniform limit of trigonometric polynomials) to deduce that (5.1.2) holds for any continuous function. Finally, uniform distribution for rectangles follows exactly as in the one-dimensional case by finding continuous functions $\varphi_1 \leq \chi_\Delta \leq \varphi_2$ such that $\int (\varphi_2 - \varphi_1) < \epsilon$. It is also easy to see within this framework that if 1, γ_1, and γ_2 are rationally dependent, then the translation T_γ is not minimal, as was pointed out at the end of Section 5.1.2: If $m_1 \gamma_1 + m_2 \gamma_2 = k$ with $m_1, m_2, k \in \mathbb{Z}$, and $m_1^2 + m_2^2 > 0$, then $e^{2\pi i (m_1 \gamma_1 + m_2 \gamma_2)} = 1$ and the values of the nonconstant character c_{m_1, m_2} do not change under translation.

The use of the Kronecker–Weyl method allows us to bypass a comparatively subtle argument, which was required in Section 5.1.2 to establish the condition for minimality. With this approach uniform distribution is deduced directly and rather straightforwardly from the rational independence of γ_1, γ_2, and 1. Also, the extension of the proof to arbitrary dimension using this method is completely routine.

■ **EXERCISES**

■ **Exercise 5.1.1** Show that 1, $\sqrt{3}$, and $\sqrt{5}$ are rationally independent.

■ **Exercise 5.1.2** Suppose $n, m \in \mathbb{Z}$ are such that 1, $\sqrt{n}$, and $\sqrt{m}$ are rationally dependent. What does this imply about n and m?

■ **Exercise 5.1.3** Describe the orbit closures for the translation by $(\alpha, 1/4 + 2\alpha)$ on $\mathbb{T}^2$, where $\alpha \notin \mathbb{Q}$.

■ **Exercise 5.1.4** (This explains the origin of the term "rational independence.") The real line $\mathbb{R}$ can be viewed as a linear space over the field $\mathbb{Q}$ of rational numbers (that is, the rationals are the scalars). Show that a set of numbers in $\mathbb{R}$ is rationally independent if and only if it is linearly independent in the linear space $\mathbb{R}$ over $\mathbb{Q}$.

■ **Exercise 5.1.5** Show that if all coordinates of the vector γ are rational numbers, say $\gamma_i = p_i/q_i$ with relatively prime p_i and q_i for each $i = 1, \ldots, n$, then T_γ is periodic and its minimal period is the least common multiple of the denominators $q_1, \ldots, q_n$.

■ **Exercise 5.1.6** Show that the closure of the orbit of a nonminimal translation on $\mathbb{T}^2$ is either a finite set or a finite union of circles.

■ **Exercise 5.1.7** Show that every closed proper subgroup Γ of $\mathbb{R}^2$ is equivalent to one of the following by a linear coordinate change: $\mathbb{R}, \mathbb{Z}, \mathbb{Z} \times \mathbb{Z}, \mathbb{Z} \times \mathbb{R}$.

■ **PROBLEMS FOR FURTHER STUDY**

■ **Problem 5.1.8** Generalize Exercise 5.1.6 to $\mathbb{R}^n$.

■ **Problem 5.1.9** Generalize Exercise 5.1.7 to $\mathbb{R}^n$.

■ **Problem 5.1.10** Write a detailed proof of the minimality criterion as in Section 5.1.5 in n dimensions.

■ **Problem 5.1.11** Formulate the uniform distribution property for a translation on the group $\mathbb{Z}_2$ of d-adic integers (Problem 4.1.15) and prove that minimality implies uniform distribution.

5.2 APPLICATIONS OF TRANSLATIONS AND LINEAR FLOWS

We now give several examples of dynamical systems where the dynamics can naturally be analyzed in terms of linear flows (or translations) on tori.

5.2.1 Linear Maps and Flows

An understanding of linear maps and flows on tori provides a tool for describing the dynamics of an important class of linear systems, namely, maps with eigenvalues of absolute value one and linear differential equations with constant coefficients whose coefficient matrix has purely imaginary eigenvalues (and whose time-T-maps thus have eigenvalues of absolute value 1). Section 4.2.4 gave particular examples of flows in $\mathbb{R}^4$ that arose from linear differential equations with constant coefficients. There tori naturally occurred as invariant sets on which we observe a linear flow. More generally, consider a linear map of $\mathbb{R}^{2m}$ whose eigenvalues form m distinct complex conjugate pairs $e^{\pm iv_j}$. As before, each pair corresponds to a two-dimensional invariant subspace in which the map acts as a rotation with respect to proper coordinates. The eigenspace and these coordinates are obtained by taking a complex eigenvector w_i and then choosing the real vectors $v_j = w_j + \bar{w}_j$ and $v'_j = i(w_j - \bar{w}_j)$ as a basis. Doing this for each pair of eigenvalues gives a basis of $\mathbb{R}^{2m}$ with respect to which the map has a block diagonal matrix representation in which each block is a 2×2 block representing a rotation. This map then leaves invariant the sets given by the equations $x_{2j-1}^2 + x_{2j}^2 = r_j^2$ for $j = 1 \ldots m$. These sets are tori whose dimension depends on the number of r_j's that are zero. Specifically, such a torus can be parametrized by polar coordinates $x_{2j-1} = r_j \cos\varphi_j$, $x_{2j} = r_j \sin\varphi_j$, and the map then acts by rotations that shift φ_j to $\varphi_j + v_j$. Clearly any $r_j = 0$ reduces the dimension of the torus.

Therefore the minimality criterion Proposition 5.1.2 (applied to $\mathbb{T}^k$) tells us that the restriction of the flow to such an invariant torus is minimal when $\{v_j \mid r_j \neq 0\} \cup \{1\}$ is rationally independent.

More generally, one can draw conclusions about the action of a linear map inside its neutral space E^0 defined in (3.3.3) when the restriction to this subspace has sufficiently many distinct eigenvalues.

5.2.2 Free Particle Motion on the Torus

The motion of a point mass on the flat torus $\mathbb{T}^n = \mathbb{R}^n / \mathbb{Z}^n$ without external forces is described by the second-order ordinary differential equation $\ddot{x} = 0$, where x is defined modulo $\mathbb{Z}^n$. Alternatively we can write

$$\dot{x} = v,$$

$$\dot{v} = 0$$

to see that the motion is along straight lines with constant speed, since v is preserved. This means that the n components of v are *integrals* (or constants) of motion. For any given v the motion corresponds to the linear flow T_v^t. Thus the phase space is $\mathbb{R}^n \times \mathbb{T}^n$ with dynamics described as follows: The tori $\{v\} \times \mathbb{T}^n$ are invariant and the motion on $\{v\} \times \mathbb{T}^n$ is given by $\{v\} \times T_v^t$. This flow is also called the *geodesic flow* on $\mathbb{T}^n$. The geodesics are the paths traced out on $\mathbb{T}^n$ by the orbits. They are projections of straight lines in $\mathbb{R}^n$ to $\mathbb{T}^n$. While for different initial velocity vectors v these curves may be variously dense, periodic, or neither, the orbits of the flow are never dense in the phase space.

One way of studying this flow via a discrete-time dynamical system is to restrict attention to vectors with footpoint on the circle $y = 0$ and pointing upward. Each of these vectors defines an orbit of the flow that returns to this set. If α is the cotangent of the angle of such a vector, then the return map is given by $(x, \alpha) \mapsto (x + \alpha, \alpha)$. This integrable twist will reappear several times (Example 6.1.2, Section 6.3.4.1).

5.2.3 Many-Particle System on the Interval

A straightforward generalization of the simple mechanical model discussed in Section 4.2.5 is given by a finite number of point particles with equal masses moving on the interval with elastic collisions among themselves and with the endpoints. Since the order of the particles cannot change, their positions $x_1, \ldots, x_n$ satisfy $0 \leq x_1 \leq \cdots \leq x_n \leq 1$; that is, the configuration space of this mechanical system is the simplex $T_n := \{(x_1, \ldots, x_n) \mid 0 \leq x_1 \leq \cdots \leq x_n \leq 1\}$, and the phase space is the space of tangent vectors with footpoints in T_n with appropriate conventions on the boundary. (See Figure 5.2.1.)

The n-dimensional analogs of the geometric considerations from Section 4.2.5 show that the system can be described as the motion of a single point particle bouncing off the faces of T_n with an n-dimensional analog of the reflection law "angle of incidence equals angle of reflection." This means that one determines the continuation of a trajectory after an impact on a face by taking the plane spanned by the incoming trajectory and the normal vector to the face and applies the two-dimensional reflection law in this plane. This prescription

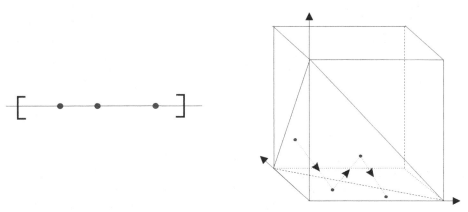

Figure 5.2.1. Three particles on a line and their configuration space.

does not determine motions that involve collisions with an edge or vertex, that is, multiple or simultaneous collisions.

The partial unfolding, which helped describe the billiard in the triangle in terms of the linear flow on the 2-torus, works here as well, with the fundamental domain being the n-dimensional cube of twice the linear size, that is, $\max |x_i| \leq 1$. The $n!2^n$ reflected copies of T_n tile this cube, and, in turn, the translated copies of this cube tile $\mathbb{R}^n$. Thus the complete unfolding of this motion on T_n produces the free particle motion on $\mathbb{R}^n$. After reducing this motion to the fundamental domain (the cube, which we identify with the n-torus) we obtain the free particle motion on the n-torus. Hence we can describe this motion in terms of the linear flow on the n-torus.

The mechanical equivalent of the geometric unfolding is the observation that, upon collision, any two particles exchange momenta and therefore one can consider only the transfer of momenta, which makes it appear as if the particles go through each other and only reverse direction at the boundary.

◼ EXERCISES

◼ Exercise 5.2.1 Consider the billiard ball motion in the unit cube. How many directions can the velocity vector take along one orbit?

◼ Exercise 5.2.2 Reduce the billiard ball motion in the unit cube I to the free particle motion on the torus and decompose it into toral translations.

◼ PROBLEMS FOR FURTHER STUDY

◼ Problem 5.2.3 Describe the reduction to a billiard problem of the motion of several particles with unequal masses on the interval.

◼ Problem 5.2.4 Describe the distribution of values of the function $\sin n + \cos \sqrt{2}n + \sin \sqrt{3}n$.

CHAPTER 6

Conservative Systems

6.1 PRESERVATION OF PHASE VOLUME AND RECURRENCE

We will see that preservation of phase volume is a natural property that holds, for example, in dynamical systems arising from mechanics, and that this property is a direct cause for ubiquitous nontrivial recurrence.

6.1.1 Criteria for Preservation of Volume

So far we have been concerned with the asymptotic behavior of individual orbits of a dynamical system. The basic examples discussed in Chapter 4 and Chapter 5 exhibit *recurrent* behavior of orbits: All orbits are either periodic, that is, return exactly to the initial position, or come arbitrarily close to the initial position, as in an irrational rotation of the circle, a nonperiodic translation of the torus, or in free particle motion on the torus. This type of behavior is different from most of the phenomena observed in Chapter 2 and Chapter 3. There, typically a nonperiodic orbit was attracted to periodic ones, and recurrence appeared only for periodic orbits, which in all nonlinear and most linear examples were few in number.

A key to understanding this difference is given by a property that is not directly observed by looking at individual orbits but by considering the evolution of large sets of initial conditions simultaneously, the *preservation of phase volume*.

1. Preservation of Phase Volume. This property is simply that the map defining a discrete-time dynamical system (or, in the case of flows, each time-t map) preserves the volume of sets in the phase space. It is quite obvious why this property is not compatible with some of the simple types of behavior observed in earlier chapters. For example, if x is a contracting fixed (or periodic) point of a map f, then a small enough ball around x is mapped by f (or an iterate of f) inside an even smaller ball. Hence the volume of that ball *decreases*. This justifies another name for the preservation of volume, *incompressibility*.

Example 6.1.1 Any translation of the torus is an isometry; it preserves the size and shape of sets. In particular, the image of any rectangle Δ is a rectangle with

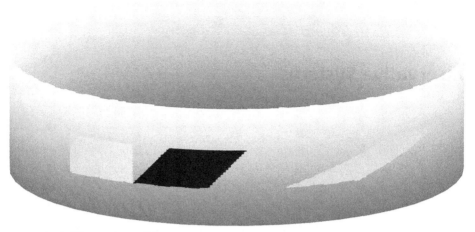

Figure 6.1.1. Distorted parallelogram.

sides of the same length; hence it has the same volume as Δ. Since every Riemann measurable set can be approximated by a finite union of rectangles, we see that volumes of such sets are also preserved.

Example 6.1.2 The *linear twist* $T\colon S^1 \times [0,1] \to S^1 \times [0,1]$ of the cylinder (Section 2.6.3), $T(x, y) = (x + y, y)$, where $x + y$ is defined modulo 1, appears naturally as a section map for the free particle motion on the 2-torus discussed in Section 5.2.2. It is not an isometry. In fact, the image of a rectangle Δ gets tilted and becomes a parallelogram with angles $\pi/4$ and $3\pi/4$. (See Figure 6.1.1.) Nevertheless, the dynamics of individual orbits can be understood in terms of isometries because we can restrict attention to any horizontal circle, where the twist acts as a rotation. When the map is iterated the parallelogram $f^n(\Delta)$ gets longer and longer and more "horizontal". Thus, the shape of Δ changes beyond recognition. However, the volume (that is, the area in this case) of $f(\Delta)$ [and of $f^n(\Delta)$ for any $n \in \mathbb{N}$] is the same as for Δ because the base and height of the parallelogram are unchanged. Alternatively, note that $\text{area}(f(\Delta)) = \int_0^1 l(f(\Delta) \cap S^1 \times \{t\})\, dt$ and that on every circle $C_t = S^1 \times \{t\}$ the map T acts as a rotation and hence $l(f(\Delta) \cap C_t) = l(\Delta \cap C_t)$, and by integration $\text{area}(f(\Delta)) = \text{area}(\Delta)$.

This argument also holds for more complicated sets.

2. The Linear Case. The previous arguments were, of course, ad hoc, and we need a more systematic method of checking whether the phase volume is preserved. As is usual in analysis, we develop an *infinitesimal* condition corresponding to the desired property. Since such conditions are based on the linear approximation of nonlinear objects, it is instructive to look at linear maps first. For linear maps the answer is provided by elementary linear algebra. If a linear map of $\mathbb{R}^n$ is represented in Euclidean coordinates by a matrix L, then the image of the standard unit cube $\Delta = \{(x_1, \ldots, x_n) \mid 0 \le x_i \le 1\}$ is a parallelepiped of volume $|\det L|$. More generally, the ratio of the volume of the image of a parallelepiped P to the volume of P itself is equal to $|\det L|$. The same follows for more general sets by approximation. Thus, a linear map preserves volume if and only if it is represented by a matrix with determinant ± 1. Notice that this property does not depend on the choice of basis: With respect to a different basis the same map is represented by the matrix $L' = C^{-1}LC$, with an invertible matrix C, and $\det L' = \det C^{-1} \det L \det C = \det L$. Alternatively note that the determinant is the product of the eigenvalues, which do not change under conjugation.

3. The Criterion. Now consider a (nonlinear) differentiable map f defined near a point $x_0 \in \mathbb{R}^n$. We have $f(x) = f(x_0) + Df_{x_0}(x - x_0) + R_{x_0}(x)$, where Df_{x_0} is the derivative of f at x_0, that is, the linear map represented in standard coordinates by the matrix of partial derivatives of f, and $R_{x_0}(x) = o(\|x - x_0\|)$. Thus, if one fixes $\epsilon > 0$ and takes a small enough parallelepiped Δ centered at x_0, then its image under f lies inside the parallelepiped $f(x_0) + (1 + \epsilon)Df_{x_0}(\Delta - x_0)$ and contains the parallelepiped $f(x_0) + (1 - \epsilon)Df_{x_0}(\Delta - x_0)$. Hence, the volume of a small parallelepiped is approximately preserved if and only if $|\det Df_{x_0}| = 1$. The determinant of Df is sometimes called the *Jacobian* of f and is denoted by Jf.

Now it is easy to deduce the criterion for preservation of phase volume.

Proposition 6.1.3 *Let $O \subset \mathbb{R}^n$ be an open set. A differentiable map $f \colon O \to \mathbb{R}^n$ preserves volume if and only if $|Jf| = \pm 1$.*

Proof If $|Jf| \neq 1$ at some point x_0, then, by the above argument for any sufficiently small parallelepiped, the volume must change. If, on the other hand, $|Jf| = 1$, then, by approximating a set A by a union of parallelepipeds, we can ensure that the volume does not change by more than a fraction ϵ. Since such ϵ can be taken arbitrarily small, volume must be preserved. $\square$

Of course, what we have derived is a particular case of the well-known *change of variables* formula from multidimensional calculus. If we treat the map f as a change of variables and take the characteristic function χ_A, then

$$\operatorname{vol} f(A) = \int_{\mathbb{R}^n} \chi_{f^{-1}(A)}(x)\,dx = \int_{\mathbb{R}^n} \chi_A(y) \det\left(\frac{\partial f}{\partial x}\right)^{-1}(y)\,dy.$$

Thus, if $\det \partial f/\partial x \equiv \pm 1$, then $\operatorname{vol}(f(A)) = \operatorname{vol}(A)$.

Definition 6.1.4 Let $O \subset \mathbb{R}^n$ be an open set. A differentiable map $f \colon O \to \mathbb{R}^n$ is said to preserve orientation if $|Jf| > 0$.

4. Differential Equations. Now consider the case of differential equations. To find the condition for incompressibility of the solutions φ^t of the system $\dot{x} = f(x)$ in $\mathbb{R}^n$, consider how the volume of small parallelepipeds is changed by a small shift $\varphi_{\Delta t}$. As before, we write $\varphi_{\Delta t}(x) = x + \Delta t f(x) + R(x, \Delta t)$, where $R(x, \Delta t) = o(\Delta t)$. Thus, if we are only interested in the change to order Δ, we should consider the Jacobian

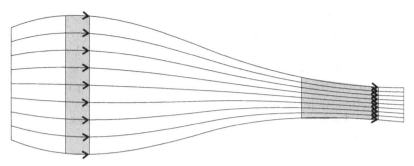

Figure 6.1.2. Incompressibility.

of the map $\tilde{\varphi}(x) = x + \Delta t f(x)$. But then we have

$$\frac{\partial \tilde{\varphi}_i(x)}{\partial x_j} = \delta_i^j + \Delta t \frac{\partial f_i}{\partial x_j},$$

where $(\delta_i^j)_{ij}$ is the identity matrix.

For a vector field $u(x) = (u_1(x_1, \ldots, x_m), u_2(x_1, \ldots, x_m), \ldots, u_m(x_1, \ldots, x_m))$ the *divergence* is defined as

$$(6.1.1) \qquad\qquad \operatorname{div}(u) = \frac{\partial u_1}{\partial x_1} + \frac{\partial u_2}{\partial x_2} + \cdots + \frac{\partial u_m}{\partial x_m}.$$

This appears when we look for terms of order Δt:

$$J\tilde{\varphi} = 1 + \Delta t \sum_{i=1}^{n} \frac{\partial f_i}{\partial x_i} + o(\Delta t) = 1 + \Delta t \operatorname{div} f + o(\Delta t).$$

By differentiating with respect to t at $t = 0$ we obtain

$$\frac{d\operatorname{vol}(\varphi^t(A))}{dt}\Big|_{t=0} = \int_A \operatorname{div} f \, dx,$$

a well-known formula from vector calculus. In particular,

Proposition 6.1.5 *If $\dot{x} = f(x)$ with* $\operatorname{div} f \equiv 0$, *then the flow generated by the vector field f preserves the phase volume.*

Corollary 6.2.3 uses this to provide us with an important natural class of examples of volume-preserving systems.

6.1.2 The Poincaré Recurrence Theorem

Now we show that the preservation of volume by a dynamical system whose phase space has finite total volume implies recurrent behavior. First, we prove a special case of a famous result by Poincaré.[1] We state it for maps on a *domain*, that is, an open set or the closure of an open set.

---

[1] This appeared in the prize memoir by Henri Poincaré, Sur le problème des trois corps et les equations de la dynamique, *Acta Mathematica* **13** (1890), 1–270.

Theorem 6.1.6 *Let X be a domain of finite volume in $\mathbb{R}^n$ or $\mathbb{T}^n$ and $f: X \to X$ an invertible differentiable volume-preserving map. Then for any $x \in X$ and $r > 0$ there exists an $n \in \mathbb{N}$ such that*

$$(6.1.2) \qquad\qquad f^n(B_r(x)) \cap B_r(x) \neq \varnothing.$$

Proof If there are $x \in X$, $r > 0$ such that $f^n(B_r(x)) \cap B_r(x) = \varnothing$ for all $n \in \mathbb{N}$, then $f^{n+k}(B_r(x)) \cap f^k(B_r(x)) = \varnothing$ for all $n, kj \in \mathbb{N}$ since f^k is invertible. Thus all images of $B_r(x)$ are pairwise disjoint and

$$\mathrm{vol}(X) \geq \mathrm{vol}\left(\bigcup_{k=0}^{n-1} f^k(B_r(x))\right) = \sum_{k=0}^{n-1} \mathrm{vol}(f^k(B_r(x))) = n\,\mathrm{vol}(B_r(x))$$

for all $n \in \mathbb{N}$ because f^k preserves volume. Then $\mathrm{vol}(X) = \infty$, since $\mathrm{vol}(B_r(x)) > 0$. This proves the contrapositive. $\square$

Corollary 6.1.7 *Let $f: X \to X$ be an invertible volume-preserving differentiable map, where X is a domain of finite volume in $\mathbb{R}^n$ or $\mathbb{T}^n$. Then for any $x \in X$ there exists a sequence of points $y_k \in X$ and a sequence $m_k \to \infty$ such that $y_k \to x$ and $f^{m_k}(y_k) \to x$ as $k \to \infty$.*

Proof Let $m_0 = 1$. Define y_k and m_k inductively. Applying Theorem 6.1.6 to $f^{-2m_{k-1}}$ yields an $m \in \mathbb{N}$ for which there exists a $y_k \in f^{-2mm_{k-1}}(B_{1/k}(x)) \cap B_{1/k}(x)$. Let $m_k := 2mm_{k-1}$. Then $m_k \to \infty$, $d(x, y_k) < 1/k$ and $d(x, f^{m_k}(y_k)) < 1/k$. $\square$

Now we study the recurrent behavior of individual orbits.

Definition 6.1.8 Let X be a metric space and $f: X \to X$ a continuous map. A point $x \in X$ is said to be *positively recurrent* with respect to f if there exists a sequence $n_k \to \infty$ such that $f^{n_k}(x) \to x$. If f is invertible, then x is said to be negatively recurrent if it is positively recurrent for f^{-1} and *recurrent* if it is both positively and negatively recurrent.

Alternatively one can describe recurrence in terms of the notion of an ω-limit set from Definition 4.3.18: A point is positively recurrent if and only if it is in its own ω-limit set. In general, one cannot expect that, for a volume-preserving map, all points are recurrent, even though this happens in all of the examples that we considered so far, such as rotations of the circle, translations of the torus, or the linear twist. In Section 6.2.2 we consider the mathematical pendulum, which has some nonrecurrent orbits, namely, those on the homoclinic loop, which converge to the unstable equilibrium. In Chapter 7 we encounter volume-preserving systems with much more complicated orbit structure, where many types of behavior, including nonrecurrent ones, coexist. In general, one can only assert the existence of many recurrent orbits in a volume-preserving systems.

Theorem 6.1.9 *Let X be a closed domain of finite volume in $\mathbb{R}^n$ or $\mathbb{T}^n$ and $f: X \to X$ an invertible volume-preserving map. Then the set of recurrent points for f is dense in X.*

Proof Given $\epsilon > 0$ and $N \in \mathbb{N}$ a point, $x \in X$ is said to be (ϵ, N)-*recurrent* if there exists $n > N$ such that $d(f^n(x), x) < \epsilon$. By Corollary 6.1.7, the set of (ϵ, N)-recurrent orbits is dense in X for any $N \in \mathbb{N}$ and $\epsilon > 0$: Given $x \in X$, $\delta, \epsilon > 0$, and $N \in \mathbb{N}$, take k such that $d(y_k, x) < \delta$, $n_k > N$, and $d(y_k, x) + d(f^{n_k}(y_k), x) < \epsilon$. It is also open by continuity of f and its iterates. On the other hand, x is positively recurrent if and only if it is $(1/n, k)$-recurrent for all $n, k \in \mathbb{N}$ [take $n_k > k$ such that $d(f^{n_k}(x), x) < 1/k$ to get $n_k \to \infty$ and $f^{n_k}(x) \to x$]. Thus, the set of all positively recurrent points is the intersection of the open dense sets of $(1/n, k)$-recurrent points for $n, k \in \mathbb{N}$. That this intersection is dense follows from the Baire Category Theorem (Lemma A.1.15). $\square$

We have established the density of positively recurrent points. In the same way we obtain the density of negatively recurrent points by considering $(2^{-k}, N)$-recurrent points for f^{-1}. Finally, the density of recurrent points follows by considering points that are $(2^{-k}, N)$-recurrent for both f and f^{-1}. This is an open dense set, and the (countable) intersection over all $k, N \in \mathbb{N}$ is therefore also dense by Lemma A.1.15.

6.1.3 Uniformity of Recurrence

The kind of recurrence established by Theorem 6.1.9 is irregular in two different ways. First, as we already noticed, it may not be uniform in space: Some points may be recurrent, while others are not. Second, it says nothing about the set of moments for which a point returns approximately. One way to sharpen and specify a notion of recurrence is to ask how regular (or uniform) it is with respect to *time*. So far the only case where we can answer this is that of periodic points. If $x \in X$ is a periodic point with minimal period n, then the orbit $\mathcal{O}(x) = \{x, f(x), \ldots, f^{n-1}(x)\}$ of x is a finite (hence discrete) set consisting of n points; and if r is so small that $d(f^k(x), f^l(x)) > r$ for $k \neq l \in \{0, \ldots, n-1\}$, then $d(f^n(y), y) < r$ implies $f^n(y) = y$ for any point $y \in \mathcal{O}(x)$, and hence the set $\{n \in \mathbb{Z} \mid d(f^n(y), y) < r\}$ is the arithmetic progression $n\mathbb{Z}$.

One of the properties of an arithmetic progression is that it overlaps with any set of consecutive integers that has more elements than the difference n defining the progression. Put differently, all gaps have the same length. One way of relaxing this condition is to impose a bound on the length of the gaps.

Definition 6.1.10 A subset S of $\mathbb{N}$ or $\mathbb{Z}$ is said to be *syndetic* if there is an $N \in \mathbb{N}$ such that $\{n + k \mid 1 \leq k \leq N\} \cap S \neq \varnothing$ for all n.

This property provides a key to studying the uniformity of recurrence for nonperiodic points.

Definition 6.1.11 Let $f : X \to X$ be a continuous map of a metric space X. A point $x \in X$ is said to be *uniformly recurrent* if, for any $r > 0$, the set $\{n \mid d(x, f^n(x)) < r\}$ is syndetic, that is, there exists $N = N(r)$ such that, among any N successive iterates $f^{n+k}(x)$, $k = 0, \ldots, N-1$, there is at least one for which $d(x, f^{n+k}(x)) < r$.

Obviously any periodic point is uniformly recurrent. The proof of Proposition 4.1.1 shows that this holds for an irrational rotation with $N = \lfloor 1/r \rfloor + \epsilon$. Since periodic points are uniformly recurrent, we have

Proposition 6.1.12 *For a circle rotation all points are uniformly recurrent.*

While the periodic case shows that the uniform recurrence of all points does not imply minimality, there is a close connection between the two notions.

Theorem 6.1.13 *Suppose X is compact and $f: X \to X$ is a homeomorphism. Then a point is uniformly recurrent if and only if its orbit closure is a compact minimal set (see Definition 4.1.4).*

Proof Suppose x is uniformly recurrent and let U be a neighborhood of x with compact closure. Then $R := \{n \in \mathbb{Z} \mid f^n(x) \in U\}$ is syndetic, so we can take $N \in \mathbb{N}$ such that among any N successive iterates $f^{n+k}(x)$, $k = 0, \ldots, N-1$, there is at least one for which $f^{n+k}(x) \in U$. Then $\mathcal{O}(x) = \{f^n(x) \mid n \in \mathbb{Z}\} = \{f^{n+k}(x) \mid n \in R, 0 \le k < N\} \subset \bigcup_{k=0}^{N-1} f^n(U) =: U_N$. Since the closure of U_N is compact, so is that of $\mathcal{O}(x)$. Furthermore, for $y \in \mathcal{O}(x)$ this gives $y \in U_N$; hence $\mathcal{O}(y) \cap U \ne \emptyset$. Since U is arbitrary, this implies $x \in \overline{\mathcal{O}(y)}$, which then gives $\overline{\mathcal{O}(y)} = \overline{\mathcal{O}(x)}$.

Conversely, suppose $\overline{\mathcal{O}(x)}$ is a compact minimal set and take a neighborhood U of x. Since $\overline{\mathcal{O}(x)} \smallsetminus \{f^n(U) \mid n \in \mathbb{Z}\}$ is a closed proper invariant subset of $\overline{\mathcal{O}(x)}$, it must be empty by minimality. This shows that $\overline{\mathcal{O}(x)} \subset \{f^n(U) \mid n \in \mathbb{Z}\}$, and the definition of compactness implies that there is a finite subcover, so $\mathcal{O}(x) \subset \overline{\mathcal{O}(x)} \subset f^m(U_N)$ for some $N \in \mathbb{N}$ and $m \in \mathbb{Z}$.

As before, let $R := \{n \in \mathbb{Z} \mid f^n(x) \in U\}$. Now $\mathcal{O}(x) \subset f^m(U_N)$, so for any $i \in \mathbb{Z}$ there is a $y \in U$ and a nonnegative $k < N$ such that $f^i(x) = f^m(f^k(y))$. Thus $f^{i-m-k}(x) \in U$ and $i - m - k \in R$. Since m is fixed and $0 \le k < N$, this proves that R is syndetic, as required. $\square$

Uniform recurrence implies that closures partition the phase space:

Proposition 6.1.14 *Suppose X is compact and $f: X \to X$ is a homeomorphism. Then the orbit closures define a partition of X by compact sets if and only if every point is uniformly recurrent.*

Proof By Theorem 6.1.13 and Exercise 4.1.8, uniform recurrence of all points implies that orbit closures are disjoint or equal (as well as compact). Conversely, the disjoint-or-equal dichotomy implies the minimality of orbit closures, so compactness and Theorem 6.1.13 imply the uniform recurrence of all orbits. $\square$

The linear twist (Example 6.1.2) illustrates these results nicely.

■ EXERCISES

■ **Exercise 6.1.1** Prove that an orientation- and volume-preserving (that is, length-preserving) homeomorphism of the circle is a rotation.

■ **Exercise 6.1.2** Show that a volume-preserving map does not have an attracting fixed point.

■ **Exercise 6.1.3** Decide whether the twist $T: S^1 \times [0, 1] \to S^1 \times [0, 1]$, $T(x, y) = (x + f(y), y)$, where the sum is defined modulo 1 and f is differentiable, preserves area.

■ **Exercise 6.1.4** Decide whether the flow generated by the differential equation

$$\begin{pmatrix} \dot{x} \\ \dot{y} \end{pmatrix} = \begin{pmatrix} y \\ -x \end{pmatrix}$$

preserves area.

■ **Exercise 6.1.5** Decide whether the flow generated by the differential equation

$$\begin{pmatrix} \dot{x} \\ \dot{y} \end{pmatrix} = \begin{pmatrix} y \\ -\sin x \end{pmatrix}$$

preserves area.

■ **Exercise 6.1.6** Decide whether the flow generated by the differential equation

$$\begin{pmatrix} \dot{x} \\ \dot{y} \end{pmatrix} = \begin{pmatrix} y \\ -y - \sin x \end{pmatrix}$$

preserves area.

■ **Exercise 6.1.7** Let X be a closed domain of finite volume in $\mathbb{R}^n$ or $\mathbb{T}^n$ and $f : X \to X$ an invertible map with an attracting fixed point. Show that the set of recurrent points for f is not dense in X.

■ **Exercise 6.1.8** Give an example of a metric space and of countably many open dense sets in it whose intersection is empty.

■ **Exercise 6.1.9** Prove the statement of Theorem 6.1.9 for an open domain with compact closure.

■ PROBLEMS FOR FURTHER STUDY

■ **Problem 6.1.10** Let X be a metric space and $f : X \to X$ topologically transitive. Show that the set of points whose orbit is not dense is the union of countably many nowhere dense sets.

■ **Problem 6.1.11** If an interval is represented as a countable union of closed sets, then one of those sets contains an interval.

6.2 NEWTONIAN SYSTEMS OF CLASSICAL MECHANICS

The discovery that mechanical systems are described by differential equations (the force affects the second derivative of the position) and the development of calculus constituted one of the most profound revolutions of human thought and in particular created an enterprise of describing, predicting, and designing physical systems that has been spectacularly successful over the past three centuries. We present some of the methods of describing and solving such systems.

In a mechanical system a set of data such as positions (or configurations; these may include angles) and velocities of its parts describe its state completely in the sense of the following determinacy principle: The present state of a mechanical system determines its future evolution uniquely. [In our terminology: A mechanical system defines a dynamical system on its state space.] If, for example, our mechanical system consists only of a single point mass, then the state is given by the position x in Euclidean space and the velocity $v = \dot{x}$, the derivative of x with respect to time t. In particular, if the entire evolution is determined by these data, that is, x is determined as a function of t, then so is the second derivative $\ddot{x}$ of x. Therefore $\ddot{x}$ is given in terms of t, x, and $\dot{x}$: $\ddot{x} = f(t, x, \dot{x})$. Thus, such mechanical systems are described by differential equations. [The existence and uniqueness of solutions of differential equations (Theorem 9.4.1) in turn then implies the determinacy principle.] In fact, the "state" of a mechanical system is always described in terms of positions and velocities, and accordingly the differential equations that arise in mechanics are always of second order. The set of positions, or configurations, is called the *configuration space*, and the space of states the *state space* or *phase space*.[2]

This section is an excursion into mechanics. This is an important subject, and it belongs here because there are two aspects of mechanical systems that can make their dynamics simpler than systems of other differential equations with as many variables may be: The confinement of orbits to energy levels in conservative systems effectively reduces the dimension, and, on the other hand, dissipation of energy due to friction may make the dynamics asymptotically simple.

We begin with the Newton equation and its basic properties, and introduce some mechanical ideas using the mathematical pendulum. We also discuss the central force problem (and Kepler's second law), which is at the heart of celestial mechanics, from which in turn dynamics derived some of its most important motivation. Finally, we introduce the Lagrangian approach to mechanics, which is related to the principle that the path of a mechanical system through its phase space solves an optimization problem.

6.2.1 The Newton Equation

The central law of classical mechanics is Newton's Law that an external force acting on a mechanical system, such as a point mass, a rigid body, a planet, and so on, causes a proportional change in velocity:

$$f = ma.$$

It describes, for example, the motion in $\mathbb{R}^n$ of a point of mass m under the influence of a force f by giving the acceleration a. It also describes a pendulum, where constraint forces are present.

1. Second-Order Differential equations. The Newton equation gives rise to a second-order differential equation: If the position of the point is taken to be a point $x \in \mathbb{R}^n$, then the acceleration is $a := \ddot{x} = d^2x/dt^2$. If the force f is a function of x only (this rules out friction), then we get the equation

$$m\frac{d^2x}{dt^2} = f(x).$$

---

[2] See also Chapter 6 of Poincare's Science and Hypothesis: *The Foundations of Science; Science and Hypothesis, The Value of Science, Science and Method*, translated by George Bruce Halsted, The Science Press, Lancaster, PA, 1946.

Example 6.2.1 For an apple falling off a tree, the force is the constant gravitational force mg (where g is some $10m/s^2$ or $32ft/s^2$) and hence $\ddot{x} = -g$, where x is the height. (Apple trees are not tall enough for air resistance to matter.)

Integrating twice we get $x(t) = -gt^2/2 + v(0)t + x(0)$.

Example 6.2.2 (Harmonic Oscillator) For a mass attached to a spring and displaced from the equilibrium (rest) position by x, Hooke's law states that the force exerted by the spring is $-kx$, where k is the spring constant (which measures the stiffness of the spring). Thus we obtain $\ddot{x} = -kx$.

We get the solutions $x(t) = a\sin(\sqrt{k}t) + b\cos(\sqrt{k}t)$ by educated guessing and linear combination.

2. Conversion to First Order. To study such second-order systems of differential equations in any generality it makes sense to convert the equations to first order by defining the velocity as an extra independent variable, that is, by setting $v := \dot{x} = (dx/dt) \in \mathbb{R}^n$. Then $md^2x/dt^2 = f(x)$ becomes

$$\frac{d}{dt}x = v$$

$$\frac{d}{dt}mv = f(x).$$

This is an autonomous first-order system of differential equations in the new variables $\binom{x}{v}$. Its general solution defines a dynamical system on $\mathbb{R}^n \times \mathbb{R}^n$ (or a subset) in coordinates (x, v) (Section 9.4.7). These equations have a number of special properties that set them apart from general autonomous systems of differential equations in $\mathbb{R}^{2n}$.

3. Volume Preservation. Since the pertinent derivatives are all zero, the vector field in $\mathbb{R}^{2n}$ defined by a Newtonian equation is divergence free, that is, it has zero divergence [see (6.1.1)].

By Proposition 6.1.5, this implies that Newtonian systems preserve phase volume:

Corollary 6.2.3 *Newtonian systems preserve the volume given by $dx\, dv$ in the phase space.*

4. Energy and Momentum. The quantity $p := mv$ is called *momentum*. The *kinetic energy* is given by $\frac{1}{2}m\langle v, v\rangle$.

If the force f is a gradient vector field $f = -\nabla V := -(\partial V/\partial x_1, \ldots, \partial V/\partial x_m)$, then

(6.2.1) $$\frac{d}{dt}mv = -\nabla V.$$

The function $V: \mathbb{R}^n \to \mathbb{R}$ is called *potential energy* and the *total energy* $H = \frac{1}{2}m\langle v, v\rangle + V$ is preserved because

$$\frac{dH}{dt} = \langle v, m\dot{v}\rangle + \frac{dV}{dt} = \langle v, m\dot{v}\rangle + \langle \dot{x}, \nabla V\rangle = \langle v, m\dot{v} + \nabla V\rangle = 0,$$

and a function with zero derivative along a curve is constant on that curve. This is a useful simple principle in analysis, and it can be used to advantage in connection with continuous-time systems. Energy is also conserved in constrained systems (although momentum may not be). Because of the conservation of energy, these systems are said to be *conservative*.

5. Geodesic Flow. The entire preceding discussion can be carried out for *free particle motion*, that is, when the force is zero. In Euclidean space this yields constant velocity motion along a line. In spaces other than Euclidean space, the resulting motions have constant speed, but the concept of a line has to be generalized to that of a *geodesic*. Therefore, this is also known as the *geodesic flow*. Examples are the sphere whose geodesic flow is constant speed motion along great circles, these being the geodesics on the sphere, and the torus, where the motion is along projections of lines. Geometrically and physically one can determine geodesic on a surface in $\mathbb{R}^3$ from a direction by intersecting the surface with the plane spanned by the normal vector and the desired direction. This geometry corresponds to the fact that, for motion along a geodesic with constant speed, the acceleration is perpendicular to the surface because the only available force is the constraint force that keeps a particle on the surface. This constraint force is orthogonal to the surface because any tangent component would result in a net force on the particle in the surface.

Geodesic flows appear in a few other places in this book, such as Section 5.2.2 and Section 6.2.8.

6.2.2 The Mathematical Pendulum

As an example consider a pendulum consisting of a point mass in the plane attached by a rod to a fixed joint, like the pendulum in a grandfather clock.

1. The Model. If we take $2\pi x$ to be the angle of deviation from the vertical, then the pendulum is subject to a downward gravitational force mg (where m is the mass of the pendulum and g is the gravitational acceleration of $9.81 m/s^2$) whose angular component is $-mg \sin 2\pi x$. Equating this with mass times acceleration, that is, with $m \cdot 2\pi L\ddot{x}$, shows that the pendulum is described by the differential equation

$$2\pi mL\ddot{x} + mg \sin 2\pi x = 0.$$

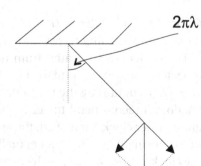

Figure 6.2.1. The mathematical pendulum.

2. Nondimensionalizing. It is often useful to unclutter such differential equations by *nondimensionalizing* them, that is, by choosing, for example, a time scale in such a way that the coefficients of the differential equation are dimensionless, and maybe fewer in number as well. The first step to this end is to pick some time T to be determined later in such a way that the derivatives of x in τ are roughly of size one when we replace the time t by a dimensionless time $\tau := t/T$. Note that by the chain rule $dx/dt = dx/d\tau \, d\tau/dt = (1/T)dx/d\tau$ and likewise $d^2x/dt^2 = (1/T^2)d^2x/d\tau^2$. Thus, the differential equation becomes

$$2\pi \, mL/T^2 \frac{d^2x}{d\tau^2} + mg \sin 2\pi x = 0.$$

Both terms are forces and will therefore become dimensionless if we divide by the force mg, yielding

(6.2.2)
$$\frac{2\pi L}{gT^2} \frac{d^2x}{d\tau^2} + \sin 2\pi x = 0.$$

The dimensionless coefficient $2\pi L/(gT^2)$ should be of order one if $d^2x/d\tau^2$ is to be of order one, so we take $T = \sqrt{2\pi L/g}$ and finally obtain the differential equation

$$\ddot{x} + \sin 2\pi x = 0,$$

where the dot now denotes differentiation with respect to τ. Physically, the choice of T is natural, because it is directly related to the period of the harmonic oscillator that arises from linearizing the mathematical pendulum (Section 6.2.2.7).

3. Conversion to First Order. This differential equation is equivalent to the system of first-order differential equations

(6.2.3)
$$\dot{x} = v,$$
$$\dot{v} = -\sin 2\pi x$$

for $x \in S^1$, $v \in \mathbb{R}$. A particular interest of this elementary example lies in the fact that, due to the periodicity of the angular coordinate (but not the velocity!), the phase space is a cylinder $S^1 \times \mathbb{R}$.

The total energy is given by $H(x, v) = (1/2)v^2 - (1/2\pi) \cos 2\pi x$. Since there is a constraint, it is useful to explicitly verify energy conservation: $dH/dt = v\dot{v} + (1/2\pi) \sin 2\pi x \dot{x} = 0$. Thus the orbits are on level curves $H = $ const.

4. Orbits. For $-1/2\pi < H < 1/2\pi$, each energy level consists of a single closed curve corresponding to oscillations around the stable equilibrium $(x, v) = (0, 0)$. Those orbits are separated from higher-energy orbits corresponding to full and repeated rotations around the joint by a *homoclinic loop* (see Definition 2.3.4) with $H = 1/2\pi$ containing the unstable equilibrium $(x, v) = (1/2, 0)$. (See Figure 6.2.2.) We do not recommend to verify these homoclinic orbits with the pendulum in a grandfather clock. First of all, these orbits are so unstable that (because of friction, however small) one does not actually observe them, and, second, you would break the clock. For $H > 1/2\pi$, each energy level consists of two orbits, corresponding to rotation in opposite directions. Thus almost all orbits of this system are periodic,

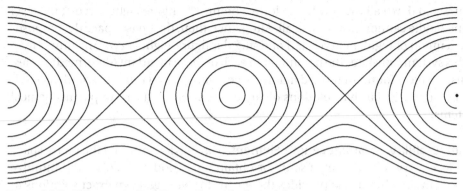

Figure 6.2.2. Phase portrait of the mathematical pendulum.

but there are two distinct families of these. The lower-energy orbits encircle the equilibrium $(0, 0)$. Any two of these can be continuously deformed into each other. The high-energy orbits go around the cylinder (like a rubber band around a rolled-up poster), and these can be deformed into each other also. But none of the lower-energy orbits can be deformed into any high-energy orbit. Accordingly, these two families are separated by the *singular* orbits homoclinic to the unstable equilibrium $(1/2, 0)$.

5. The Global Picture. Qualitatively, this situation corresponds to the situation of a cylinder bent into the shape of a U-tube (like the drain pipe under a sink) with H being the height function (Figure 6.2.3). The level curves are then horizontal slices. The stable equilibrium $(0, 0)$ corresponds to the lowest point on this tube, and the unstable equilibrium is the saddle of the pipe. The energy levels above $1/2\pi$ consist of pairs of closed curves above the saddle that can be slid up and down freely, but not to below the saddle. The lower-energy levels are below the saddle and can be

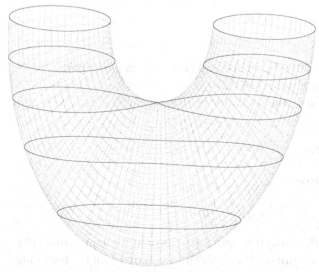

Figure 6.2.3. Energy as height function.

contracted down all the way to the lowest point. The figure eight curve at the level of the saddle cannot be moved at all. Altogether, then, the phase space decomposes into a union of regular curves and one singular figure-eight curve. Any time-t map of this system translates the parameter in each of these invariant (deformed) circles. Thus, we have found that the dynamics decomposes into rotations of (deformed) circles, in particular that rotations of circles occur naturally in simple mechanical systems.

6. Integrability, Invariant Length Element. Having decomposed the phase space into level curves is not only useful for obtaining a good qualitative and intuitive understanding, but it also provides the means for writing down exact solutions in terms of functions obtainable from elementary functions by integration, inversion, and algebraic operations. For example, in this case the solutions involve elliptic integrals, which cannot be expressed directly in terms of elementary functions.

Proposition 6.2.4 provides a mechanism for producing the solutions. To briefly describe it in our context, we can write the standard area element in the phase space as $dHdl$, where dH is flow invariant since H is flow invariant, and dl is the length element along the curves $H = $ const., divided by $\|\nabla H\|$. By (6.2.3), $\|\nabla H\|$ is the speed of motion along $H = $ const., so dl is also flow invariant and hence the flow also preserves the area. As we noted in Section 6.1, the last fact is equivalent to the vanishing of divergence and hence is common to all Newtonian systems.

7. Linearization. This is a good opportunity to see how linearization affects a phase portrait, or how well the linearized system reflects the actual dynamics. Since linearization is supposed to provide a good local approximation of the behavior of the nonlinear system, it is of most interest near an equilibrium. We consider motions near the stable equilibrium, that is, *small oscillations*.

We mean to compare our picture to the explicit solutions of the differential equation

$$\dot{x} = v$$

$$\dot{v} = -2\pi x$$

obtained from (6.2.3) by replacing the right-hand side with its linear part at $(0, 0)$. This is the *harmonic oscillator* (Example 6.2.2).

The total energy of the harmonic oscillator is $H(x, v) = v^2/2 + \pi x^2$. Note that for the Hamiltonian

$$H(x, v) = \frac{1}{2}v^2 - \frac{1}{2\pi}\cos 2\pi x$$

the second-order Taylor expansion is

$$\frac{1}{2}v^2 - \frac{1}{2\pi}(1 - \frac{1}{2}(2\pi x)^2) = \frac{1}{2}v^2 - \frac{1}{2\pi} + \pi x^2,$$

that is, the same as that of the linearized system (up to the additive constant $-1/2\pi$, which changes neither level sets nor derivatives). The level curves for the harmonic oscillator are $v^2/2 + \pi x^2 = $ const., that is, ellipses centered at the origin. From the explicit solution

$$A \begin{pmatrix} \sin(\sqrt{2\pi}\,(t+c))/\sqrt{\pi} \\ \cos(\sqrt{2\pi}\,(t+c))\sqrt{2} \end{pmatrix}$$

we know that all solutions have the same period.

Qualitatively, this simple phase portrait looks much like that for the mathematical pendulum near the origin and accordingly the linearized system provides a sensible approximation here. Note, however, two differences. Away from the origin the phase portraits become qualitatively different due to the presence of a second equilibrium and the homoclinic loop for the mathematical pendulum. Furthermore, the periodic solutions encircling $(0, 0)$ have longer periods for larger energy, because the speed near the unstable equilibrium is small. In particular, the nonlinear system does not have constant periods of the solutions near $(0, 0)$, that is, the period depends on the amplitude of the oscillation. This is one of the issues that affects the design of pendulum clocks. The amplitude has to be kept constant (or the pendulum needs to be equipped with cycloidal cheeks to remove the dependence on amplitude).

Note that the phase space of the mathematical pendulum and harmonic oscillator decomposes into invariant (deformed) circles and the time-one map of the dynamics simply shifts the parameter on each circle by a constant amount depending on the circle. Thus, a detailed qualitative description of this time-1 map is naturally obtained by studying each of these circles separately, hence by studying circle rotations (Section 4.1.1).

6.2.3 Invariant Volume on Energy Levels

In addition to preserving phase volume, as we just noted, Newtonian systems also preserve the level sets of the Hamiltonian energy function H. In Section 6.2.2.6 we observed that in the case of the mathematical pendulum the flow also preserves a length parameter on energy level curves. We obtained this directly from the equations of motion (6.2.3), and used it to deduce area preservation. Now we show that any map that preserves volume and the level sets of a function also preserves a volume on these level sets. To see the idea most clearly let us first study the two-dimensional case with some care, which will in particular illuminate our conclusions about the pendulum.

Proposition 6.2.4 *Suppose $f \colon \mathbb{R}^2 \to \mathbb{R}^2$ preserves area and H is an invariant function, that is, $H(f(p)) = H(p)$ for all $p \in \mathbb{R}^2$. Then each level set that is non-critical, that is, contains no critical points, can be decomposed into curves $c_z :=$ $\{p \mid H(p) = z\}$ each of which can be parametrized as $c_z(t)$ such that $f(c_z(t)) = c_z(t + s(z))$ for some function s depending only on the level set. In other words, with respect to this parameter f acts on each curve like a translation.*

Proof We parametrize c_z in such a way that $\|c_z'(t)\|\,\|DH_{c_z(t)}\| = 1$ for a all t. Consider a unit tangent vector v and the unit normal vector $w = DH/\|DH\|$ to c_z at a point $p = c_z(t)$. Then the area of the rectangle P spanned by ϵv and ϵw is ϵ^2. As illustrated in Figure 6.2.4, its image $f(P)$ under f has area

Figure 6.2.4. Motion on level sets.

(up to a small relative error that goes to zero as $\epsilon \to 0$) equal to that of the parallelogram at $f(p) = c_z(\tilde{t})$ spanned by the images $\epsilon v'$ and $\epsilon w'$ of ϵv and ϵw under Df. With $\epsilon v'$ as the base of the parallelogram, its height δ is the length of the projection of $\epsilon w'$ to the normal to c_z at $f(p)$. By linear approximation

$$H(p + \epsilon w) \approx H(p) + \epsilon \|DH_p\|$$

$$H(f(p) + \epsilon w') \approx H(f(p)) + \delta \|DH_{f(p)}\|.$$

Since f preserves H, the left sides are equal, as are the first terms on the right. Therefore $\delta \approx \epsilon \|DH_p\|/\|DH_{f(p)}\|$ and $f(P)$ has area $\delta \epsilon \|v'\| \approx \|v'\| \epsilon^2 \|DH_p\|/ \|DH_{f(p)}\|$, which by area preservation must equal ϵ^2. This implies that $\|v'\| \|DH_{f(p)}\| = \|DH_p\|$ and hence $v'\|DH_{f(p)}\| = v\|DH_p\|$. (Note that ϵ does not appear in these last equations, and that they are therefore exact.) If we take the unit vector $v = c_z'(t)\|DH_p\|$, then $v' = Df(c_z'(t))\|DH_p\|$ and thus

$$c_z'(\tilde{t})\frac{d\tilde{t}}{dt} = \frac{d}{dt}c_z(\tilde{t}) = Df(c_z'(t)) = \frac{v'}{\|DH_p\|} = \frac{v}{\|DH_{f(p)}\|} = \frac{c_z'(t)\|DH_p\|}{\|DH_{f(p)}\|} = c_z'(\tilde{t})$$

because $c_z'\|DH\| = 1$ everywhere. This evidently implies that $d\tilde{t}/dt = 1$ and hence $\tilde{t} = t + s(z)$ for some function s depending only on z, that is, depending only on the invariant curve. $\square$

We note again that this description corresponds exactly to the situation encountered with the mathematical pendulum. A further common feature is that here, too, the calculations are entirely local. This shows that it is not necessary to have a map defined on all of $\mathbb{R}^2$, nor a function H defined on the entire phase space. With appropriate qualifications, then, our result is valid for a map of an open subset of $\mathbb{R}^2$, say, that preserves area and a function H on this open set. Likewise, we can also apply this result to maps of the cylinder $S^1 \times \mathbb{R}$, for example, because these local calculations work just as well in that setting. The basic reason can be explained geometrically: Consider a *flow box*, that is, a small area swept out by flowing a local transversal for a small amount of time. If one follows this flow box further along to an area where level sets are close together (large gradient of H), then these level sets "squeeze" the flow box in the transverse direction, and so it has to elongate by area preservation (see Figure 6.1.2). This elongation corresponds to increased speed in regions of large gradient of H. This picture works in higher dimension as well, but now the elongation corresponds to volume expansion on level sets.

Theorem 6.2.5 *If* $f: \mathbb{R}^n \to \mathbb{R}^n$ *is volume-preserving and* H *is an invariant function, then on each level set one can define a volume function* Ivol *that is invariant under* f. *That is, if for any open subset* O *of a noncritical level set we define* $\mathrm{Ivol}(O) := \int_O 1/\|\operatorname{grad} H\|$, *then* $\mathrm{Ivol}(O) = \mathrm{Ivol}(f(O))$ *for any such* O.

Proof Analogously to before pick a point p (not critical for H) in a level set $A_h := \{p \mid H(p) = h\}$ of H and take an orthonormal set $v_1, \ldots, v_{n-1}$ of tangent vectors to A_h at p. Let v_n be a unit normal vector to A_h at p (in the same direction as $\operatorname{grad} H$). Consider the parallelepiped P spanned by $\epsilon v_1, \ldots, \epsilon v_n$ of volume ϵ^n and denote the parallelepiped spanned by $\epsilon v_1, \ldots \epsilon v_{n-1}$ by Q. The image $f(P)$ under f of P is essentially the parallelepiped spanned by $\epsilon Df v_1, \ldots, \epsilon Df v_n$. The volume of this latter parallelepiped does not change if we replace $\epsilon Df v_n$ by its projection to the normal vector v'_n to A_h at $f(p)$ (because volume equals base times height). Denote the length of this projection by δ. By linear approximation we have

$$H(p + \epsilon v_n) \approx H(p) + \epsilon\|\operatorname{grad} H_p\|$$

$$H(f(p) + v'_n) \approx H(f(p)) + \delta\|\operatorname{grad} H_{f(p)}\|.$$

The left sides and the first terms on the right are equal, so (up to small error) we have $\delta = \epsilon\|\operatorname{grad} H_p\|/\|\operatorname{grad} H_{f(p)}\|$. Since the volume $\delta \operatorname{vol}(f(Q))$ of $f(P)$ is equal to the volume $\epsilon \operatorname{vol}(Q)$ of P, we must have

$$\mathrm{Ivol}(Q) = \frac{\operatorname{vol}(Q)}{\|\operatorname{grad} H_p\|} = \frac{\operatorname{vol}(f(q))}{\|\operatorname{grad} H_{f(p)}\|} = \mathrm{Ivol}(f(Q)). \qquad \square$$

6.2.4 Constants of Motion

The key observation for our analysis of the mathematical pendulum was the fact that the total energy was preserved, that is, it is a *constant of motion*, or a *first integral*. Hence the two-dimensional phase space of the system splits into invariant one-dimensional-level curves of constant energy. On each such curve there are only several simple possibilities for the behavior of solutions since we are essentially dealing with first-order autonomous differential equations. Namely, for *regular* energy levels, that is, noncritical values of the total energy, the vector field does not vanish. Hence, if a solution is bounded, it must be periodic; and if it is unbounded, it goes to infinity along the particular noncritical level curve of energy. As we saw in the pendulum case, the nonconstant solutions on a critical energy level are attracted asymptotically, as time goes to $+\infty$ and to $-\infty$, to (possibly different) constant solutions (compare with the discussion in Section 2.3). While a pathological situation is also possible in principle when a critical solution does not converge to a fixed point but wanders near a whole curve of fixed solutions instead, this situation does not appear in natural models. Thus for Newtonian systems with one degree of freedom the simple description above gives a fairly complete qualitative analysis of the orbit behavior.

6.2.5 Central Forces

Among the main subjects of classical mechanics is that of celestial mechanics, that is, the description of the motion of the planets around the sun, or of the moons

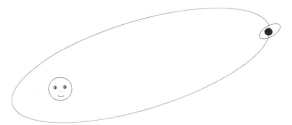

Figure 6.2.5. Central force.

around the planets and such like. Its simplest model has two bodies moving freely, but subject to mutual gravitational attraction. One may either pass to coordinates centered at the center of mass of the system or assume that one of them (the sun) is much heavier than the other and hence essentially stationary (or rather, moving with constant velocity). Either way one can write the position of the second body (the planet) as $x \in \mathbb{R}^3 \setminus \{0\}$ and its velocity as $v \in \mathbb{R}^3$. The potential energy of the gravitational field is given by $V(x) = -1/\|x\|$, so Newton's equation becomes

$$\ddot{x} = \nabla \frac{1}{\|x\|} = -\frac{x}{\|x\|^3} \qquad \text{or} \qquad \begin{aligned} \dot{x} &= v, \\ \dot{v} &= -\frac{x}{\|x\|^3}. \end{aligned}$$

The kinetic energy is $\langle v, v \rangle / 2$, as usual. Thus the total energy is $E(x, v) = \langle v, v \rangle / 2 - 1/\|x\|$. It is conserved since our equations have the form (6.2.1). There are other constants of motion, namely, the components of *angular momentum* $x \times v = (x_2 v_3 - x_3 v_2, x_3 v_1 - x_1 v_3, x_1 v_2 - x_2 v_1)$. To check this, note, for example, that

$$(6.2.4) \qquad \frac{d}{dt}(x_1 v_2 - x_2 v_1) = \dot{x}_1 v_2 + x_1 \dot{v}_2 - \dot{x}_2 v_1 - x_2 \dot{v}_1$$

$$= v_1 v_2 - \frac{x_1 x_2}{\|x\|^3} - v_2 v_1 + \frac{x_2 x_1}{\|x\|^3} = 0.$$

(See also Lemma 6.2.6.) We will describe the dynamics by explicitly solving the equations of motion. Since $v \perp x \times v$, the motion is in a plane perpendicular to $x \times v$. Thus for any given direction of $x \times v$ the problem reduces to a problem in $\mathbb{R}^2 \setminus \{0\}$, that is, with $x_3 = v_3 = 0$ after a suitable coordinate change.

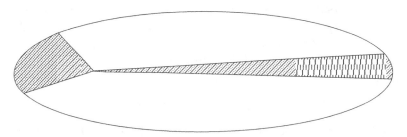

Figure 6.2.6. Kepler's Second Law.

In passing we note that $x_1 v_2 - x_2 v_1$ is twice the area of the triangle with vertices 0, x, $x + v$. Thus $x_1 v_2 - x_2 v_1$ is twice the derivative of the area swept out by x. The fact of this being constant is known as Kepler's Second Law: The ray from the sun to a planet sweeps out equal areas in equal amounts of time. If $A := x_1 v_2 - x_2 v_1 \neq 0$, then we can show that the orbits are on conics. Recall from analytic geometry that in polar coordinates conics are given by $r = ed/(1 + e \cos(\theta - \theta_0))$ with eccentricity $e \in (0, 1)$ for ellipses, $e = 1$ for parabolas, and $e > 1$ for hyperbolas. If we write $r = \|x\|$, then

$$\frac{d}{dt}\left(\frac{x_1}{r}\right) = \frac{v_1 r^2 - x_1 \langle x, v \rangle}{r^3} = -(x_1 v_2 - x_2 v_1)\frac{x_2}{r^3} = A\dot{v}_2,$$

so $A v_2 = x_1/r + C$ for some $C \in \mathbb{R}$. Likewise, $A v_1 = -x_2/r - D$. Then

$$C x_1 + D x_2 + r = A x_1 v_2 - \frac{x_1^2}{r} - A x_2 v_1 - \frac{x_2^2}{r} + r = A(x_1 v_2 - x_2 v_1) = A^2,$$

and in polar coordinates $x_1 = r \cos\alpha$, $x_2 = r \sin\alpha$ one has

(6.2.5)
$$r(\alpha) = \frac{r A^2}{r + C x_1 + D x_2} = \frac{A^2}{1 + C \cos\alpha + D \sin\alpha}$$

$$= \frac{A^2}{1 + \sqrt{C^2 + D^2}\cos(\alpha - \beta)},$$

where $\cos\beta = C/\sqrt{C^2 + D^2}$ and $\sin\beta = D/\sqrt{C^2 + D^2}$, that is, β is such that $r(\beta)$ is minimal (the *perihelion angle*). Equation (6.2.5) is the equation of a conic with eccentricity $e = \sqrt{C^2 + D^2}$, which is determined by E and A, that is, the values of energy and angular momentum:

(6.2.6)
$$e^2 = C^2 + D^2 = \left(\frac{A v_2 - x_1}{r}\right)^2 + \left(\frac{A v_1 + x_2}{r}\right)^2$$

$$= \frac{x_1^2 + x_2^2}{r^2} + 2A^2 \frac{v_1^2 + v_2^2}{2} - 2A\frac{x_1 v_2 - x_2 v_1}{r} = 1 + 2EA^2.$$

Thus the orbit is an ellipse if $E < 0$, a hyperbola if $E > 0$, and a parabola if $E = 0$. In qualitative terms we should emphasize two main properties of the solutions to the central force problem. All bounded orbits are periodic (elliptic orbits). All unbounded orbits go to infinity in both positive and negative time (hyperbolic and parabolic orbits). This simple dichotomy is a specific property of the gravitational potential; that is, it depends crucially on the power of $r = \|x\|$ that appears in the potential V. For other powers of r bounded orbits tend to be nonperiodic. In fact, as a consequence of the theory of general relativity, the planet Mercury effectively experiences a slightly different power of r in the potential. Accordingly, its perihelion angle changes slowly over time, that is, its orbit is essentially an ellipse but does not quite close up. This slow drift of the perihelion angle is called precession. In fact, the interaction with Venus produces some drift that can be perfectly accounted for by Newtonian theory. However, observational accuracy was good enough in the nineteenth century to detect a further drift: There is a difference between observed precession (5.70"/year) versus precession attributable to planetary interaction by Newtonian gravity (5.27"/year). (It helps that the orbit is more elliptic than others in the solar system: The eccentricity is 0.2056, and the perihelion and aphelion distances are $4.59 \cdot 10^7$km and $6.97 \cdot 10^7$km, respectively.) General relativity produces the minute required correction.[3]

6.2.6 Harmonic Oscillator

A simple central force problem is given by the potential $V(x) = \|x\|^2$ in the plane. Newton's equation becomes

$$\ddot{x} = -\nabla \|x\|^2 = -x,$$

and the components decouple into a harmonic oscillators. Therefore the solutions are independent oscillations with the same frequency in either coordinate. The planar orbits accordingly trace out ellipses centered at the origin.

Only the inverse-square and square potential allow periodic solutions. This was one ingredient in Newton's deduction that gravity is (at least very nearly) given by an inverse-square potential force.

6.2.7 Spherical Pendulum

A simple-looking central force system is given by the spherical pendulum, that is, a point mass attached to a point by a rod and subject to gravity. (See Figure 6.2.7.) The equations of motion are easy to write if we note that we can use the potential energy V as in Section 6.2.1.4. The potential energy is given by the height of the mass above its rest position, which is $U(x) = 1 - \sqrt{1 - x_1^2 - x_2^2}$. In this case a second integral of motion independent of energy is the angular momentum with respect to the vertical axis, that is, the third coordinate of the angular momentum. (This is related to the natural rotational symmetry of the system.) To describe the motion for fixed values of both integrals we use polar coordinates. These are adapted to the rotational symmetry and to the fact that the force $-\nabla U$ is radial, that is, directed toward the origin. We write $x = (x_1, x_2) = (r\cos\theta, r\sin\theta)$.

Lemma 6.2.6 *In a central force field, angular momentum is preserved.*

Proof Angular momentum is defined as the cross product $M := x \times \dot{x}$. By the product rule, $\dot{M} = \dot{x} \times \dot{x} + x \times \ddot{x} = x \times \ddot{x} = 0$, because in a central force field x and $\ddot{x}$ are collinear. $\square$

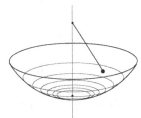

Figure 6.2.7. The spherical pendulum.

[3] Albert Einstein, Erklärung der Perihelbewegung des Merkur aus der allgemeinen Relativitätstheorie, *Sitzungsberichte der königlich preußischen Akademie der Wissenschaften* **XLVII** (1915), 831–839.

To express angular momentum in polar coordinates we choose a time-dependent basis of a radial unit vector v_r and an angular unit vector v_θ perpendicular to it and pointing in the direction of increasing θ. Then $\dot{v}_r = \dot{\theta} v_\theta$ and $\dot{v}_\theta = -\dot{\theta} v_r$, so $\dot{x} = (d/dt)(\|x\| v_r) = d\|x\|/dt\, v_r + \|x\| \dot{v}_r = \dot{r} v_r + r\dot{\theta} v_\theta$, and hence the angular momentum is

$$M = x \times \dot{x} = x \times \dot{r} v_r + x \times r\dot{\theta} v_\theta = r\dot{\theta} x \times v_\theta = r^2 \dot{\theta} v_r \times v_\theta.$$

By Lemma 6.2.6, $r^2 \dot{\theta}$ is constant.

We can use this to reduce the problem by finding an equation of motion for r that does not involve θ.

Differentiating $\dot{x} = \dot{r} v_r + r\dot{\theta} v_\theta$ by using $\dot{v}_r = \dot{\theta} v_\theta$ and $\dot{v}_\theta = -\dot{\theta} v_r$ gives

$$-\frac{\partial U}{\partial r} v_r = -\nabla U = \ddot{x} = (\ddot{r} - r\dot{\theta}) v_r + (2\dot{r}\dot{\theta} + r\ddot{\theta}) v_\theta,$$

so $\ddot{r} - r\dot{\theta} = -\partial U/\partial r$ and $2\dot{r}\dot{\theta} + r\ddot{\theta} = 0$. Inserting $\dot{\theta} = \|M\|/r^2$ (angular momentum) gives $\ddot{r} = -\partial U/\partial r + \|M\|/r^3$, the promised equation without θ. Since $U = 1 - \sqrt{1 - r^2}$, we have $\partial U/\partial r = r/\sqrt{1 - r^2}$, and consequently $\ddot{r} = (\|M\|/r^3) - (r/\sqrt{1 - r^2})$, where $\|M\|$ is determined from initial conditions.

This is the desired equation of motion for r alone.

6.2.8 The Lagrange Equation and the Variational Approach
Using

$$(6.2.7) \qquad\qquad L(x, v) = \frac{1}{2} m \langle v, v \rangle - V(x),$$

the Newton equation (6.2.1) becomes

$$(6.2.8) \qquad\qquad \frac{d}{dt} \frac{\partial L}{\partial v} = \frac{\partial L}{\partial x}.$$

This is called the *Lagrange equation* or the *Euler–Lagrange equation*. One reason Lagrange introduced his formalism was that using $f = ma$ as described earlier can become rather laborious when one considers constrained systems. For example, a three-dimensional mathematical pendulum consists of a mass attached by a rod to a fixed point and thus constrains the mass point to a sphere (see Section 6.2.7). To deal with this, one has to develop notions of constraint forces – forces that are at all times just such that the system will obey the constraint. Here Lagrange's approach greatly simplifies the problem, because it is coordinate-independent.

Theorem 6.2.7 *If L is a smooth function of $(x, v) \in \mathbb{R}^n \times \mathbb{R}^n$, $x, y \in \mathbb{R}^n$, and $T > 0$, define the Lagrange action functional*

$$(6.2.9) \qquad\qquad F(c) := \int_0^T L(c(t), \dot{c}(t))\, dt$$

on parametrized smooth curves $c \colon [0, T] \to \mathbb{R}^n$ with $c(0) = x$, $c(T) = y$. A curve c is a critical point of F if and only if c satisfies (6.2.8).

Proof If L is a smooth function of $(x, v) \in \mathbb{R}^n \times \mathbb{R}^n$, $x, y \in \mathbb{R}^n$, and $T > 0$, consider smooth curves $c \colon [0, T] \to \mathbb{R}^n$ with $c(0) = x$, $c(T) = y$. Then the

Lagrange action functional (6.2.9) is well defined. To find a curve c so that $F(c)$ is minimal consider curves $c_s : [0, T] \to \mathbb{R}^n$ depending smoothly on $s \in (-\epsilon, \epsilon)$ such that $c_0 = c$ and $c_s(0) = x$, $c_s(T) = y$. Then $F(c_s)$ is a real-valued function of s and, if $F(c_0)$ is minimal, then c is a critical point of F because for any such curves c_s integration by parts gives

$$
\begin{aligned}
0 = \frac{d}{ds} F(c_s)|_{s=0} &= \frac{d}{ds}|_{s=0} \int_0^T L(c_s(t), \dot{c}_s(t)) dt \\
&= \int_0^T \left(\frac{\partial L}{\partial x} \frac{dc_s}{ds}|_{s=0} + \frac{\partial L}{\partial v} \frac{d}{ds}|_{s=0} \dot{c}_s(t) \right) dt \\
&= \left[\frac{\partial L}{\partial v} \frac{dc_s}{ds}|_{s=0} \right]_0^T - \int_0^T \left(\frac{d}{dt} \frac{\partial L}{\partial v} - \frac{\partial L}{\partial x} \right) \frac{dc_s}{ds}|_{s=0} dt \\
&= - \int_0^T \left(\frac{d}{dt} \frac{\partial L}{\partial v} - \frac{\partial L}{\partial x} \right) \frac{dc_s}{ds}|_{s=0} dt,
\end{aligned}
$$

using $(dc_s/ds)|_{s=0} = 0$ for $t = 0, T$. The last integral vanishes regardless of the values of dc_s/ds along c_0. Then

$$
\frac{d}{dt} \frac{\partial L}{\partial v} - \frac{\partial L}{\partial x} = 0
$$

because otherwise there is a time $t \in (0, T)$ such that this expression is nonzero at $c_0(t)$; choosing c_s such that

$$
\frac{dc_s}{ds}(t) = \frac{d}{dt} \frac{\partial L}{\partial v}(c_0(t)) - \frac{\partial L}{\partial x}(c_0(t)) \qquad \text{and} \qquad \frac{dc_s}{ds} = 0
$$

outside a small neighborhood of t makes this integral nonzero, contrary to our choice of c_0 as a critical point.

Thus the Lagrange equation (6.2.8) arises from minimizing integrals along curves, and, if L is chosen as in (6.2.7), then the critical points are exactly the solutions of Newton's equation. $\square$

Thus, solving the Lagrange equation (6.2.8) – and hence describing Newtonian dynamics – amounts to solving a variational problem, that is, finding critical points of a certain functional. This corresponds to the heuristic principle that many natural processes are optimized in some way. The natural action functional that gives (6.2.8) is defined on an infinite-dimensional space. That leads to considerable technical complications. Therefore, we do not put this approach to use in this setting. In the discrete-time situation of billiards, which were introduced in Section 6.3, however, we are led to consider an action function of finitely many variables, and we will gather significant information from this variational approach. This theme is developed further in Chapter 14.

Consider a freely moving particle where the Lagrangian has no term corresponding to potential energy. The Lagrange equation (6.2.8) implies that orbits minimize action, which is related to energy. This happens to imply that orbits also minimize the length between any of its points (so long as these are not too far apart). Therefore the orbits are *geodesics*, that is, curves that locally minimize length. Intuitively this corresponds to the fact that a freely moving particle follows the next best thing to a straight line, and geodesics are the "straight lines" of a curved space. The resulting flow is called the *geodesic flow*.

■ EXERCISES

■ **Exercise 6.2.1** A pebble is dropped in a well and hits the bottom in one second. How deep is the well?

■ **Exercise 6.2.2** Consider a particle that is subject to gravity but no other force. Set up the Newton equation for its coordinates (x, y, z) as functions of time (where z is the height) and solve the resulting differential equation.

■ **Exercise 6.2.3** A football is punted with an initial vertical velocity of 30 m/s. How high does it fly?

■ **Exercise 6.2.4** Prove Lemma 6.2.6 using computations in coordinates like (6.2.4).

■ **Exercise 6.2.5** Using elementary spherical geometry, describe the dynamics of the geodesic flow on the round sphere (that is, the unit sphere in $\mathbb{R}^3$).

■ PROBLEMS FOR FURTHER STUDY

■ **Problem 6.2.6** Consider the system of n point masses in $\mathbb{R}^3$ whose pairwise interaction depends only on their mutual distances, that is, $V(x) = \sum V_{ij}(\|x_i - x_j\|)$. Show that the coordinates of the velocity of the center of gravity and of the angular momentum are first integrals (that is, constants of motion).

■ **Problem 6.2.7** (Two-Body Problem in the Plane)   Show that for the system of two point masses in the plane with interaction as in the previous exercise the four integrals (energy, angular momentum, and coordinates of the velocity of the center of gravity) are independent. Describe the motion relative to the center of gravity.

■ **Problem 6.2.8** Obtain the solutions for the mathematical pendulum using the recipe described in Section 6.2.2.6.

■ **Problem 6.2.9** Prove that the linear and inverse-square central forces are the only ones for which all orbits are closed.

6.3 BILLIARDS: DEFINITION AND EXAMPLES

In Section 4.2.5 we studied a class of dynamical systems that can be viewed as either mechanical or optical. The mechanical model is that of a particle moving in a confined region and bouncing elastically off the walls. Hence such systems are called billiard flows. In Section 4.2.5 they arose from a simple two-particle system on an interval. There are many more situations when billiards arise; but independently of whether they are obtained from a concrete model, studying

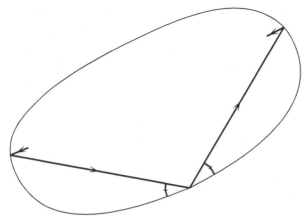

Figure 6.3.1. A billiard.

billiards is very illuminating for the following reason:

> In this problem the formal side, usually so formidable in dynamics, almost completely disappears, and only the interesting qualitative questions need to be considered.[4]

Thus they are more feasible to study in detail, yet they are representative of the dynamical complexity of less tractable systems. While in Section 4.2.5 a billiard system arose in connection with a physical system of two particles on a line segment, we saw in Section 5.2.3 that this works for any number of particles in an interval.

The main object of this and the subsequent section is to study billiards of a type different from those in Section 4.2.5 or even Section 5.2.3. These are convex billiards, that is, billiards where the table has a smooth convex boundary, such as a circle or ellipse (Figure 2.2.2). Not all remarks in this section, however, depend on convexity.

6.3.1 The Billiard Flow

Consider the motion of a point mass (or a light ray) in a bounded region D in the plane with boundary B. In the case of a traditional (pool) billiard, the region would be a rectangle, in the case of the example from Section 4.2.5 it is a triangle. The orbits of the motion are sequences of line segments in D where each two successive segments share a boundary point, and at this point the two segments make the same angle with the tangent to the boundary. So the angle of incidence equals the angle of reflection, just as with a mirror (see Figure 6.3.1). If the orbit encounters a corner of the boundary, then it ends there (because reflection is not well defined in that case). One can think of the table having pockets in the corners. The speed of the motion is constant (no friction). Every orbit is completely determined by specifying an initial location as well as an initial direction of motion; that is, the phase space of this system is the set of all tangent vectors of fixed length (for example, unit length) supported at points of the interior of D together with vectors at boundary points

[4] George David Birkhoff, *Dynamical Systems*, American Mathematical Society Colloquium Publications **9**, American Mathematical Society, Providence, RI, 1966, Section VI.6, p. 170

pointing inward. We can describe such points by using the Euclidean coordinates (x_1, x_2) of the base point and the cyclic angular coordinate α of the direction vector.

6.3.2 The Billiard Map

A billiard flow is a system with a continuous time-parameter, but at those times where a reflection occurs there is a discontinuous change in direction. In Section 4.2.5.3 this motivated the unfolding construction for polygonal billiards. For convex billiards this is not possible, and it is better to describe the system in a different way by ignoring the times between collisions with the boundary and using discrete time, that is, by constructing a *section map*, which assigns to a collision configuration (boundary point with inward vector) the next collision configuration that it determines. There is no loss of information because two successive collision sites determine the line between them. Therefore we only consider boundary points with attached inward vectors and define a map ϕ on the set C of these by assigning to one of these initial conditions the place of the next collision together with the reflected direction. This description is reasonable even if there are corners in the boundary, but is undefined at those places.

The map $\phi\colon C \to C$ is usually called the *billiard-ball map* or simply the *billiard map* and can be more carefully described as follows: A vector $v \in C$ supported at $p \in B$ determines an oriented line l, which intersects B in two points p and p'. Then $\phi(v)$ is a vector at p' pointing inward in the direction of the line obtained by reflecting l in the line tangent to B at p'. The natural coordinates in the phase space C are the cyclic length parameter $s \in [0, L)$ on B, where L is the total length of B (recall from Section 2.6.2 that the identification of L with 0 makes this interval into a circle), and the angle $\theta \in (0, \pi)$ with the direction of the positively oriented tangent direction. Thus, the phase space is a cylinder (Section 2.6.3). Note that keeping p fixed while increasing the angle of a vector results in monotonically increasing p' (see Figure 6.3.2). This means that the map thus defined on the cylinder has the *twist* property, which is introduced in Definition 14.2.1. We saw a special instance of this in Example 6.1.2.

The billiard map on the cylinder does not describe the billiard flow completely because it does not provide the times between collisions. But these can be calculated as the lengths of the segments between impacts.

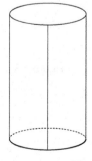 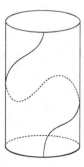

Figure 6.3.2. Phase space of the billiard.

6.3.3 Billiard Models

In Section 4.2.5 we first encountered a billiard system when it arose from the system of two particles on a line segment. Convex billiards, the subject of this chapter, also arise as pertinent models for other problems. Birkhoff's statement quoted at the beginning of Section 6.3 was made in connection with the following analogy. Consider a particle moving freely on a convex surface, without external forces. That is, the particle is constrained only to the surface and otherwise moves solely according to its inertial forces. A physical realization of this could be given (expensively) by a cavity in the shape of the surface, which is stationary in a gravity-free environment. A drop of mercury in this cavity moves in the described manner (being constrained to the cavity wall by centrifugal force). A different way of describing this mixture of constrained and free motion is to say that the acceleration is always normal to the surface (because the only force acting on the particle is the constraint force).

If the surface in question is a triaxial ellipsoid and we squash it flat by making one axis ever shorter, then the limiting dynamics, as the shortest axis shrinks to length zero, is the same as that in the resulting elliptic billiard table. While this is not an exact model of the free motion on any ellipsoid, our description of the dynamics of the elliptic billiard has many similarities to that of the free particle motion on the ellipsoid, yet it is more easily obtained. There are similar analogies between other billiard tables and free particle motion on corresponding surfaces. Billiard models have led to discoveries of which analogs could afterwards be proved for the corresponding surfaces.

6.3.4 The Circle

The simplest convex billiard is the circle. Let D be the unit disk with boundary $B = \{(x, y) \mid x^2 + y^2 = 1\}$. The billiard map can be written explicitly in terms of the cyclic length parameter s along the circle and the angle $\theta \in (0, \pi)$ with the positive tangent direction. Thus, the phase space of the billiard map is a cylinder $C = S^1 \times (0, \pi)$ with s playing the role of the angle coordinate on C.

1. The Billiard Map. The billiard map ϕ is given by $(s', \theta') = (s + 2\theta, \theta)$, so the angle θ is a constant of motion (that is, constant along each orbit). Note that this is essentially the linear twist from Example 6.1.2 that is shown in Figure 6.1.1 and that also arose from free particle motion on the torus in Section 5.2.2.

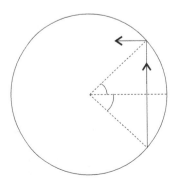

Figure 6.3.3. The billiard map in the circle.

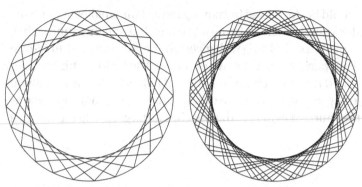

Figure 6.3.4. Rational orbit and segment of an irrational orbit of the circle billiard.

This means that the phase cylinder C splits into ϕ-invariant circles $\theta = \theta_0$. The dynamics on this invariant circle is a rotation by $2\theta_0$, and for any billiard orbit the successive encounters with the boundary B lie on the orbit of a rotation of B by $2\theta_0$. Consequently, if θ_0 is commensurable with 2π (that is, $\theta_0/\pi \in \mathbb{Q}$ or θ_0 is a rational number of degrees), then the billiard map on the circle is periodic and the orbits correspond to inscribed star-shaped polygons. If θ_0 is incommensurable with 2π, then by Proposition 4.1.1 all orbits are dense on the circle. (See Figure 6.3.4.)

2. Caustics. The invariant circle $\theta = \theta_0$ corresponds to all those rays making an angle θ_0 with the boundary $B = \{(x, y) \mid x^2 + y^2 = 1\}$. The union of these rays is the annulus $\cos^2 \theta \leq x^2 + y^2 \leq 1$. Its inner boundary $x^2 + y^2 = \cos^2 \theta$ is called the *caustic* associated to the invariant circle. The complement of this annulus is the intersection of all the left half-planes to these rays. A caustic is in particular the envelope of the rays defining it; that is, it is a smooth curve tangent to every ray in the family, or, in this case, with the property that if a ray is tangent to it, then so is its reflection in the boundary of the billiard table.

An extreme case of this is the point in the center of the circular billiard table. Any ray through the center is reflected right back, and to the extent that there is any caustic at all, it is this single point, that is, it is a focus. This suggests that caustics are natural generalizations of a focus in case of imperfect focussing. The corresponding invariant circle for the billiard map on the cylinder, by the way, is $\theta = \pi/2$ and consists of period-2 orbits. This is apparent from both the geometry and the formula for the billiard map.

3. The Variational Approach. It is well to note at this point an alternative description of how successive points of impact of a billiard trajectory on the boundary are related. If two of them are given, with the intermediate one only known approximately, then one can find the intermediate point exactly by using the law of reflection (without knowing the approximate location of the intermediate point there would be two opposite choices). A way of describing the rule of the equality of angles at the intermediate point is that this choice of intermediate point minimizes the sum of the lengths of the two resulting rays. Indeed, if the angles were not equal, the sum of the lengths could be decreased by moving toward the smaller angle. Note that this observation does not depend on being in a circle. Indeed, this approach of finding an orbit by minimizing something with given endpoints occurred already in the context of Lagrangian mechanics, where we minimized an action. This is no coincidence at all, but related to the mechanical nature of billiards. Viewing a billiard system as an optical arrangement, we can also describe this variational approach as Fermat's principle of the "hurried light ray" that reaches its target along the shortest route.

6.3.5 **The Ellipse**
Consider an elliptic region D with boundary

$$B = \left\{ (x, y) \mid \frac{x^2}{a^2} + \frac{y^2}{b^2} = 1 \right\}.$$

1. Periodic Points. Unlike the circle, the elliptic billiard does not have an invariant circle of period-2 orbits going through the center, but there are two special orbits of period 2 on the symmetry axes of the ellipse. These are the only two lines that intersect the ellipse with two right angles. Accordingly, the endpoints of the longer symmetry axis can be characterized by being the only pair of boundary points with maximum mutual distance. Likewise, the endpoints of the shorter symmetry axis can be characterized by being saddle points for the distance between the endpoints. The length of the longer axis is equal to the *diameter* of the ellipse, that is, the maximum distance between any two points in the domain. The length of the shorter axis is equal to the *width*, which is defined to be the minimal width of a strip (between two parallel lines) that contains the ellipse, that is, the width of the narrowest hallway through which this elliptic table can be pushed.

2. The Generating Function. The extremal property of these special orbits will come up again and suggests the following definition: Parametrize the boundary B by the arc length parameter s and consider two points p and p' with corresponding coordinates s and s' on B. Let $H(s, s')$ to be the negative of the Euclidean distance between p and p'. H is called the *generating function* for the billiard (and will be discussed in general in Section 6.4.2). (See Figure 6.3.5.) Thus, the long period-2 orbit corresponds to a minimum of H, whereas the short orbit corresponds to a saddle of H. We will see that any convex billiard has at least two orbits of period 2

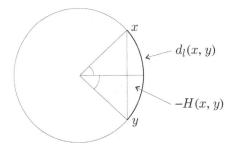

Figure 6.3.5. The generating function of the circle.

that can be described similarly in terms of diameter and width. Note again the similarity to the variational approach in Lagrangian mechanics.

For the circle, by the way, the generating function is $H(s, s') = -2\sin\frac{1}{2}(s' - s)$. As should be expected, it has many critical points, namely, any (s, s') with $s' - s = \pi$, corresponding to all diameters.

3. Caustics. The elliptic billiard table has many caustics:

Proposition 6.3.1 *Every smaller confocal ellipse (that is, having the same foci) is a caustic.*

Proof To see this, consider Figure 6.3.6. It shows an elliptic billiard table with foci f_1 and f_2 as well as a ray $p_0 p_1$ disjoint from the segment connecting the foci and its image $p_1 p_2$ under the billiard map. These two rays thus make the same angle with the tangent line at p_1. The same goes for the rays $f_1 p_1$ and $p_1 f_2$, being part of an orbit as well. Therefore the angles $p_0 p_1 f_1$ and $f_2 p_1 p_2$ are equal, as indicated. Now reflect $f_1 p_1$ in $p_0 p_1$ to get $f_1' p_1$, and $f_2 p_1$ in $p_2 p_1$ to get $f_2' p_1$. By construction the two new angles are equal to the two previously studied. Therefore the triangle $f_1 p_1 f_2'$ is obtained from the triangle $f_1' p_1 f_2$ by rotation around p_1, and hence $l(f_1 f_2') = l(f_1' f_2) =: L$.

Now $p_0 p_1$ is tangent to a confocal ellipse at the point a because the reflection of af_1' in $p_0 p_1$ is af_1, which occurs only for reflection in the tangent to the confocal ellipse defined by $l(f_2 x) + l(x f_1) = l(f_2 f_1') = L$ on which a lies. Likewise, b is a point of tangency with the same ellipse $l(f_1 x) + l(x f_2) = l(f_1 f_2') = L$. $\square$

Therefore there is a family of invariant circles in the phase space C of this billiard, corresponding to the families of rays tangent to a given confocal ellipse. These circles can be parametrized, for example by the (positive) eccentricity of the corresponding elliptic caustic.

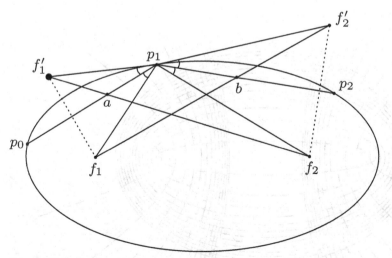

Figure 6.3.6. Confocal ellipses are caustics.

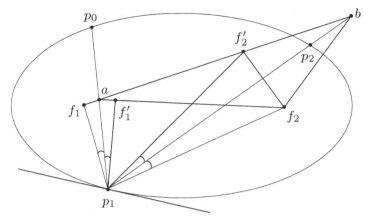

Figure 6.3.7. Confocal hyperbolae are caustics.

This is only half the picture:

Proposition 6.3.2 *There is a caustic corresponding to any ray that goes between the foci. It consists of (both pieces of) a hyperbola with the same foci.*

Proof The proof is almost the same, and Figure 6.3.7 shows the required diagram in which the same construction is performed. Note that again $l(f_1 f_2') = l(f_1' f_2) =: \Delta$ by rotating the corresponding triangles around p_1, and that a and b both are points of tangency with the hyperbola $l(f_1 x) - l(f_2 x) = \pm \Delta$ (a and b correspond to opposite signs here). $\square$

Successive tangencies are always with opposite branches of the hyperbola. (See Figure 6.3.8) Correspondingly, then, each of these caustics produces a pair of closed invariant arcs in C [parametrized by the (negative) eccentricity of the corresponding hyperbola], which are interchanged by the billiard map. This family of invariant sets is separated from those corresponding to positive

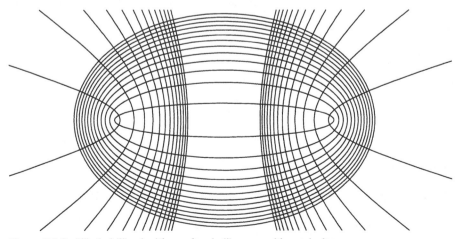

Figure 6.3.8. Elliptic billiard with confocal ellipses and hyperbolas.

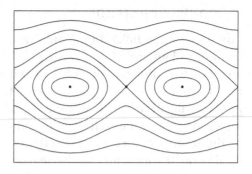

Figure 6.3.9. Phase portrait of the elliptic billiard.

eccentricity by the curve corresponding to the family of rays going through the foci. (See Figure 6.3.9.) The only orbit not accounted for by this classification is the period-2 orbit corresponding to the minor axis of the ellipse.

4. Invariant Circles. To study the motion on an invariant circle corresponding to a confocal elliptic caustic we use the fact that the billiard map preserves area if we take $-\cos\theta$ instead of θ as the second coordinate (Proposition 6.4.2). Because invariant circles come in a family corresponding to elliptic caustics, the corresponding eccentricity is an invariant function on this part of the phase space. As explained after Proposition 6.2.4, area preservation and the presence of an invariant function allow us to parametrize these invariant curves in such a way that the motion (under the billiard map) along each of them is a circle rotation. Therefore the dynamics of the billiard map can be understood completely on this set: It is an open set of pairwise disjoint invariant circles each of which is rotated rigidly with respect to an appropriate parametrization.

Remark 6.3.3 The rotation number varies (by checking extreme cases and by continuity).

■ **EXERCISES**

■ **Exercise 6.3.1** Describe the billiard map in a rectangle.

■ **Exercise 6.3.2** Describe the billiard map in a right triangle.

■ **Exercise 6.3.3** Describe the billiard-ball motion in the annulus between two concentric circles.

■ **Exercise 6.3.4** Describe the billiard-ball motion in the quarter-circle $\{(x, y) \in \mathbb{R}^2 \mid x > 0,\ y > 0,\ x^2 + y^2 \le 1\}$.

■ **Exercise 6.3.5** Show that a billiard orbit through the foci of an ellipse accumulates on the major axis. (Since the same holds for the reversed orbit, these orbits make up a heteroclinic loop, or separatrix, for the period-2 orbit through both foci.)

■ **PROBLEMS FOR FURTHER STUDY**

■ **Problem 6.3.6** Find a first integral for the billiard flow in the ellipse that is a quadratic function of both the coordinates and the velocities.

6.4 CONVEX BILLIARDS

The study of the billiard in a circular and in an elliptic table has shown several features to look for when studying other billiard tables, notably periodic points and caustics, and has provided us with several notions to help organize our investigations. For example, we will be studying the orbit structure to some extent by discussing the qualitative properties of the billiard map on the phase cylinder, as opposed to reasoning exclusively by directly apparent geometry. This was presaged by the absence of an explicit description of the billiard map for the ellipse, where we described important qualitative features of the map, namely, a decomposition into invariant sets that are easily studied individually.

6.4.1 Smooth Convexity

The billiards we want to study now are those whose boundary is given by a smooth closed curve B that is convex in a stronger sense than that of Definition 2.2.13. We require the boundary to have nonzero curvature. An equivalent way of expressing this is that, if B is parametrized by arc length, then the second derivative is never zero.

This implies (strict) convexity as in Definition 2.2.13; that is there are no turning points and therefore no "inward bulges", and we have the defining property that every line that enters the interior of the table enters and exits transversely [with an angle in $(0, \pi)$] and intersects the boundary in exactly two points. Often it suffices to satisfy this latter geometric assumption and to allow isolated points of zero second derivative.

There are occasions where the derivative condition is necessary, and we call billiards satisfying it *strictly differentiably convex*. This is a stronger notion of convexity than the "strict convexity" introduced in Definition 2.2.13.

Thus, as in the case of circles and ellipses, the phase space is a cylinder C parametrized by a parameter s on the boundary (usually arc-length) and the angle $\theta \in (0, \pi)$.

6.4.2 The Generating Function

As in the case of the circle and the ellipse, define a function H by taking two points p and p' on B with corresponding arc-length coordinates s and s' and letting $H(s, s')$ be the negative of the Euclidean distance between p and p'. H is called the *generating function* for the billiard. Although we do not usually have an explicit formula H, as in the case of the circular billiard, it is nevertheless amenable to analytic treatment.

Lemma 6.4.1 *If θ' is the angle of the segment joining p and p' with the negative tangent at p' and θ is the angle of the segment joining p and p' with the positive tangent at p, then*

(6.4.1)
$$\frac{\partial}{\partial s'} H(s, s') = -\cos\theta',$$

$$\frac{\partial}{\partial s} H(s, s') = \cos\theta.$$

(See Figure 6.4.1.)

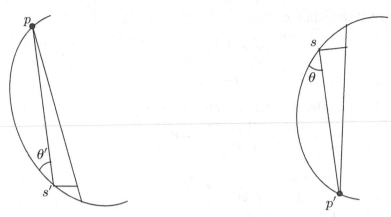

Figure 6.4.1. Derivatives of the generating function.

Proof

$$\frac{\partial}{\partial s'} H(s, s') = -\frac{d}{dt} d(p, c(t)) = -\frac{d}{dt} \sqrt{\langle c(t) - p, c(t) - p \rangle}$$

$$= -\frac{1}{2\sqrt{\langle c(t) - p, c(t) - p \rangle}} 2\langle c'(t), c(t) - p \rangle = -\frac{\langle c(t) - p, c(t) - p \rangle}{\|c(t) - p\|}.$$

For $t = s'$, the last expression is exactly $-\cos\theta'$ because c' is a unit vector. The second equation is proved in the same way. $\square$

The generating function helps to decide when a sequence of boundary points lies on an orbit. Any two certainly do, but, unlike a pair of points, a triple does not always lie on part of an orbit. Those triples that do can be described as critical points of a certain functional. Consider three points p_{-1}, p_0, and p_1 with corresponding coordinates s_{-1}, s_0, and s_1 on B. If they are part of a billiard orbit, then by definition the segments joining p_{-1} with p_0 and p_0 with p_1 make the same angles with the tangent at p_0. Consequently, by Lemma 6.4.1,

(6.4.2) $$\frac{d}{ds} H(s_{-1}, s) + \frac{d}{ds} H(s, s_1) = 0 \text{ at } s = s_0,$$

that is, p_0 is obtained as a *critical point* of the functional $s \mapsto H(s_{-1}, s) + H(s, s_1)$ on triples of boundary points. As in the Lagrange formulation, this describes an orbit segment of a dynamical system as a critical point of a functional defined on a space of "potential" orbit segments of the dynamical system. The procedure can be iterated to produce orbit segments as critical points of functionals depending on several variables.

6.4.3 Area Preservation

The explicit form (6.4.1) of the derivatives of the generating function is in turn directly useful for the study of the billiard map. It implies that the billiard map preserves area on the phase cylinder C if instead of θ we use the coordinate $r = -\cos\theta$.

Proposition 6.4.2 *In the coordinates (s, r) the billiard ball map $\phi(s, r) = (S(s, r), R(s, r))$ is area- and orientation-preserving (see Section 6.1.1.3).*

Proof Equation (6.4.1) simplifies to

$$\frac{\partial}{\partial s'} H(s, s') = r',$$

(6.4.3)

$$\frac{\partial}{\partial s} H(s, s') = -r,$$

where $r' = -\cos\theta'$. Define $\tilde{H}(s, r) := H(s, S(s, r))$. Then

$$\frac{\partial \tilde{H}}{\partial s} = \frac{\partial H}{\partial s} + \frac{\partial H}{\partial s'}\frac{\partial S}{\partial s} = -r + R\frac{\partial S}{\partial s}$$

and

$$\frac{\partial \tilde{H}}{\partial r} = \frac{\partial H}{\partial s'}\frac{\partial S}{\partial r} = R\frac{\partial S}{\partial r};$$

so by calculating $\partial^2 \tilde{H}/\partial s\partial r$ two ways we get

$$-1 + \frac{\partial R}{\partial r}\frac{\partial S}{\partial s} + R\frac{\partial^2 S}{\partial s\partial r} = \frac{\partial^2 \tilde{H}}{\partial s\partial r} = \frac{\partial^2 \tilde{H}}{\partial r\partial s} = \frac{\partial R}{\partial s}\frac{\partial S}{\partial r} + R\frac{\partial^2 S}{\partial r\partial s}$$

and

$$\frac{\partial R}{\partial r}\frac{\partial S}{\partial s} - \frac{\partial R}{\partial s}\frac{\partial S}{\partial r} = 1.$$

This means that the Jacobian determinant of ϕ is 1 and hence that ϕ preserves area and orientation (see Proposition 6.1.3 and Definition 6.1.4). $\square$

6.4.4 Smoothness of the Billiard Map

The equations (6.4.3) are not only useful for showing that area is preserved. They also code the dynamics because they locally determine the functions S and R. This has various uses. The first is smoothness of the billiard map:

Proposition 6.4.3 *Assume the curve B is C^k, that is, the Euclidean coordinates are C^k functions of the length parameter. Then the functions S and R are C^{k-1} for $0 < r < 1$.*

Proof Apply the Implicit-Function Theorem 9.2.3 to

$$0 = F(s, s', r, r') := \begin{pmatrix} \frac{\partial}{\partial s'} H(s, s') - r' \\ \frac{\partial}{\partial s} H(s, s') + r \end{pmatrix}.$$

The hypotheses of the Implicit-Function Theorem 9.2.3 are satisfied because the total derivative

$$\begin{pmatrix} \frac{\partial^2}{\partial s'^2} H(s, s') & -1 \\ \frac{\partial^2}{\partial s\partial s'} H(s, s') & 0 \end{pmatrix}$$

of F with respect to (s', r') is invertible: Its determinant $(\partial^2/\partial s\partial s')H(s, s') = \partial r'/\partial s$ is clearly nonzero for geometrical reasons: Increasing s while s' is constant decreases θ' (see Figure 6.4.2) and hence r', so

(6.4.4) $$\frac{\partial^2}{\partial s\partial s'} H(s, s') = \frac{\partial r'}{\partial s} < 0.$$

If B is C^k, then the generating function is also C^k, and by the Implicit-Function Theorem 9.2.3 the functions S and R are C^{k-1} for $0 < r < 1$. $\square$

Figure 6.4.2. Increasing s while s' is constant.

6.4.5 Special Period-2 Orbits in Convex Billiards

Now we generalize our previous observation that in an elliptic billiard two orbits of period 2 can be obtained by a geometric description via the diameter and width of the region (see Section 6.3.5.1). This uses only a bit of knowledge about the generating function. Intuitively, the following result is quite obvious.

Proposition 6.4.4 *Let D be a convex bounded region whose boundary B is C^2 with nonzero curvature. Then the associated billiard map has at least two distinct period-2 orbits that are described as follows: For one of them, the distance between the corresponding boundary points is the diameter of D; for the other, it is the width of D.*

Proof The generating function $H(s, s')$ is defined and continuous on the torus $B \times B$ and differentiable except on the diagonal. Since it is zero on the diagonal and negative elsewhere, it attains its minimum d away from the diagonal. Let (s, s') be such that $H(s, s') = d$. Since it is a critical point, (6.4.1) implies that $\theta = \theta' = \pi/2$, so we obtain the first of these period-2 orbits. (The preceding argument only depends on convexity, by the way, and can easily be made to work for C^1 curves.) Now consider the curve $(s, g(s))$ on the torus, where $s' = g(s)$ is the coordinate of the boundary point other than s on the line through s for which $\theta = -\theta'$. (This line is the one connecting two points with parallel tangents, so the minimal length of such lines is the width; see Figure 6.4.3.) On this curve H is bounded from above by a negative number and thus attains a negative maximum w. Parametrize s and s' by the angle α

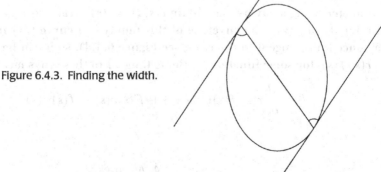

Figure 6.4.3. Finding the width.

that the connecting line makes with some reference direction. Then $\theta = -\theta'$ implies

$$\frac{\partial H(s(\alpha), s'(\alpha))}{\partial \alpha} = \frac{\partial H}{\partial s}\frac{ds}{d\alpha} + \frac{\partial H}{\partial s'}\frac{ds'}{d\alpha} = \cos\theta\left(\frac{ds}{d\alpha} + \frac{ds'}{d\alpha}\right).$$

If s is the arc length parameter, then $d\alpha/ds$ is the curvature of B at the point corresponding to s; hence it is nonzero (and finite). Therefore $ds/d\alpha$ is positive. The same goes for $ds'/d\alpha$, and therefore at a critical point of $H(s(\alpha), s'(\alpha))$ we have $\cos\theta = 0$. Consequently, there is a line of the desired kind for which $\theta = \pi/2$, and we have obtained the second point of period 2. $\square$

A period-3 counterpart of the first orbit from Proposition 6.4.4 can be constructed by considering the inscribed triangle with the largest perimeter. A similar construction works for orbits of period 4. For higher periods there are different types of orbits, for example inscribed pentagons versus pentagrams.

The construction of these orbits in the more general setting of area-preserving twist maps is carried out in Section 14.1. There are also counterparts of the second type of orbit.

In the case of an ellipse (and, of course, also of a circle) all orbits of period higher than 2 come in continuous families corresponding to their invariant circle, but this is a rather exceptional property.

6.4.6 The Mirror Equation of Geometric Optics

We now look for caustics in convex billiards. Recall that these are envelopes of families of orbits for which the reflected family of rays has the same envelope. We presently define these a little more carefully, but it is clear that in order to be able to study caustics it is important that we understand how the envelope of a family of rays and the envelope of the reflected family are related. Or, alternatively, given a smooth arc in the billiard table and a family of tangent rays, consider all of the arcs obtained from reflecting each tangent ray in the boundary of the billiard table. What is their envelope? To obtain a caustic, it is, of course, necessary that the new envelope be another part of the same curve.

To bring elementary differential geometry to bear on this question we need to define an envelope given a parametrization of a family of rays. To parametrize a family of rays in the plane choose a curve c parametrized by $s \in (-\epsilon, \epsilon)$ and a family $v(\cdot)$ of unit vectors also parametrized by $s \in (-\epsilon, \epsilon)$. Denote the parameter along a ray by t to obtain $r(s, t) = c(s) + tv(s)$ as a parametrization of a family of rays. The envelope of this family is a curve that intersects each ray once (and tangent to the rays; see Figure 6.4.4), so it can be parametrized as $r(s, f(s))$ for some function f. Being tangent to these rays means that

$$\frac{d}{ds}r(s, f(s)) = c'(s) + f'(s)v(s) + f(s)v'(s)$$

Figure 6.4.4. An envelope.

is parallel to v, that is, it has no component normal to v. We can express this by taking v' as the normal vector to v (because v is a family of unit vectors). Thus, if $v' \neq 0$ (the rays are not supposed to be parallel in order to admit an envelope), the tangency condition is

$$(6.4.5) \quad 0 = \left\langle \frac{d}{ds}r(s, f(s)), v'(s) \right\rangle = \langle c'(s) + f'(s)v(s) + f(s)v'(s), v'(s) \rangle$$

$$= \langle c'(s) + f(s)v'(s), v'(s) \rangle = \langle c'(s), v'(s) \rangle + f(s)\langle v'(s), v'(s) \rangle.$$

Thus f is uniquely determined:

$$f(s) = -\frac{\langle c'(s), v'(s) \rangle}{\langle v'(s), v'(s) \rangle}.$$

Notice as a special case that if c is constant, then c is the focus of the family of rays and indeed, by the above formula, $f \equiv 0$ also parametrizes the focus: $r(s, f(s)) = r(s, 0) = c(s)$. To relate an envelope with that of the family of rays reflected in (a part of) the boundary of a billiard table it is convenient to parametrize the family of rays in this fashion and to take the curve c to be the segment of the boundary to be studied. The advantage is that the family of reflected rays can be parametrized using the same curve c.

Thus, we parametrize a piece of the boundary of a billiard table by a curve c using the arc length parameter s, that is, such that $T := c'$ is a unit vector. If we choose the normal vector N to c to point to the inside of the billiard table, then we can define the *curvature* κ of c by $T'(s) = \kappa(s)N(s)$. This implies, by the way, that $N'(s) = -\kappa(s)T(s)$. With these choices we have $\kappa \geq 0$ for a convex billiard. If $c''(s) \neq 0$, then $\kappa(s) > 0$. In particular, strictly differentiably convex billiard tables have nonzero boundary curvature everywhere.

Parametrize a family of rays by

$$r(s, t) = c(s) + tv(s),$$

with v pointing inward and making an angle $\alpha(s) \in [0, \pi]$ with $T(s)$. Then the reflected family can be parametrized by

$$\bar{r}(s, t) = c(s) + t\bar{v}(s),$$

with $\bar{v}$ pointing inward and making an angle $\pi - \alpha(s)$ with $T(s)$. The effect of reflection in the boundary on an envelope is succinctly described by the *mirror equation* in geometric optics (see Figure 6.4.5):

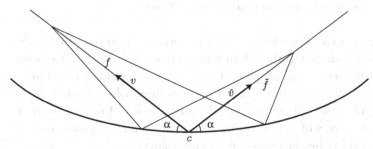

Figure 6.4.5. The mirror equation.

Theorem 6.4.5 *If f and $\bar{f}$ are the envelopes of these two families, respectively, then*

$$\frac{1}{f} + \frac{1}{\bar{f}} = \frac{2\kappa}{\sin\alpha}.$$

Proof As in the statement we drop the variable s. We can write

$$v = \quad \cos\alpha\, T + \sin\alpha\, N$$

$$\bar{v} = -\cos\alpha\, T + \sin\alpha\, N.$$

and differentiate to get

$$v' = -\sin\alpha \cdot \alpha'\, T + \cos\alpha\, T' + \cos\alpha \cdot \alpha'\, N + \sin\alpha\, N'$$

$$= -(\alpha' + \kappa)\sin\alpha\, T + (\alpha' + \kappa)\cos\alpha\, N$$

$$\bar{v}' = \quad \sin\alpha \cdot \alpha'\, T - \cos\alpha\, T' + \cos\alpha \cdot \alpha'\, N + \sin\alpha\, N'$$

$$= \quad (\alpha' - \kappa)\sin\alpha\, T + (\alpha' - \kappa)\cos\alpha\, N.$$

Then

$$f = -\frac{\langle T, v'\rangle}{\langle v', v'\rangle} = -\frac{-(\alpha'+\kappa)\sin\alpha}{(\alpha'+\kappa)^2} = \frac{\sin\alpha}{\alpha'+\kappa}$$

and likewise $\bar{f} = -\sin\alpha/(\alpha' - \kappa)$, and hence

$$\frac{1}{f} + \frac{1}{\bar{f}} = \frac{(\alpha'+\kappa) - (\alpha'-\kappa)}{\sin\alpha} = \frac{2\kappa}{\sin\alpha}. \qquad \square$$

This result still holds in the extreme case where an envelope degenerates to a single point, as in a circular billiard table with a family of rays through the center being reflected back to the center. In this case, $f = \bar{f} = \rho$, the radius, and $\sin\alpha = 1$, so $\kappa = 1/\rho$. Another extreme test case is that of a parallel bundle of lines encountering the boundary of the circular billiard. In this case we can take $1/f = 0$ in the mirror equation (taking the point p where the ray through the center meets the boundary) to find that $\bar{f} = \rho/2$, that is, this bundle focuses (approximately) at the midpoint between the center and p.

6.4.7 Caustics

Now we can study caustics using the mirror equation. For circles and ellipses we found them to be sometimes associated with invariant circles, and other times not. It is worthwhile to maintain this distinction. Thus we now carefully define caustics and also delineate the notion of an invariant circle.

Definition 6.4.6 An *invariant circle* Γ for ϕ is a ϕ-invariant set in C that is the graph of a continuous function (other than 0 or π) from B to $[0, \pi]$. A *caustic* is a piecewise smooth curve γ, all of whose tangents are part of a billiard orbit and such that, whenever a ray of a billiard orbit defines a line tangent to γ, so does its image under the billiard map ϕ. A caustic is said to come from an invariant circle if the family of rays defining it constitutes an invariant circle of ϕ in C. A caustic is said to be convex if it is a convex curve.

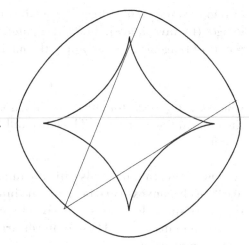

Figure 6.4.6. A nonconvex caustic.

A caustic does not necessarily have to be inside the billiard table (the hyperbolas for the elliptic billiard are not), but that this is certainly the case for convex caustics (otherwise they would have tangent lines that do not meet the billiard table).

A convex caustic comes from the invariant circle defined by its tangent lines, and it can be alternatively described as follows: The intersection of all of the left half-planes to these rays or the intersection of the right half-planes is nonempty and the boundary of this region is the caustic.

There are nonconvex caustics, and they may be contained in the billiard table. An example is shown in Figure 6.4.6. It is obtained by perturbing a circular billiard and relies on the fact that the center of a circular billiard is a degenerate caustic. The existence of convex caustics restricts the geometry of a billiard table.

Theorem 6.4.7 *A convex C^2 billiard table with a point of zero curvature has no convex caustics.*

Proof To see this, suppose that in Figure 6.4.7 γ is a convex caustic, and consider two rays tangent to it with a common point $p \in B$ and such that one of these rays is the image of the other under the billiard map. The caustic is the

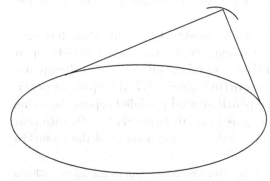

Figure 6.4.7. A convex caustic.

envelope of the family of rays defined by the invariant circle, as well as for their images (by invariance). Thus, if f_p and $\bar{f}_p$ denote the distances from p to the points of tangency, we can apply the mirror equation to conclude that

$$\frac{1}{f} + \frac{1}{\bar{f}} = \frac{2\kappa}{\sin\alpha},$$

where κ is the curvature of B at p and α is the angle that the two rays make with the tangent line to B at p. The left-hand side of this last equation is positive, so $\kappa \neq 0$. $\square$

This shows that billiards with a boundary point of zero curvature (those that are "barely convex") are far from the integrable ones such as the ellipse, where the phase space decomposes nicely into invariant curves on which the dynamics is easy to understand. These billiards are therefore likely to be dynamically far more complicated.

It is a remarkable fact that this is the only way to avoid convex caustics. A strictly differentiably convex billiard always has infinitely many caustics, indeed a nonnull set of them.[5]

6.4.8 The String Construction

On the other hand, the same considerations enable us to produce many tables with convex caustics. In fact, we can prescribe a convex curve and construct a family of billiard tables having this caustic. To that end we denote the length of the caustic between the points of tangency (and on the side away from p) in Figure 6.4.7 by ℓ_p. Then we have

Proposition 6.4.8 $S(\gamma) := f_p + \bar{f}_p + \ell_p$ *is independent of* p.

Proof Differentiate the right-hand side with respect to the length parameter s on B that parametrizes p. Denoting the length parameter on γ by t and its values at the points of tangency by t_p and $\bar{t}_p$, we get

$$\frac{d}{ds}f_p = \cos\alpha - \frac{d}{ds}t_p \qquad \frac{d}{ds}\bar{f}_p = -\cos\alpha + \frac{d}{ds}\bar{t}_p \qquad \frac{d}{ds}\ell_p = \frac{d}{ds}t_p - \frac{d}{ds}\bar{t}_p.$$

The sum of these is zero. $\square$

The number $L(\gamma) := S(\gamma) - l(\gamma)$ is called the *Lazutkin parameter* of the caustic γ.

The preceding proposition enables us to construct billiard tables having a given convex curve γ as a caustic. This *string construction* uses a closed loop of string of some length $S > l(\gamma)$ pulled tight around the curve γ using the tip of a pencil to pull the loop away from the curve (in Figure 6.4.7, the tip of the pencil would be at the top right). Moving the pencil around γ while keeping the string taut traces out a billiard table that has γ as a caustic (with $S(\gamma) = S$). Different values of S give different billiard tables with γ as a caustic (and the Lazutkin

---

[5] This is a related to KAM theory. See Vladimir F. Lazutkin, The Existence of a Caustics for a Billiard Problem in a Convex Domain, *Mathematics of the USSR, Isvestia* **7** (1973), 185–214.

parameter measures the excess length of the loop of string). A familiar case of this is the prescription of obtaining an ellipse by pulling a loop of string tautly around a line segment (whose endpoints are then the foci). Of course this is not a smoothly convex caustic. Strings of different lengths give confocal ellipses.

This string construction enables us to find many billiard tables with a given caustic. It does not give us any billiard table for which we know that there are many caustics. There is, in fact, a long-standing open problem raised by Birkhoff: Suppose a billiard table has an open set of caustics (as opposed to only isolated ones). Must it necessarily be an elliptic table?

▨ EXERCISES

▨ **Exercise 6.4.1** Show that, if a convex billiard table has two perpendicular symmetry axes, then the billiard ball map has a period-4 orbit.

▨ **Exercise 6.4.2** Generalize the statement of the previous exercise to the case of two symmetry axes subtending an angle of $2\pi/n$.

▨ **Exercise 6.4.3** For an equilateral triangle, a square, and a regular pentagon describe the billiard tables obtained from the string construction.

▨ **Exercise 6.4.4** Write a functional of several variables whose critical points produce periodic orbits of a billiard.

▨ **Exercise 6.4.5** Give an example of a convex billiard table other than a circle that has a continuous family of period-2 orbits.

▨ PROBLEMS FOR FURTHER STUDY

▨ **Problem 6.4.6** Give an example of a billiard table other than a circle that has a period-2 orbit in any direction.

▨ **Problem 6.4.7** Show that the family from the previous problem defines a nonconvex caustic as its envelope.

▨ **Problem 6.4.8** Develop the ideas of Section 6.2.3 to prove area preservation of the billiard map from considerations of the *flux* of the billiard flow through the section defined by the boundary. The flux through a surface is the integral of the normal velocity.

▨ **Problem 6.4.9** Construct a smooth curve for which the astroid is a nonconvex caustic as in Figure 6.4.6.

CHAPTER 7

Simple Systems with Complicated Orbit Structure

This chapter presents a rich array of properties of a collection of examples. Its coherence derives from the fact that it is part of a general theory we outline in Chapter 10. The examples (other than the quadratic map f_4) are instances of hyperbolic dynamical systems (or symbolic dynamical systems), and the properties we derive here are largely properties common to hyperbolic and symbolic dynamical systems.

7.1 GROWTH OF PERIODIC POINTS

Periodic orbits represent the most distinctive special class of orbits. So far we have mostly encountered maps with few periodic orbits or, as in the case of a rational rotation, only periodic orbits. In these basic examples different periods did not appear for the same map. Even the most complex situations so far still involve periodic orbits neatly organized by period in families such as invariant curves in plane rotations, linear twists, the time-1 map for the mathematical pendulum, or billiards. There we placed more emphasis on coherence than complexity. Now we encounter the first examples with a different periodic pattern. In these cases, when periodic points of different periods are present, we want to count them.

Definition 7.1.1 For a map $f: X \to X$, let $P_n(f)$ be the number of periodic points of f with (not necessarily minimal) period n, that is, the number of fixed points for f^n.

This section introduces numerous new examples of dynamical systems. For now they are introduced with a view to their periodic orbit structure, but in due time numerous other fascinating features of their orbit structure will emerge.

7.1.1 Linear Expanding Maps
Consider the noninvertible map E_2 of the circle given in multiplicative notation by

$$E_2(z) = z^2, \qquad |z| = 1,$$

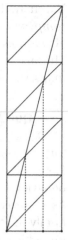

Figure 7.1.1. Periodic points for an expanding map.

and in additive notation by

(7.1.1) $E_2(x) = 2x \pmod{1}$.

Proposition 7.1.2 $P_n(E_2) = 2^n - 1$ and periodic points for E_2 are dense in S^1.

Proof If $E_2^n(z) = z$, then $z^{2^n} = z$, and $z^{2^n - 1} = 1$. Thus every root of unity of order $2^n - 1$ is a periodic point for E_2 of period n. There are exactly $2^n - 1$ of these, and they are uniformly spread over the circle with equal intervals. In particular, when n becomes large these intervals become small. (See Figure 7.1.1) □

We see from Proposition 7.1.2 that a natural measure of asymptotic growth of the number of periodic points is the exponential growth rate $p(f)$ for the sequence $P_n(f)$:

(7.1.2) $$p(f) = \varlimsup_{n \to \infty} \frac{\log_+ P_n(f)}{n},$$

where $\log_+(x) = \log(x)$ for $x \geq 1$, 0 otherwise. In particular, Proposition 7.1.2 shows that $p(E_2) = \lim_{n \to \infty} (\log 2^n + \log(1 - 2^{-n}))/n = \log 2$.

The maps

$$E_m: x \mapsto mx \pmod{1},$$

where m is an integer of absolute value greater than one, represent a straightforward generalization of the map E_2. Not surprisingly, these maps also have dense sets of periodic orbits. The proof of Proposition 7.1.2 holds verbatim with the replacement of 2 by m:

Proposition 7.1.3 $P_n(E_m) = |m^n - 1|$ and periodic points for E_m are dense.

Proof $z = E_m^n(z) = z^{m^n}$ has $|m^n - 1|$ solutions that are evenly spaced. □

See also Section 7.1.3.

Another property of the maps E_m worth noticing is preservation of length similar to the property of preservation of phase volume discussed in Section 6.1.

Naturally, the length of an image of any arc increases; however, if one considers the *complete preimage* of an arc Δ under E_m, one immediately sees that it consists of $|m|$ arcs of length $l(\Delta)/|m|$ each, placed along the circle at equal distances. The analysis in Section 6.1.2 can be extended to noninvertible volume-preserving maps, so recurrent points are dense in this situation as well.

7.1.2 Quadratic and Quadratic-Like Maps

For $\lambda \in \mathbb{R}$, let $f_\lambda : \mathbb{R} \to \mathbb{R}$, $f_\lambda(x) := \lambda x(1 - x)$. For $0 \leq \lambda \leq 4$, the f_λ map the unit interval $I = [0, 1]$ into itself. The family f_λ is referred to as the *quadratic family*. For $\lambda \leq 3$, this family was discussed in detail in Section 2.5, and the asymptotic behavior for any such λ is fairly simple and changes with λ only a few times. As it turns out, for the remaining interval of parameter values the quadratic family develops a bewildering array of complicated but different types of behavior, which change with caleidoscopic speed (see Figure 7.1.2 and Chapter 11). Note that $P_n(f_\lambda) \leq 2^n$ because the nth iterate of f_λ is a polynomial of degree 2^n, and hence the equation $(f_\lambda)^n(x) = x$ has at most 2^n solutions. While one may expect that in the complex plane this equation would indeed have exactly 2^n solutions for most values of the parameter λ, this is certainly not the case for real solutions.

Here we consider the behavior of the quadratic family for large values of the parameter, namely, $\lambda \geq 4$. While for $\lambda > 4$ the interval $[0, 1]$ is not preserved, the set of points that remains in that interval is still quite interesting.

The analysis of the behavior of the quadratic family on the unit interval for $0 \leq \lambda \leq 3$ carried out in Section 2.5 showed simple periodic patterns: Only points of periods 1 and 2 appear, and their number is small. With moderate effort this analysis can be extended as far as $\lambda = 1 + \sqrt{6}$ (Proposition 11.2.1). On the other hand, we have:

Proposition 7.1.4 *For $\lambda \geq 4$ we have $P_n(f_\lambda) = 2^n$.*

Proof Since $P_n(f_\lambda) \leq 2^n$, it suffices to prove the reverse inequality. To that end we use the following observation: If $f : \mathbb{R} \to \mathbb{R}$ is continuous and $\Delta \subset [0, 1]$ is an interval such that one endpoint is mapped to 0 and the other to 1, then by the Intermediate-Value Theorem there is a fixed point of f in Δ. Now $[0, 1] \subset [f_\lambda(0), f_\lambda(1/2)]$ and $[0, 1] \subset [f_\lambda(1/2), f_\lambda(1)]$, so there are intervals $\Delta_0 \subset [0, 1/2]$ and $\Delta_1 \subset [1/2, 1]$ whose images under f_λ are exactly $[0, 1]$, giving us two fixed points for f. The nonzero fixed

Figure 7.1.2. Bifurcation diagram.

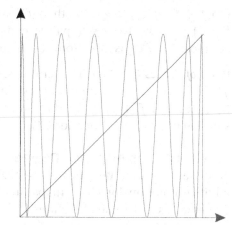

Figure 7.1.3. Periodic points of f_4.

point is indeed in the interior of Δ_1 because the right endpoint of Δ_1 is 1 and hence is mapped to 0, so the other endpoint is mapped to 1 and therefore neither are fixed.

Furthermore, the preimages of Δ_0 and Δ_1 under f consist of two intervals each, so there are four intervals whose images under f^2 are exactly $[0, 1]$. Each contains a fixed point of f_λ^2, again every one except 0 being in the interior of the corresponding interval, so no two of these fixed points coincide.

Repeating this argument successively for higher iterates of f_λ we obtain 2^n intervals whose images under f_λ^n are $[0, 1]$, and each of which therefore contains at least one fixed point, giving 2^n distinct orbits of period n for f_λ. $\square$

It is useful that the argument to show that $P_n(f_\lambda) \geq 2^n$ applies to any continuous map $f: [0, 1] \to \mathbb{R}$ with $f(0) = f(1) = 0$ and such that there is a $c \in [0, 1]$ with $f(c) \geq 1$. In this more general case it is somewhat more convenient, however, to talk about intervals whose images under f^n contain $[0, 1]$ rather than being exactly $[0, 1]$.

In the quadratic case (for $\lambda > 4$) one can refine the preceding argument slightly to show that there are exactly 2^n periodic points (rather than using that the degree of f^n is 2^n). This also works for some continuous maps f of this more general nature, which are monotone on $[0, c]$ as well as $[c, 1]$. A continuous map defined on an interval that is increasing to the left of an interior point and decreasing thereafter is said to be *unimodal*. Thus we have found

Proposition 7.1.5 *If $f: [0, 1] \to \mathbb{R}$ is continuous, $f(0) = f(1) = 0$, and there exists $c \in [0, 1]$ such that $f(c) > 1$, then $P_n(f) \geq 2^n$. If, in addition, f is unimodal and expanding (that is, $|f(x) - f(y)| > |x - y|$) on each interval of $f^{-1}((0, 1))$, then $P_n(f) = 2^n$.*

The heart of the proof is the following lemma:

Lemma 7.1.6 *Denote by $\mathcal{M}_k$ the collection of continuous maps $f: [0, 1] \to \mathbb{R}$ such that $f^{-1}((0, 1)) = \bigcup_{i=1}^{k} I_i$ with $I_i \subset [0, 1]$ open intervals, f monotonic on I_i, and $f(I_i) = (0, 1)$. Then $f \circ g \in \mathcal{M}_{kl}$ whenever $f \in \mathcal{M}_k$ and $g \in \mathcal{M}_l$.*

Proof If $f \in \mathcal{M}_k$ and $g \in \mathcal{M}_l$, then $f^{-1}((0,1)) = \bigcup_{i=1}^{k} I_i$ and $g^{-1}(I_i) = \bigcup_{j=1}^{l} J_{ij}$ with $\{J_{ij} \mid 1 \le i \le k, \ 1 \le j \le l\}$ pairwise disjoint and $(f \circ g)^{-1}((0,1)) = \bigcup_{ij} J_{ij}$. The composition $f \circ g$ is monotonic on J_{ij} and $f \circ g(J_{ij}) = (0,1)$. $\square$

Proof of Proposition 7.1.5 The lemma shows that $P_n(f) \ge k^n$ for $f \in \mathcal{M}_k$ because $f^n \in \mathcal{M}_{k^n}$. If f is expanding on every interval of $f^{-1}((0,1))$, then the same holds for iterates of f. This shows that on each of those intervals there is at most one solution of $f^n(x) = x$. Therefore, $P_n(f) \le k^n$, proving equality. $\square$

7.1.3 Expanding Maps and Degree

Next we consider a nonlinear generalization of the expanding maps E_m. We use additive notation for circle maps. In this notation derivatives of maps can be expressed as real-valued functions.

Definition 7.1.7 A continuously differentiable map $f \colon S^1 \to S^1$ is said to be an *expanding* map if $|f'(x)| > 1$ for all $x \in S^1$.

Since f' is continuous and periodic, the minimum of $|f'|$ is attained and hence is greater than 1.

Proposition 4.3.1 gives us a function $F \colon \mathbb{R} \to \mathbb{R}$ that satisfies $[F(x)] = f([x])$ and $F(s+1) = F(s) + \deg(f)$, where $\deg(f)$ is the *degree* of f. It has the following simple property:

Lemma 7.1.8 *If $f, g \colon S^1 \to S^1$ are continuous, then* $\deg(g \circ f) = \deg(f) \deg(g)$, *in particular* $\deg(f^n) = \deg(f)^n$.

Proof If F, G are lifts of f and g, respectively, then $G(s+k) = G(s+k-1) + \deg(g) = \cdots = G(s) + k \deg(g)$ and $G(F(s+1)) = G(F(s) + \deg(f)) = G(F(s)) + \deg(g) \deg(f)$. $\square$

This property is useful for counting periodic points.

Proposition 7.1.9 *If $f \colon S^1 \to S^1$ is an expanding map, then* $|\deg(f)| > 1$ *and* $P_n(f) = |\deg(f)^n - 1|$.

Proof $|f'| > 1$ implies $|F'| > 1$ for any lift, so, by the Mean-Value Theorem A.2.3, $|\deg(f)| = |F(x+1) - F(x)| > 1$. By the chain rule an iterate of an expanding map is itself expanding, so by Lemma 7.1.8 it suffices to consider the case $n = 1$. Take a lift F of f and consider it on the interval $[0,1]$. The fixed points of f are the projections of the points x for which $F(x) - x \in \mathbb{Z}$. The function $g(x) := F(x) - x$ satisfies $g(1) = g(0) + \deg(f) - 1$, so by the Intermediate-Value Theorem there are at least $|\deg(f) - 1|$ points x where $g(x) \in \mathbb{Z}$. If $g(0) \in \mathbb{Z}$, then there are $|\deg(f) - 1| + 1$ such points, but 0 and 1 project to the same point on S^1. Now $g'(x) \ne 0$, so g is strictly monotone and hence takes every value at most once. Thus there are exactly $|\deg(f) - 1|$ fixed points on S^1. $\square$

This proposition in particular establishes part of the analog of Proposition 7.1.2 for E_m.

Similarly to quadratic maps, the argument that shows $P_n(f) \geq |\deg(f)^n - 1|$ works for any continuous map. It is trivial for maps of degree 1 because the assertion is vacuous. Indeed, irrational rotations do not have any fixed or periodic points. For maps of degree 0 it merely guarantees a fixed point. For maps f with $|\deg(f)| > 1$, however, this result gives exponential growth of the number of periodic points: $p(f) \geq \log_+(|\deg(f)|)$.

7.1.4 Hyperbolic Linear Map of the Torus

The previous examples were all one-dimensional, but the patterns of the growth and distribution of periodic points observed in those examples also appear in higher dimension.

A convenient model to demonstrate this is built from the following linear map of $\mathbb{R}^2$:

$$L(x, y) = (2x + y, x + y) = \begin{pmatrix} 2 & 1 \\ 1 & 1 \end{pmatrix} \begin{pmatrix} x \\ y \end{pmatrix}.$$

If two vectors (x, y) and (x', y') represent the same element of $\mathbb{T}^2$, that is, if $(x - x', y - y') \in \mathbb{Z}^2$, then $L(x, y) - L(x', y') \in \mathbb{Z}^2$, so $L(x, y)$ and $L(x', y')$ also represent the same element of $\mathbb{T}^2$. Thus L defines a map $F_L: \mathbb{T}^2 \to \mathbb{T}^2$:

$$F_L(x, y) = (2x + y, x + y) \pmod 1.$$

The map F_L is, in fact, an automorphism of the torus viewed as an additive group. It is invertible because the matrix $\begin{pmatrix} 2 & 1 \\ 1 & 1 \end{pmatrix}$ has determinant one, so L^{-1} also has integer entries [in fact $\begin{pmatrix} 2 & 1 \\ 1 & 1 \end{pmatrix}^{-1} = \begin{pmatrix} 1 & -1 \\ -1 & 2 \end{pmatrix}$] and hence defines a map $F_{L^{-1}} = F_L^{-1}$ on $\mathbb{T}^2$ by the same argument. The eigenvalues of L are

$$(7.1.3) \qquad \lambda_1 = \frac{3 + \sqrt{5}}{2} > 1 \quad \text{and} \quad \lambda_1^{-1} = \lambda_2 = \frac{3 - \sqrt{5}}{2} < 1.$$

Figure 7.1.4 gives an idea of the action of F_L on the fundamental square $I = \{(x, y) \mid 0 \leq x < 1, \ 0 \leq y \leq 1\}$. The lines with arrows are the eigendirections. For any matrix L with determinant ± 1, the map F_L preserves the area of sets on the torus.

Proposition 7.1.10 *Periodic points of F_L are dense and $P_n(F_L) = \lambda_1^n + \lambda_1^{-n} - 2$.*

Proof To obtain density we show that points with rational coordinates are periodic points. Let $x, y \in \mathbb{Q}$. Taking the common denominator write $x = s/q, y = t/q$, where $s, t, q \in \mathbb{Z}$. Then $F_L(s/q, t/q) = ((2s + t)/q, (s + t)/q)$ is a rational point whose coordinates also have denominator q. But there are only q^2 different points on $\mathbb{T}^2$ whose coordinates can be represented as rational numbers with denominator q, and all iterates $F_L^n(s/q, t/q)$, $n = 0, 1, 2 \ldots$, belong to that finite set. Thus they must repeat, that is, $F_L^n(s/q, t/q) = F_L^m(s/q, t/q)$ for some $n, m \in \mathbb{Z}$. But since F_L is invertible, $F_L^{n-m}(s/q, t/q) = (s/q, t/q)$ and $(s/q, t/q)$ is a periodic point, as required. This gives density. (By contrast, not all rational points are periodic for E_m. See Exercise 7.1.1.)

Next we show that points with rational coordinates are the only periodic points for F_L. Write $F_L^n(x, y) = (ax + by, cx + dy) \pmod 1$, where $a, b, c, d \in \mathbb{Z}$. If

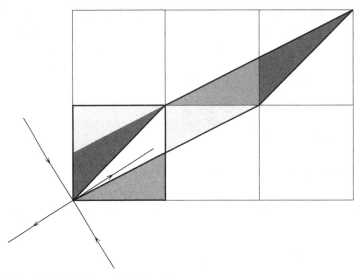

Figure 7.1.4. The hyperbolic toral map.

$F_L^n(x, y) = (x, y)$, then

$$ax + by = x + k,$$

$$cx + dy = y + l$$

for $k, l \in \mathbb{Z}$. Since 1 is not an eigenvalue for L^n, we can solve for (x, y):

$$x = \frac{(d-1)k - bl}{(a-1)(d-1) - cb}, \qquad y = \frac{(a-1)l - ck}{(a-1)(d-1) - cb}.$$

Thus $x, y \in \mathbb{Q}$.

Now we calculate $P_n(F_L)$. The map

$$G = F_L^n - \mathrm{Id}\colon (x, y) \mapsto ((a-1)x + by, \ cx + (d-1)y) \pmod 1$$

is a well-defined noninvertible map of the torus onto itself. As before, if $F_L^n(x, y) = (x, y)$, then $(a-1)x + by$ and $cx + (d-1)y$ are integers; hence $G(x, y) = 0 \pmod 1$, that is, the fixed points of F_L^n are exactly the preimages of the point $(0, 0)$ under G. Modulo 1 these are exactly the points of $\mathbb{Z}^2$ in $(L^n - \mathrm{Id})([0, 1) \times [0, 1))$. We presently show that their number is given by the area of $(L^n - \mathrm{Id})([0, 1) \times [0, 1))$, which is $|\det(L^n - \mathrm{Id})| = |(\lambda_1^n - 1)(\lambda_1^{-n} - 1)| = \lambda_1^n + \lambda_1^{-n} - 2$. $\square$

Lemma 7.1.11 *The area of a parallelogram with integer vertices is the number of lattice points it contains, where points on the edges are counted as half, and all vertices count as a single point.*

Proof Denote the area of the parallelogram by A. Adding the number of lattice points it contains in the prescribed way gives an integer N, which is the same for any translate of the parallelogram.

Now consider the canonical tiling of the plane by copies of this parallelogram translated by integer multiples of the edges. Denote by l the longest diagonal. The area of the tiles can be determined in a backward way by determining how many tiles lie in the square $[0, n) \times [0, n)$ for $n > 2l$. Those that lie inside cover the smaller square $[l, n - l) \times [l, n - l)$ completely, so there are at least

$$\frac{(n - 2l)^2}{A} \geq \frac{n^2}{A}\left(1 - \frac{4l}{n}\right).$$

Since any tile that intersects the square is contained in $[l, n - l) \times [l, n - l)$, there are at most

$$\frac{(n + 2l)^2}{A} = \frac{n^2}{A}\left(1 + \frac{4l}{n}\left(1 + \frac{l}{n}\right)\right) < \frac{n^2}{A}\left(1 + \frac{6l}{n}\right).$$

The number n^2 of integer points in the square is at least the number of points in tiles in the square and at most the number of points in tiles that intersect the square. Therefore

$$N \cdot \frac{n^2}{A}\left(1 - \frac{4l}{n}\right) \leq n^2 \leq N \cdot \frac{n^2}{A}\left(1 + \frac{6l}{n}\right) \qquad \text{and} \qquad 1 - \frac{4l}{n} \leq \frac{A}{N} \leq 1 + \frac{6l}{n}$$

for all $n > 2l$.

This shows that $N = A$. $\square$

7.1.5 Inverse Limits

The closest invertible analog to E_2 so far is the toral automorphism induced by $\left(\begin{smallmatrix} 2 & 1 \\ 1 & 1 \end{smallmatrix}\right)$. We digress briefly to describe a general construction that "makes a map invertible", that is, it that produces an invertible map from a noninvertible one in a standard way. The way to overcome noninvertibility is to replace the points of the given space by sequences $(x_n)_{n \in \mathbb{Z}}$ with $f(x_n) = x_{n+1}$. This way the ambiguity about preimages is resolved by listing the entire orbit explicitly. Indeed, the map $F((x_n)_{n \in \mathbb{Z}}) := (x_{n+1})_{n \in \mathbb{Z}}$ is clearly invertible.

Definition 7.1.12 If X is a metric space and $f \colon X \to X$ continuous, then the *inverse limit* is defined on the space

$$X' := \{(x_n)_{n \in \mathbb{Z}} \mid x_n \in X \text{ and } f(x_n) = x_{n+1} \text{ for all } n \in \mathbb{Z}\}$$

by $F((x_n)_{n \in \mathbb{Z}}) := (x_{n+1})_{n \in \mathbb{Z}}$.

Consider explicitly $f = E_2$ on S^1. Then the inverse limit is the space

$$\mathbb{S} := \{(x_n)_{n \in \mathbb{Z}} \mid x_n \in X \text{ and } f(x_n) = x_{n+1} \text{ for all } n \in \mathbb{Z}\}$$

with the map $F((x_n)_{n \in \mathbb{Z}}) := (x_{n+1})_{n \in \mathbb{Z}} = (2x_n)_{n \in \mathbb{Z}} \pmod 1$. This is called the *solenoid*.

Compared to listing the whole sequence, there is a more economical way to identify a point in $\mathbb{S}$. Once an entry, such as x_0, is specified, all subsequent entries are uniquely determined (by the orbit of $x_0 \in S^1$ under E_2). In order to specify all previous members of the sequence, one need only choose (recursively) one of two preimages at each step. For any given x_0 this can be coded by a one-sided 0-1-sequence. Since the space Ω_2 of these is a Cantor set (Section 7.3.5), the solenoid $\mathbb{S}$ is locally the product of an interval (points on S^1 near x_0) and a Cantor set.

There is a beautiful way to visualize the inverse limit construction. Beginning with a circle of "initial conditions" x_0, there are "twice as many" possible preimages x_{-1}, so the circle has to be doubled up like a rubber band around a newspaper. But there are twice as many second preimages, and so on, so it is necessary to double up ad infinitum. This is analogous to the construction of the ternary Cantor set, where an interval becomes two, then four, and so on.

The definitive geometric realization is carried out in Section 13.2 and illustrated in Figure 13.2.1 and on the cover of this book. This picture is representative of a great wealth of ideas in dynamics and deserves to be an icon for chaotic dynamics. Together with the horseshoe and linear toral automorphisms, the expanding map E_2 and the solenoid are the most tractable representatives of hyperbolic dynamical systems, and these have provided the framework of concepts and techniques within which each chaotic dynamical system is studied and described. This framework is developed in this chapter and the next, and it is described further in Chapter 10.

▨ EXERCISES

▨ **Exercise 7.1.1** Prove that for the expanding map $E_m(|m| \geq 2)$ rational points are preimages of periodic points ("eventually periodic").

▨ **Exercise 7.1.2** Find a necessary and sufficient condition for a rational point to be periodic under E_m.

▨ **Exercise 7.1.3** Carry out the proof of Proposition 7.1.3 for the case $m < -1$.

▨ **Exercise 7.1.4** Prove that for any $n \in \mathbb{N}$ and $\lambda \geq 4$ the quadratic map f_λ has a periodic point whose *minimal* period (Definition 2.2.6) is n.

▨ **Exercise 7.1.5** Give an example of a continuous map $f \colon [0, 1] \to \mathbb{R}$ with $f(0) = f(1) = 0$ for which there exists $c \in [0, 1]$ such that $f(c) > 1$, and such that $P_n(f) > 2^n$.

▨ **Exercise 7.1.6** Give an example of a smooth unimodal map f such that $P_n(f) < 2^n$.

▨ **Exercise 7.1.7** Show that a continuous map f of S^1 can be deformed to $E_{\deg(f)}$, that is, that there is a continuous map $F \colon [0, 1] \times S^1 \to S^1$ with $F(0, \cdot) = E_{\deg(f)}$ and $F(1, \cdot) = f$.

▨ **Exercise 7.1.8** Show that maps of different degrees cannot be deformed into each other, that is, that there is no continuous map $F \colon [0, 1] \times S^1 \to S^1$ such that $\deg(F(0, \cdot)) \neq \deg(F(1, \cdot))$.

▨ **Exercise 7.1.9** Suppose $f \colon S^1 \to S^1$ has degree 2 and 0 is an attracting fixed point. Show that $P_n(f) > 2^n$.

▨ **Exercise 7.1.10** Consider the Fibonacci sequence from Section 1.2.2, Example 2.2.9, and Section 3.1.9. Show that the sequence obtained from taking the last digit of each Fibonacci number is periodic.

▨ **Exercise 7.1.11** Apply the inverse limit construction to a homeomorphism and prove that the result is naturally equivalent to the original system.

▨ PROBLEMS FOR FURTHER STUDY

▨ **Problem 7.1.12** Prove that the solenoid in Section 7.1.5 is connected but not path-connected.

7.2 TOPOLOGICAL TRANSITIVITY AND CHAOS

We will show that some of the examples considered in the previous section are topologically transitive in the sense of Definition 4.1.3, that is, they have dense orbits. That there are at the same time infinitely many periodic points makes these examples different from irrational rotations and the other topologically transitive examples of Chapter 4 and Chapter 5. In expanding maps and hyperbolic linear maps of the torus we even found that the union of the periodic points is dense, which means that dense and periodic orbits are inextricably intertwined.

Thus, the global orbit structure is far more complex in these examples. This intertwining of density and periodicity is an essential feature of the complexity of the orbit structure. It causes sensitive dependence of any orbit on its initial conditions (see Definition 7.2.11 and Theorem 7.2.12), which is regarded as an essential ingredient of *chaos*.

Definition 7.2.1 A continuous map $f: X \to X$ of a metric space is said to be *chaotic* if it is topologically transitive and its periodic points are dense.[1]

Circle rotations show that neither condition alone gives much complexity.

We will show presently that expanding and hyperbolic maps are chaotic. In fact, we show the stronger property of topological mixing (Definition 7.2.5), which is absent in the minimal examples of Chapter 4 and Chapter 5. Before introducing the mixing property, we give an alternative definition of topological transitivity.

7.2.1 A Criterion for Topological Transitivity

We defined topological transitivity as the existence of a dense orbit. However, it is useful to have an alternate characterization in terms of subsets of phase space. In order to include noninvertible maps, we say that a sequence $(x_i)_{i \in \mathbb{Z}}$ is an orbit of f if $f(x_i) = x_{i+1}$ for all $i \in \mathbb{Z}$. However, we simply write $f^i(x)$ for $i \in \mathbb{Z}$ anyway to keep the notations more familiar.

Proposition 7.2.2 *Let X be a complete separable (that is, there is a countable dense subset) metric space with no isolated points. If $f: X \to X$ is a continuous map, then the following four conditions are equivalent:*

    (1) *f is topologically transitive, that is, it has a dense orbit.*
    (2) *f has a dense positive semiorbit.*
    (3) *If $\varnothing \neq U, V \subset X$, then there exists an $N \in \mathbb{Z}$ such that $f^N(U) \cap V \neq \varnothing$.*
    (4) *If $\varnothing \neq U, V \subset X$, then there exists an $N \in \mathbb{N}$ such that $f^N(U) \cap V \neq \varnothing$.*

Of course, the implications (4) $\Rightarrow$ (3) and (2) $\Rightarrow$ (1) are clear. To show which hypotheses are needed for which of the remaining directions, we prove Proposition 7.2.2 in the following form.

---

[1] There is no universally accepted definition of chaos, but this definition is equivalent to the one most commonly found in expository literature, which was put forward by Robert Devaney.

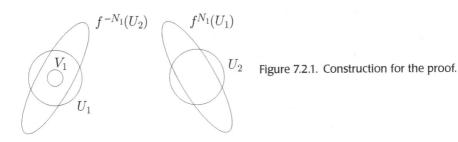

Figure 7.2.1. Construction for the proof.

Lemma 7.2.3 *Let X be a metric space and $f: X \to X$ a continuous map. Then (1) implies (3). If X has no isolated points, then (1) implies (4). If X is separable, then (3) implies (1) and (4) implies (2).*

Proof Let f be topologically transitive and suppose the orbit of $x \in X$ is dense. Then there exists an $n \in \mathbb{Z}$ such that $f^n(x) \in U$, and there is an $m \in \mathbb{Z}$ such that $f^m(x) \in V$; hence $f^{m-n}(U) \cap V \neq \varnothing$. This implies (3).

If we can choose $m > n$, then by taking $N := m - n$ we have even established (4). Otherwise we use the assumption that X has no isolated points, so $f^m(x)$ is not an isolated point and therefore there are $n_k \in \mathbb{Z}$ such that $|n_k| \to \infty$, $f^{n_k}(x) \in V$, and $f^{n_k}(x) \to f^m(x)$ as $k \to \infty$. Indeed, $n_k \to -\infty$ since $n_k \leq n$ by assumption (otherwise we are in the first case), so we can choose an $m' < 2m - n$ from among the n_k such that $f^{m'}(x) \in f^{m-n}(U)$. Then $x' := f^{n-m}(f^{m'}(x)) \in U$ and $f^{2m-n-m'}(x') = f^m(x) \in V$, so $f^N(U) \cap V \neq \varnothing$ with $N := 2m - n - m' \in \mathbb{N}$. Thus (1) $\Rightarrow$ (4) if X has no isolated points.

Now assume separability and one of the intersection conditions (3) and (4). We give one argument to prove both that (3) implies (1) and (4) implies (2). For a countable dense subset $S \subset X$, let $U_1, U_2, \ldots$ be the countable collection of balls centered at points of S with rational radius. We need to construct an orbit or semiorbit that intersects every U_n. By (3) there exists $N_1 \in \mathbb{Z}$ such that $f^{N_1}(U_1) \cap U_2 \neq \varnothing$. If (4) holds, we can take $N_1 \in \mathbb{N}$. Let V_1 be an open ball of radius at most $1/2$ such that $\bar{V}_1 \subset U_1 \cap f^{-N_1}(U_2)$. (See Figure 7.2.1.) There exists $N_2 \in \mathbb{Z}$ such that $f^{N_2}(V_1) \cap U_3$ is nonempty, and, if (4) holds, we can take $N_2 \in \mathbb{N}$. Again, take an open ball V_2 of radius at most $1/4$ such that $\bar{V}_2 \subset V_1 \cap f^{-N_2}(U_3)$. By induction, we construct a nested sequence of open balls V_n of radii at most 2^{-n} such that $\bar{V}_{n+1} \subset V_n \cap f^{-N_{n+1}}(U_{n+2})$. The centers of these balls form a Cauchy sequence whose limit x is the unique point in the intersection $V = \bigcap_{n=1}^{\infty} \bar{V}_n = \bigcap_{n=1}^{\infty} V_n$. Then $f^{N_{n-1}}(x) \in U_n$ for every $n \in \mathbb{N}$, and all $N_n \in \mathbb{N}$ if (4) holds.

If f is noninvertible, the last step may involve choices for negative values of N_n: Take i_k such that $N_{i_k} < 0$ for all k and $N_{i_{k+1}} < N_{i_k}$. Choose $x_0 = x$ and $x_{N_{i_k}} \in U_{i_k+1}$. Together with $f(x_k) = x_{k+1}$, this defines an orbit of x. $\square$

Corollary 7.2.4 *A continuous open (Definition A.1.16) map f of a complete separable metric space without isolated points is topologically transitive if and only if there are no two disjoint open nonempty f-invariant sets.*

Proof If $U, V \subset X$ are open, then the invariant sets $\tilde{U} := \bigcup_{n \in \mathbb{Z}} f^n(U)$ and $\tilde{V} := \bigcup_{n \in \mathbb{Z}} f^n(V)$ are open because f is an open map, and therefore not disjoint by assumption, so $f^n(U) \cap f^m(V) \neq \varnothing$ for some $n, m \in \mathbb{Z}$. Then $f^{n-m}(U) \cap V \neq \varnothing$ and f is topologically transitive by Proposition 7.2.2. The other direction is obvious: A dense orbit visits every open set. $\square$

7.2.2 Topological Mixing
There is a property of a dynamical system that immediately implies this criterion but is indeed much stronger:

Definition 7.2.5 A continuous map $f \colon X \to X$ is said to be *topologically mixing* if for any two nonempty open sets $U, V \subset X$ there is an $N \in \mathbb{N}$ such that $f^n(U) \cap V \neq \varnothing$ for every $n > N$.

By Proposition 7.2.2, every topologically mixing map is topologically transitive. On the other hand, our simple examples are not mixing. No translation T_y, in particular no circle rotation, is topologically mixing. This follows from the fact that translations preserve the natural metric on the torus induced by the standard Euclidean metric on $\mathbb{R}^n$ and from the following general criterion.

Lemma 7.2.6 *Isometries are not topologically mixing.*

Proof Let $f \colon X \to X$ be an isometry (that is, a map that preserves the metric on X). Take distinct points $x, y, z \in X$, and let $\delta := \min(d(x, y), d(y, z), d(z, x))/4$. Let U, V_1, V_2 be δ-balls around x, y, z correspondingly. Since f preserves the diameter of any set, the diameter of $f^n(U)$ is at most 2δ whereas the distance between any $p \in V_1$ and $q \in V_2$ is greater than 2δ. Thus for each n either $f^n(U) \cap V_1 = \varnothing$ or $f^n(U) \cap V_2 = \varnothing$. $\square$

7.2.3 Expanding Maps
For expanding maps we prove topological mixing by showing the stronger fact that, for any open set, its image under some iterate of the map contains S^1. For the linear expanding maps E_m this is obvious: Every open set contains an interval of the form $[l/|m|^k, (l+1)/|m|^k]$ for some integers k and $l \leq |m|^k$. The image of this interval under E_m^k is S^1.

Proposition 7.2.7 *Expanding maps of S^1 are topologically mixing.*

Proof Let $f \colon S^1 \to S^1$ such that $|f'(x)| \geq \lambda > 1$ for all x. Consider a lift F of f to $\mathbb{R}$. Then $|F'(x)| \geq \lambda$ for $x \in \mathbb{R}$. If $[a, b] \subset \mathbb{R}$ is an interval, then by the Mean-Value Theorem A.2.3 there exists a $c \in (a, b)$ such that $|F(b) - F(a)| = |F'(c)(b - a)| \geq \lambda(b - a)$ and so the length of any interval is increased by a factor at least λ^n under F^n. Consequently, for every interval I there exists $n \in \mathbb{N}$ such that the length of $F(I)$ exceeds 1. Thus the image of the projection of I to S^1 under f^n contains S^1. Since every open set of S^1 contains an interval, this shows that every open set has an image under an iterate of f that contains S^1. $\square$

Corollary 7.2.8 *Linear expanding maps of S^1 are chaotic.*

Proof Transitivity follows from Proposition 7.2.7 and the density of periodic points from Proposition 7.1.3. ☐

For nonlinear expanding maps, this result also holds by invoking Theorem 7.4.3 (which is only stated for degree 2 in the following, but holds for any expanding map).

7.2.4 Hyperbolic Linear Map on the Torus

The hyperbolic linear map F_L of the torus induced by the linear map L with matrix $\left(\begin{smallmatrix}2&1\\1&1\end{smallmatrix}\right)$ was introduced in Section 7.1.4. The eigenvectors corresponding to the first eigenvalue belong to the line $y = (\sqrt{5} - 1/2)x$. The family of lines parallel to it is invariant under L, and L uniformly expands distances on those lines by a factor λ_1. Similarly, there is an invariant family of contracting lines $y = (-\sqrt{5} - 1/2)x + \text{const}$.

Proposition 7.2.9 *The automorphism F_L is topologically mixing.*

Proof Fix open sets $U, V \subset \mathbb{T}^2$. The L-invariant family of lines

$$(7.2.1) \qquad\qquad y = \frac{\sqrt{5} - 1}{2}x + \text{const}.$$

projects to $\mathbb{T}^2$ as an F_L-invariant family of orbits of the linear flow T_ω^t with irrational slope $\omega = (1, (\sqrt{5} - 1)/2)$. By Proposition 5.1.3, this flow is minimal. Thus the projection of each line is everywhere dense on the torus, and hence U contains a piece J of an expanding line; furthermore, for any $\epsilon > 0$, there exists $T = T(\epsilon)$ and a segment of an expanding line of length T that intersects any ϵ-ball on the torus. Since all segments of a given length are translations of one another, this property holds for all segments. Now take ϵ such that V contains an ϵ-ball and $N \in \mathbb{N}$ such that $f^N(J)$ has length at least T. Then $f^n(J) \cap V \neq \varnothing$ for $n \geq N$ and thus $f^n(U) \cap V \neq \varnothing$ for $n \geq N$. ☐

Corollary 7.2.10 *The automorphism F_L is chaotic.*

Proof Combine Proposition 7.2.9, and Proposition 7.1.10. ☐

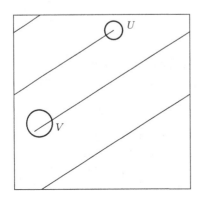

Figure 7.2.2. Topological mixing.

7.2.5 Chaos

At the outset of this section we motivated our definition of a chaotic map by saying that it implies sensitive dependence on initial conditions. We now justify this claim by defining and verifying sensitive dependence.

Definition 7.2.11 A map $f: X \to X$ of a metric space is said to exhibit *sensitive dependence* on initial conditions if there is a $\Delta > 0$, called a sensitivity constant, such that for every $x \in X$ and $\epsilon > 0$ there exists a point $y \in X$ with $d(x, y) < \epsilon$ and $d(f^N(x), f^N(y)) \geq \Delta$ for some $N \in \mathbb{N}$.

This means that the slightest error (ϵ) in any initial condition (x) can lead to a macroscopic discrepancy (Δ) in the evolution of the dynamics. Accordingly, Δ tells us at what scale these errors show up. Suppose I start a dynamical system in a state x, let it evolve for a while, and try to reproduce this experiment. Even if I reproduce x to within a billionth of an inch, the initial minuscule error may magnify to a large difference in behavior in finite (often relatively short) time, that is, I may find that the second run of the same experiment bears little resemblance to the first. This is what Poincaré meant by his comment quoted in Section 1.1.1.

For linear expanding maps this property is clearly true: Any initial error of an orbit for E_m grows exponentially (by a factor of $|m|$ at every iteration) until it has grown to more than $1/2|m|$. In particular, $\delta = 1/2|m|$ is a sensitivity constant. On the other hand, this property clearly fails for isometries because points do not move apart at all under iteration.

It is important for the definition that Δ does not depend on x, nor on ϵ, but only on the system. Thus, the *smallest* error *anywhere* can lead to discrepancies of size Δ eventually.[2]

Theorem 7.2.12 *Chaotic maps exhibit sensitive dependence on initial conditions, except when the entire space consists of a single periodic orbit.*

Proof Unless the entire space consists only of a single periodic orbit, the density of periodic points implies that there are two distinct periodic orbits. Since they have no common point, there are periodic points p, q such that $\Delta := \min\{d(f^n(p), f^m(q)) \mid n, m \in \mathbb{Z}\}/8 > 0$. (Note that n and m need not agree.) We now show that Δ is a sensitivity constant.

If $x \in X$, the orbit of one of these two points keeps a distance at least 4Δ from x: If they were both within less than 4Δ of x, then their mutual distance would be less than 8Δ. Suppose this point is q.

Take any $\epsilon \in (0, \Delta)$. By the density of periodic points, there is a periodic point $p \in B(x, \epsilon)$ whose period we call n. Then the set

$$V := \bigcap_{i=0}^{n} f^{-i}(B(f^i(q), \Delta))$$

---

[2] The meteorologist Edward Lorenz described this as the "butterfly effect": Weather appears to be a chaotic dynamical system, so it is conceivable that a butterfly that flutters by in Rio may cause a typhoon in Tokyo a few days later.

of points whose first n iterates track those of q up to Δ is an open neighborhood of q. By Proposition 7.2.2 (used in the direction that does not require completeness) there exists a $k \in \mathbb{N}$ such that $f^k(B(x, \epsilon)) \cap V \neq \varnothing$, that is, there exists a $y \in B(x, \epsilon)$ such that $f^k(y) \in V$. If $j := \lfloor k/n \rfloor + 1$, then $k/n < j \le (k/n) + 1$ and

$$k = n \cdot \frac{k}{n} < nj \le n\left(\frac{k}{n} + 1\right) = k + n.$$

If we take $N := nj$, then this shows that $0 < N - k \le n$. Since $f^N(p) = p$, the triangle inequality gives

$$d(f^N(p), f^N(y)) = d(p, f^N(y))$$

(7.2.2)
$$\ge d(x, f^{N-k}(q)) - d(f^{N-k}(q), f^N(y)) - d(p, x)$$

$$\ge 4\Delta - \Delta - \Delta = 2\Delta$$

because $p \in B(x, \epsilon) \subset B(x, \Delta)$ and

$$f^N(y) = f^{N-k}(f^k(y)) \in f^{N-k}(V) \subset B(f^{N-k}(q), \Delta)$$

by definition of V. Both p and y are in $B(x, \epsilon)$ and either $d(f^N(p), f^N(x)) \ge \Delta$ or $d(f^N(y), f^N(x)) \ge \Delta$ by (7.2.2). $\square$

Remark 7.2.13 There are maps exhibiting sensitive dependence that are not chaotic, such as the linear twist from Section 6.1.1. Here, any point x has arbitrarily nearby points (on a vertical segment through x) that move a considerable distance away after sufficiently many iterates. The set of periodic points consists of those points whose second coordinate is rational and is hence dense. On the other hand, this map is clearly not topologically transitive.

Sensitive dependence can be derived from topological mixing alone, without an assumption on periodic points:

Proposition 7.2.14 *A topologically mixing map (on a space with more than one point) has sensitive dependence.*

Proof Take $\Delta > 0$ such that there are points x_1, x_2 with $d(x_1, x_2) > 4\Delta$. We show that Δ is a sensitivity constant.

Let $V_i = B_\Delta(x_i)$ for $i = 1, 2$. Suppose $x \in X$ and U is a neighborhood of x. By topological mixing there are $N_1, N_2 \in \mathbb{N}$ such that $f^n(U) \cap V_1 \neq \varnothing$ for $n \ge N_1$ and $f^n(U) \cap V_2 \neq \varnothing$ for $n \ge N_2$. For $n \ge N := \max(N_1, N_2)$, there are points $y_1, y_2 \in U$ with $f^n(y_1) \in V_1$ and $f^n(y_2) \in V_2$; hence $d(f^n(y_1), f^n(y_2)) \ge 2\Delta$. By the triangle inequality $d(f^n(y_1), f^n(x)) \ge \Delta$ or $d(f^n(y_2), f^n(x)) \ge \Delta$. $\square$

▦ EXERCISES

▦ **Exercise 7.2.1** Find the maximal sensitivity constant for E_2.

▦ **Exercise 7.2.2** Find the supremum of sensitivity constants for F_L in Section 7.2.4.

▦ **Exercise 7.2.3** Prove that, for a topologically mixing map, any number less than the diameter $\sup\{d(x, y) \mid x, y \in X\}$ of the space X is a sensitivity constant.

■ **Exercise 7.2.4** Consider the linear twist $T: S^1 \times [0, 1] \to S^1 \times [0, 1]$, $T(x, y) = (x + y, y)$ from Section 6.1.1 that was remarked upon in Remark 7.2.13. Prove that it has the following property of *partial topological mixing*: Let $U, V \subset S^1$ be nonempty open sets. Then there exists $N(U, V) \in \mathbb{N}$ such that $T^n(U \times [0, 1]) \cap (V \times [0, 1]) \neq \varnothing$ for any $n \geq N$.

■ **Exercise 7.2.5** Show that for a compact space sensitive dependence is a topological invariant (see Section 7.3.6).

■ **Exercise 7.2.6** Prove that for any two periodic points of F_L the set of heteroclinic points (see Definition 2.3.4) is dense.

■ **Exercise 7.2.7** Consider a 2×2 integer matrix L without eigenvalues of absolute value 1 and with $|\det L| > 1$. Prove that the induced noninvertible hyperbolic linear map $F_L: \mathbb{T}^2 \to \mathbb{T}^2$ is topologically mixing.

7.3 CODING

One of the most important ideas for studying complicated dynamics sounds strange at first. It involves throwing away some information by tracking orbits only approximately. The idea is to divide the phase space into finitely many pieces and to follow an orbit only to the extent of specifying which piece it is in at a given time. This is a bit like the itinerary of the harried tourist in Europe, who decides that it is Tuesday, so the place must be Belgium. A more technological analogy would be to look at the records of a cell phone addict and track which local transmitters were used at various times.

In these analogies one genuinely loses information, because the sequence of European countries or of local cellular stations does not pinpoint the traveller at any given moment. However, orbits in a dynamical system do not move around at whim, and the deterministic nature of the dynamics has the effect that a complete *itinerary* of this sort may (and often does) give all the information about a point. This is the process of *coding* of a dynamical system.

7.3.1 Linear Expanding Maps
The linear expanding maps

$$E_m: S^1 \to S^1, \quad E_m(x) = mx \pmod 1$$

from Section 7.1.1 are chaotic (Corollary 7.2.8), that is, they exhibit coexistence of dense orbits (Proposition 7.2.7) with a countable dense set of periodic orbits (Proposition 7.1.3). Thus the orbit structure is both complicated and highly nonuniform. Now we look at these maps from a different point of view, which in turn gives a deeper appreciation of just how complicated their orbit structure really is. To simplify notations, assume as before that $m = 2$.

Consider the binary intervals

$$\Delta_n^k := \left[\frac{k}{2^n}, \frac{k+1}{2^n} \right] \quad \text{for } n = 1, \dots \quad \text{and} \quad k = 0, 1, \dots, 2^n - 1.$$

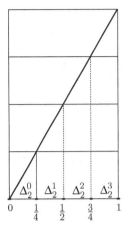

Figure 7.3.1. Linear coding.

Figure 7.3.1 illustrates this for $n = 2$. Let $x = 0.x_1 x_2 \ldots$ be the binary representation of $x \in [0, 1]$. Then $2x = x_1.x_2 x_3 \ldots = 0.x_2 x_3 \ldots \pmod 1$. Thus

(7.3.1) $$E_2(x) = 0.x_2 x_3 \ldots \pmod 1.$$

This is the first and easiest example of *coding*, which we will discuss in greater detail shortly.

7.3.2 Implications of Coding

We briefly derive a few new facts about linear expanding maps that are best seen via this coding.

1. Proof of Transitivity via Coding. First, we use this representation to give another proof of topological transitivity by describing explicitly the binary representation of a number whose orbit under the iterates of E_2 is dense. Consider an integer k, $0 \le k \le 2^n - 1$. Let $k_0 \ldots k_{n-1}$ be the binary representation of k, maybe with several zeroes at the beginning. Then $x \in \Delta_n^k$ if and only if $x_i = k_i$ for $i = 0, \ldots n - 1$. Therefore we write $\Delta_{k_0 \ldots k_{n-1}} := \Delta_n^k$ from now on. Now put the binary representations of all numbers from 0 to $2^n - 1$ (with zeroes in front if necessary) one after another and form a finite sequence, which we denote by ω_n, that is, ω_n is obtained by concatenating all 2^n binary sequences of length n. Having done this for every $n \in \mathbb{N}$, put the sequences ω_n, $n = 1, 2, \ldots$ in that order, call the resulting infinite sequence ω, and consider the number x with the binary representation $0.\omega$. Since by construction moving ω to the left and cutting off the first digits produces at various moments binary representations of any n-digit number, this means that the orbit of the point x under the iterates of the map E_2 intersects every interval $\Delta_{k_0 \ldots k_{n-1}}$ and hence is dense.

This construction extends to any $m \ge 2$. To construct a dense orbit for E_m with $m \le -2$, we notice that $E_m^2 = E_{m^2}$. Obviously the orbit of any point under the iterates of a square of a map is a subset of the orbit under the iterates of the map itself; thus if the former is dense, so is latter. So we apply our construction to the map E_{m^2} and obtain a point with dense orbit under E_m.

2. Exotic Asymptotics. Next we use this approach to show that besides periodic and dense orbits there are other types of asymptotic behavior for orbits of expanding maps. One can construct such orbits for E_2, but the simplest and most elegant example appears for the map E_3.

Proposition 7.3.1 *There exists a point $x \in S^1$ such that the closure of its orbit with respect to the map E_3 in additive notation coincides with the standard middle-third Cantor set K. In particular, K is E_3-invariant and contains a dense orbit.*

Proof The middle-third Cantor set K can be described as the set of all points on the unit interval that have a representation in base 3 with only 0's and 2's as digits (see Section 2.7.1). Similarly to (7.3.1), the map E_3 acts as the shift of digits to the left in the base 3 representation. This implies that K is E_3-invariant. It remains to show that E_3 has a dense orbit in K.

Every point in K has a unique representation in base 3 without 1's. Let $x \in K$ and

(7.3.2)
$$0.x_1 x_2 x_3 \ldots$$

be such a representation. Let $h(x)$ be the number whose representation in base 2 is

$$0.\frac{x_1}{2} \frac{x_2}{2} \frac{x_3}{2} \ldots,$$

that is, it is obtained from (7.3.2) by replacing 2's by 1's. Thus we have constructed a map $h \colon K \to [0, 1]$ that is continuous, nondecreasing [that is, $x > y$ implies $h(x) \ge h(y)$], and one-to-one, except for the fact that binary rationals have two preimages each (compare Section 2.7.1 and Section 4.4.1). Furthermore, $h \circ E_3 = E_2 \circ h$. Let $D \subset [0, 1]$ be a dense set of points that does not contain binary rationals. Then $h^{-1}(D)$ is dense in K because, if Δ is an open interval such that $\Delta \cap K \ne \varnothing$, then $h(\Delta)$ is a nonempty interval open, closed, or semiclosed and hence contains points of D. Now take any $x \in [0, 1]$ whose E_2-orbit is dense; the E_3-orbit of $h^{-1}(x) \in K$ is dense in K. $\square$

3. Nonrecurrent Points. Another interesting example is the construction of a nonrecurrent point, that is, such a point x that for some neighborhood U of x all iterates of x avoid U (see Definition 6.1.8). In fact, there is a dense set of nonrecurrent points for the map E_2.

Pick any fixed sequence $(\omega_0, \ldots, \omega_{n-1})$ of 0's and 1's and add a tail of 0's if $\omega_{n-1} = 1$, or of 1's if $\omega_{n-1} = 0$. Call the resulting infinite sequence ω. As before, let x be the number with binary representation $0.\omega$. Thus, x lies in a prescribed interval $\Delta_{\omega_0 \ldots \omega_{n-1}}$ and by construction $x \ne 0$. On the other hand, $E_2^n x = 0$ and hence $E_2^m x = 0$ for all $m \ge n$, so x is a nonrecurrent point.

Thus, we have found that E_m is chaotic and topologically mixing, that its periodic and nonrecurrent orbits are dense, and that E_3 has orbits whose closure is a Cantor set.

7.3.3 A Two-Dimensional Cantor Set
We now describe a map in the plane that naturally gives rise to a two-dimensional Cantor set (previously encountered in Problem 2.7.5) on which ternary expansion of the coordinates provides all information about the dynamics. This *horseshoe map* plays a central role in our further development.

Consider a map defined on the unit square $[0, 1] \times [0, 1]$ by the following construction: First apply the linear transformation $(x, y) \mapsto (3x, y/3)$ to get a horizontal strip whose left and right thirds will be rigid in the next transformation. Holding the left third fixed, bend and stretch the middle third such that the right third falls rigidly on the top third of the original unit square. This results in a "G"-shape. For points that are in and return to the unit square, this map is given analytically by

$$(x, y) \mapsto \begin{cases} (3x, y/3) & \text{if } x \leq 1/3 \\ (3x - 2, (y + 2)/3) & \text{if } x \geq 2/3. \end{cases}$$

The inverse can be written as

$$(x, y) \mapsto \begin{cases} (x/3, 3y) & \text{if } y \leq 1/3 \\ ((x + 2)/3, 3y - 2) & \text{if } y \geq 2/3. \end{cases}$$

Geometrically, the inverse looks like an "e"-shape rotated counterclockwise by $90°$.

To iterate this map one triples the x-coordinate repeatedly and always assumes that the resulting value is either at most $1/3$ or else at least $2/3$, that is, that the first ternary digit is 0 or 2, but not 1. (If the expansion is not unique, one requires such a choice to be possible.) Comparing with the construction of the ternary Cantor set in Section 2.7.1, one sees that the x-coordinate lies in the ternary Cantor set C. Looking at the inverse one sees likewise that, in order for all preimages to be defined, the y-coordinate lies in the Cantor set as well. Therefore this map is defined for all positive and negative iterates on the two-dimensional Cantor set $C \times C$. There is a straightforward way of using ternary expansion to code the dynamics. For a point (x, y) the map shifts the ternary expansion of x one step to the left, dropping the first term, and shifts the ternary expansion of y to the right. It is natural to fill in the now-ambiguous first digit of the shifted y-coordinate with the entry from the x-coordinate that was just dropped. This retains all information, and the best way of vizualizing the result is to write the expansion of the y-coordinate in reverse and in front of that of the x-coordinate. This gives a bi-infinite string of 0's and 2's (remember, no 1's allowed), which is shifted by the map. Of course, one should verify that the inverse acts by shifting in the opposite direction.

7.3.4 Sequence Spaces

Now we are ready to discuss the concept of coding in general. We mean by coding a representation of points in the phase space of a discrete-time dynamical system or an invariant subset by sequences (not necessarily unique) of symbols from a certain "alphabet," in this case the symbols $0, \ldots, N - 1$. So we should acquaint ourselves with these spaces.

Denote by Ω_N^R the space of sequences $\omega = (\omega_i)_{i=0}^\infty$ whose entries are integers between 0 and $N - 1$. Define a metric by

(7.3.3)
$$d_\lambda(\omega, \omega') := \sum_{i=0}^\infty \frac{\delta(\omega_i, \omega_i')}{\lambda^i},$$

where $\delta(k, l) = 1$ if $k \neq l$, $\delta(k, k) = 0$, and $\lambda > 2$. The same definition can be made for two-sided sequences by summing over $i \in \mathbb{Z}$:

$$(7.3.4) \qquad d_\lambda(\omega, \omega') := \sum_{i \in \mathbb{Z}} \frac{\delta(\omega_i, \omega'_i)}{\lambda^{|i|}},$$

for some $\lambda > 3$. This means that two sequences are close if they agree on a long stretch of entries around the origin.

Consider the symmetric *cylinder* defined by

$$C_{\alpha_{1-n}\ldots\alpha_{n-1}} := \{\omega \in \Omega_N \mid \omega_i = \alpha_i \text{ for } |i| < n\}.$$

Fix a sequence $\alpha \in C_{\alpha_{1-n}\ldots\alpha_{n-1}}$. If $\omega \in C_{\alpha_{1-n}\ldots\alpha_{n-1}}$, then

$$d_\lambda(\alpha, \omega) = \sum_{i \in \mathbb{Z}} \frac{\delta(\alpha_i, \omega_i)}{\lambda^{|i|}} = \sum_{|i| \geq n} \frac{\delta(\alpha_i, \omega_i)}{\lambda^{|i|}} \leq \sum_{|i| \geq n} \frac{1}{\lambda^{|i|}} = \frac{1}{\lambda^{n-1}} \frac{2}{\lambda - 1} < \frac{1}{\lambda^{n-1}}.$$

Thus $C_{\alpha_{1-n}\ldots\alpha_{n-1}} \subset B_{d_\lambda}(\alpha, \lambda^{1-n})$, the λ^{1-n}-ball around α. If $\omega \notin C_{\alpha_{1-n}\ldots\alpha_{n-1}}$, then

$$d_\lambda(\alpha, \omega) = \sum_{i \in \mathbb{Z}} \frac{\delta(\alpha_i, \omega_i)}{\lambda^{|i|}} \geq \lambda^{1-n}$$

because $\omega_i \neq \alpha_i$ for some $|i| < n$. Thus $\omega \notin B_{d_\lambda}(\alpha, \lambda^{1-n})$, and the symmetric cylinder is the ball of radius λ^{1-n} around any of its points:

$$(7.3.5) \qquad C_{\alpha_{1-n}\ldots\alpha_{n-1}} = B_{d_\lambda}(\alpha, \lambda^{1-n}).$$

Therefore, balls in Ω_N are described by specifying a symmetric stretch of entries around the initial one.

For one-sided sequences this discussion works along the same lines [one only needs $\lambda > 2$ in (7.3.4)] and λ^{1-n}-balls are described by specifying a string of n initial entries.

Our examples [see (7.3.1)] suggest to represent points in the phase space by sequences in such a way that the sequences representing the image of a point are obtained from those representing the point itself by the shift (translation) of the symbols. In this way the given transformation corresponds to the *shift transformation*

$$(7.3.6) \qquad \begin{array}{ll} \sigma: \Omega_N \to \Omega_N, & (\sigma\omega)_i = \omega_{i+1} \\ \sigma^R: \Omega_N^R \to \Omega_N^R, & (\sigma^R\omega)_i = \omega_{i+1}. \end{array}$$

We often write σ_N for the shift σ on Ω_N and likewise σ_N^R for σ^R on Ω_N^R. For invertible discrete-time systems, any coding involves sequences of symbols extending in both directions; while for noninvertible systems, one-sided sequences do the job. Section 7.3.7 studies these shifts as dynamical systems.

Among the shift transformations that arise from coding there is also a new kind of combinatorial model for a dynamical system that is described by the possibility or impossibility of certain successions of events.

Definition 7.3.2 Let $A = (a_{ij})_{i,j=0}^{N-1}$ be an $N \times N$ matrix whose entries a_{ij} are either 0's or 1's. (We call such a matrix a 0-1 matrix.) Let

$$(7.3.7) \qquad \Omega_A := \{\omega \in \Omega_N \mid a_{\omega_n\omega_{n+1}} = 1 \text{ for } n \in \mathbb{Z}\}.$$

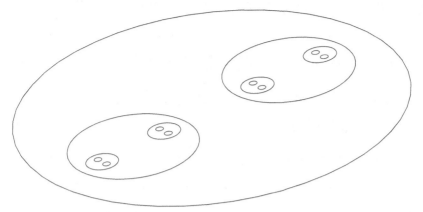

Figure 7.3.2. Obtaining a Cantor set.

The space Ω_A is closed and shift-invariant, and the restriction

$$\sigma_N\!\restriction_{\Omega_A} =: \sigma_A$$

is called the *topological Markov chain*determined by A.

This is a particular case of a *subshift of finite type*.

7.3.5 Coding

Sequences representing a given point of the phase space are called the *codes* of that point. We have several examples of coding: for the map E_m on the whole circle by sequences from the *alphabet* $\{0, \dots, |m| - 1\}$; for the restriction of the map E_3 to the middle-third Cantor set K by one-sided sequences of 0's and 1's; and for the ternary horseshoe in Section 7.3.3 by bi-infinite sequences of 0's and 2's. In both cases we used one-sided sequences, all sequences appeared as codes of some points, and each code represented only one point. There was, however, an important difference: In the first case, which involved for positive m a representation in base m, a point could have either one or two codes; in the latter there was only one code.

This shows that the space of binary sequences is a Cantor set (Definition 2.7.4). In fact, this also holds for the other sequence spaces.

7.3.6 Conjugacy and Factors

This situation can be roughly described by saying that the shift (Ω_2^R, σ^R) "contains" the map f up to a continuous coordinate change. (We already encountered such a situation in Theorem 4.3.20.)

Definition 7.3.3 Suppose that $g \colon X \to X$ and $f \colon Y \to Y$ are maps of metric spaces X and Y and that there is a continuous surjective map $h \colon X \to Y$ such that $h \circ g = f \circ h$. Then f is said to be a *factor* of g via the *semiconjugacy* or *factor map* h. If this h is a homeomorphism, then f and g are said to be *conjugate* and h is said to be a *conjugacy*.

These notions made a brief appearance in Section 4.3.5 in connection with modeling an arbitrary homeomorphism of the circle by a rotation. The notion of conjugacy is natural and central; two conjugate maps are obtained from one another by a continuous change of coordinates. Hence all properties that are independent of such changes of coordinates are unchanged, such as the numbers of periodic orbits for each period, sensitive dependence (Exercise 7.2.5), topological transitivity, topological mixing, and hence also being chaotic. Such properties are said to be topological invariants. Later in this book we will encounter further important topological invariants such as topological entropy (Definition 8.2.1).

7.3.7 Dynamics of Shifts and Topological Markov Chains

We now study the properties of shifts and topological Markov chains introduced in (7.3.6) and Definition 7.3.2 in more detail. These are important because many interesting dynamical systems are coded by shifts or topological Markov chains. To such dynamical systems the results of this section have immediate applications.

Proposition 7.3.4 *Periodic points for the shifts σ_N and σ_N^R are dense in Ω_N and Ω_N^R, correspondingly, $P_n(\sigma_N) = P_n(\sigma_N^R) = N^n$, and both σ_N and σ_N^R are topologically mixing.*

Proof Periodic orbits for a shift are periodic sequences, that is, $(\sigma_N)^m \omega = \omega$ if and only if $\omega_{n+m} = \omega_n$ for all $n \in \mathbb{Z}$. In order to prove density of periodic points, it is enough to find a periodic point in every ball (symmetric cylinder), because every open set contains a ball. To find a periodic point in $C_{\alpha_{-m},\dots,\alpha_m}$, take the sequence ω defined by $\omega_n = \alpha_{n'}$ for $|n'| \le m$, $n' = n \pmod{2m+1}$. It lies in this cylinder and has period $2m+1$.

Every periodic sequence ω of period n is uniquely determined by its coordinates $\omega_0, \dots, \omega_{n-1}$. There are N^n different finite sequences $(\omega_0, \dots, \omega_{n-1})$.

To prove topological mixing, we show that $\sigma_N^n(C_{\alpha_{-m},\dots,\alpha_m}) \cap C_{\beta_{-m},\dots,\beta_m} \ne \varnothing$ for $n > 2m+1$, say, $n = 2m+k+1$ with $k > 0$. Consider any sequence ω such that

$$\omega_i = \alpha_i \text{ for } |i| \le m, \qquad \omega_i = \beta_{i-n} \quad \text{for } i = m+k+1, \dots, 3m+k+1.$$

Then $\omega \in C_{\alpha_{-m},\dots,\alpha_m}$ and $\sigma_N^n(\omega) \in C_{\beta_{-m},\dots,\beta_m}$.

The arguments for the one-sided shift are analogous. $\square$

There is a useful geometric representation of topological Markov chains. Connect i with j by an arrow if $a_{ij} = 1$ to obtain a *Markov graph* G_A with N vertices and several oriented edges. We say that a finite or infinite sequence of vertices of G_A is an *admissible path* or *admissible sequence* if any two consecutive vertices in the sequence are connected by an oriented arrow. A point of Ω_A corresponds to a doubly infinite path in G_A with marked origin; the topological Markov chain σ_A corresponds to moving the origin to the next vertex. The following simple combinatorial lemma is a key to the study of topological Markov chains:

Lemma 7.3.5 *For every $i, j \in \{0, 1, \dots, N-1\}$, the number N_{ij}^m of admissible paths of length $m+1$ that begin at x_i and end at x_j is equal to the entry a_{ij}^m of the matrix A^m.*

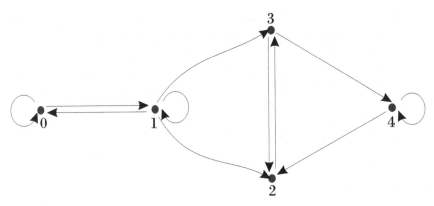

Figure 7.3.3. A Markov graph.

Proof We use induction on m. First, it follows from the definition of the graph G_A that $N_{ij}^1 = a_{ij}$. To show that

$$(7.3.8) \qquad\qquad N_{ij}^{m+1} = \sum_{k=0}^{N-1} N_{ik}^m a_{kj},$$

take $k \in \{0, \dots, N-1\}$ and an admissible path of length $m+1$ connecting i and k. It can be extended to an admissible path of length $m+2$ connecting i to j (by adding j) if and only if $a_{kj} = 1$. This proves (7.3.8). Now, assuming by induction that $N_{ij}^m = a_{ij}^m$ for all ij, we obtain $N_{ij}^{m+1} = a_{ij}^{m+1}$ from (7.3.8). $\square$

Corollary 7.3.6 $P_n(\sigma_A) = \mathrm{tr}\, A^n$.

Proof Every admissible closed path of length $m+1$ with marked origin, that is, a path that begins and ends at the same vertex of G_A, produces exactly one periodic point of σ_A of period m. $\square$

Because the eigenvalue of largest absolute value dominates the trace, it determines the exponential growth rate:

Proposition 7.3.7 $p(\sigma_A) = r(A)$, where $r(A)$ is the spectral radius.

Proof "$\leq$" is clear. To show "$\geq$" we need to avoid cancellations: If $\lambda_j = re^{2\pi i \varphi_j}(1 \leq j \leq k)$ are the eigenvalues of maximal absolute value then there is a sequence $m_n \to \infty$ such that $m_n \varphi_j \to 0 \,(\mathrm{mod}\ 1)$ for all j (recurrence for toral translations, Section 5.1), so $\sum_i \lambda_i^{m_n} \sim r^{m_n}$. $\square$

Example 7.3.8 The Markov graph in Figure 7.3.3 produces three fixed points, $\overline{0}$, $\overline{1}$, and $\overline{4}$. $\overline{01}$ and $\overline{23}$ give four periodic points with period 2. The period-3 orbits are generated by $\overline{011}$, $\overline{001}$, $\overline{234}$.

Topological Markov chains can be classified according to the recurrence properties of various orbits they contain. Now we concentrate on those topological Markov chains that possess the strongest recurrence properties.

Definition 7.3.9 A matrix A is said to be *positive* if all its entries are positive. A 0-1 matrix A is said to be *transitive* if A^m is positive for some $m \in \mathbb{N}$. A topological Markov chain σ_A is said to be *transitive* if A is a transitive matrix.

Lemma 7.3.10 *If A^m is positive, then so is A^n for any $n \geq m$.*

Proof If $a_{ij}^n > 0$ for all i, j, then for each j there is a k such that $a_{kj} = 1$. Otherwise, $a_{ij}^n = 0$ for every n and i. Now use induction. If $a_{ij}^n > 0$ for all i, j, then $a_{ij}^{n+1} = \sum_{k=0}^{N-1} a_{ik}^n a_{kj} > 0$ because $a_{kj} = 1$ for at least one k. $\square$

Lemma 7.3.11 *If A is transitive and $\alpha_{-k}, \ldots, \alpha_k$ is admissible, that is, $a_{\alpha_i \alpha_{i+1}} = 1$ for $i = -k, \ldots, k-1$, then the intersection $\Omega_A \cap C_{\alpha_{-k},\ldots,\alpha_k} =: C_{\alpha_{-k},\ldots,\alpha_k, A}$ is nonempty and moreover contains a periodic point.*

Proof Take m such that $a_{\alpha_k, \alpha_{-k}}^m > 0$. Then one can extend the sequence α to an admissible sequence of length $2k + m + 1$ that begins and ends with α_{-k}. Repeating this sequence periodically, we obtain a periodic point in $C_{\alpha_{-k},\ldots,\alpha_k, A}$. $\square$

Proposition 7.3.12 *If A is a transitive matrix, then the topological Markov chain σ_A is topologically mixing and its periodic orbits are dense in Ω_A; in particular, σ_A is chaotic and hence has sensitive dependence on initial conditions.*

Proof The density of periodic orbits follows from Lemma 7.3.11. To prove topological mixing, pick open sets $U, V \subset \Omega_A$ and nonempty symmetric cylinders $C_{\alpha_{-k},\ldots,\alpha_k, A} \subset U$ and $C_{\beta_{-k},\ldots,\beta_k, A} \subset V$. Then it suffices to show that $\sigma_A^n(C_{\alpha_{-k},\ldots,\alpha_k, A}) \cap C_{\beta_{-k},\ldots,\beta_k, A} \neq \varnothing$ for any sufficiently large n. Take $n = 2k + 1 + m + l$ with $l \geq 0$, where m is as in Definition 7.3.9. Then $a_{\alpha_k \beta_{-k}}^{m+l} > 0$ by Lemma 7.3.10, so there is an admissible sequence of length $4k + 2 + m + l$ whose first $2k + 1$ symbols are identical to $\alpha_{-k}, \ldots, \alpha_k$ and the last $2k + 1$ symbols to $\beta_{-k}, \ldots, \beta_k \beta$. By Lemma 7.3.11, this sequence can be extended to a periodic element of Ω_A which belongs to $\sigma_A^n(C_{\alpha_{-k},\ldots,\alpha_k, A}) \cap C_{\beta_{-k},\ldots,\beta_k, A}$. $\square$

Example 7.3.13 The matrix $\left(\begin{smallmatrix} 1 & 1 \\ 0 & 1 \end{smallmatrix}\right)$ is not transitive because all its powers are upper triangular and hence there is no path from 1 to 0. In fact, the space Ω_A is countable and consists of two fixed points $(\ldots, 0, \ldots, 0, \ldots)$ and $(\ldots, 1, \ldots, 1, \ldots)$, and a single heteroclinic orbit connecting them (consisting of the sequences that are 1 up to some place and 0 thereafter).

Example 7.3.14 For the matrix

$$\begin{pmatrix} 0 & 0 & 1 & 1 \\ 0 & 0 & 1 & 1 \\ 1 & 1 & 0 & 0 \\ 1 & 1 & 0 & 0 \end{pmatrix}$$

every orbit alternates between entries from the first group $\{0, 1\}$ on the one hand and from the second group $\{2, 3\}$ on the other hand, that is, the parity (even–odd) must alternate. Therefore no power of the matrix has all entries positive.

■ **EXERCISES**

■ **Exercise 7.3.1** Prove that E_2 has a nonperiodic orbit all of whose even iterates lie in the left half of the unit interval.

■ **Exercise 7.3.2** Prove that E_2 has a uncountably many orbits for which no segment of length 10 has more than one point in the left half of the unit interval.

■ **Exercise 7.3.3** Prove that linear maps that are conjugate in the sense of linear algebra are topologically conjugate in the sense of Definition 7.3.3.

■ **Exercise 7.3.4** Write down the Markov matrix for Figure 7.3.3 and check Corollary 7.3.6 up to period 3.

■ **Exercise 7.3.5** Consider the metric

$$(7.3.9) \qquad\qquad d_\lambda'(\alpha, \omega) := \sum_{i \in \mathbb{Z}} \frac{|\alpha_i - \omega_i|}{\lambda^{|i|}}$$

on Ω_N. Show that for $\lambda > 2N - 1$ the cylinder $C_{\alpha_{1-n}\ldots\alpha_{n-1}}$ is a λ^{1-n}-ball for d_λ'.

■ **Exercise 7.3.6** Repeat the previous exercise for one-sided shifts (with $\lambda > N$).

■ **Exercise 7.3.7** Consider the metric

$$(7.3.10) \qquad\qquad d_\lambda''(\alpha, \omega) := \lambda^{-\max\{n \in \mathbb{N} \,|\, \alpha_i = \omega_i \text{ for } |i| \leq n\}}$$

[and $d_\lambda''(\alpha, \alpha) = 0$] on Ω_N. Show that the cylinder $C_{\alpha_{1-n}\ldots\alpha_{n-1}}$ is a ball for d_λ''.

■ **Exercise 7.3.8** Find the supremum of sensitivity constants for a transitive topological Markov chain with respect to the metric d_λ''.

■ **Exercise 7.3.9** Find the supremum of sensitivity constants for a transitive topological Markov chain with respect to the metric d_λ'.

■ **Exercise 7.3.10** Show that for $m < n$ the shift on Ω_m is a factor of the shift on Ω_n.

■ **Exercise 7.3.11** Prove that the quadratic map f_4 on $[0, 1]$ is not conjugate to any of the maps f_λ for $\lambda \in [0, 4)$.

■ **Exercise 7.3.12** Show that the topological Markov chains determined by the matrices

$$\begin{pmatrix} 1 & 1 \\ 1 & 0 \end{pmatrix} \qquad \text{and} \qquad \begin{pmatrix} 1 & 1 & 0 \\ 0 & 0 & 1 \\ 1 & 1 & 0 \end{pmatrix}$$

are conjugate.

■ **Exercise 7.3.13** Find the smallest positive value of $p(\sigma_A)$ for a transitive topological Markov chain with two states (that is, with a 2×2 matrix A).

■ **PROBLEMS FOR FURTHER STUDY**

■ **Problem 7.3.14** Find all factors of an irrational rotation R_α of the circle.

■ **Problem 7.3.15** Find the smallest value of $p(\sigma_A)$ for a transitive topological Markov chain with three states (that is, with a 3×3 matrix A).

7.4 MORE EXAMPLES OF CODING

We now carry out a coding construction for several familiar dynamical systems.

7.4.1 Nonlinear Expanding Maps

There is a correspondence between general (not necessarily linear) expanding maps of the circle (Section 7.1.3) and a shift on a sequence space. The construction is similar to the one from Section 7.3.1. There is some effort involved, but there is a beautiful prize at the end: We obtain a complete classification of a large class of maps in terms of a simple invariant.

To keep notations simple, we consider an expanding map $f: S^1 \to S^1$ of degree 2. By Proposition 7.1.9, f has exactly one fixed point p. (For maps of higher degree, we could pick any one of the fixed points.) Since $\deg(f) = 2$, there is exactly one point $q \neq p$ such that $f(q) = p$. The points p and q divide the circle into two arcs. Starting from p in the positive direction, denote the first arc by Δ_0 and the second arc by Δ_1. Define the coding for $x \in S^1$ as follows: x is represented by the sequence $\omega \in \Omega_2^R$ for which

(7.4.1) $$f^n(x) \in \Delta_{\omega_n}.$$

This representation is unique unless $f^n(x) \in \{p, q\} = \Delta_0 \cap \Delta_1$. This lack of uniqueness is similar to the case of binary rationals for the map E_2. Suppose a point x has an iterate in $\{p, q\}$. Then either $x = p$ and $f^n(x) = p$ for all $n \in \mathbb{N}$, or else the point q must appear before p in the sequence of iterates, that is, $f^n(x) \notin \{p, q\}$ for all n less than some k and then $f^k(x) = q$ and $f^{k+1}(x) = p$. In this case we make the following convention. p has two codes, all 0's and all 1's, and q has two codes, $01111111\ldots$ and $1000000\ldots$, and any x such that $F^k(x) = q$ has two codes given by the first $k - 1$ digits uniquely defined by (7.4.1), followed by either of the codes for q.

Actually, going the other way around is better:

Proposition 7.4.1 If $f: S^1 \to S^1$ is an expanding map of degree 2, then f is a factor of σ^R on Ω_2^R (Definition 7.3.3), that is, there is a surjective continuous map $h: \Omega_2^R \to S^1$ such that $f^n(h(\omega)) \in \Delta_{\omega_n}$ for all $n \in \mathbb{N}_0$, that is, $h \circ \sigma^R = f \circ h$.

Proof That the domain of h is Ω_2^R requires that every sequence of 0's and 1's appears as the code of some point. First, f maps each of the two intervals Δ_0 and Δ_1 onto S^1 almost injectively, the only identification being at the ends. Let

$$\Delta_{00} \text{ be the core of } \Delta_0 \cap f^{-1}(\Delta_0),$$
$$\Delta_{01} \text{ be the core of } \Delta_0 \cap f^{-1}(\Delta_1),$$
$$\Delta_{10} \text{ be the core of } \Delta_1 \cap f^{-1}(\Delta_0),$$
$$\Delta_{11} \text{ be the core of } \Delta_1 \cap f^{-1}(\Delta_1).$$

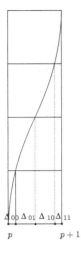

Figure 7.4.1. Nonlinear coding.

What we mean by "core" is that each indicated intersection consists of an interval as well as an isolated point (p or q), and we discard this extraneous point. Each of these four intervals is mapped onto S^1 by f^2, again the only identification being at the ends. By definition, any point from Δ_{ij} has ij as the first two symbols of its code. Proceeding inductively we construct for any finite sequence $\omega_0, \ldots, \omega_{n-1}$ the interval

$$(7.4.2) \qquad \Delta_{\omega_0,\ldots,\omega_{n-1}} := \text{the core of } \Delta_{\omega_0} \cap f^{-1}(\Delta_{\omega_1}) \cdots \cap f^{1-n}(\Delta_{\omega_{n-1}}),$$

which is mapped by f^n onto S^1 with identification of the endpoints. Now take any infinite sequence $\omega = \omega_1, \cdots \in \Omega_2^R$. The intersection $\bigcap_{n=1}^{\infty} \Delta_{\omega_0,\ldots,\omega_{n-1}}$ of the nested closed intervals $\Delta_{\omega_0,\ldots,\omega_{n-1}}$ is nonempty, and any point in this intersection has the sequence ω as its code.

So far we have only used the fact that f is a monotone map of degree 2. To show that h is well defined, we use the expanding property to check that $\bigcap_{n=1}^{\infty} \Delta_{\omega_0,\ldots,\omega_{n-1}}$ consists of a single point, hence a point with a given code is unique.

If $g: I \to S^1$ is an injective map of an open interval I with a nonnegative derivative, then by the Mean-Value Theorem A.2.3 $l(g(I)) = \int_I g'(x)\,dx = g'(\xi)l(I)$ for some $\xi \in I$. Thus, in our case, there is a ξ_n such that

$$1 = l(S^1) = \int_{\Delta_{\omega_0,\ldots,\omega_{n-1}}} (f^n)'(x)\,dx = (f^n)'(\xi_n) \cdot l(\Delta_{\omega_0,\ldots,\omega_{n-1}}).$$

Since f is expanding $|(f^n)'| > \lambda^n$ for some $\lambda > 1$, hence $l(\Delta_{\omega_0,\ldots,\omega_{n-1}}) < \lambda^{-n} \to 0$ as $n \to \infty$ and $\bigcap_{n=1}^{\infty} \Delta_{\omega_0,\ldots,\omega_{n-1}}$ consists of a single point x_ω. This gives a well-defined surjective map $h: \Omega_2^R \to S^1$, $\omega \mapsto x_\omega$.

Give Ω_2^R the metric d_4 from (7.3.3). We showed in Section 7.3.4 that if $\epsilon = \lambda^{-n}$ and $\delta = 4^{-n}$, then $d(\omega, \omega') < \delta$ implies that $\omega_i = \omega_i'$ for $i < n$ and hence $|x_\omega - x_{\omega'}| \le l(\Delta_{\omega_0,\ldots,\omega_{n-1}}) < \lambda^{-n} = \epsilon$. Thus h is continuous.

That $h(\sigma^R(\omega)) = f(h(\omega))$ is clear from the construction. $\square$

7.4.2 Classification via Coding

Proposition 7.4.1 and the discussion preceding it established a semiconjugacy between the one-sided 2-shift and the expanding map f on S^1, that is,

Proposition 7.4.2 *Let* $f\colon S^1 \to S^1$ *be an expanding map of degree 2. Then* f *is a factor of the one-sided 2-shift* (Ω_2^R, σ_R) *via a semiconjugacy* $h\colon \Omega_2^R \to S^1$. *If* $h(\omega) = h(\omega') =: x$, *then there exists an* $n \in \mathbb{N}_0$ *such that* $f^n(x) \in \{p, q\}$, *where* $p = f(p) = f(q), q \neq p$.

The last sentence of this proposition says that h is "very close" to being a conjugacy: There are only countably many image points where injectivity fails.

An important feature of this coding is that it is obtained in a uniform way for all expanding maps, and that the absence of injectivity occurs at points defined by their dynamics, namely, the fixed point and its preimages. This leads us to the prize promised at the beginning:

Theorem 7.4.3 *If* $f, g\colon S^1 \to S^1$ *are expanding maps of degree 2, then* f *and* g *are topologically conjugate; in particular, every expanding map of* S^1 *of degree 2 is conjugate to* E_2.

Proof We have semiconjugacies $h_f, h_g\colon \Omega_2^R \to S^1$ for f and g. For $x \in S^1$, consider the set $H_x := h_g(h_f^{-1}(\{x\}))$. If x is a point of injectivity of h_f, that is, $h_f^{-1}(\{x\})$ is a single point, then so is H_x. Otherwise, x is a preimage of the fixed point under some iterate of f and $h_f^{-1}(\{x\})$ consists of a collection of sequences that are mapped under h_g to a single point. Therefore, H_x always consists of precisely one point $h(x)$. The bijective map $h\colon S^1 \to S^1$ thus defined is clearly a conjugacy: $h \circ f = g \circ h$. It is continuous because h_f sends open sets to open sets, that is, the image of a sequence and all sufficiently closeby sequences contains a small interval. Exchanging f and g shows that h^{-1} is also continuous. $\square$

This holds for any degree via an appropriate coding. It is the first major conjugacy result that establishes conjugacy with a specific model for all maps from a certain class. The Poincaré Classification Theorem 4.3.20 comes close, but requires extra assumptions (such as the existence of the second derivative; see Section 4.4.3) to produce a conjugacy with a rotation.

7.4.3 Quadratic Maps

For $\lambda > 4$ consider the quadratic map

$$f\colon \mathbb{R} \to \mathbb{R}, \quad x \to \lambda x(1 - x).$$

If $x < 0$, then $f(x) < x$ and $f'(x) > \lambda > 4$, so $f^n(x) \to -\infty$. When $x > 1$, $f(x) < 0$ and hence $f^n(x) \to -\infty$. Thus the set of points with bounded orbits is $\bigcap_{n \in \mathbb{N}_0} f^{-n}([0, 1])$.

Proposition 7.4.4 *If* $\lambda > 2 + \sqrt{5}$ *and* $f\colon \mathbb{R} \to \mathbb{R}, x \to \lambda x(1 - x)$, *then there is a homeomorphism* $h\colon \Omega_2^R \to \Lambda := \bigcap_{n \in \mathbb{N}_0} f^{-n}([0, 1])$ *such that* $h \circ \sigma^R = f \circ h$, *that is,* $f_{\restriction_\Lambda}$ *is conjugate to the 2-shift.*

Proof Let

$$\Delta_0 = \left[0, \frac{1}{2} - \sqrt{\frac{1}{4} - \frac{1}{\lambda}} \right] \quad \text{and} \quad \Delta_1 = \left[\frac{1}{2} + \sqrt{\frac{1}{4} - \frac{1}{\lambda}}, 1 \right].$$

Then $f^{-1}([0, 1]) = \Delta_0 \cup \Delta_1$ by solving the quadratic equation $f(x) = 1$. Likewise, $f^{-2}([0, 1]) = \Delta_{00} \cup \Delta_{01} \cup \Delta_{11} \cup \Delta_{10}$ consists of four intervals, and so forth. Consider the partition of Λ by Δ_0 and Δ_1. These pieces do not overlap and

$$|f'(x)| = |\lambda(1 - 2x)| = 2\lambda \left| x - \frac{1}{2} \right| \geq 2\lambda \sqrt{\frac{1}{4} - \frac{1}{\lambda}}$$

$$= \sqrt{\lambda^2 - 4\lambda} > \sqrt{(2 + \sqrt{5})^2 - 4(2 + \sqrt{5})} = 1$$

on $\Delta_0 \cup \Delta_1$. Thus, for any sequence $\omega = (\omega_0, \omega_1, \dots)$, the diameter of the intersections

$$\bigcap_{n=0}^{N} f^{-n}(\Delta_{\omega_n})$$

decreases (exponentially) as $N \to \infty$. This shows that for a sequence $\omega = (\omega_0, \omega_1, \dots)$ the intersection

(7.4.3) $$h(\{\omega\}) = \bigcap_{n \in \mathbb{N}_0} f^{-n}(\Delta_{\omega_n})$$

consists of exactly one point and this map $h: \Omega_2^R \to \Lambda$ is a homeomorphism. $\square$

Remark 7.4.5 It turns out that Proposition 7.4.4 holds whenever $\lambda > 4$ (Proposition 11.4.1), but this is significantly less straightforward to prove than the present result. The situation present in either case, where a map folds an interval entirely over itself, is referred to as a one-dimensional *horseshoe*, in analogy to the geometry seen in the next subsection.

7.4.4 Linear Horseshoe

We now describe Smale's original "*horseshoe*," which provides one of the best examples of perfect coding. (In Section 7.3.3 a special case was constructed, in which ternary expansion provides the coding.)

Let Δ be a rectangle in $\mathbb{R}^2$ and $f: \Delta \to \mathbb{R}^2$ a diffeomorphism of Δ onto its image such that the intersection $\Delta \cap f(\Delta)$ consists of two "horizontal" rectangles Δ_0 and Δ_1 and the restriction of f to the components $\Delta^i := f^{-1}(\Delta_i)$, $i = 0, 1$, of $f^{-1}(\Delta)$ is a hyperbolic linear map, contracting in the vertical direction and expanding in the horizontal direction. This implies that the sets Δ^0 and Δ^1 are "vertical" rectangles. One of the simplest ways to achieve this effect is to bend Δ into a "horseshoe", or rather into the shape of a permanent magnet (Figure 7.4.2), although this method produces some inconveniences with orientation. Another way, which is better from the point of view of orientation, is to bend Δ roughly into a paper clip shape (Figure 7.4.3). This is an exaggerated version of the ternary horseshoe in Section 7.3.3, which also leaves some extra margin. If the horizontal and vertical rectangles lie strictly inside Δ, then the maximal invariant subset $\Lambda = \bigcap_{n=-\infty}^{\infty} f^{-n}(\Delta)$ of Δ is contained in the interior of Δ.

Proposition 7.4.6 $f_{\restriction_\Lambda}$ *is topologically conjugate to* σ_2.

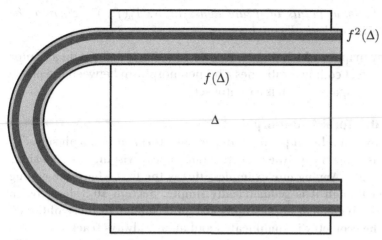

Figure 7.4.2. The horseshoe.

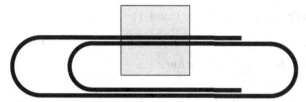

Figure 7.4.3. The paper clip.

Proof We use Δ^0 and Δ^1 as the "pieces" in the coding construction and start with positive iterates. The intersection $\Delta \cap f(\Delta) \cap f^2(\Delta)$ consists of four thin horizontal rectangles: $\Delta_{ij} = \Delta_i \cap f(\Delta_j) = f(\Delta^i) \cap f^2(\Delta^j)$, $i, j \in \{0, 1\}$ (see Figure 7.4.2). Continuing inductively, one sees that $\bigcap_{i=0}^{n} f^i(\Delta)$ consists of 2^n thin disjoint horizontal rectangles whose heights are exponentially decreasing with n. Each such rectangle has the form $\Delta_{\omega_1,\ldots,\omega_n} = \bigcap_{i=1}^{n} f^i(\Delta^{\omega_i})$, where $\omega_i \in \{0, 1\}$ for $i = 1, \ldots, n$. Each infinite intersection $\bigcap_{n=1}^{\infty} f^n(\Delta^{\omega_n})$, $\omega_n \in \{0, 1\}$, is a horizontal segment, and the intersection $\bigcap_{n=1}^{\infty} f^n(\Delta)$ is the product of the horizontal segment with a Cantor set in the vertical direction. Similarly, one defines and studies vertical rectangles $\Delta^{\omega_0,\ldots,\omega_{-n}} = \bigcap_{i=0}^{n} f^{-i}(\Delta^{\omega_{-i}})$, the vertical segments $\bigcap_{n=0}^{\infty} f^{-n}(\Delta^{\omega_{-n}})$, and the set $\bigcap_{n=0}^{\infty} f^{-n}(\Delta)$, which is the product of a segment in the vertical direction with a Cantor set in the horizontal direction.

The desired invariant set $\Lambda = \bigcap_{n=-\infty}^{\infty} f^{-n}(\Delta)$ is the product of two Cantor sets and hence is a Cantor set itself (Problem 2.7.5), and the map

$$h: \Omega_2 \to \Lambda, \qquad h(\{\omega\}) = \bigcap_{n=-\infty}^{\infty} f^{-n}(\Delta^{\omega_n})$$

is a homeomorphism that conjugates the shift σ_2 and the restriction of the diffeomorphism f to the set Λ. $\square$

Since periodic points and topological mixing are invariants of topological conjugacy, Proposition 7.4.6 and Proposition 7.3.4 immediately give substantial information about the behavior of f on Λ.

Corollary 7.4.7 *Periodic points of f are dense in* Λ, $P_n(f_{\restriction_\Lambda}) = 2^n$, *and the restriction of f to Λ is topologically mixing.*

Remark 7.4.8 Any map for which there is a perfect coding is defined on a Cantor set, because the perfect coding establishes a homeomorphism between the phase space and a sequence space, which is a Cantor set.

7.4.5 Coding of the Toral Automorphism

The idea of coding can be applied to hyperbolic toral automorphisms. To simplify notations and keep the construction more visual, we consider the standard example. Among our examples, this is the first where the coding is ingenious, even though it is geometrically simple. Section 10.3 describes a construction whose dynamical implications are quite similar to those obtained here, but where the geometry is complicated and almost always fractal.

Theorem 7.4.9 *For the map*

$$F(x, y) = (2x + y, x + y) \qquad (\text{mod } 1)$$

of the 2-torus from Section 7.1.4 there is a semiconjugacy $h\colon \Omega_A \to \mathbb{T}^2$ with

$$F \circ h = h \circ \sigma_5 {\restriction_{\Omega_A}}, \qquad where$$

(7.4.4)
$$A = \begin{pmatrix} 1 & 1 & 0 & 1 & 0 \\ 1 & 1 & 0 & 1 & 0 \\ 1 & 1 & 0 & 1 & 0 \\ 0 & 0 & 1 & 0 & 1 \\ 0 & 0 & 1 & 0 & 1 \end{pmatrix}.$$

Proof Draw segments of the two eigenlines at the origin until they cross sufficiently many times and separate the torus into disjoint rectangles. Specifically, extend a segment of the contracting line in the fourth quadrant until it intersects the segment of the expanding line twice in the first quadrant and once in the third quadrant (see Figure 7.4.4). The resulting configuration is a decomposition of the torus into two rectangles $R^{(1)}$ and $R^{(2)}$. Three pairs among the seven vertices of the plane configuration are identified, so there are only four different points on the torus that serve as vertices of the rectangles; the origin and three intersection points. Although $R^{(1)}$ and $R^{(2)}$ are not disjoint, one can apply the method used for the horseshoe, using $R^{(1)}$ and $R^{(2)}$ as basic rectangles. The expanding and contracting eigendirections play the role of the "horizontal" and "vertical" directions, correspondingly. Figure 7.4.5 shows that the image $F(R^{(i)})$ $(i = 1, 2)$ consists of several "horizontal" rectangles of full length. The union of the boundaries $\partial R^{(1)} \cup \partial R^{(2)}$ consists of the segments of the two eigenlines at the origin just described. The image of the contracting segment is a part of that segment. Thus, the images of $R^{(1)}$ and $R^{(2)}$ have to be "anchored" at parts of their "vertical" sides; that is, once one of the images "enters" either $R^{(1)}$ or $R^{(2)}$, it has to stretch all the way through it. By matching things up along the contracting direction one sees that $F(R^{(1)})$ consists of three components, two in $R^{(1)}$ and one in $R^{(2)}$. The image of $R^{(2)}$ has two components, one in each

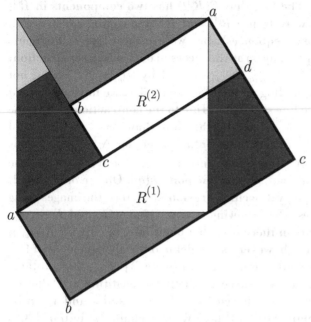

Figure 7.4.4. Partitioning the torus.

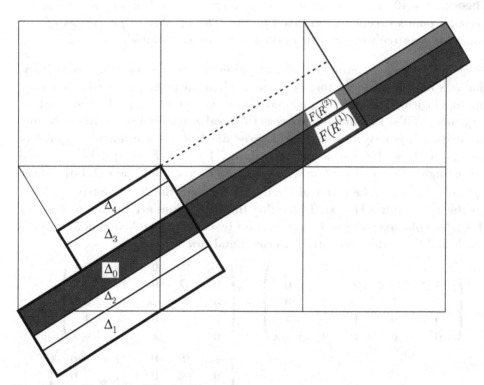

Figure 7.4.5. The image of the partition.

rectangle (see Figure 7.4.5). The fact that $F(R^{(1)})$ has two components in $R^{(1)}$ would cause problems if we were to use $R^{(1)}$ and $R^{(2)}$ for coding construction (more than one point for some sequences), but we use these five components $\Delta_0, \Delta_1, \Delta_2, \Delta_3, \Delta_4$ (or their preimages) as the pieces in our coding construction. There is exactly one rectangle $\Delta_{\omega_{-\ell}...\omega_0,\omega_1...\omega_k}$ defined by $\bigcap_{n=-\ell}^{k} F^{-n}(\Delta_{\omega_n})$, not several. (As in the case of expanding maps in Section 7.3.1, we have to discard extraneous pieces, in this case line segments.) Due to the contraction of F in the "vertical" direction, $\Delta_{\omega_{-\ell}...\omega_0,\omega_1...\omega_k}$ has "height" less than $((3-\sqrt{5})/2)^\ell$, and due to the contraction of F^{-1} in the "horizontal" direction $\Delta_{\omega_{-\ell}...\omega_0,\omega_1...\omega_k}$ has "width" less than $((3-\sqrt{5})/2)^k$. These go to zero as $\ell \to \infty$ and $k \to \infty$, so the intersection $\bigcap_{n\in\mathbb{Z}} F^{-n}(\Delta_{\omega_n})$ defines at most one point $h(\omega)$. On the other hand, because of the "Markov" property described previously, that is, the images going full length through rectangles, the following is true: If $\omega \in \Omega_5$ and $F^{-1}(\Delta_{\omega_n})$ overlaps $\Delta_{\omega_{n+1}}$ for all $n \in \mathbb{Z}$, then there is such a point $h(\omega)$ in $\bigcap_{n\in\mathbb{Z}} F^{-n}(\Delta_{\omega_n})$. Thus, we have a coding, which, however, is not defined for all sequences of Ω_5.

Instead, we have to restrict attention to the subspace Ω_A of Ω_5 that contains only those sequences where any two successive entries constitute an "allowed transition", that is, 0, 1, 2 can be followed by 0, 1, or 3, and 3 and 4 can be followed by 2 or 4. This is exactly the topological Markov chain (Definition 7.3.2) for (7.4.4). $\square$

Theorem 7.4.10 *The semiconjugacy between σ_A and F is one-to-one on all periodic points except for the fixed points. The number of preimages of any point not negatively asymptotic to the fixed point is bounded.*

Proof We describe carefully the identifications arising from our semiconjugacy, that is, what points on the torus have more than one preimage. First, obviously, the topological Markov chain σ_A has three fixed points, namely, the constant sequences of 0's, 1's, and 4's, whereas the toral automorphism F has only one, the origin. It is easy to see that all three fixed points are indeed mapped to the origin. As we have seen in Proposition 7.1.10, $P_n(F) = \lambda_1^n + \lambda_1^{-n} - 2$, and accordingly $P_n(\sigma_A) = \operatorname{tr} A^n = \lambda_1^n + \lambda_1^{-n} = P_n(F) + 2$ (Corollary 7.3.6), where $\lambda_1 = (3+\sqrt{5})/2$ is the maximal eigenvalue for both the 2×2 matrix $\left(\begin{smallmatrix} 2 & 1 \\ 1 & 1 \end{smallmatrix}\right)$ and for the 5×5 matrix (7.4.4). To see that the eigenvalues are the same, consider $A - \lambda\,\text{Id}$, subtract column 4 from the first two columns and column 5 from the third, and then add rows 1 and 2 to row 4 and row 3 to row 5:

$$\begin{pmatrix} 1-\lambda & 1 & 0 & 1 & 0 \\ 1 & 1-\lambda & 0 & 1 & 0 \\ 1 & 1 & -\lambda & 1 & 0 \\ 0 & 0 & 1 & -\lambda & 1 \\ 0 & 0 & 1 & 0 & 1-\lambda \end{pmatrix} \to \begin{pmatrix} -\lambda & 0 & 0 & 1 & 0 \\ 0 & -\lambda & 0 & 1 & 0 \\ 0 & 0 & -\lambda & 1 & 0 \\ \lambda & \lambda & 0 & -\lambda & 1 \\ 0 & 0 & \lambda & 0 & 1-\lambda \end{pmatrix}$$

$$\to \begin{pmatrix} -\lambda & 0 & 0 & 1 & 0 \\ 0 & -\lambda & 0 & 1 & 0 \\ 0 & 0 & -\lambda & 1 & 0 \\ 0 & 0 & 0 & 2-\lambda & 1 \\ 0 & 0 & 0 & 1 & 1-\lambda \end{pmatrix}.$$

Furthermore, one can see that every point $q \in \mathbb{T}^2$ whose positive and negative iterates avoid the boundaries $\partial R^{(1)}$ and $\partial R^{(2)}$ has a unique preimage, and vice versa. In particular, periodic points other than the origin (which have rational coordinates) fall into this category. The points of Ω_A whose images are on those boundaries or their iterates under F fall into three categories corresponding to the three segments of stable and unstable manifolds through 0 that define parts of the boundary. Thus sequences are identified in the following cases: They have a constant infinite right (future) tail consisting of 0's or 4's, and agree otherwise – this corresponds to a stable boundary piece – or else an infinite left (past) tail (of 0's and 1's, or of 4's), and agree otherwise – this corresponds to an unstable boundary piece. □

■ EXERCISES

■ **Exercise 7.4.1** Prove that for $\lambda \geq 1$ every bounded orbit of the quadratic map f_λ is in $[0, 1]$.

■ **Exercise 7.4.2** Give a detailed argument that (7.4.3) defines a homeomorphism.

■ **Exercise 7.4.3** Construct a Markov partition for $\left(\begin{smallmatrix} 1 & 1 \\ 1 & 0 \end{smallmatrix}\right)$ that consists of two squares.

■ **Exercise 7.4.4** Construct a Markov partition and describe the corresponding topological Markov chain for the automorphism F_L, where $L = \left(\begin{smallmatrix} 1 & 1 \\ 2 & 1 \end{smallmatrix}\right)$.

■ **Exercise 7.4.5** Given a 0-1 $n \times n$-matrix A, describe a system of n rectangles $\Delta_1, \ldots, \Delta_n$ in $\mathbb{R}^2$ and map $f: \Delta := \bigcup_{i=1}^n \Delta_i \to \mathbb{R}^2$ such that the restriction of f to the set of points that stay inside Δ for all iterates of f is topologically equivalent to the topological Markov chain σ_A.

■ **Exercise 7.4.6** Check that the process (7.4.2) of discarding extraneous points in the coding construction amounts to taking $\Delta_{\omega_0, \ldots, \omega_{n-1}} = \bigcap_{i=0}^{n-1} \text{Int}(f^{-i}(\Delta_{\omega_i}))$, and $\{h(\omega)\} := \bigcap_{n \in \mathbb{N}} \Delta_{\omega_0, \ldots, \omega_{n-1}}$.

■ PROBLEMS FOR FURTHER STUDY

■ **Problem 7.4.7** Show that the assertion of Theorem 7.4.3 remains true for any map f of degree 2 such that $f' \geq 1$ and $f' = 1$ only at finitely many points.

■ **Problem 7.4.8** Prove the assertion of Theorem 7.4.9 for some 0-1 matrix A for any automorphism

$$F_L: \mathbb{T}^2 \to \mathbb{T}^2, x \mapsto Lx \quad (\text{mod } 1),$$

where L is an integer 2×2 matrix with determinant $+1$ or -1 and with real eigenvalues different from ± 1.

7.5 UNIFORM DISTRIBUTION

We now investigate whether the notion of the uniform distribution of orbits that appeared in previous chapters for rotations of the circle and translations of the torus has any meaning for the group of examples discussed in the present chapter, such as linear or nonlinear expanding maps of the circle, shifts, and automorphisms of the torus.

7.5.1 Failure of Unique Ergodicity

We discuss linear expanding maps in more detail. One can define asymptotic frequencies of visits for an interval $\Delta \subset S^1$ as in (4.1.2) and Birkhoff averages of functions as in (4.1.5). Proposition 4.1.7 and Theorem 4.1.15 showed that for rotations these frequencies converge uniformly. This is called unique ergodicity (Definition 4.1.18).

In the present situation, those expressions do not converge uniformly to constants, that is, linear expanding maps are not uniquely ergodic: Consider a continuous function that vanishes only at 0. Then the Birkhoff average of the fixed point is 0 while all other periodic points have nonzero averages. Furthermore, there are orbits for which the limit of the average frequencies of visits to an interval does not exist, as one can see by coding. For E_2, the average frequency of visits to the interval $[0, 1/2]$ of a point $x \in S^1$ with unique binary representation is equal to the proportion of 0's among the first n digits. Let x be a point whose binary representation consists of alternating blocks of 0's and 1's in such a way that the length of the nth block is n times the total length of all preceding blocks. The proportion of 0's after the end of the nth block of 0's [which is the $(2n-1)$st block] is greater than $1 - 1/(2n-1)$; but after the end of the nth block of 1's (the $2n$th block) it is less than $1/2n$. Thus the limit points for the average frequencies cover the whole interval $[0, 1]$.

Therefore we have to investigate another mode of convergence.

7.5.2 Convergence in the Mean

The previous counterexample orbits are somewhat special, so there is still hope that a "majority" of orbits are nevertheless uniformly distributed, or that "on average" we see convergence. This is indeed true in fairly great generality. We explicitly show it for the map E_2.

Proposition 7.5.1 *If*

$$\varphi(x) := \chi_{[0,1/2]} := \begin{cases} 1 & \text{if } x \le 1/2 \\ 0 & \text{if } x > 1/2 \end{cases}$$

is the characteristic function of the interval $[0, 1/2]$ *and*

$$\mathcal{B}_n(\varphi)\,(x) := \frac{1}{n} \sum_{k=0}^{n-1} \varphi\left(E_2^k(x)\right)$$

is the Birkhoff average, then $\int_{S^1} |\mathcal{B}_n(\varphi)(x) - \int_{S^1} \varphi(t)\, dt|\, dx \xrightarrow[n\to\infty]{} 0$ *exponentially.*

Remark 7.5.2 One can prove analogous results for any binary interval, and by linear combination and approximation this gives equidistribution.

Proof In terms of binary representation $x = 0.x_1 x_2 \dots$

$$n\mathcal{B}_n(\varphi)(x) = F_{[0,1/2]}(x, n) := \text{card}\{k \mid 0 \le k \le n-1 \quad \text{and}$$

$$E_2^k x \in [0, 1/2]\} = \sum_{k=0}^{n-1} 1 - x_k.$$

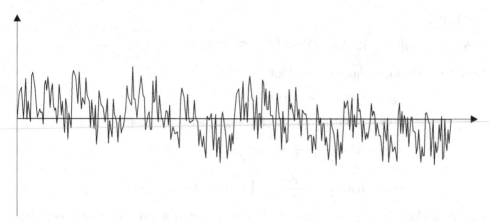

Figure 7.5.1. Distribution after 10 iterates.

Thus, the Birkhoff average is constant on each binary interval

$$\Delta_n^i = \left[\frac{i}{2^n}, \frac{i+1}{2^n}\right] \qquad i = 0, 1, \ldots, 2^n - 1.$$

As noted before, this is not uniformly close to any constant; in fact, it takes all values $i/n, i = 0, 1, \ldots, n$. Nevertheless, for most intervals the average frequency of visits is close to $\ell([0, 1/2]) = 1/2$. (See Figure 7.5.1.) To show this we use that the number of sequences of 0's and 1's of length n that have exactly k 0's (this corresponds to the average frequency k/n) is the binomial coefficient

$$\binom{n}{k} = \frac{n!}{k!(n-k)!},$$

and, accordingly, the proportion of such sequences is $\binom{n}{k}2^{-n}$. Thus, if $\epsilon > 0$, then

(7.5.1) $$\operatorname{card}\left\{i \colon |\mathcal{B}_n(\varphi)(x) - 1/2| < \epsilon \ \text{ for } x \in \Delta_n^i\right\} = \sum_{k=\lfloor(1/2-\epsilon)n\rfloor+1}^{\lfloor(1/2+\epsilon)n\rfloor} \binom{n}{k},$$

where $\lfloor\cdot\rfloor$ denotes integer part. To give a lower bound we estimate the sum of the remaining binomial coefficients from above. Since $\binom{n}{k} = \binom{n}{n-k}$, this sum is

$$2\sum_{k=0}^{\lfloor(1/2-\epsilon)n\rfloor} \binom{n}{k}.$$

The binomial coefficients are increasing with k for $k < n/2$ because the ratio of successive coefficients is $(k+1)/(n-k)$. Thus the largest term in the sum is the last. Since the number of terms does not exceed $n/2$, we find that

$$2\sum_{k=0}^{\lfloor(1/2-\epsilon)n\rfloor} \binom{n}{k} \le n\binom{n}{\lfloor(1/2-\epsilon)n\rfloor} = n\binom{n}{\lfloor\alpha n\rfloor},$$

where $\alpha = (1/2) - \epsilon$. Dividing by 2^n to get at the proportion of "bad" sequences shows that we need an upper bound for $n\binom{n}{\lfloor\alpha n\rfloor}2^{-n}$.

We use the classical *Stirling formula* $n! \asymp \sqrt{2\pi n}n^n e^{-n}$, where $f(n) \asymp g(n)$ means $\lim_{n\to\infty} f(n)/g(n) = 1$. Writing $l := \lfloor\alpha n\rfloor$, this gives

$$\binom{n}{\lfloor\alpha n\rfloor} = \frac{n!}{l!(n-l)!} \asymp \frac{n^n e^{-n}\sqrt{2\pi n}}{l^l(n-l)^{n-l}e^{-l}e^{l-n}\sqrt{2\pi l}\sqrt{2\pi(n-l)}}$$

and hence

(7.5.2) $n\dbinom{n}{\lfloor \alpha n \rfloor}2^{-n} \asymp ne^{n\log n - l\log l -(n-l)\log(n-l)-n\log 2}\sqrt{\dfrac{n}{2\pi l(n-l)}}.$

In order to obtain an upper bound for

$$n\log n - l\log l - (n-l)\log(n-l) - n\log 2$$

$$= (n-l)[\log n - \log(n-l) - \log 2] + l[\log n - \log l - \log 2]$$

$$= (n-l)\log\frac{n}{2(n-l)} + l\log\frac{n}{2l}$$

$$= (n-l)\log\left(1+\frac{2l-n}{2(n-l)}\right) + l\log\left(1+\frac{n-2l}{2l}\right),$$

we use that $\log(1+x)\le x$ (because the logarithm is concave down), and, in fact, that $\log(1+x) \le x - \delta$ whenever $|x| > \epsilon$. Since $2l = 2\lfloor \alpha n \rfloor = 2\lfloor n/2 - n\epsilon \rfloor \asymp (1-2n)\epsilon$ for large n, we get $2l - n/2(n-l) \asymp 2\epsilon$ and $n - 2l/2l \asymp -2\epsilon$, so

$$(n-l)\log\left(1+\frac{2l-n}{2(n-l)}\right) + l\log\left(1+\frac{n-2l}{2l}\right)$$

$$\le (n-l)\left(\frac{2l-n}{2(n-l)}-\delta\right)+l\left(\frac{n-2l}{2l}-\delta\right) = -n\delta,$$

with $\delta > 0$ depending on ϵ. Taking δ also small enough to leave a margin for the small error in (7.5.2) gives

(7.5.3) $2\displaystyle\sum_{k=0}^{\lfloor(1/2-\epsilon)n\rfloor}\binom{n}{k}2^{-n} \le ne^{-n\delta}\sqrt{\dfrac{n}{2\pi\lfloor \alpha n\rfloor(n-\lfloor\alpha n\rfloor)}} =: \Delta(n,\epsilon),$

which goes to 0 exponentially as $n\to\infty$.

Returning to (7.5.1), we see that for any fixed $\epsilon > 0$ and sufficiently large n the proportion of the binary intervals for which the average number of 0's differs from $1/2$ by less than ϵ is at least $1 - \Delta(n, \epsilon)$ and hence exponentially close to 1. Since every interval has the same length, the function $\mathcal{B}_n(\varphi)$ is close to $1/2$, except on a union of intervals the sum of whose lengths tends to zero exponentially as $n\to\infty$. Because the function is bounded (above and below), this implies that its average deviation from $1/2 = \ell([0, 1/2]) = \int_{S^1}\varphi(t)\,dt$ is exponentially small. $\square$

A crucial point was to reduce the estimate for the total length of the "bad" set (for which the Birkhoff averages deviate from the space averages) to a combinatorial calculation of a number of intervals. This relies on the fact that all binary intervals of a given rank have the same length. This, in turn, implies (but does not follow from) the fact that E_2 preserves the length in the sense of complete preimages (see Section 7.1.1).

7.5.3 Pointwise Convergence of Averages

There is an essential difference between uniform distribution in the mean (the Birkhoff averages for large n are close to the space average outside of a set of small length) and the original notion of uniform distribution as discussed in Chapter 4, where we calculated Birkhoff averages of individual points. It is natural to suppose that uniform distribution in the mean implies convergence of Birkhoff averages for "most" points. The problem is to make precise sense of the word "most". Even in the simplest case we just discussed, both the set A of points for which Birkhoff averages converge to the space average and its complement are dense, so the characteristic function of A is not Riemann integrable and there is no length of A. However, there is an adequate notion of smallness in this sense:

Definition 7.5.3 A *null set* in $\mathbb{R}$ is a set that, for any $\epsilon > 0$, can be covered by a (not necessarily finite or disjoint) collection of open intervals whose lengths add to at most ϵ. A property of points that holds for all points except those of a null set is said to hold *almost everywhere* or for *almost every* point.

Any subset of a null set is itself a null set. Easy examples of null sets are finite sets. Even countable sets are necessarily null sets because they are (by definition of countable) the range of a sequence $(x_n)_{n\in\mathbb{N}}$, which can be Covered by the intervals $(x_n - \epsilon 2^{-n-1}, x_n + \epsilon 2^{-n-1})$ whose lengths $\epsilon 2^{-n}$ sum to ϵ. This means that $\mathbb{Q}$ is a null set, as is the set of periodic points of E_2. A countable union $\bigcup_{n\in\mathbb{N}} N_n$ of null sets is itself a null set: Cover N_n by intervals whose lengths add to at most $\epsilon 2^{-n}$. The ternary Cantor set is also a null set because it is covered by the union C_n of 2^n closed intervals of length 3^{-n} each (see Section 2.7.1) and hence by 2^n open intervals of slightly larger length, for any $n \in \mathbb{N}$.

Lemma 7.5.4 *An interval that is not a point is not a null set.*

Proof Every interval that is not a point contains a closed interval that is not a point, so it suffices to prove this for closed intervals. Consider a cover of a closed interval $[a, b]$ by open intervals. There is a finite subcover (by compactness; see Definition A.1.18). Consider all endpoints of intervals from this cover that are in (a, b) and order them as $x_0 := a < x_1 < x_2 < \cdots < x_k < b =: x_{k+1}$. This way, $[a, b]$ is split into the union of $k + 1$ intervals $I_1 = [a, x_1), I_2 = [x_1, x_2), \ldots, I_{k+1} = [x_k, b]$. Each interval I_j is covered m_j times by intervals from our finite cover, and hence the lengths of these intervals sum to at least $m_j(x_{j+1} - x_j) > x_{j+1} - x_j$. Therefore, all of the lengths of intervals from this finite cover, and hence even more so all the lengths of the intervals from the original cover, add to at least $(b - x_k) + (x_k - x_{k-1}) + \cdots + (x_1 - a) = b - a$. $\square$

Corollary 7.5.5 *The complement of a null set is dense.*

Proof Otherwise the null set contains an interval that is not a point. $\square$

Theorem 7.5.6 *With the notations of Proposition 7.5.1, $\mathcal{B}_n(\varphi)(x) \xrightarrow[n\to\infty]{} \int_{S^1} \varphi(t)\, dt$ almost everywhere.*

Remark 7.5.7 Compared to Proposition 7.5.1, this result sacrifices exponential convergence to gain control of almost every point.

Proof The set of x where we have convergence is

$$\left\{ x \mid \mathcal{B}_n(\varphi)(x) \to \frac{1}{2} \right\} = \bigcap_{m=1}^{\infty} \bigcup_{N\in\mathbb{N}} \bigcap_{n\geq N} \left\{ x \mid \left| \mathcal{B}_n(\varphi)(x) - \frac{1}{2} \right| < \frac{1}{m} \right\},$$

so we need to show that the ("bad") set

$$B := \left\{ x \mid \mathcal{B}_n(\varphi)(x) \not\to \frac{1}{2} \right\} = \bigcup_{m=1}^{\infty} \bigcap_{N \in \mathbb{N}} \bigcup_{n \geq N} \left\{ x \mid \left| \mathcal{B}_n(\varphi)(x) - \frac{1}{2} \right| \geq \frac{1}{m} \right\}$$

is a null set. To that end note that

$$\{ x \mid |\mathcal{B}_n(\varphi)(x) - 1/2| \geq 1/m \}$$

can be covered by a collection of (binary) intervals whose lengths add to no more than $\Delta(n, 1/m)$, where Δ is defined by (7.5.3), and exponentially small in n for any given m. Therefore

$$\bigcup_{n \geq N} \{ x \mid |\mathcal{B}_n(\varphi)(x) - 1/2| \geq 1/m \}$$

can be covered by intervals whose lengths sum to at most $\bar{\Delta}(N, 1/m) := \sum_{n \geq N} \Delta$ $(n, 1/m)$. This series converges [geometrically; see (7.5.3)], so $\bar{\Delta}(N, 1/m) \xrightarrow{N \to \infty} 0$. There is an N_0 such that $\bar{\Delta}(N_0, 1/m) < \epsilon 2^{-m}$, so $\bigcap_{N \in \mathbb{N}} \bigcup_{n \geq N} \{ x \mid |\mathcal{B}_n(\varphi)(x) - 1/2| \geq 1/m \}$ can be covered by a union of intervals whose lengths add to no more than $\epsilon 2^{-m}$. The union of these covers gives a covering of B by intervals whose lengths add to no more than ϵ. $\square$

7.5.4 The Law of Large Numbers

The two kinds of uniform distribution, $\int_{S^1} |\mathcal{B}_n(\varphi)(x) - \int_{S^1} \varphi(t) \, dt| \, dx \to 0$ (Proposition 7.5.1) and Theorem 7.5.6 are called the weak and strong *law of large numbers*, respectively, because they describe the fact that any initial probability distribution (represented by a continuous function φ) tends to a uniform distribution after repeated applications of the map, that is, large numbers of iterations tend to make any distribution look uniform in the long run.

These two notions of uniform distribution are both weaker than the strongly uniform distribution observed for irrational rotations in Theorem 4.1.15 and found there to be related to the notion of unique ergodicity (Definition 4.1.18). Likewise, the weak and strong laws of large numbers are related to a notion of *ergodicity*, which we do not define here because this requires familiarity with measure theory. In fact, a consequence of the pertinent theory (specifically, the Birkhoff Ergodic Theorem) is that these two notions of uniform distribution are equivalent, even though they look different on the surface.

7.5.5 Distribution of Periodic Points

Another interesting conclusion about some orbit averages can be drawn from the proof of Theorem 7.5.6.

Theorem 7.5.8 *For any $\epsilon > 0$,*

$$\frac{\text{card}\{ p \mid E_2^n(p) = p, \quad |\lim_{m \to \infty} \mathcal{B}_m(\varphi)(p) - 1/2| \geq \epsilon \}}{2^n - 1}$$

converges to 0 exponentially.

Proof There are $2^n - 1$ periodic points of period n for the transformation E_2, one in each binary interval Δ_n^i, $i = 0, \ldots, 2^n - 1$ with Δ_n^0 and $\Delta_n^{2^n-1}$ sharing 0 as an endpoint. Denote the n-periodic point in Δ_n^i by p_n^i. Since $\mathcal{B}_n$ is constant on Δ_n^i, the values $\mathcal{B}_n(p_n^i)$ for all but an exponentially small proportion of these $2^n - 1$ periodic points are within ϵ of $1/2$. But then $\lim_{m\to\infty} \mathcal{B}_m(p) = \mathcal{B}_n(p)$ for any n-periodic point p since the averages are taken over a periodic sequence. $\square$

This result is not a consequence of Theorem 7.5.6 because the set of periodic points is a null set. It is rather a natural consequence of the fact that the set of periodic points is "uniformly distributed".

■ **EXERCISES**

■ **Exercise 7.5.1** Prove for the map E_2 that the set of points for which there are asymptotic frequencies is the union of countably many nowhere dense sets.

■ **Exercise 7.5.2** Prove the counterpart of Proposition 7.5.1 for the map E_3.

■ **Exercise 7.5.3** Prove the counterpart of Proposition 7.5.1 for the map E_m.

■ **Exercise 7.5.4** Prove the counterpart of Theorem 7.5.6 for the map E_3.

■ **Exercise 7.5.5** Prove the counterpart of Theorem 7.5.6 for the map E_m.

■ **Exercise 7.5.6** Prove the counterpart of Theorem 7.5.8 for the map E_m.

7.6 INDEPENDENCE, ENTROPY, MIXING

Investigating the distribution properties of orbits constitutes a probabilistic approach to dynamics. Especially the present examples, where convergence to averages is far from uniform, suggest that even in a deterministic dynamical system there are features that appear to reflect randomness. We take a closer look at dynamical systems where this is a central feature.

7.6.1 The Coin-Tossing Model
We now take a fresh look at the coding construction for expanding maps carried out in Section 7.3.1. Think of a binary representation of a number from the interval $[0, 1]$ as an account of the result of an infinite sequence of coin-tossing experiments: We write zero each time the coin lands heads up and one when it lands tails up. If the coin is *fair*, that is, heads and tails are equally likely, and our trials are *independent*, then the probability of any sequence of zeros and ones in the first n trials is 2^{-n}. We call the appearance of any fixed sequence of heads and tails during n trials an *elementary event*. The probability of any outcome during any n trials (not necessarily the first ones or even successive) is the total length of the collection of binary intervals that correspond to all elementary events compatible with the given outcome. Thus the calculations with lengths in Section 7.5 have a probabilistic interpretation. For example, to find the probability of having k tails in n trials, take the number of sequences of zeros and ones of length n with exactly k ones, which is equal to $\binom{n}{k}$, and divide by 2^n. Similarly, (7.5.1) gives the number of trials for which the deviation of the average number of tails from $1/2$ is at most ϵ. The calculations with Birkhoff averages of the characteristic function φ with respect

to the map E_2 (Section 7.5) can be reformulated to give estimates of probabilities of various outcomes in the fair coin-tossing model: The probability that the average number of tails differs from one-half by more than any fixed number converges to zero exponentially as the number of trials goes to infinity.

Section 7.3.5 explains how the coding of a map produces a shift on a sequence space. The coding for E_2 using the partition into the left half and right half of the unit interval corresponds exactly to a coin-tossing scheme; it also takes place on a space of sequences of zeros and ones.

7.6.2 Bernoulli Schemes

A more general scheme of random trials with several possible outcomes (not necessarily equally likely or independent) can still be described in terms of the shift on a symbolic space. The probabilities of elementary events may not be equal. Elements of this space correspond to possible outcomes of infinite sequences of trials, and the shift transformation corresponds to taking one step forward in time. The simplest class of such schemes is the *stationary Bernoulli scheme*. Fix the probabilities $p_0, \ldots, p_{N-1}$ of the N symbols $0, \ldots, N-1$ to occur in a trial, and suppose that successive trials are mutually independent, that is, the probability of a particular outcome of any trial does not depend on the results of the previous one. This means that the probability of any finite sequence of outcomes is the product of the probabilities of the outcomes in the sequence.

Consider the case $N = 2$ where $p := p_0 \neq p_1 = 1 - p$, that is, the tossing of a biased coin, and follow the calculations from Section 7.5. The probability of any elementary event that includes k occurrences of 0 and $n - k$ occurrences of 1 is equal to $\binom{n}{k} p^k (1 - p)^{n-k}$. Thus, if φ is the characteristic function of the set of sequences whose first entry is 0, then the space average of φ is p, and for most finite sequences the proportion of 0 entries should be approximately p. This expectation is justified by adapting the calculation from Section 7.5 to this new situation.

Proposition 7.6.1 *For the stationary Bernoulli scheme with weights p and $1 - p$,*

$$\text{probability} \left\{ i : |\mathcal{B}_n(\varphi)(x) - p| < \epsilon \text{ for } x \in \Delta_n^i \right\} \to 1$$

exponentially as $n \to \infty$.

Proof Instead of (7.5.1) we have

(7.6.1)

$$\text{probability} \left\{ i : |\mathcal{B}_n(\varphi)(x) - p| < \epsilon \text{ for } x \in \Delta_n^i \right\} = \sum_{k=\lfloor (p-\epsilon)n \rfloor + 1}^{\lfloor (p+\epsilon)n \rfloor} \binom{n}{k} p^k (1 - p)^{n-k}.$$

To find a lower bound estimate the sum of the remaining binomial coefficients from above. One part is

$$\sum_{k=0}^{\lfloor (p-\epsilon)n \rfloor} \binom{n}{k} p^k (1 - p)^{n-k},$$

and estimating the corresponding sum starting from $\lfloor (p+\epsilon)n \rfloor + 1$ is left to the reader. For sufficiently large n, the last term is the largest because passing from one term to the next is accomplished by multiplying with $(n - k/k)(p/1 - p)$, which is greater than 1 for $k < pn$. As in Section 7.5, this leads to the upper bound

$$\sum_{k=0}^{\lfloor(p-\epsilon)n\rfloor} \binom{n}{k} p^k (1-p)^{n-k} \leq (l+1) \binom{n}{l} p^l (1-p)^{n-l}$$

$$\asymp (l+1) e^{n\log n - l\log l - (n-l)\log(n-l)} \sqrt{\frac{n}{2\pi l(n-l)}} p^l (1-p)^{n-l}$$

$$= e^{n\log n - l\log l - (n-l)\log(n-l) + l\log p + (n-l)\log(1-p)} (l+1) \sqrt{\frac{n}{2\pi l(n-l)}} p^l (1-p)^{n-l}.$$

The exponent

$$(n-l)[\log n - \log(n-l) + \log(1-p)] + l[\log n - \log l + \log p]$$

$$= (n-l) \log \frac{n(1-p)}{n-l} + l \log \frac{np}{l}$$

$$= (n-l) \log \left(1 + \frac{n(1-p) - n+l}{n-l}\right) + l \log \left(1 + \frac{np-l}{l}\right)$$

can be estimated again using convexity of the logarithm: For $l = \lfloor(p-\epsilon)n\rfloor$ and $\epsilon > 0$, there is a $\delta > 0$ such that

$$(n-l) \log \left(1 + \frac{n(1-p) - n+l}{n-l}\right) + l \log \left(1 + \frac{np-l}{l}\right)$$

$$\leq (n-l) \left(\frac{n(1-p) - n+l}{n-l} - \delta\right) + l \left(\frac{np-l}{l} - \delta\right) = -n\delta.$$

Thus the exponent is negative and the upper bound goes to zero exponentially as $n \to \infty$.

Together with the analogous estimate of the other tail of the sum, this implies that the right-hand side of (7.6.1) goes to 1 exponentially, as claimed. □

7.6.3 Entropy

The elementary events that constitute the bulk of the possible outcomes, that is, those with an average number of appearances close to the space average, are not the most probable by themselves, but they have a specific exponential size. For any such event C the logarithm of the probability of C divided by n is approximately $p \log p + (1-p) \log(1-p)$. This is called the *entropy* of the probability distribution $(p, 1-p)$ and is intimately connected with the degree of uncertainty generated by the randomness of the scheme. The above exponential asymptotic for the probability of typical outcomes is the most elementary case of the celebrated Shannon–MacMillan Theorem, which is one of the cornerstones of information theory.[3]

7.6.4 Bernoulli Measures on the Interval

Returning to the expanding map E_2, we can interpret the probabilities of elementary events as a substitute of the length of the corresponding binary intervals. To

---

[3] See, for example, Theorem 2.3 in Karl Petersen, *Ergodic Theory*, Cambridge Studies in Advanced Mathematics **2**, Cambridge University Press, Cambridge, 1983.

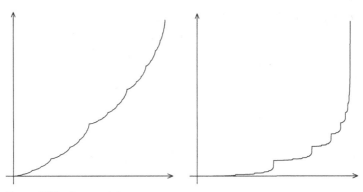

Figure 7.6.1. $\Psi_{1/3}$ and $\Psi_{1/10}$.

keep the distinction, we refer to this as the *Bernoulli measure* of these intervals, as opposed to their length. For example, the measure of $[0, 1/4]$ is $m_p([0, 1/4]) = p^2$ and $m_p([1/4, 1/2]) = m_p([1/2, 3/4]) = p(1 - p)$, $m_p([3/4, 1]) = (1 - p)^2$. Naturally the measure of a union of binary intervals is taken to be the sum of their measures. It is not hard to check that any way of approximating an interval by unions of binary intervals from inside or outside leads to the same limit of the corresponding measures, so we obtain a measure $m_p(I)$ for any interval I, and hence their finite unions. As in the definition of the usual integral, we can use this measure to define the integral of any continuous function (in fact, some other functions as well, such as functions with finitely many discontinuities). Analogously to before, a *null set* is a set such that given $\epsilon > 0$ there is a countable cover by open intervals whose (Bernoulli) measures add to less than ϵ.

Analogously to Definition 4.2.2, it is convenient to represent this measure by its *distribution function* $\Psi_p(x) = m_p([0, x])$. Evidently, $\Psi_p(0) = 0$ and $\Psi_p(1) = 1$, and Ψ_p is nondecreasing. In fact, it is increasing, since every binary interval has positive measure and between any two numbers there is a binary interval. Likewise it is continuous for $p \notin \{0, 1\}$, because the measures of short binary intervals are small. Clearly, $\Psi_{1/2}(x) = x$, but, as it turns out, for $p \neq 1/2$ the function Ψ_p is not differentiable at many points. However it has an interesting self-similarity property: $\psi_p(x/2) = p\psi(x)$. Figure 7.6.1 shows $\Psi_{1/3}$ and $\Psi_{1/10}$.

7.6.5 Mixing

Analogously to the way in which topological mixing is a stronger property than topological transitivity, there is a property, called *mixing*, corresponding in a like manner to the uniform distribution (or a law of large numbers). We develop this notion in the concrete setting of piecewise continuous maps of the circle.

To that end, start from a distribution function: Suppose $\Psi\colon [0, 1] \to [0, 1]$ is a continuous nondecreasing function with $\Psi(0) = 0$ and define the measure of $[a, b]$ to be $m([a, b]) := \Psi(b) - \Psi(a)$. This extends to finite unions of intervals by additivity. In particular, if we represent the circle as $[0, 1]$ with endpoints identified, then the measure of an arc containing 0 is defined to be the sum of the measures of the two pieces on either side of 0. The measure of an open or half-open arc is the same as that of its closure.

Consider a piecewise continuous, piecewise monotone map $f\colon S^1 \to S^1$ (or of an interval). This notion was introduced in Definition 4.2.2, where we noted that $m_f(I) := m(f^{-1}(I))$ is well defined. If $m_f = m$, then we say that the measure m is *invariant* under f. This is not the first situation of the preservation of a measure in Section 6.1 we noted that Newtonian systems preserve volume (Corollary 6.2.3), which is a natural measure in higher dimension.

Definition 7.6.2 Consider a piecewise continuous, piecewise monotone map $f\colon S^1 \to S^1$ and suppose a measure m is invariant under f. Then f is said to be *mixing* (with respect to m) if $m(\Delta_1 \cap f^{-n}(\Delta_2)) \xrightarrow[n\to\infty]{} m(\Delta_1) \cdot m(\Delta_2)$ for any two arcs Δ_1, Δ_2.

Since $\Delta_1 \cap f^{-n}(\Delta_2)$ is a finite union of intervals, its measure is well defined.

Proposition 7.6.3 *If f is mixing with respect to a measure m whose distribution function is increasing, then f is topologically mixing.*

Proof The assumption on m implies that every interval has positive measure. If U, V are open, then there are intervals $\Delta_1 \subset V$, $\Delta_2 \subset U$. Since f is mixing with respect to m, there is an $N \in \mathbb{N}$ such that $m(\Delta_1 \cap f^{-n}(\Delta_2)) > 0$ and hence $\Delta_1 \cap f^{-n}(\Delta_2) \neq \varnothing$ for $n \geq N$. Applying f^n we get $\Delta_1 \cap f^n(\Delta_2) \neq \varnothing$ for all $n \geq N$; hence $V \cap f^n(U) \neq \varnothing$ for $n \geq N$. $\square$

Proposition 7.6.4 *The Bernoulli measure m_p on S^1 from Section 7.6.4 is invariant under E_2, and E_2 is mixing with respect to m_p.*

Proof To prove invariance note that any arc can be approximated arbitrarily well by a union of nonoverlapping binary intervals. Therefore it suffices to check that a binary interval has the same measure as its preimage. A binary interval Δ is determined by a finite string $0.x_1 x_2 \ldots x_n$ of binary digits, and its measure $m(\Delta) = \prod_{i=1}^{n} x_i$ has one factor p for each $x_i = 0$ and a factor $1 - p$ for each $x_i = 1$. The preimage consists of two binary intervals with binary strings $0.0x_1 x_2 \ldots x_n$ and $0.1x_1 x_2 \ldots x_n$, whose measures sum to

$$pm(\Delta) + (1 - p)m(\Delta) = m(\Delta).$$

Mixing likewise needs to be checked only for binary intervals. Suppose Δ_1 and Δ_2 are binary intervals given by strings $0.\alpha_1 \ldots \alpha_m$ and $0.\omega_1 \ldots \omega_r$, respectively. Then $E_2^{-n}(\Delta_2)$ is the disjoint union of the 2^n binary intervals determined by the strings $0.x_1 \ldots x_n \omega_1 \ldots \omega_r$ with all possible combinations $x_1 \ldots x_n$. If $n > m$, then $\Delta_1 \cap E_2^{-n}(\Delta_2)$ is the disjoint union of the binary intervals with all possible strings $0.\alpha_1 \ldots \alpha_m x_{m+1} \ldots x_n \omega_1 \ldots \omega_n$. Its measure is

$$\sum_{x_1 \ldots x_n} \prod_{i=1}^{m} \alpha_i \prod_{j=1}^{n-m} x_{m+i} \prod_{k=1}^{r} \omega_i = m(\Delta_1) m(\Delta_2) \sum_{x_1 \ldots x_n} \prod_{j=1}^{n-m} x_{m+i} = m(\Delta_1) m(\Delta_2)$$

because the last sum is the sum of the measures of all binary intervals for strings of length $n - m$, hence the measure of the circle, that is, 1. $\square$

The two other prominent examples of chaotic systems are mixing as well: the two-sided shift and the hyperbolic linear automorphism of the torus. Of course, this statement requires definitions of mixing and of measures for these contexts.

Proposition 7.6.5 *A two-sided shift with a Bernoulli measure defined analogously to Section 7.6.4 is mixing, where mixing is understood as in Definition 7.6.2 with arcs replaced by cylinders.*

The proof uses the same argument as in the case of E_2 and is left as an exercise (Exercise 7.6.1).

Proposition 7.6.6 *The hyperbolic automorphism F of $\mathbb{T}^2$ induced by the linear map L with matrix $\left(\begin{smallmatrix} 2 & 1 \\ 1 & 1 \end{smallmatrix}\right)$ is mixing with respect to area measure, where mixing is understood as in Definition 7.6.2 with arcs replaced by parallelograms.*

Proof Since F is invertible, we can replace the condition $m(\Delta_1 \cap F^{-n}(\Delta_2)) \xrightarrow{n\to\infty} m(\Delta_1) \cdot m(\Delta_2)$ by the condition $m(B \cap F^n(A)) \xrightarrow{n\to\infty} m(A) \cdot m(B)$. For convenience, we use particular kinds of parallelograms as "test sets" to replace the arcs in the case of E_2. Instead of the arc A we use a parallelogram A whose sides are parallel to the eigendirections; we denote the length of the sides along the expanding eigenline by a_1 and the length of the other sides by a_2. Instead of the arc B we use a parallelogram with two vertical sides of length b_2 and two sides of length b_1 parallel to the eigendirection for the eigenvalue λ greater than one, denoting the cosine of their angle with the horizontal by c. Consider now $F^n(A)$ for some large n. Its bottom side of length a_1 is mapped to a line of length $\lambda^n a_1$ that intersects the vertical side of B in approximately $cb_2\lambda^n a_1$ places because the intersection points are the images of the section map for the linear flow generating the eigenlines, and this section map is an irrational rotation, which has the uniform distribution property. To determine the measure of the intersection $F^n(A) \cap B$ note that (aside from at most two strips near the edges of B and two pieces not transversing B all the way) it consists of as many strips of width $\lambda^{-n} a_2$ and of length b_1, giving a combined area of $(c\lambda^n a_1 b_2)(\lambda^{-n} a_2)b_1 = (a_1 a_2)(c b_1 b_2)$, which is precisely $m(A)m(B)$. (See Figure 7.6.2.) $\square$

This kind of argument works for any linear toral automorphism:

Proposition 7.6.7 *Any hyperbolic linear automorphism of the torus is mixing with respect to area as measure.*

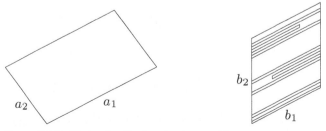

Figure 7.6.2. Mixing for the toral automorphism.

◼ EXERCISES

◼ **Exercise 7.6.1** Show that the full shift on two symbols is mixing with respect to any Bernoulli measure.

◼ **Exercise 7.6.2** Estimate the second tail in the proof of Proposition 7.6.1.

◼ **Exercise 7.6.3** Show that if $0 < p < q < 1$, then there is a set A such that A is a null set for m_p and its complement is a null set for m_q.

◼ **Exercise 7.6.4** Prove the counterpart of Theorem 7.5.6 for the Bernoulli measure m_p.

CHAPTER 8

Entropy and Chaos

In this chapter we look at two related notions that are important parameters for chaotic dynamical systems. The first is the fractal dimension of a set. By permitting noninteger values, this notion extends the topological concept of dimension to sets such as Cantor sets. While all Cantor sets are homeomorphic, they may look thicker or thinner depending on the parameters in their construction. Fractal dimension is a measure of the thickness of these sets. When the Cantor set in question arises as an invariant set of a hyperbolic dynamical system its dimension is related in deep ways to other dynamically important quantities, notably the contraction and expansion rates in the system. This is an active research topic, and we illustrate it with the Smale horseshoe.

The other notion is entropy. It measures the global orbit complexity on an exponential scale and is intimately related to the growth rate of periodic points and contraction and expansion rates. As an invariant of topological conjugacy, it also provides a means for telling apart dynamical systems that are not conjugate.

The values of dimension and entropy of an invariant set of a dynamical system are related, and so are the constructions involved in defining them. The common root is the notion of *capacity* of a set, with which we begin the chapter.

8.1 DIMENSION OF A COMPACT SPACE

8.1.1 Capacity

For a compact metric space there is a notion of the "size" or capacity inspired by the notion of volume. Suppose X is a compact space with metric d. Then a set $E \subset X$ is said to be *r-dense* if $X \subset \bigcup_{x \in E} B_d(x, r)$, where $B_d(x, r)$ is the r-ball with respect to d around x (see Section 2.6.1). Define the *r-capacity* of (X, d) to be the minimal cardinality $S_d(r)$ of an r-dense set.

For example, if $X = [0, 1]$ with the usual metric, then $S_d(r)$ is approximately $1/2r$ because it takes over $1/2r$ balls (that is, intervals) to cover a unit length, and the $\lfloor 2 + 1/2r \rfloor$-balls centered at $ir(2 - r)$, $0 \leq i \leq \lfloor 1 + 1/2r \rfloor$ suffice. As another example, if $X = [0, 1]^2$ is the unit square, then $S_d(r)$ is roughly r^{-2}

because it takes at least $1/\pi r^2$ r-balls to cover a unit area, and, on the other hand, the $(1 + 1/r)^2$-balls centered at points (ir, jr) provide a cover. Likewise, for the unit cube $(1 + 1/r)^3$, r-balls suffice.

In the case of the ternary Cantor set with the usual metric we have $S_d(3^{-i}) = 2^i$ if we cheat a little and use closed balls for simplicity; otherwise, we could use $S_d((3 - 1/i)^{-i}) = 2^i$ with honest open balls.

8.1.2 Box Dimension

One interesting aspect of capacity is the relation between its dependence on r [that is, with which power of r the capacity $S_d(r)$ increases] and dimension.

If $X = [0, 1]$, then

$$\lim_{r \to 0} -\frac{\log S_d(r)}{\log r} \geq \lim_{r \to 0} -\frac{\log(1/2r)}{\log r} = \lim_{r \to 0} \frac{\log 2 + \log r}{\log r} = 1$$

and

$$\lim_{r \to 0} -\frac{\log S_d(r)}{\log r} \leq \lim_{r \to 0} -\frac{\log\lfloor 2 + 1/2r \rfloor}{\log r} \leq \lim -\frac{\log(1/r)}{\log r} = 1,$$

so $\lim_{r \to 0} -\log S_d(r)/\log r = 1 = \dim X$. If $X = [0, 1]^2$, then $\lim_{r \to 0} -\log S_d(r)/\log r = 2 = \dim X$; and if $X = [0, 1]^3$, then $\lim_{r \to 0} -\log S_d(r)/\log r = 3 = \dim X$. This suggests that $\lim_{r \to 0} -\log S_d(r)/\log r$ defines a notion of dimension.

Definition 8.1.1 If X is a totally bounded metric space (Definition A.1.20), then

$$\mathrm{bdim}(X) := \lim_{r \to 0} -\frac{\log S_d(r)}{\log r}$$

is called the *box dimension* of X.

8.1.3 Examples

Let us test this notion on less straightforward spaces.

1. The Ternary Cantor Set. If C is the ternary Cantor set, then

$$\mathrm{bdim}(C) = \lim_{r \to 0} -\frac{\log S_d(r)}{\log r} = \lim_{n \to \infty} -\frac{\log 2^i}{\log 3^{-i}} = \frac{\log 2}{\log 3}.$$

If C_α is constructed by deleting a middle interval of relative length $1 - (2/\alpha)$ at each stage, then $\mathrm{bdim}(C_\alpha) = \log 2/\log \alpha$. This increases to 1 as $\alpha \to 2$ (deleting ever smaller intervals), and it decreases to 0 as $\alpha \to \infty$ (deleting ever larger intervals). Thus we get a small box dimension if in the Cantor construction the size of the remaining intervals decreases rapidly with each iteration.

This illustrates, by the way, that the box dimension of a set may change under a homeomorphism, because these Cantor sets are pairwise homeomorphic.

2. The Sierpinski Carpet. It is easy to handle other Cantor-like sets, such as the Sierpinski carpet S from Section 2.7.2. For the square Sierpinski carpet we can cheat as in the capacity calculation for the ternary Cantor set and use closed balls (sharing their center with one of the small remaining cubes at a certain stage) for covers. Then $S_d(3^{-i}/\sqrt{2}) = 8^i$ and

$$\text{bdim}(S) = \lim_{n \to \infty} - \frac{\log 8^i}{\log 3^{-i}/\sqrt{2}} = \frac{\log 8}{\log 3} = \frac{3 \log 2}{\log 3},$$

which is three times that of the ternary Cantor set (but still less than 2, of course). For the triangular Sierpinski carpet we similarly get box dimension $\log 3/\log 2$.

3. The Koch Snowflake.

The Koch snowflake K from Section 2.7.2 has $S_d(3^{-i}) = 4^i$ by covering it with (closed) balls centered at the edges of the ith polygon. Thus

$$\text{bdim}(K) = \lim_{n \to \infty} - \frac{\log 4^i}{\log 3^{-i}} = \frac{\log 4}{\log 3} = \frac{2 \log 2}{\log 3},$$

which is less than that of the Sierpinski carpet, corresponding to the fact that the iterates look much "thinner". Notice that this dimension exceeds 1, however, so it is larger than the dimension of a curve. All of these examples have (box) dimension that is not an integer, that is, fractional or "fractal". This has motivated calling such sets *fractals*.

4. The Smale Horseshoe.

Suppose that in the construction of the Smale horseshoe (Section 7.4.4) the expansion rate on the linear pieces is $\lambda > 2$ and the contraction rate is $\mu < 1/\lambda$ (without loss of generality). Given $n \in \mathbb{N}$, the invariant set $\Lambda = \bigcap_{n=-\infty}^{\infty} f^{-n}(\Delta)$ is contained in $\Lambda = \bigcap_{i=-n}^{n} f^{-i}(\Delta)$, which consists of 4^n rectangles with sides λ^{-n} and μ^n and can therefore be covered by about $4^n/(\lambda^n \mu^n)$ squares with sides μ^n. Thus $S_d(\mu^{-n}) \asymp 4^n/(\lambda^n \mu^n)$ and

$$\text{bdim}(\Lambda) = \lim_{n \to \infty} - \frac{\log S_d(\mu^{-n})}{\log \mu^{-n}} = \lim_{n \to \infty} - \frac{n(\log 4 - \log \lambda - \log \mu)}{n \log \mu}$$

$$= 1 + \frac{\log 4 - \log \lambda}{- \log \mu}.$$

5. Sequence Spaces.

Consider the two-sided sequence space Ω_N with the metric d_λ of (7.3.4). According to (7.3.5), there is a disjoint cover by N^{2n-1} balls of radius λ^{1-n}, namely, the cylinders $C_{\alpha_{1-n} \dots \alpha_{n-1}} = \{\omega \in \Omega_N \mid \omega_i = \alpha_i \text{ for } |i| < n\}$. Therefore $S_{d_\lambda}(\lambda^{1-n}) = N^{2n-1}$ and hence the box dimension is

$$\text{bdim}(\Omega_N, d_\lambda) = \lim_{r \to 0} - \frac{\log S_d(r)}{\log r} = \lim_{n \to \infty} - \frac{\log N^{2n-1}}{\log \lambda^{1-n}} = \lim_{n \to \infty} \frac{2n-1}{n-1} \frac{\log N}{\log \lambda} = 2 \frac{\log N}{\log \lambda}.$$

Analogously to the Cantor set example, the box dimension decreases as λ increases, corresponding to the rapid decrease of the radius of cylinders (as a function of the length of the specified string) for large λ.

8.1.4 Dependence on the Metric

A different issue related to capacity is the dependence of $S_d(r)$ on the metric for a given r. If one replaces a metric by a larger one (with finer resolution, as it were), then balls become smaller and hence $S_d(r)$ increases. The rate at which it does so is a new measure of the rate of refinement of the metrics. A simple example is scaling of the metric, that is, multiplying by a positive factor a. Clearly $S_{ad}(ar) = S_d(r)$ and

$$\lim_{r \to 0} -\frac{\log S_{ad}(r)}{\log r} = \lim_{r \to 0} -\frac{\log S_{ad}(ar)}{\log ar} = \lim_{r \to 0} -\frac{\log S_d(r)}{\log ar}$$

$$= \lim_{r \to 0} -\frac{\log S_d(r)}{\log a + \log r} = \lim_{r \to 0} -\frac{\log S_d(r)}{\log r}.$$

Thus, scaling does not affect the box dimension. However, one may study the asymptotic behavior of $S_{d_i}(r)$ for a sequence d_i of metrics as $i \to \infty$ for fixed r. We presently do this in our study of entropy.

■ EXERCISES

■ **Exercise 8.1.1** Prove that the cardinality of a minimal cover is not always the same as the minimal cardinality of a cover.

■ **Exercise 8.1.2** Compute the box dimension of $\mathbb{Q} \cap [0, 1]$.

■ **Exercise 8.1.3** For the Smale horseshoe show that $0 < \text{bdim}(\Lambda) < 2$.

■ **Exercise 8.1.4** For an S-shaped horseshoe with three crossings compute the box dimension of the invariant set and prove that it lies between 0 and 2.

■ **Exercise 8.1.5** Find the dimension of the metric d''_λ (7.3.10) on Ω_N and Ω_N^R.

■ **Exercise 8.1.6** Show that the dimension $\text{bdim}(\Omega_N^R, d_\lambda)$ of the one-sided shift space Ω_N^R with the metric d_λ is $\log N / \log \lambda$.

■ **Exercise 8.1.7** Show that the triangular Sierpinski carpet has box dimension $\log 3 / \log 2$.

■ **Exercise 8.1.8** Construct Cantor sets on the interval with box dimension 0 and 1.

■ **Exercise 8.1.9** Determine the box dimension of the set of points in $[0, 1]$ that have a binary expansion with no consecutive 0's.

8.2 TOPOLOGICAL ENTROPY

8.2.1 Measures of Complexity and Invariants

We have encountered several indicators of the complexity of a dynamical system: topological transitivity, minimality, density of the set of periodic points, chaos, and topological mixing. Especially the latter indicate the presence of intertwining and separation of different orbits. These are all qualitative ("yes–no") measures of complexity. So far the only quantitative measure of complexity is the growth rate of periodic orbits. While the otherwise simple rational rotations have infinitely many periodic points, it is chaotic examples that are distinguished by the exponential growth of finite numbers of periodic points.

1. Entropy. A step beyond the periodic orbit growth is to measure the growth of all orbits in some sense. This is done by the most important numerical invariant related to the orbit growth, the topological entropy. It represents the exponential growth rate for the number of orbit segments distinguishable with arbitrarily fine but finite precision. In a sense, the topological entropy describes in a crude but suggestive way the total exponential complexity of the orbit structure with a single number. Indeed, we will see that the chaotic systems from among our examples are distinguished by having positive entropy, and the topological entropy is no less than the growth rate of periodic orbits. Therefore it is appropriate to view entropy as a quantitative measure of the amount of chaos in a dynamical system.

2. Invariants. At this point it might be useful to give another motivation for studying invariants of dynamical systems. Invariants are quantities associated with a dynamical system that agree for two dynamical systems that are equivalent in the sense of conjugacy (Definition 7.3.3). When one encounters a new dynamical system it is natural to wonder whether it is equivalent to a previously studied one, which would save a lot of work; or one may try to see whether certain collections of dynamical systems are pairwise equivalent or can be subdivided neatly into equivalence classes (under topological conjugacy). Either way, one needs to decide whether there is a conjugacy between two given systems. If one is unable, after much trying, to find one, the need becomes apparent for methods to show that there can be no conjugacy. Invariants provide a means to do this: If one system is transitive and the other one is not, then they cannot be conjugate. If one circle homeomorphism has rotation number α and another has rotation number $\beta \neq \alpha$, then these two homeomorphisms are not topologically conjugate. Similarly, entropy is an attractive invariant (Corollary 8.2.3) not least for the reason that it takes on real values (as opposed to "yes–no" only) and hence gives a finer distinction between different dynamical systems than transitivity, mixing, and so on. On the other hand, it is defined for a broad class of dynamical systems rather than only circle maps.

8.2.2 First Definition of Entropy

To define entropy we measure the rate of increase of the capacity $S_d(r)$ for fixed r as the metric is refined in a dynamically significant way. This is different from the definition of box dimension, where we study the change in capacity as a function of r for a fixed metric. Suppose $f : X \to X$ is a continuous map of a compact metric space X with distance function d and define an increasing sequence of metrics d_n^f, $n = 1, 2, \ldots$, starting from $d_1^f = d$ by

$$(8.2.1) \qquad\qquad d_n^f(x, y) = \max_{0 \leq i \leq n-1} d(f^i(x), f^i(y)).$$

In other words, d_n^f is the distance between the orbit segments $\mathcal{O}_n(x) = \{x, \ldots, f^{n-1}x\}$ and $\mathcal{O}_n(y)$. We denote the open ball $\{y \in X \mid d_n^f(x, y) < r\}$ by $B_f(x, r, n)$.

Definition 8.2.1 Let $S_d(f, r, n)$ be the r-capacity of d_n^f. Explicitly, a set $E \subset X$ is r-dense with respect to d_n^f, or (n, r)-dense, if $X \subset \bigcup_{x \in E} B_f(x, r, n)$. Then $S_d(f, r, n)$ is the minimal cardinality of an (n, r)-dense set or, equivalently, the cardinality of a *minimal (n, r)-dense set*. This is the minimal number of initial conditions whose behavior up to time n approximates the behavior of *any* initial condition up to r.

Consider the exponential growth rate

(8.2.2) $$h_d(f, r) := \overline{\lim_{n \to \infty}} \frac{1}{n} \log S_d(f, r, n)$$

of $S_d(f, r, n)$. Obviously $h_d(f, r)$ does not decrease with r, so we can define

(8.2.3) $$h_d(f) := \lim_{r \to 0} h_d(f, r).$$

Call $h(f) := h_{\text{top}}(f) := h_d(f)$ the *topological entropy* of f.

Note that we take a double limit, first with respect to n and then with respect to r. The important limit is the one in n, because it is there that the dynamics enters. In many interesting cases the limit in r is, in fact, trivial, because $h_d(f, r)$ is independent of r (for small r) to begin with.

A priori, $h_d(f)$ might depend on the metric d. Actually it does not, so long as one changes to a homeomorphic metric (Definition A.1.17). This justifies dropping the reference to the metric in (8.2.3).

Proposition 8.2.2 *If d' is a metric on X equivalent to d, then $h_{d'}(f) = h_d(f)$.*

Proof The identity map Id: $(X, d) \to (X, d')$ is a homeomorphism by assumption and uniformly continuous in both directions by the compactness of X. Thus, given $r > 0$, there exists a $\delta(r) > 0$ such that, if $d'(x_1, x_2) < \delta$, then $d(x_1, x_2) < r$, that is, any δ-ball in the metric d' is contained in an r-ball in the metric d. By (8.2.1) this also holds for d'^f_n and d^f_n. Thus $S_{d'}(f, \delta, n) \geq S_d(f, r, n)$ for every n, so $h_{d'}(f, \delta) \geq h_d(f, r)$ and $h_{d'}(f) \geq \lim_{\delta \to 0} h_{d'}(f, \delta) \geq \lim_{r \to 0} h_d(f, r) = h_d(f)$. Interchanging d and d' one obtains $h_d(f) \geq h_{d'}(f)$, and hence equality. $\square$

Corollary 8.2.3 *Topological entropy is an invariant of topological conjugacy.*

Proof Let $f: X \to X$, $g: Y \to Y$ be topologically conjugate via a homeomorphism $h: X \to Y$ (see Definition 7.3.3). Fix a metric d on X and define d' on Y as the pullback of d, that is, $d'(y_1, y_2) = d(h^{-1}(y_1), h^{-1}(y_2))$ (Section 2.6.1). Then h becomes an isometry, so $h_d(f) = h_{d'}(g)$. $\square$

8.2.3 Subexponential Growth
As a first example of how to apply this concept, consider situations with relatively simple dynamics.

Proposition 8.2.4 *The topological entropy of contractions and isometries is zero. In particular, any translation T_γ of the torus or any linear flow T^t_ω on the torus (see Section 5.1) has zero entropy.*

Proof If X is a compact metric space and $f: X \to X$ is 1-Lipschitz, then $d^f_n = d$ for all n and consequently $S_n(f, r, n)$ does not depend on n; so $h(f) = 0$. The situation with isometric flows is completely similar to that of maps. $\square$

This absence of any growth is most removed from the case of positive topological entropy. Between these two extreme cases there is a variety of situations of "moderate", that is, subexponential, growth for those quantities. An example is given by the linear twist $T \colon S^1 \times [0,1] \to S^1 \times [0,1]$, $T(x,y) = (x+y, y)$ in Section 6.1.1. In this case we can give a d_n^f-r-dense set of nr^2 balls with centers spaced uniformly r apart along the horizontal and uniformly nr apart on the vertical. The centers are then also $r/2$-separated.

8.2.4 Entropy via Covers

Topological entropy is not always easy to calculate, and it helps to have alternative definitions in order to be able to choose a convenient one as the situation requires (this comes in handy already in Proposition 8.2.9).

There are several quantities similar to $S_d(f, r, n)$ that can be used to define topological entropy. Let $D_d(f, r, n)$ be the minimal number of sets whose diameter in the metric d_n^f is less than r and whose union covers X.

Lemma 8.2.5 $\tilde{h}_d(f, r) := \lim_{n \to \infty} (1/n) \log D_d(f, r, n)$ *exists for any $r > 0$.*

Proof If A is a set of d_n^f-diameter less than r and B is a set of d_m^f-diameter less than r, then $A \cap f^{-n}(B)$ has d_{m+n}^f-diameter less than r. Thus if $\mathfrak{A}$ is a cover of X by $D_d(f, r, n)$ sets of d_n^f-diameter less than r and $\mathfrak{B}$ is a cover of X by $D_d(f, r, m)$ sets of d_m^f-diameter less than r, then the cover by all sets $A \cap f^{-n}(B)$, where $A \in \mathfrak{A}$, $B \in \mathfrak{B}$, contains at most $D_d(f, r, n) \cdot D_d(f, r, m)$ sets and is a cover by sets of d_{m+n}^f-diameter less than r. Thus

$$D_d(f, r, m+n) \le D_d(f, r, n) \cdot D_d(f, r, m)$$

for all m, n. For $a_n = \log D_d(f, r, n)$, this means $a_{m+n} \le a_n + a_m$ and hence $\lim_{n \to \infty} a_n / n$ exists by Lemma 4.3.7. $\square$

Proposition 8.2.6 *If* $\underline{h}_d(f, r) := \varliminf_{n \to \infty} (1/n) \log S_d(f, r, n)$, *then*

$$(8.2.4) \qquad \lim_{r \to 0} \tilde{h}_d(f, r) = \lim_{r \to 0} \underline{h}_d(f, r) = \lim_{r \to 0} h_d(f, r) = h(f).$$

Proof The diameter of an r-ball is at most $2r$, so every covering by r-balls is a covering by sets of diameter $\le 2r$, that is,

$$(8.2.5) \qquad\qquad D_d(f, 2r, n) \le S_d(f, r, n).$$

On the other hand, any set of diameter $\le r$ is contained in the r-ball around each of its points, so

$$(8.2.6) \qquad\qquad S_d(f, r, n) \le D_d(f, r, n).$$

Thus

$$\tilde{h}_d(f, 2r) \le \underline{h}_d(f, r) \le h_d(f, r) \le \tilde{h}_d(f, r). \quad \square$$

8.2.5 Topological Entropy via Separated Sets

Another way to define topological entropy is via the maximal number $N_d(f, r, n)$ of points in X with pairwise d_n^f-distances at least r. We say that such a set of points is (n, r)-*separated*. (See Figure 8.2.1.) Such points generate the maximal number of orbit segments of length n that are distinguishable with precision r.

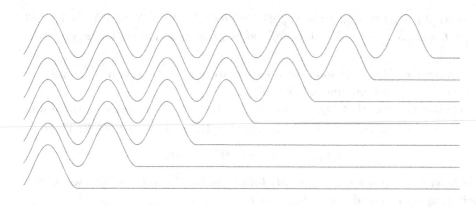

Figure 8.2.1. A separated set.

Proposition 8.2.7

$$(8.2.7) \qquad h_{top}(f) = \lim_{r\to 0} \overline{\lim_{n\to\infty}} \frac{1}{n} \log N_d(f, r, n) = \lim_{r\to 0} \underline{\lim_{n\to\infty}} \frac{1}{n} \log N_d(f, r, n).$$

Remark 8.2.8 This justifies the verbal description of entropy as the exponential growth rate for the number of orbit segments distinguishable with arbitrarily fine but finite precision that we gave at the beginning of this section.

Proof A maximal (n, r)-separated set is (n, r)-dense, that is, for any such set of points the r-balls around them cover X, because otherwise it would be possible to increase the set by adding any point not covered. Thus

$$(8.2.8) \qquad S_d(f, r, n) \le N_d(f, r, n).$$

On the other hand, no $r/2$-ball can contain two points r apart. Thus

$$(8.2.9) \qquad N_d(f, r, n) \le S_d\left(f, \frac{r}{2}, n\right).$$

Using (8.2.8) and (8.2.9) we obtain

$$(8.2.10) \qquad \underline{h}_d(f, r) \le \overline{\lim_{n\to\infty}} \frac{1}{n} \log N_d(f, r, n) \le \overline{\lim_{n\to\infty}} \frac{1}{n} \log N_d(f, r, n) \le h_d\left(f, \frac{r}{2}\right).$$

The result follows by Proposition 8.2.6. □

8.2.6 Some Properties of Entropy

The following proposition contains some standard elementary properties of topological entropy. The proofs demonstrate the usefulness of switching back and forth from one of the three definitions to another.

Proposition 8.2.9

(1) If Λ is a closed f-invariant set, then $h_{top}(f_{\restriction\Lambda}) \le h_{top}(f)$.
(2) If $X = \bigcup_{i=1}^m \Lambda_i$, where Λ_i, $(i = 1, \ldots, m)$ are closed f-invariant sets, then $h_{top}(f) = \max_{1\le i\le m} h_{top}(f_{\restriction\Lambda_i})$.
(3) $h_{top}(f^m) = |m| h_{top}(f)$.
(4) If g is a factor of f, then $h_{top}(g) \le h_{top}(f)$.

(5) $h_{top}(f \times g) = h_{top}(f) + h_{top}(g)$, where $f: X \to X$, $g: Y \to Y$ and $f \times g: X \times Y \to X \times Y$ is defined by $(f \times g)(x, y) = (f(x), g(y))$.

Proof Statement (1) is obvious since every cover of X by sets of d_n^f-diameter less than r is at the same time a cover of Λ.

To prove (2) note that $D_d(f, r, n) \leq \sum_{i=1}^m D_d(f_{\restriction \Lambda_i}, r, n)$, because the union of covers of $\Lambda_1, \ldots, \Lambda_m$ by sets of diameter less than r is a cover of X. Thus

$$D_d(f_{\restriction \Lambda_i}, r, n) \geq \frac{1}{m} D_d(f, r, n)$$

for at least one i. Since there are only finitely many i's, at least one i works for infinitely many n. For this $i \in \{1, \ldots, m\}$

$$\varlimsup_{n \to \infty} \frac{\log D_d(f_{\restriction \Lambda_i}, r, n)}{n} \geq \varlimsup_{n \to \infty} \frac{\log D_d(f, r, n) - \log m}{n} = \tilde{h}_d(f, r).$$

Together with (1) this proves (2).

If m is positive, then (3) follows from two remarks. First

$$d_n^{f^m}(x, y) = \max_{0 \leq i \leq n-1} d(f^{im}(x), f^{im}(y)) \leq \max_{0 \leq i \leq mn-1} d(f^i(x), f^i(y)) = d_{nm}^f(x, y),$$

so any $d_n^{f^m}$ r-ball contains a d_{mn}^f r-ball and

(8.2.11) $S_d(f^m, r, n) \leq S_d(f, r, mn).$

Hence $h_{top}(f^m) \leq m h_{top}(f)$. On the other hand, for every $r > 0$ there is a $\delta(r) > 0$ such that $B(x, \delta(r)) \subset B_f(x, r, m)$ for all $x \in X$. Thus

$$B_{f^m}(x, \delta(r), n) = \bigcap_{i=0}^{n-1} f^{-im} B(f^{im}(x), \delta(r))$$

$$\subset \bigcap_{i=0}^{n-1} f^{-im} B_f(f^{im}(x), r, m) = B_f(x, r, mn).$$

Consequently,

$$S_d(f, r, mn) \leq S_d(f^m, \delta(r), n)$$

and $m h_{top}(f) \leq h_{top}(f^m)$. If f is invertible, then $B_f(x, r, n) = B_{f^{-1}}(f^{n-1}(x), r, n)$ and $S_d(f, r, n) = S_d(f^{-1}, r, n)$; so $h_{top}(f) = h_{top}(f^{-1})$.

If m is negative, then (3) follows from the statement for $m > 0$ and $n = -1$.

Statement (4) deals with $f: X \to X$, $g: Y \to Y$, $h: X \to Y$ such that $h \circ f = g \circ h$ and $h(X) = Y$ (Definition 7.3.3). Denote by d_X, d_Y the distance functions in X and Y, correspondingly.

h is uniformly continuous, so for any $r > 0$ there is $\delta(r) > 0$ such that, if $d_X(x_1, x_2) < \delta(r)$, then $d_Y(h(x_1), h(x_2)) < r$. Thus the image of any $(d_X)_n^f$ ball of radius $\delta(r)$ lies inside a $(d_Y)_n^f$ ball of radius r, that is,

$$S_{d_X}(f, \delta(r), n) \geq S_{d_Y}(g, r, n).$$

Taking logarithms and limits, we obtain (4).

To prove (5) use the product metric $d((x_1, y_1), (x_2, y_2)) = \max(d_X(x_1, x_2), d_Y(y_1, y_2))$ in $X \times Y$. Balls in the product metric are products of balls on X and Y. The same is true for balls in $d_n^{f \times g}$. Thus $S_d(f \times g, r, n) \leq S_{d_X}(f, r, n) S_{d_Y}(g, r, n)$

and $h_{top}(f \times g) \leq h_{top}(f) + h_{top}(g)$. On the other hand, the product of any (n, r)-separated set in X for f and any (n, r)-separated set in Y for g is an (n, r)-separated set for $f \times g$. Thus

$$N_d(f \times g, r, n) \geq N_{d_X}(f, r, n) \times N_{d_Y}(g, r, n)$$

and hence $h_{top}(f \times g) \geq h_{top}(f) + h_{top}(g)$. $\square$

■ **EXERCISES**

■ **Exercise 8.2.1** Compute the topological entropy of $f(x) = x(1 - x)$ on $[0, 1]$.

■ **Exercise 8.2.2** Compute the topological entropy of the linear horseshoe.

■ **Exercise 8.2.3** Suppose $f: S^1 \to S^1$ is an orientation-preserving C^2-diffeo-morphism without periodic points. Find $h_{top}(f)$.

■ **Exercise 8.2.4** Let $f: \mathbb{T}^3 \to \mathbb{T}^3$, $f(x, y, z) = (x, x + y, y + z)$. Find $h_{top}(f)$.

■ **Exercise 8.2.5** Suppose $X = \bigcup_i X_i$ is compact, $f: X \to X$ such that each X_i is closed and f-invariant. Show that $h_{top}(f) = \sup h_{top}(f_{\restriction X_i})$.

■ **PROBLEMS FOR FURTHER STUDY**

■ **Problem 8.2.6** Given $f: X \to X, g: Y \to Y$, suppose $h \circ f = g \circ h$, where $h: X \to Y$ is a continuous surjective map such that every $y \in Y$ has finitely many preimages. Show that $h_{top}(f) = h_{top}(g)$.

8.3 APPLICATIONS AND EXTENSIONS

8.3.1 Expanding Maps

The expanding maps E_m represent the first situation in our survey where a really complicated orbit structure appears. Since one of the features of this structure is the exponential growth of periodic orbits (Proposition 7.1.2), it is natural to expect the total exponential orbit complexity, measured by the topological entropy, to be positive too.

Proposition 8.3.1 *If $m \in \mathbb{N}$, $|m| \geq 2$, then $h_{top}(E_m) = \log |m| = p(E_m)$.*

Proof For the map E_m, and in fact for any expanding map, the distance between iterates of any two points grows until it becomes greater than a certain constant depending on the map $(1/2|m|$ for the map E_m). To simplify notations, assume $m > 0$. If $d(x, y) < m^{-n}/2$, then $d_n^{E_m}(x, y) = d(E_m^{n-1}(x), E_m^{n-1}(y))$; so if $d_n^{E_m}(x, y) \geq r$, then $d(x, y) \geq rm^{-n}$. Taking $r = m^{-k}$, this shows that $\{im^{-n-k} \mid i = 0, \ldots, m^{n+k} - 1\}$ is a maximal set of points whose pairwise $d_n^{E_m}$-distances are at least m^{-k}, that is,

$$N_d(E_m, m^{-k}, n) = m^{n+k},$$

and consequently

$$h(E_m) = \lim_{k \to \infty} \overline{\lim_{n \to \infty}} \frac{\log N_d(E_m, m^{-k}, n)}{n} = \lim_{k \to \infty} \lim_{n \to \infty} \frac{n+k}{n} \log m = \log m.$$

The case $m < 0$ is completely parallel. $\square$

Since topological entropy is invariant under topological conjugacy (Corollary 8.2.3) and every expanding map of degree m is topologically conjugate to the map E_m (Theorem 7.4.3), we obtain from Proposition 8.3.1

Corollary 8.3.2 *If $f: S^1 \to S^1$ is an expanding map of degree m, then*

$$h_{top}(f) = p(f) = \log |m|.$$

8.3.2 Shifts and Topological Markov Chains

Proposition 8.3.3 $h_{top}(\sigma_A) = p(\sigma_A) = \log |\lambda_A^{\max}|$ *for any topological Markov chain σ_A.*

Proof Analogously to Section 7.3.4, any cylinder

$$(8.3.1) \qquad C_{\alpha_{-m},\ldots,\alpha_{n+m}}^{-m,\ldots,n+m} := \{\omega \in \Omega_N \mid \omega_i = \alpha_i \text{ for } -m \le i \le m+n\}$$

is at the same time the ball of radius $r_m = \lambda^{-m}/2$ around each of its points with respect to the metric $d_n^{\sigma_N}$ associated with the shift σ_N (because $\lambda > 3$). Thus, any two $d_n^{\sigma_N}$ balls of radius r_m are either identical or disjoint, and there are exactly N^{n+2m+1} different ones of the form (8.3.1); so $S_{d_\lambda}(\sigma_N, r_m, n) = N^{n+2m+1}$ and

$$h(\sigma_N) = \lim_{m\to\infty} \lim_{n\to\infty} \frac{1}{n} \log N^{n+2m+1} = \log N.$$

Similarly, if σ_A is a topological Markov chain, then $S_d(\sigma_A, r_m, n)$ is the number of those cylinders (8.3.4) that have nonempty intersection with Ω_A. Assume that each row of A contains at least one 1. Since the number of admissible paths of length n that begin with i and end with j is the entry a_{ij}^n of A^n (see Lemma 7.3.5), the number of nonempty cylinders of rank $n+1$ in Ω_A is $\sum_{i,j=0}^{N-1} a_{ij}^n < C \cdot \|A^n\|$ for some constant C. On the other hand, $\sum_{i,j=0}^{N-1} a_{ij}^n > c\|A^n\|$ for another constant $c > 0$ because all numbers a_{ij}^n are nonnegative and hence the left-hand side is the norm $\sum_{i,j=0}^{N-1} a_{ij}^n$ of A^n, which is equivalent to the usual norm because all norms on $\mathbb{R}^{N^2}$ are equivalent. Thus, we have

$$(8.3.2) \qquad S_{d_\lambda}(\sigma_A, r_m, n) = \sum_{i,j=0}^{N-1} a_{ij}^{n+2m}$$

and

$$(8.3.3) \quad h(\sigma_A) = \lim_{m\to\infty} \overline{\lim_{n\to\infty}} \frac{1}{n} \log S_{d_\lambda}(\sigma_A, r_m, n)$$

$$= \lim_{m\to\infty} \lim_{n\to\infty} \frac{1}{n} \log \|A^{n+2m}\| = \lim_{n\to\infty} \frac{1}{n} \log \|A^n\| = \log r(A) = \log |\lambda_A^{\max}|,$$

where $r(A)$ is the spectral radius of the matrix A (Definition 3.3.1). Equation (8.3.3) and Proposition 7.3.7 now give the claim. $\square$

8.3.3 The Hyperbolic Toral Automorphism

In calculating the entropy of the toral automorphism we use both coding and our knowledge of the growth rate of periodic points.

Proposition 8.3.4 *If $F_L: \mathbb{T}^2 \to \mathbb{T}^2$ is given by $F_L(x, y) = (2x + y, x + y)$ (mod 1), then*

$$h(F_L) = p(F_L) = \frac{3 + \sqrt{5}}{2}.$$

Proof In Section 7.4.5 we showed that

$$F_L(x, y) = (2x + y, x + y) \quad \text{(mod 1)}$$

is a factor of the topological Markov chain σ_A, where

$$A = \begin{pmatrix} 1 & 1 & 0 & 1 & 0 \\ 1 & 1 & 0 & 1 & 0 \\ 1 & 1 & 0 & 1 & 0 \\ 0 & 0 & 1 & 0 & 1 \\ 0 & 0 & 1 & 0 & 1 \end{pmatrix},$$

and that the maximal eigenvalue of A is $\lambda_A^{\max} = 3 + \sqrt{5}/2$. Proposition 8.2.9(4) shows that

(8.3.4) $$h(F_L) \le h(\sigma_A) = \log \frac{3 + \sqrt{5}}{2}.$$

On the other hand, we next show that the set of n-periodic points of F_L is $(n, 1/4)$-separated for any $n \in \mathbb{N}$. This implies $N_d(F_L, 1/4, n) \ge P_n(F_L)$ and

$$h(F_L) \ge p(F_L) = \log \frac{3 + \sqrt{5}}{2}$$

by Proposition 7.1.10. By (8.3.4), the result then follows.

If p, q are n-periodic points and $d(p, q) < 1/4$, then there is a uniquely defined minimal rectangle R with vertices p, s, q, t formed by segments of expanding and contracting lines passing through p and q. (See Figure 8.3.1.) Under the action of F_L the sides ps and qt expand with coefficient $\lambda_1 = (3 + \sqrt{5}/2) > 2$ while the other two sides contract with coefficient λ_1^{-1}.

This implies $F_L^n(R) \ne R$ because F_L^n cannot leave all four sides invariant while also expanding and contracting them. Equivalently, $F_L^{-n}(R) \ne R$. Therefore, there is a $k \le n$ for which $F_L^k(R)$ is not a minimal rectangle. For the smallest such k we then

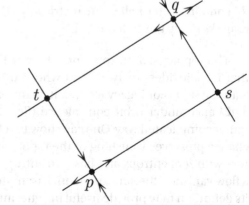

Figure 8.3.1. Heteroclinic points.

have $d(F_L^k(p), F_L^k(q)) > 1/4$ because a rectangle with diagonals shorter than $1/4$ is minimal. Thus the periodic points of period n form an $(n, 1/4)$-separated set. $\square$

Remark 8.3.5 In the case of expanding maps E_m and for topological Markov chains σ_A one can also show that periodic points form (n, r_0)-separated sets for some r_0. This allows us to produce the inequality $h_{\text{top}} \geq p$ in a uniform way for all three cases.

8.3.4 Periodic Points and Entropy

Our examples show an interesting pattern. For both smooth examples with complicated exponentially growing orbit structure, namely, expanding maps (Proposition 8.3.1) and hyperbolic toral automorphisms (Proposition 8.3.4), the two natural measures of the exponential orbit growth – the growth rate p of periodic points and the topological entropy h_{top} – coincide. This is a rather widespread phenomenon, although not universal. It is related to the local hyperbolic structure, that is, the stretching and folding common to these examples. (This is systematically introduced in Chapter 10.) For topological Markov chains the growth rate of periodic points and topological entropy also coincide (Proposition 8.3.3). Hyperbolicity is a relevant explanation here, too, since by Proposition 7.4.6 topological Markov chains are topologically conjugate to the restriction of some smooth systems to special invariant sets that possess hyperbolic behavior.

8.3.5 Topological Entropy for Flows

The definition of topological entropy $h_{\text{top}}(\Phi)$ for a flow $\Phi = (\varphi^t)_{t \in \mathbb{R}}$ is completely parallel to that for the discrete-time case. The only change is that the metrics in (8.2.1) are replaced by the nondecreasing family

$$d_T^\Phi(x, y) = \max_{0 \leq t \leq T} d(\varphi^t(x), \varphi^t(y))$$

of metrics. This parallelism has a particularly useful consequence analogous to Proposition 8.2.9.(3).

Proposition 8.3.6 $h_{top}(\Phi) = h_{top}(\varphi^1)$.

Proof If $r > 0$, then by compactness and continuity for there is a $\delta(r) > 0$ such that $d(x, y) \leq \delta(r)$ implies $\max_{0 \leq t \leq 1} d(\varphi^t(x), \varphi^t(y)) < r$. Then any r-ball in the metric d_T^Φ contains a $\delta(r)$-ball in the metric $d_{\lfloor T \rfloor}^{\varphi^1}$. On the other hand, $d_n^\Phi \geq d_n^{\varphi^1}$. These two remarks imply the statement. $\square$

The topological entropy for a flow is thus invariant under *flow equivalence*, that is, coincides for two flows whose time-t maps are topologically conjugate with the same conjugacy for all t. It changes under time change (Definition 9.4.12) and hence under orbit equivalence (flow equivalence with a time change) in a rather complicated way. One can show that for a flow without fixed points any time change preserves vanishing of the topological entropy, that is, a time change of a flow with zero entropy also has zero entropy. If the topological entropy for a map or a flow vanishes, the subexponential asymptotic of any of the quantities involved in its definition may provide useful insights into the complexity of the orbit structure.

8.3.6 Local Entropy as a Measure of Sensitive Dependence

As we mentioned in the introduction to this section, entropy can be viewed as a measure of the amount of chaos in a system. We now explicitly show how entropy provides a quantitative measure of the amount of sensitive dependence in a dynamical system. To that end we introduce a closely related notion of *local entropy*, explain how it measures sensitive dependence on the one hand, and how it is related to topological entropy on the other.

Fix a point x and a "microscopic" ϵ as well as a "macroscopic" r, and let $N_d(f, r, n, x, \epsilon)$ be the maximal number of points in $B_d(x, \epsilon)$ with pairwise d_n^f-distances at least r. A large such number would certainly indicate rather sensitive dependence on initial conditions.

Definition 8.3.7 If

$$h_{d,x,r}(f) := \lim_{\epsilon \to 0} \overline{\lim_{n \to \infty}} \frac{1}{n} \log N_d(f, r, n, x, \epsilon),$$

then

$$h_{d,x}(f) := \lim_{r \to 0} h_{d,x,r}(f)$$

is called the *local entropy* of f at x.

Remark 8.3.8 The limits exist because the dependence on ϵ is increasing and on r decreasing.

Proposition 8.3.9 $h_{d,x}(f) \leq h_{top}(f)$.

Proof Topological entropy corresponds to the case of leaving ϵ fixed at a size for which $B(x, \epsilon)$ is the entire space. Therefore any point with strong sensitive dependence in this sense necessarily produces large topological entropy. $\square$

On the other hand, there is a relation with $h_d(f, r)$ [see (8.2.2)]:

Proposition 8.3.10 *For $r > 0$ there exists an x such that*

$$h_{d,x,r}(f) \geq h_d(f, r).$$

Proof If $S_d(f, r, n, x, \epsilon)$ is the minimal number of d_n^f-r-balls covering $B_d(x, \epsilon)$, then there is an x such that

(8.3.5) $$S_d(f, r, n) \leq S_d(\epsilon) S_d(f, r, n, x, \epsilon)$$

because we can take a cover of the space by $S_d(\epsilon)$-balls of radius ϵ and, denoting their centers by x_j, we have

$$S_d(f, r, n) \leq \sum_{j=1}^{S_d(\epsilon)} S_d(f, r, n, x_j, \epsilon);$$

hence
$$S_d(f, r, n, x, \epsilon) \geq S_d(f, r, n)/S_d(\epsilon)$$
for x being one of the x_j.

As $n \to \infty$ we obtain a sequence of such points x_n satisfying (8.3.2) for the respective values of n. Take an accumulation point x of this sequence and consider the 2ϵ-ball around it. For sufficiently large n we have $B_d(x_n, \epsilon) \subset B_d(x, 2\epsilon)$ and hence
$$S_d(f, r, n, x, 2\epsilon) \geq S_d(f, r, n)/S_d(\epsilon)$$
for all n, which implies
$$\varlimsup_{n \to \infty} \frac{1}{n} \log S_d(f, r, n, x, 2\epsilon) \geq \varlimsup_{n \to \infty} \frac{1}{n} \log)(S_d(f, r, n)/S_d(\epsilon)) = h_d(f, r).$$
Using arguments as before we can replace $S_d(f, r, n, x, 2\epsilon)$ by the corresponding number of r-dense points and let $\epsilon \to 0$ to get
$$h_{d,x,r}(f) \geq h_d(f, r)$$
for all r. $\square$

Remark 8.3.11 Since $h_d(f, r) \xrightarrow[r \to 0]{} h(f)$, this shows that there are points with $h_{d,x,r}(f)$ arbitrarily close to the topological entropy. Thus the supremum of local entropies over the space is the topological entropy, and topological entropy indeed measures the amount of sensitive dependence on initial conditions.

■ EXERCISES

■ **Exercise 8.3.1** Prove Corollary 8.3.2 without reference to topological conjugacy.

■ **Exercise 8.3.2** Construct a map with positive topological entropy that has no periodic points.

■ **Exercise 8.3.3** Construct a topologically transitive map of a compact metric space that has infinite topological entropy.

■ **Exercise 8.3.4** Prove that the local entropy of E_m is independent of the point and equals topological entropy.

■ **Exercise 8.3.5** Prove that the local entropy of the shift on m symbols is independent of the point and equals topological entropy.

■ **Exercise 8.3.6** Prove that the local entropy of the toral automorphism induced by $\begin{pmatrix} 2 & 1 \\ 1 & 1 \end{pmatrix}$ is independent of the point and equals topological entropy.

■ **Exercise 8.3.7** Consider the closed unit disk in $\mathbb{R}^2$ and the map f_λ on it defined in polar coordinates by $f_\lambda(re^{i\theta}) = \lambda re^{2i\theta}$, where $0 \leq \lambda \leq 1$. Show that $h_{\text{top}}(f_1) \geq \log 2$ and that $h_{\text{top}}(f_\lambda) = 0$ for $\lambda < 1$.

■ PROBLEMS FOR FURTHER STUDY

■ **Problem 8.3.8** Prove that $h_{\text{top}}(\varphi^t) = |t| h_{\text{top}}(\varphi^1)$ for any flow φ^t.

■ **Problem 8.3.9** Give an example of a topologically transitive map for which local entropy is not constant.

PART 2

PANORAMA OF DYNAMICAL SYSTEMS

This part of the book develops results and themes from the course and in doing so presents several strands of research in dynamics. The exposition relies on and refers to the course, but it stands on its own as a careful description of selected modern achievements. The choice of subjects was influenced by the topics of the course and the degree to which these results have been digested in the literature.

Starting with Chapter 10, we adopt a style less technically thorough than that of the course. Many results come with at least a proof outline, but not all arguments labeled as proofs are as complete and self-contained as in the course. The focus is on making clear how ideas are elaborated and used. Many references are given in the text.

The chapters are largely independent of each other, and can be read selectively and in any order.

CHAPTER 9

Simple Dynamics as a Tool

9.1 INTRODUCTION

9.1.1 Applications of the Contraction Principle

The collection of simple dynamical systems with complicated orbit structure presented in Chapter 7 and revisited in Section 8.3 is representative of *hyperbolic* dynamical systems. Much of the core theory of hyperbolic dynamics consists of results that are obtained (more or less) directly from the Contraction Principle, which first appeared as an example of a dynamical system with simple dynamics in Chapter 2. Although we already used it in Section 2.5 as a tool that can tell us much about other dynamical systems, its pervasive role in hyperbolic dynamics motivates a more thorough presentation of its uses. Accordingly, the main theme of this chapter is to present case studies of using the Contraction Principle, that is, of putting one important insight about a specific class of simple dynamics to use in an auxiliary space to tell us about analysis as well as (complicated) dynamical systems. Since the results we obtain are rather important, we take some time to develop them further, notably when it comes to the basic theory of differential equations. In this chapter we maintain the same standard of proof as in the course.

As in the preceding chapters, this intrinsically interesting development has a utilitarian undercurrent. The results obtained here are important for the study of dynamical systems. In the case of existence and uniqueness of solutions of differential equations this is evident, but all other results presented here also figure in our development and are standard tools in dynamics. This chapter does not present nearly all such applications; some others are presented elsewhere, such as the Anosov Closing Lemma (Theorem 10.2.2), which follows from the Contraction Principle by way of the Hyperbolic Fixed Point Theorem 9.5.4. The Stable Manifold Theorem 9.5.2 is the foremost example and is featured prominently here.

9.1.2 Overview

We begin by deriving two important results in analysis, the Inverse- and Implicit-Function Theorems. The latter immediately tells us something new about the

Contraction Principle itself: The fixed point of a contraction depends smoothly on the contraction. A first and straightforward application of these results is the persistence of transverse fixed points in Section 9.3, where we show that a simple condition on the linear part of a map at a fixed point can guarantee that the fixed point persists when the map is perturbed. This is similar to the situation for contractions (Proposition 2.2.20) and very much in the spirit of linearization (which is discussed in Section 2.1, the beginning of Chapter 3, and, for example, Section 6.2.2.7). In these first applications the space on which the contraction is defined is the same space in which the problem is posed. However, the applications of the Contraction Principle to existence and uniqueness of solutions of differential equations (Section 9.4) and in the theorem on stable manifolds (Section 9.5), like many other important applications in analysis, use the Contraction Principle by reducing the situation at hand to a search for a fixed point in a space of functions, which has infinite dimension, rather than in a Euclidean space.

9.1.3 Creating a Context for the Contraction Principle
While the common feature is the application of the Contraction Principle in some auxiliary space, the degree of cleverness required to set this up varies in these examples. Picard iteration (Section 9.4) is a straightforward application, even though the space in question is not as simple as in the earlier applications. Of course, this is also the oldest example. The initial step in the proof of the Inverse-Function Theorem 9.2.2 requires more creativity but is close to the Newton method. The proof of persistence of transverse fixed points (Proposition 9.3.1) has no equally obvious motivation for the initial step, but it exhibits some features common to other applications of the Contraction Principle in dynamics. The central point is the combination of transversality and closeness (smallness of a perturbation), which is being used to produce an invertible map by transversality whose inverse is composed with a strongly contracting map arising from the perturbation. (The trick is to do this in such a way that the desired object is a fixed point of the resulting contraction.)

Except for the Picard iteration, all applications of the Contraction Principle in this chapter depend on linearization. This, too, is typical of applications in the theory of smooth dynamical systems.

9.2 IMPLICIT- AND INVERSE-FUNCTION THEOREMS IN EUCLIDEAN SPACE

9.2.1 The Inverse-Function Theorem
The inverse function theorem says that, if a differentiable map has invertible derivative at some point, then the map is invertible near that point. This result is related to linearization: If we assume a certain qualitative ("yes–no") fact about the linear part (invertibility), then it holds for the nonlinear map itself – at least in a neighborhood. The version for the real line is familiar from calculus:

Theorem 9.2.1 *Suppose $I \subset \mathbb{R}$ is an open interval and $f: I \to \mathbb{R}$ a differentiable function. If $a \in I$ is such that $f'(a) \neq 0$ and f' is continuous at a, then f is invertible on a neighborhood U of a and $(f^{-1})'(y) = 1/f'(x)$, where $y = f(x)$.*

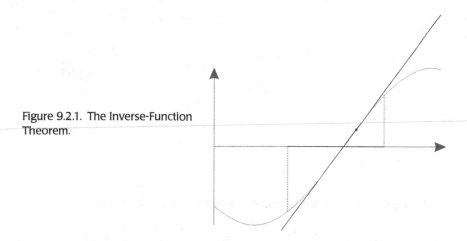

Figure 9.2.1. The Inverse-Function Theorem.

Usually one thinks of invertibility as the easy part and the derivative formula as the hard one, because the basic calculus examples of invertible real-valued functions are given by formulas where invertibility is rather apparent. However, the main content of this result is to conclude invertibility from knowledge only of the linear part of a map *at one point*, without any such extra information. The derivative formula is then an easy postlude. We even get higher derivatives easily. This result is fairly easy in $\mathbb{R}^n$ as well, however, we first give a proof for the simple case of a single variable.

Proof Given y, we want to solve the equation $f(x) = y$ for x, which is the same as finding a root of $F_y(x) := y - f(x)$. To this end we first set up a suitable contracting map.

The space. The space on which the contraction acts is the real line.

Defining the contraction. The Newton method in Section 2.2.8 suggests making an initial guess x (where y is fixed for the moment) and improve the guess by repeatedly applying the map

$$F_y(x) = x - \frac{F_y(x)}{F_y'(x)} = x + \frac{y - f(x)}{f'(x)}.$$

To verify that this is a contraction involves taking and estimating the second derivative of f, but we don't assume it exists. It is convenient to instead consider the map

$$\varphi_y(x) := x + \frac{y - f(x)}{f'(a)}$$

on I. Its fixed points are solutions of our problem because $\varphi_y(x) = x$ if and only if $f(x) = y$.

The contraction property. Now we show that φ_y is a contraction of some closed subset of O. Then, by the Contraction Principle, it has a unique fixed point and hence there exists a unique x such that $f(x) = y$.

To that end let $A := f'(a)$ and $\alpha = |A|/2$. By continuity of f' at a there is an $\epsilon > 0$ such that $W := (a - \epsilon, a + \epsilon) \subset I$ and $|f'(x) - A| < \alpha$ for x in the closure $\overline{W}$ of W.

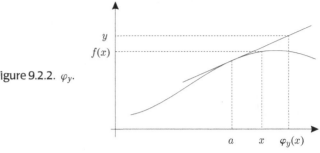

Figure 9.2.2. φ_y.

To see that φ_y is a contraction on $\ddot{W}$ note that, if $x \in \ddot{W}$, then

$$|\varphi_y'(x)| = \left|1 - \frac{f'(x)}{A}\right| = \left|\frac{A - f'(x)}{A}\right| < \frac{\alpha}{|A|} = 1/2.$$

Using Proposition 2.2.3 we get $|\varphi_y(x) - \varphi_y(x')| \le |x - x'|/2$ for $x, x' \in W$.

We also need to show that $\varphi_y(\ddot{W}) \subset \ddot{W}$ for y sufficiently close to $b := f(a)$. Let $\delta = A\epsilon/2$ and $V = (b - \delta, b + \delta)$. Then for $y \in V$ we have

$$|\varphi_y(a) - a| = \left|a + \frac{y - f(a)}{A} - a\right| = \left|\frac{y - b}{A}\right| < \left|\frac{\delta}{a}\right| = \frac{\epsilon}{2};$$

so if $x \in \ddot{W}$, then

$$|\varphi_y(x) - a| \le |\varphi_y(x) - \varphi_y(a)| + |\varphi_y(a) - a| < \frac{x - a}{2} + \frac{\epsilon}{2} \le \epsilon$$

and hence $\varphi_y(x) \in W$.

Therefore Proposition 2.2.21 applied to $\varphi_y \colon \ddot{W} \to \ddot{W}$ for $y \in V$ produces a unique fixed point $g(y) \in W$, which depends continuously on y.

Next we prove that the inverse is differentiable. For $y = f(x) \in V$ we want to show that $g'(y)$ exists and is the reciprocal of $B := f'(g(y))$.

Let $U := g(V) = W \cap f^{-1}(V)$ (the preimage under f), so U is open. Take $y + k = f(x + h) \in V$. Then

$$\frac{|h|}{2} \ge |\varphi_y(x + h) - \varphi_y(x)| = \left|h + \frac{f(x) - f(x + h)}{A}\right| = \left|h - \frac{k}{A}\right| \ge |h| - \left|\frac{k}{A}\right|,$$

and hence

$$\frac{|h|}{2} \le \left|\frac{k}{A}\right| < \frac{|k|}{\alpha} \quad \text{and} \quad \frac{1}{|k|} < \frac{2}{\alpha|h|}.$$

Since $g(y + k) - g(y) - k/B = h - k/B = -(f(x + h) - f(x) - Bh)/B$, we therefore get

$$\frac{|g(y + k) - g(y) - k/B|}{|k|} < \frac{2}{|B|\alpha} \frac{|f(x + h) - f(x) - Bh|}{|h|} \xrightarrow[|h| \le 2|k|/\alpha \to 0]{} 0,$$

which proves $g'(y) = 1/B = 1/f'(g(y))$.

Finally, suppose $f \in C^r$. We show inductively that $g \in C^r$. To that end assume $g \in C^k$ for some $k < r$ (we start the induction with $k = 0$). Then $f'(g(y)) \in C^k$ and so is its reciprocal g'. Thus, $g \in C^{k+1}$. $\square$

Now we adapt this argument to $\mathbb{R}^n$:

Theorem 9.2.2 (Inverse-Function Theorem) *Suppose* $O \subset \mathbb{R}^m$ *is open,* $f\colon O \to \mathbb{R}^m$ *is differentiable, and* Df *is invertible at a point* $a \in O$ *and continuous at a. Then there exist neighborhoods* $U \subset O$ *of a and* V *of* $b := f(a) \in \mathbb{R}^m$ *such that* f *is a bijection from* U *to* V *[that is,* f *is one-to-one on* U *and* $f(U) = V$*]. The inverse* $g\colon V \to U$ *of* f *is differentiable with* $Dg(y) = (Df(g(y)))^{-1}$*. Furthermore, if* f *is* C^r *(that is, all partial derivatives of* f *up to order* r *exist and are continuous) on* U*, then so is its inverse.*

Proof The proof is actually the same as before. We only need to replace various numbers by linear maps and some absolute values by norms.

The space. The contraction acts in $\mathbb{R}^m$.

The map. For any given $y \in \mathbb{R}^m$, consider the map

$$\varphi_y(x) := x + Df(a)^{-1}(y - f(x))$$

on O. Notice that $\varphi_y(x) = x$ if and only if $f(x) = y$, so we try to find a unique fixed point for φ_y. We need a set W on which it is a contraction.

The contraction property. Let $A := Df(a)$, $\alpha < \|A^{-1}\|^{-1}/2$, and, using continuity of Df at a, take $\epsilon > 0$ such that $\|Df(x) - A\| < \alpha$ for x in the closure of $W := B(a, \epsilon)$. To see that φ_y is a contraction, note that

$$\|D\varphi_y(x)\| = \|\operatorname{Id} - A^{-1}Df(x)\| = \|A^{-1}(A - Df(x))\| < \|A^{-1}\|\alpha =: \lambda < 1/2$$

for $x \in W$ and apply Corollary 2.2.15 to get $\|\varphi_y(x) - \varphi_y(x')\| \le \lambda\|x - x'\|$ for $x, x' \in W$. Therefore, by Proposition 2.2.20 there is a neighborhood V of b such that φ_y is a contraction of $\overline{W}$ for all $y \in V$ and has a unique fixed point $g(y) \in W$ (which depends continuously on y). $U := g(V) = W \cap f^{-1}(V)$ is open.

The determinant of $Df(x)$ depends continuously on Df and hence is continuous at a as a function of x. Thus, by taking V (and hence U) smaller, if necessary, we may assume $\det Df \ne 0$ on U and therefore that $Df(x)$ is invertible on U.

For $y = f(x) \in V$ we want to show that $Dg(y)$ exists and is the inverse of $B := Df(g(y))$. Take $y + k = f(x + h) \in V$. Then

$$(9.2.1) \qquad \frac{\|h\|}{2} \ge \|\varphi_y(x + h) - \varphi_y(x)\| = \|h + A^{-1}(f(x) - f(x + h))\|$$

$$= \|h - A^{-1}k\| \ge \|h\| - \|A^{-1}\|\|k\|,$$

so

$$\frac{\|k\|}{\alpha} > \|A^{-1}\|\|k\| \ge \frac{\|h\|}{2} \qquad \text{and} \qquad \frac{1}{\|k\|} < \frac{2}{\alpha\|h\|}.$$

Since $g(y + k) - g(y) - B^{-1}k = h - B^{-1}k = -B^{-1}(f(x + h) - f(x) - Bh)$, we get

$$\frac{\|g(y + k) - g(y) - B^{-1}k\|}{\|k\|} < \frac{\|B^{-1}\|}{\alpha/2}\frac{\|f(x + h) - f(x) - Bh\|}{\|h\|} \xrightarrow[\|h\| \le 2\|k\|/\alpha \to 0]{} 0,$$

which proves $Dg(y) = B^{-1}$.

Finally, suppose $f \in C^r$ and $g \in C^k$ for some $k < r$. Then $Df(g(y)) \in C^k$ and so is its inverse Dg by using the formula for matrix inverses (the entries of A^{-1} are polynomials in those of A divided by $\det A \ne 0$). Thus, $g \in C^{k+1}$. $\square$

9.2.2 The Implicit-Function Theorem

A result closely related to the Inverse-Function Theorem is the Implicit-Function Theorem. It follows easily from the Inverse-Function Theorem and is therefore indirectly an application of the Contraction Principle. Furthermore, as we will see in the next subsection, it immediately tells us more about the Contraction Principle itself regarding the dependence of the fixed point of a contraction on the contraction (see also Figure 2.2.3).

Like the Inverse-Function Theorem, the Implicit-Function Theorem transfers information about the linear part of a map to the map itself. To see how, consider the question answered by the Implicit-Function Theorem in the case of a linear map. Suppose $A \colon \mathbb{R}^n \times \mathbb{R}^m \to \mathbb{R}^n$ is a linear map and write it as $A = (A_1, A_2)$, where $A_1 \colon \mathbb{R}^n \to \mathbb{R}^n$ and $A_2 \colon \mathbb{R}^m \to \mathbb{R}^n$ are linear. Suppose we pick $k \in \mathbb{R}^m$ and want to find $h \in \mathbb{R}^n$ such that $A(h, k) = 0$. To see when this can be done, rewrite this as $A_1 h + A_2 k = 0$ to conclude that if A_1 is invertible, then

$$(9.2.2) \qquad\qquad A(h, k) = 0 \Leftrightarrow h = -(A_1)^{-1} A_2 k.$$

One can interpret this as saying that the equation $A(h, k) = 0$ implicitly defines a map $h = Lk$ such that $A(Lk, k) = 0$. The Implicit-Function Theorem says that if this is true for the linear part of a map, then it is true for the map itself: Under some assumptions corresponding to invertibility of A_1, the equation $f(x, y) = 0$ implicitly defines a map $x = g(y)$ such that $f(g(y), y) = 0$. To state those assumptions for a map $f \colon \mathbb{R}^n \times \mathbb{R}^m \to \mathbb{R}^n$ we write $Df = (D_1 f, D_2 f)$ analogously to the above, with $D_1 f \colon \mathbb{R}^n \to \mathbb{R}^n$ and $D_2 f \colon \mathbb{R}^m \to \mathbb{R}^n$.

The Implicit-Function Theorem then tells us that if we can solve an equation given a particular value of a parameter, then there is a solution for nearby parameter values as well.

Theorem 9.2.3 (Implicit-Function Theorem) *Let $O \subset \mathbb{R}^n \times \mathbb{R}^m$ be open and $f \colon O \to \mathbb{R}^n$ a C^r map. If there is a point $(a, b) \in O$ such that $f(a, b) = 0$ and $D_1 f(a, b)$ is invertible, then there are open neighborhoods $U \subset O$ of (a, b) and $V \subset \mathbb{R}^m$ of b such that for every $y \in V$ there exists a unique $x =: g(y) \in \mathbb{R}^n$ with $(x, y) \in U$ and $f(x, y) = 0$. Furthermore, g is C^r and $Dg(b) = -(D_1 f(a, b))^{-1} D_2 f(a, b)$.*

Proof $F(x, y) := (f(x, y), y) \colon O \to \mathbb{R}^n \times \mathbb{R}^m$ is C^r, and if $A = Df(a, b)$, then $DF(a, b)(h, k) = (A(h, k), k)$ by the chain rule. This gives zero only if $k = 0$ and $A(h, k) = 0$; hence $(h, k) = 0$ by (9.2.2). Therefore DF is invertible and, by the Inverse-Function Theorem 9.2.2, there are open neighborhoods $U \subset O$ of (a, b) and $W \subset \mathbb{R}^n \times \mathbb{R}^m$ of $(0, b)$ such that $F \colon U \to W$ is invertible with C^r inverse $G = F^{-1} \colon W \to U$. Thus, for any $y \in V := \{y \in \mathbb{R}^m \mid (0, y) \in W\}$ there exists an $x =: g(y) \in \mathbb{R}^n$ such that $(x, y) \in U$ and $F(x, y) = (0, y)$, that is, $f(x, y) = 0$.

Now $(g(y), y) = (x, y) = G(0, y)$ and hence g is C^r. To find $Dg(b)$, let $\gamma(y) := (g(y), y)$. Then $f(\gamma(y)) \equiv 0$ and hence $Df(\gamma(y)) D\gamma(y) = 0$ by the chain rule. For $y = b$, this gives $D_1 f(a, b) Dg(b) + D_2 f(a, b) = Df(a, b) D\gamma(b) = 0$, completing the proof. $\square$

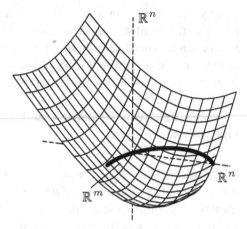

Figure 9.2.3. The Implicit-Function Theorem.

9.2.3 The Smooth Contraction Principle

Returning to dynamics we apply smoothness of the implicit function g to the Contraction Principle to show that the fixed point of a contraction depends smoothly on the contraction itself (see Figure 2.2.3). To express this we write our contractions as maps with a parameter.

Theorem 9.2.4 *Suppose* $f: \mathbb{R}^n \times \mathbb{R}^m \to \mathbb{R}^n$ *is* C^r *and there exists a* $\lambda < 1$ *such that* $d(f(x, y), f(x', y)) \le \lambda d(x, x')$ *for* $x, x' \in X$. *Then for every* $y \in Y$ *there is a unique fixed point* $g(y)$ *of* $x \mapsto f(x, y)$ *and* g *is* C^r.

Proof Existence of the fixed point $g(y)$ follows from the contraction principle. Now write $F(x, y) := f(x, y) - x$ and notice that this is a C^r function that satisfies the hypotheses of the Implicit-Function Theorem 9.2.3: It is zero at $(a, b) = (g(y), y)$ (any choice of y is fine here) and $\|D_1 F v\| = \|D_1 f v - v\| \ge \|v\| - \|D_1 f v\| \ge (1 - \lambda)\|v\| > 0$ for $v \ne 0$, so $D_1 F$ is invertible. Thus $g \in C^r$. $\square$

Remark 9.2.5 Instead of the domain $\mathbb{R}^n \times \mathbb{R}^m$, one can take $A \times O$, where $O \subset \mathbb{R}^m$ is open and A is the closure of an open set, say. (One needs a closed set to apply the contraction principle, but also a good enough one to be able to differentiate r times.)

Remark 9.2.6 Suppose f_λ depends smoothly on λ and $f := f_0$ is as in Proposition 2.2.20. Show that there is a smooth family $\lambda \mapsto x_\lambda$ with x_0 as in Proposition 2.2.20 and $f_\lambda(x_\lambda) = x_\lambda$.

9.3 PERSISTENCE OF TRANSVERSE FIXED POINTS

The fixed point of a contraction simultaneously exhibits two kinds of stability. As an attracting fixed point it is asymptotically stable. Proposition 2.2.20 and Proposition 2.6.14 (as well as Theorem 9.2.4) state that it is also stable under perturbations of the map, that is, perturbations of the map have a unique fixed point nearby. This is an important robustness property of the local dynamics, and we now use the Contraction Principle to describe a general condition under which an analogous conclusion holds. This is a straightforward and simple illustration of the use of the Contraction Principle and the Implicit-Function Theorem in dynamics where the Contraction Principle is applied to a derived system in the same space.

Recall that two C^1-maps f and g are C^1-close if $|f - g| + \|Df - Dg\|$ is uniformly small.

Proposition 9.3.1 *If p is a periodic point of period m for a C^1 map f and the differential Df_p^m does not have one as an eigenvalue (in this case p is said to be a* transverse periodic point*), then for every map g sufficiently C^1-close to f there is a unique periodic point of period m close to p.*

Note that in dimension one the assumption on the derivative simply means that it is not one. Accordingly, in the example of the basic bifurcation of Section 2.3.2 (see Figure 2.3.2) the single fixed point appears or disappears exactly when there is a tangency with the diagonal, that is, the derivative of the map is one. Figure 9.3.1 illustrates this. The axis of the independent variable points right, the vertical axis is for the "output", and the axis toward the rear gives a parameter with which the map changes. The plane shows the diagonal for various parameters, and the graphs of perturbed maps combine to a surface that intersects the diagonals in the family of fixed points.

Proof *The space.* We define a contraction in a neighborhood of p.

The map. Introduce local coordinates near p with p as the origin. In these coordinates, Df_0^m becomes a matrix. Since 1 is not among its eigenvalues, the map $F = f^m - \text{Id}$ defined locally in these coordinates is locally invertible by the Inverse-Function Theorem 9.2.2. Now let g be a map C^1-close to f. Near 0 one can write $g^m = f^m - H$, where H is small together with its first derivatives. A fixed point

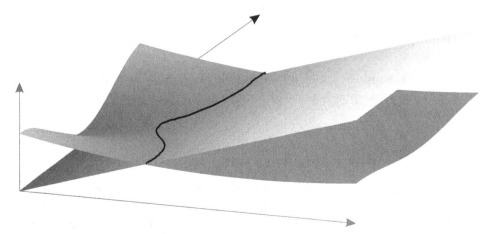

Figure 9.3.1. Persistence of a fixed point.

for g^m can be found from the equation $x = g^m(x) = (f^m - H)(x) = (F + \mathrm{Id} - H)(x)$ or $(F - H)(x) = 0$ or

$$x = F^{-1}H(x).$$

The contraction property. Since F^{-1} has bounded derivatives and H has very small first derivatives, one can show that $F^{-1}H$ is a contracting map. More precisely, let $\|\cdot\|_0$ denote the C^0-norm, $\|dF^{-1}\|_0 = L$, and

$$\max(\|H\|_0, \|dH\|_0) \le \epsilon.$$

Then, since $F(0) = 0$, we get $\|F^{-1}H(x) - F^{-1}H(y)\| \le \epsilon L\|x - y\|$ for every x, y close to 0 and $\|F^{-1}H(0)\| \le L\|H(0)\| \le \epsilon L$, and hence $\|F^{-1}H(x)\| \le \|F^{-1}H(x) - F^{-1}H(0)\| + \|F^{-1}H(0)\| \le \epsilon L\|x\| + \epsilon L$. Thus, if $\epsilon \le (R/L(1 + R))$, the disc $X = \{x \mid \|x\| \le R\}$ is mapped by $F^{-1}H$ into itself and the map $F^{-1}H \colon X \to X$ is contracting. By the contraction principle it has a unique fixed point in X, which is thus a unique fixed point for g^m near 0. $\square$

Remark 9.3.2 It is easy to show that a transverse fixed point is isolated.

Remark 9.3.3 If f is a C^1 map with a hyperbolic fixed point p, that is, $Df|_p$ has no eigenvalues on the unit circle, and g is sufficiently C^1-close to f, then g has a unique fixed point near p and this fixed point is a hyperbolic fixed point of g. Theorem 9.5.4 quantifies "near."

9.4 SOLUTIONS OF DIFFERENTIAL EQUATIONS

Differential equations are a natural setting in which dynamical issues arise, and they appear in several important contexts. At the basis of the use of differential equations in science is that they describe a system in a deterministic way. This means that for any allowed initial condition there is a solution, which then describes the evolution from that initial condition onward. In addition, determinacy requires that the solution be unique – if it were not, then the initial data would not determine the evolution uniquely and the model would have no predictive value.

Right now we examine only the basic fact of existence and uniqueness of solutions by itself. While it can conveniently be taken for granted in the sequel, it is appealing to derive it here as yet another application of the Contraction Principle. Obtaining existence of solutions in this way has the advantage that smooth dependence of the fixed point of a contraction on the contraction has beautiful and useful implications about the behavior of solutions of a differential equation as the initial condition is varied: Small changes in initial condition change the solution only slightly.

9.4.1 The Uniform Case

The present use of the Contraction Principle is called *Picard iteration*. It is the first time we use the Contraction Principle in a function space. The idea here is that we can write a differential equation with initial condition as an integral equation and then apply the integral to continuous functions as candidates for solutions. This operation turns out to be a contraction and hence to improve our guesses at a solution iteratively.

Theorem 9.4.1 *Suppose* $f\colon \mathbb{R} \times \mathbb{R}^n \to \mathbb{R}^n$ *is a continuous function that is Lipschitz-continuous in* $y \in \mathbb{R}^n$ *with Lipschitz constant M. Given any* $(a, b) \in \mathbb{R} \times \mathbb{R}^n$ *and* $\delta < 1/M$, *there is a unique solution* $\varphi_{a,b}\colon (a - \delta, a + \delta) \to \mathbb{R}^n$ *of the differential equation* $\dot{y} = f(t, y)$ *with* $\varphi_{a,b}(a) = b$.

Proof **The space.** We use a contraction defined on the space of continuous functions (candidate solutions). Specifically, the hypothesis on f means that $\| f(t, y') - f(t, y) \| \le M \| y' - y \|$ for $t \in \mathbb{R}$, $y, y' \in \mathbb{R}^n$. Consider the set of continuous functions $\varphi\colon [a - \delta, a + \delta] \to \mathbb{R}^n$ and let $\|\varphi\| := \max_{|t-a| \le \delta} \|\varphi(t)\|$. This is a complete metric space by Theorem A.1.13.

The map. We apply the Contraction Principle to the Picard operator defined by

$$\mathcal{P}_{a,b}(\varphi)(t) := b + \int_a^t f(x, \varphi(x))\, dx.$$

The contraction property. Note that

$$\|\mathcal{P}_{a,b}(\varphi_1) - \mathcal{P}_{a,b}(\varphi_2)\| = \max_{|t-a| \le \delta} \left\| \int_0^t f(x, \varphi_1(x)) - f(x, \varphi_2(x))\, dx \right\| \le M\delta \|\varphi_1 - \varphi_2\|,$$

that is, $\mathcal{P}_{a,b}$ is a contraction and hence has a unique fixed point. It remains to show that fixed points of $\mathcal{P}_{a,b}$ are solutions of $\dot{y} = f(t, y)$ with $\varphi(a) = b$ (and vice versa). To that end, differentiate the fixed-point condition $\varphi_{a,b}(t) = b + \int_a^t f(x, \varphi_{a,b}(x))\, dx$ with respect to t to get $\dot{\varphi}_{a,b}(t) = f(t, \varphi_{a,b}(t))$ by the Fundamental Theorem of Calculus. Evidently fixed points $\varphi_{a,b}$ of $\mathcal{P}_{a,b}$ satisfy $\varphi_{a,b}(a) = b$. To see conversely that solutions are fixed points, insert a solution into the fixed-point condition and observe that the integrand is $\dot{\varphi}$, yielding a fixed point by the Fundamental Theorem. Thus existence and uniqueness of the fixed point gives existence and uniqueness of solutions. □

In fact, the solutions are defined for all time in this case (Proposition 9.4.7) by piecing together the local ones obtained here.

Example 9.4.2 One can explicitly carry out this iteration scheme for the differential equation $\dot{y} = y$, $y(0) = 1$ with $y_0(x) = 1$ as the initial guess. Then $y_1(x) = 1 + \int_0^x y(x)\, dx = 1 + \int_0^x dx = 1 + x$ and $y_2(x) = 1 + \int_0^x 1 + x\, dx = 1 + x + x^2/2$. Inductively, $y_k(x) = \sum_{n=0}^k x^n/n!$, so $y(x) = \sum_{n=0}^\infty x^n/n! = e^x$.

Picard invented this scheme well before the Contraction Principle was available, and this method of successive approximation was carried out by verifying that the errors shrink sufficiently fast.

9.4.2 The Nonuniform Case

It may happen that the Lipschitz constant of the right-hand side of the differential equation depends on t and that the right-hand side is not even defined for all time and not on all of $\mathbb{R}^n$ either. In that case there is still a result like Theorem 9.4.1, but some care must be taken that the solutions do not leave the domain of the right-hand side:

Theorem 9.4.3 *Suppose $I \in \mathbb{R}$ is an open interval, $O \subset \mathbb{R}^n$ open, $f: I \times O \to \mathbb{R}^n$ a continuous function that is an M-Lipschitz-continuous function of $y \in O$ for any fixed $t \in I$. Given any $(a, b) \in I \times O$, there exists a $\delta > 0$ such that there is a unique solution $\varphi_{a,b}: (a - \delta, a + \delta) \to \mathbb{R}^n$ of the differential equation $\dot{y} = f(t, y)$ with $\varphi_{a,b}(a) = b$.*

Proof *The space.* The hypothesis on f means that $\| f(t, y') - f(t, y) \| \leq M \| y' - y \|$ for $t \in I$, y, $y' \in O$. Take a closed bounded subset K of O and a closed interval $I' \subset I$ containing a. Let $B > \sup_{t \in I', x \in K} \| f(t, x) \|$ and take $\delta \in (0, 1/M)$ such that $[a - \delta, a + \delta] \subset I'$ and the ball $B(b, B\delta)$ is contained in K. Now consider the set $\mathcal{C}$ of continuous functions $\varphi: [a - \delta, a + \delta] \to O$ such that $\| \varphi - b \| < B\delta$, where again $\| \varphi \| := \max_{|t-a| \leq \delta} \| \varphi(t) \|$. $\mathcal{C}$ is a closed subset of the complete metric space of all continuous functions on $[a - \delta, a + \delta]$ (with this norm) and hence is itself complete.

The map. The Picard operator is again defined by

$$\mathcal{P}_{a,b}(\varphi)\,(t) := b + \int_a^t f(x, \varphi(x))\, dx.$$

Then $\| \mathcal{P}_{a,b}(\varphi) - b \| \leq \max_{|t-a| \leq \delta} \| \int_a^t f(x, \varphi(x))\, dx \| < B\delta$, so $\mathcal{P}_{a,b}(\varphi) \in \mathcal{C}$ for $\varphi \in \mathcal{C}$, that is, $\mathcal{P}_{a,b}$ is well defined.

The contraction property. Since

$$\| \mathcal{P}_{a,b}(\varphi) - \mathcal{P}_{a,b}(\varphi') \| = \max_{|t-a| \leq \delta} \left\| \int_0^t f(x, \varphi(x)) - f(x, \varphi'(x))\, dx \right\| \leq M\delta \| \varphi - \varphi' \|,$$

$\mathcal{P}_{a,b}$ is a contraction of $\mathcal{C}$ and hence has a unique fixed point. As before, fixed points correspond to solutions. $\square$

Remark 9.4.4 Note that we only obtain local solutions here. Global ones can be obtained by piecing together local ones; by uniqueness, any two local solutions must agree on the intersection of their domains. In fact, the only obstacle to extending solutions is that they may run into the boundary of O, beyond which the ordinary differential equation makes no sense. We carry this out explicitly in Section 9.4.7.

9.4.3 Continuous Dependence

Since $\mathcal{P}_{a,b}$ depends continuously on a and b and $\mathcal{P}_{a,b'}(\mathcal{C}) \subset \mathcal{C}$ for b' sufficiently close to b, the solutions depend continuously on the initial value b by Proposition 2.6.14.

Proposition 9.4.5 *Under the hypotheses of Theorem 9.4.3, solutions depend continuously on the initial value; that is, given $\epsilon > 0$ there exists an $\eta > 0$ such that, if $\| b' - b \| < \eta$, then $\max_{|t-a| \leq \delta} \| \varphi_{a,b'}(t) - \varphi_{a,b}(t) \| < \epsilon$.*

Proof We clearly need to pick η such that $B(b', B\delta) \subset K$ (see the beginning of the previous proof) whenever $\| b' - b \| < \eta$, to make sure that $\varphi_{a,b'}$ is defined for $|t - a| < \delta$. Once this is the case, however, the conclusion (for possibly smaller η) is simply a restatement of the continuous dependence of the fixed point of a contraction on a parameter, in this case with respect to the norm $\| \varphi \| := \max_{|t-a| \leq \delta} \| \varphi(t) \|$. $\square$

9.4.4 Smooth Dependence

The map $\mathcal{P}\colon \mathcal{C} \times \mathbb{R} \times O \to \mathcal{C}$ goes into a linear space, where differentiation makes sense (Definition A.2.1). It depends linearly (hence smoothly) on $b \in O$, and the dependence on $\varphi \in \mathcal{C}$ is through f and hence is as smooth as f. To indicate how one sees this consider the first derivative. The Mean-Value Theorem gives

$$\mathcal{P}_{a,b}(\varphi)(t) - \mathcal{P}_{a,b}(\psi)(t) = \int_a^t f(x, \varphi(x))\, dx - \int_a^t f(x, \psi(x))\, dx$$

$$= \int_a^t f(x, \varphi(x)) - f(x, \psi(x))\, dx$$

$$= \int_a^t (\partial f/\partial y)(x, c_x)(\varphi(x) - \psi(x))\, dx$$

$$\approx \int_a^t (\partial f/\partial y)(x, \varphi(x))(\varphi(x) - \psi(x))\, dx.$$

The first derivative is thus given by $D\mathcal{P}_{a,b}(\varphi)(\eta)(t) = \int_a^t (\partial f/\partial y)(x, \varphi(x))\eta(x)\, dx$.
    Corollary 2.2.15 implies

Proposition 9.4.6 *If in Proposition 9.4.5 the function f is C^r, then the solutions are C^{r+1} and depend C^r on the initial value b, that is, $b \mapsto \varphi_{a,b}(a + t)$ is a C^r map for all $t \in (-\delta, \delta)$.*

    The fact that the solutions themselves are C^{r+1} follows inductively from the differential equation, which shows that $\dot{y}$ is C^k whenever y and f are C^k.

9.4.5 Nonexistence and Nonuniqueness

To see that the hypotheses are really needed, consider Figure 9.4.1. It shows the solutions $x = ct^2$ for $t\dot{x} = 2x$, where uniqueness fails for the initial condition $a = b = 0$ and existence fails for any initial condition $a = 0$, $b \neq 0$. The right portion of the picture shows that solutions do not always extend to all t where the right-hand side $f(t, x)$ is defined for all t: The solutions $x = -1/(t + c)$ of $\dot{x} = x^2$ have singularities for finite t. Existence of solutions can be proved using only continuity of the right-hand side of the differential equation. However, the possible failure of uniqueness shows that continuous dependence on the initial value cannot be expected without a Lipschitz condition.

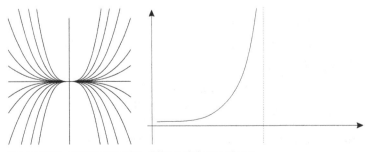

Figure 9.4.1. Problems with differential equations.

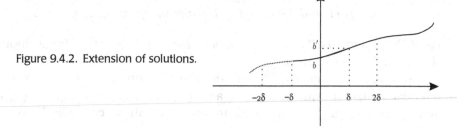

Figure 9.4.2. Extension of solutions.

9.4.6 Extension of Solutions

For reasons that were previewed in Section 3.2.6 and are fully justified by Proposition 9.4.11, we restrict attention to differential equations of the form $\dot{x} = f(x)$, that is, differential equations whose right-hand side does not depend on time. (These are said to be *autonomous* differential equations, and the right-hand side is then called the *vector-field* generating the flow.) Physically this reflects the fixed laws of nature that we assume. We would prefer not to have to worry about the possibility of solutions being defined only up to some time, and we usually don't:

Proposition 9.4.7 *If f is defined on all of $\mathbb{R}^n$ and is Lipschitz-continuous, then the solutions of $\dot{x} = f(x)$ are defined for all t.*

Proof For any initial condition $y(0) = b$ there is a solution $\varphi_{0,b} \colon [-\delta, \delta] \to \mathbb{R}^n$ with $\varphi_{0,b}(0) = b$ by Theorem 9.4.1. For the initial condition $\varphi_{0,b}(\delta) =: b'$ there is a solution (see Figure 9.4.2) $\varphi_{\delta, \varphi_b(\delta)}$ on $[0, 2\delta]$, that is,

$$\dot{\varphi}_{\delta, \varphi_b(\delta)}(t) = f\big(\varphi_{\delta, \varphi_b(\delta)}\big) \text{ and } \varphi_{\delta, \varphi_b(\delta)}(0) = b.$$

At the same time, $\dot{\varphi}_{0,b}(t) = f(\varphi_{0,b})$ and $\varphi_{0,b}(0) = b$, so $\varphi_{\delta, \varphi_b(\delta)}(t) = \varphi_{0,b}(t)$ for $t \in [0, \delta]$ by uniqueness. Therefore there is a unique solution on $[-\delta, 2\delta]$. Extending similarly from $-\delta$ gives a solution on $[-2\delta, 2\delta]$, which can in turn be continued to $[-3\delta, 3\delta]$, and so on. Thus solutions are defined for all time, independently of the initial condition. $\square$

Applying Proposition 9.4.5 about T/δ times gives:

Proposition 9.4.8 *Solutions depend continuously on the initial value for any finite amount of time; that is, given $T, \epsilon > 0$ there exists a $\delta > 0$ such that, if $\|b' - b\| < \delta$, then $\max_{|t-a|\leq T} \|\varphi_{a,b'}(t) - \varphi_{a,b}(t)\| < \epsilon$.*

9.4.7 Flows

We now study the maps arising from solutions of differential equations. The first lemma establishes that the evolution over a given length of time is independent of the initial time. Then we conclude that these evolutions determine a *one-parameter group* of invertible differentiable maps.

Lemma 9.4.9 *The map $\phi_a^t: b \mapsto \varphi_{a,b}(a+t)$ of Proposition 9.4.6 is defined on all of $\mathbb{R}^n$ for any value of t, and it is C^r if f is. It is also independent of a.*

Proof Proposition 9.4.7 shows that ϕ_a^t is defined on $\mathbb{R}^n$ for any t. Proposition 9.4.6 shows that it is as smooth as f.

Given $a, a' \in \mathbb{R}$ and $b \in \mathbb{R}^n$, consider the solutions $\varphi_{a,b}$ and $\varphi_{a',b}$ of the differential equation with $\varphi_{a,b}(a) = b$ and $\varphi_{a',b}(a') = b$. Then $\phi_a^t(b) = \varphi_{a,b}(a+t)$ and $\phi_{a'}^t(b) = \varphi_{a',b}(a'+t)$. We need to show that these coincide. If we define $\psi_1(t) := \varphi_{a,b}(t+a)$ and $\psi_2(t) := \varphi_{a',b}(t+a')$, then

$$\dot{\psi}_1(t) = f(\psi_1(t)), \qquad \psi_1(0) = b \qquad \text{and} \qquad \dot{\psi}_2(t) = f(\psi_2(t)), \qquad \psi_2(0) = b.$$

By uniqueness, $\phi_{a'}^t(b) = \varphi_{a',b}(t+a') = \psi_2(t) = \psi_1(t) = \varphi_{a,b}(a+t) = \phi_a^t(b)$. $\square$

We drop the subscript a henceforth and write $\phi^t(b) = \varphi_{a,b}(a+t)$ from now on (and make $a = 0$ our default choice).

Definition 9.4.10 A family $(\phi^t)_{t \in \mathbb{R}}$ of maps for which $(t, x) \mapsto \phi^t(x)$ is C^r is said to be a C^r *flow* if $\phi^{s+t} = \phi^s \circ \phi^t$ for all $s, t \in \mathbb{R}$.

This "group property" holds in our situation:

Proposition 9.4.11 *A differential equation $\dot{x} = f(x)$ with $f: \mathbb{R}^n \to \mathbb{R}^n$ a C^r function and $\|Df\|$ bounded defines a C^r flow on $\mathbb{R}^n$.*

Proof Given $t \in \mathbb{R}$, the functions $\psi_1(s) := \varphi_{0,b}(s+t)$ and $\psi_2(s) := \varphi_{0,\varphi_{0,b}(t)}(s)$ are solutions of the differential equation and $\psi_2(0) = \varphi_{0,\varphi_{0,b}(t)}(0) = \varphi_{0,b}(t) = \psi_1(0)$, so $\psi_1 = \psi_2$ by uniqueness. Consequently,

$$\phi^s \circ \phi^t(b) = \phi^s(\varphi_{0,b}(t)) = \varphi_{0,\varphi_{0,b}(t)}(s) = \varphi_{0,b}(s+t) = \phi^{s+t}(b).$$

Taking $s = -t$ shows in particular that ϕ^t is invertible with inverse ϕ^{-t}. Thus these maps ϕ^t are C^r diffeomorphisms. $\square$

The concept of a smooth flow is central to the theory of dynamical systems with continuous time. It is the bridge between dynamics and differential equations. With this concept one can describe dynamics as the study of the asymptotic behavior of one-parameter groups (discrete or continuous) of transformations.

Section 2.4.1 gave a complete description of the dynamics of the flow generated by the differential equation $\dot{x} = f(x)$ on the line, where f is a Lipschitz-continuous function: There is a closed set of fixed points and the flow is monotone on every complementary interval with all orbits asymptotic to one endpoint and asymptotic to the other endpoint in negative time.

Changing the size of the right-hand side without changing the direction does not change orbits, only the speed along them.

Definition 9.4.12 The flows generated by $\dot{x} = f(x)$ and $\dot{x} = a(x) f(x)$ for some continuous nowhere zero scalar function a are said to be related by a *time change*.

9.5 HYPERBOLICITY

We start this section with a nice instance of linearization relating directly to the description of qualitative features of a nonlinear map in terms closely related to those of its linear part.

9.5.1 Hyperbolic Fixed Points

Recall our description of the dynamics of a hyperbolic linear map of $\mathbb{R}^2$ in Section 3.1. A hyperbolic linear map of $\mathbb{R}^2$ is a linear map with an eigenvalue $\lambda \in (-1, 1)$ and an eigenvalue μ outside $[-1, 1]$. A nonzero eigenvector v for λ then spans a line E^s through the origin in $\mathbb{R}^2$, and likewise we obtain a line E^u from an eigenvector w for μ.

These two lines are the main building blocks for reconstructing the dynamics of the linear map. On E^s the map is a linear contraction; on E^u its inverse is a contraction. The orbits of points not on either line lie on curves asymptotic to the stable and unstable lines obtained.

Consider now a differentiable map $f \colon \mathbb{R}^2 \to \mathbb{R}^2$ with a fixed point x_0 such that $Df(x_0)$ is a hyperbolic linear map. Remarkably, the presence of a contracting and expanding line for the linear map $Df(x_0)$ has an exact counterpart for the map f itself, except that instead of lines we get curves W^s and W^u. Each of these curves is invariant, that is, $f(W^s) = W^s$ and $f(W^u) = W^u$, and f is a contraction on W^s and f^{-1} a contraction on W^u. (See Figure 9.5.1) Strictly speaking, this statement is not perfectly accurate, but it is true in a neighborhood of the fixed point. In order not to have to worry about how large this neighborhood might be, we prove a statement that is formulated a bit differently (Theorem 9.5.2). That statement also contains only half of what we just promised, namely, only the contracting curve. But by applying it to the inverse map as well we also get the expanding curve for f.

Note also, by the way, that the contracting line E^s for a hyperbolic linear map is exactly the set of points whose successive images form a bounded sequence, that is, whose positive semiorbit is bounded. Analogously, a map with a hyperbolic fixed point (that is, one where the differential of the map is a hyperbolic linear map) has a smooth curve through the fixed point that consists of exactly those points

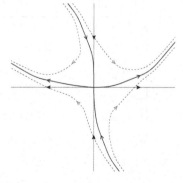

Figure 9.5.1. Stable and unstable manifolds.

whose iterates stay sufficiently near the fixed point. We prove this result for maps of $\mathbb{R}^2$, but it is not hard to see that the same argument with reinterpreted notations works in higher dimension. As in Section 9.4, the proof is a nice application of the Contraction Principle in a properly constructed "infinite-dimensional" metric space.

9.5.2 The Stable Manifold Theorem

Let us call the graph $c \subset \mathbb{R}^2$ of a C^1 function $\varphi: \{0\} \times \mathbb{R} \to \mathbb{R} \times \{0\}$ (that is, x as a function of y) with $|D\varphi| \leq \gamma$ a *vertical γ-curve*. When φ gives y as a function of x we speak of a *horizontal γ-curve*. For our considerations it is useful to have the following

Lemma 9.5.1 *Suppose $A: \mathbb{R}^2 \to \mathbb{R}^2$ is the linear map given by the matrix $\begin{pmatrix} \kappa_u & 0 \\ 0 & \kappa_s \end{pmatrix}$ with $0 < \kappa_s < \kappa_u$. For $\gamma > 0$ and $\epsilon < (\kappa_u - \kappa_s/\gamma + 2 + (1/\gamma))$ and any C^1 map $f: \mathbb{R}^2 \to \mathbb{R}^2$ for which $\|Df - A\| \leq \epsilon$, the inverse of f preserves vertical γ-curves, that is, the preimage of a γ-curve under f is again a γ-curve.*

Proof It is convenient in this proof to use the norm $\|(x, y)\| := |x| + |y|$ on $\mathbb{R}^2$ and to write $f(x, y) = (f_1(x, y), f_2(x, y))$ and $D_1 := \partial/\partial x$, $D_2 := \partial/\partial y$.

We want to show that for any γ-curve given by $x = c(y)$ we can solve the equation

$$f_1(x, y) = c(f_2(x, y)) \qquad \text{or} \qquad 0 = F(x, y) := f_1(x, y) - c(f_2(x, y))$$

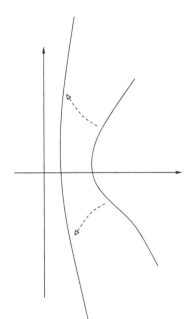

Figure 9.5.2. Preservation of vertical γ-curves.

for an implicitly defined function $x = g(y)$ with $|Dg| \leq \gamma$. Since the preimage of our γ-curve is nonempty, there exists a pair (a, b) for which $F(a, b) = 0$. Let us check that $D_1 F \neq 0$. Note that $|D_1 f_2| < \epsilon$, $|D_2 f_1| < \epsilon$ and

(9.5.1)
$$|D_1 f_1| \geq \kappa_u - |D_1 f_1 - \kappa_u| \geq \kappa_u - \epsilon.$$

Therefore

$$|D_1 F| = |D_1 f_1 - Dc \circ f_2 \, D_1 f_2| \geq \kappa_u - (1 + \gamma)\epsilon > 0.$$

Thus we can, at least locally, solve for a function giving $x = g(y)$. To estimate its derivative note that

$$|D_2 F| = |D_2 f_1 - Dc \circ f_2 \, D_2 f_2| \leq \gamma(\kappa_s + \epsilon) - \epsilon = \gamma\kappa_s - (1 + \gamma)\epsilon.$$

Thus $|Dg| < \gamma$, as required:

$$|Dg| = \left| -\frac{D_2 F}{D_1 F} \right| \leq \frac{\gamma\kappa_s - (1 + \gamma)\epsilon}{\kappa_u - (1 + \gamma)\epsilon} < \gamma.$$

[To check the last inequality clear fractions, divide by γ and use $(1 + \gamma)(1 + 1/\gamma)\epsilon < \kappa_u - \kappa_s$.]

It remains to show that g is defined on $\mathbb{R}$ and not only locally. To that end note from above that $|D_2 f_2| \leq \kappa_s + \epsilon$, that is, f^{-1} stretches curves in the vertical direction. To be specific, take any $y \in \mathbb{R}$. We need to show that the preimage of our curve contains a point (x, y). Consider the graph of c over the interval from a to $a + (y - b)$. The y-coordinates of the preimage extend from b to $b + (y - b)/(\kappa_s + \epsilon)$ and in particular include y. Thus, in this setting we have a globally defined implicit function with the right estimate on the derivative. $\square$

Note that $\gamma + 2 + (1/\gamma) \geq 4$, so we certainly need $\epsilon < (\kappa_u - \kappa_s)/4$ above. Now we can prove the result about stable curves.

Theorem 9.5.2 (Stable Manifold Theorem) *Suppose* $A: \mathbb{R}^2 \to \mathbb{R}^2$ *is the linear map given by the matrix* $\begin{pmatrix} \kappa_u & 0 \\ 0 & \kappa_s \end{pmatrix}$ *with* $0 < \kappa_s < 1 < \kappa_u$. *There exists* $\epsilon > 0$ *such that, for any* C^r *map* $f: \mathbb{R}^2 \to \mathbb{R}^2$ *with* $\|Df - A\| \leq \epsilon$, *the set of points with bounded positive semiorbits is the graph of a* C^r *function of* y.

Proof *The space.* We define a contraction on the space l^∞ of bounded sequences $(x_n)_{n \in \mathbb{N}_0}$ with the sup-norm $\|(x_n)_{n \in \mathbb{N}_0}\|_\infty := \sup_{n \in \mathbb{N}_0} |x_n|$. (This is a complete metric space by Theorem A.1.14.)

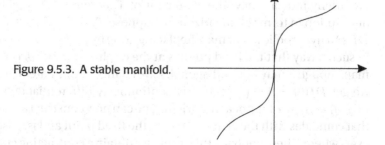

Figure 9.5.3. A stable manifold.

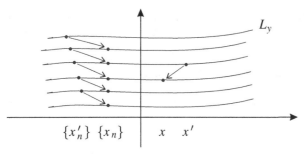

Figure 9.5.4. Proof of Theorem 9.5.2.

The map. Take $\epsilon > 0$ such that we can apply Lemma 9.5.1 to f^{-1} and A^{-1} with the roles of x and y reversed; that is, ϵ is also small enough for f to preserve *horizontal* γ-curves for some γ. For a given $y \in \mathbb{R}$ consider the successive images $f^n(L_y)$ of the line $L_y := \{(x, y) \mid x \in \mathbb{R}\}$ (these are horizontal C^r γ-curves). Given any $x \in \mathbb{R}$ and $n \in \mathbb{N}$ there is a unique $z \in \mathbb{R}$ such that $(x, z) \in f^n(L_y)$, and so we can associate with x an $x' \in \mathbb{R}$ by taking $y' \in \mathbb{R}$ such that $(x', y') \in f^{n-1}(L_y)$ and $f(x', y') = (x, y)$. Thus we define a map $\mathcal{F}_y \colon l^\infty \to l^\infty$ by $\mathcal{F}_y((x_n)_{n \in \mathbb{N}_0}) = (x'_n)_{n \in \mathbb{N}_0}$. (See Figure 9.5.4.)

The contraction property. Each of these sequences corresponds to a bounded sequence of points in the plane by adding y-coordinates y_n in such a way as to have $(x_n, y_n) \in f^n(L_y)$. The map $\mathcal{F}_y$ reflects the action of f on such sequences of points (dropping the first one and reindexing the others). This shows that a fixed point of $\mathcal{F}_y$ is the sequence of x-coordinates of an orbit of f with initial y-coordinate y, and vice versa, every bounded semiorbit gives rise to a fixed point of $\mathcal{F}_y$. $\mathcal{F}_y$ is a contraction because f expands x-coordinates by (9.5.1); that is, entrywise differences between l^∞-sequences are divided by a factor $\kappa_u - \epsilon > 1$, and the same therefore applies to differences in the sup-norm. Therefore there is a unique fixed point $(g(y), y)$ depending continuously on $y \in \mathbb{R}$.

Thus, the graph of the continuous function $g(y)$ gives the set of points with bounded positive semiorbits.

The fact that l^∞ is a normed linear space allows us to talk about differentiability (Definition A.2.1), and the fact that f and the curves $f^n(L_y)$ are C^r implies (with some work that we prefer not to do here in detail) that $\mathcal{F} \colon ((x_n)_{n \in \mathbb{N}_0}, y) \mapsto \mathcal{F}_y((x_n)_{n \in \mathbb{N}_0})$ is C^r; hence, by Theorem 9.2.4, g is also C^r. $\square$

9.5.3 Localization

We briefly indicate how the statement of Theorem 9.5.2 is related to our earlier description in terms of linearization. Suppose $f \colon \mathbb{R}^2 \to \mathbb{R}^2$ has a fixed point p with $Df(p)$ a hyperbolic linear map. Replacing (x, y) by $(x, y) - p$, we change coordinates in such a way that the fixed point is at the origin. Now $Df(p)$ is a hyperbolic linear map, though it may not be diagonal. If we change coordinates to diagonalize $Df(p)$, we get $Df(0) = A = \left(\begin{smallmatrix} \lambda & 0 \\ 0 & \mu \end{smallmatrix} \right)$. If f is continuously differentiable, then by definition $\|Df(q) - A\| < \epsilon$ for q near the origin. One can now construct a globally defined map that coincides with the given map near the fixed point and is close to the derivative everywhere. Then, applying this result, we obtain a contracting curve C, at least near the origin, and likewise an expanding curve by considering the inverse of f. One can easily show that $\bigcup_{n\in\mathbb{N}} f^{-n}(C)$ is a simple curve, that is, it has no self-intersections.

9.5.4 The Hyperbolic Fixed-Point Theorem

Note that in Theorem 9.5.2 the origin is a transverse fixed point of A and hence persists by Proposition 9.3.1. However, this persistence also follows from our arguments here. Since f is C^1-close to A, hyperbolicity also persists. This gives us the Hyperbolic Fixed-Point Theorem:

Theorem 9.5.3 *If p is a hyperbolic m-periodic point of a C^1 map f, then for every map g sufficiently C^1-close to f there is a unique m-periodic point close to p, and it is hyperbolic.*

The following "globalized" version is also useful:

Theorem 9.5.4 *Suppose $0 < \kappa_s < 1 < \kappa_u$ and $A\colon \mathbb{R}^n \to \mathbb{R}^n$ is a linear map such that $A\!\restriction_{\mathbb{R}^m \times \{0\}}$ is a κ_s-contraction and $A^{-1}\!\restriction_{\{0\} \times \mathbb{R}^{n-m}}$ is a $1/\kappa_u$-contraction. If $F\colon \mathbb{R}^n \to \mathbb{R}^n$ is a map with $f := F - A$ bounded and $\mathrm{Lip}(f) \le \epsilon := \min\{1 - \kappa_s, 1 - \kappa_u^{-1}\}$ (see Definition 2.2.1), then F has a unique fixed point p with $\|p\| < \|F(0)\|/(\epsilon - \mathrm{Lip}(f))$.*

Remark 9.5.5 Clearly p is a hyperbolic fixed point of F.

Using only methods available at this point (stable leaves or the Contraction Principle), one can strengthen Theorem 9.5.3 to give a local conjugacy. This is a simple instance of *structural stability* (see also Section 10.2.6), of which a compelling example is given by Theorem 7.4.3.

Theorem 9.5.6 (**Hartman–Grobman Theorem**)[1] *Let $U \subset \mathbb{R}^n$ be open, $f\colon U \to \mathbb{R}^n$ continuously differentiable, and $O \in U$ a hyperbolic fixed point of f. Then there exist neighborhoods U_1, U_2, V_1, V_2 of O and a homeomorphism $h\colon U_1 \cup U_2 \to V_1 \cup V_2$ such that $f = h^{-1} \circ Df_0 \circ h$ on U_1.*

In fact, the topological character of f near O is determined already by the orientation of f on stable and unstable manifolds and by their dimensions, analogous to Theorem 7.4.3, which has no closeness assumption:

Theorem 9.5.7 *Suppose $f\colon U \to \mathbb{R}^n$, $g\colon V \to \mathbb{R}^n$ have hyperbolic fixed points $p \in U$ and $q \in V$, respectively, and $\dim E^+(Df_p) = \dim E^+(Dg_q)$, $\dim E^-(Df_p) = \dim E^-(Dg_q)$, $\mathrm{sign}\det Df_p\!\restriction_{E^+(Df_p)} = \mathrm{sign}\det Dg_q\!\restriction_{E^+(Dg_q)}$, and $\mathrm{sign}\det Df_p\!\restriction_{E^-(Df_p)} = \mathrm{sign}\det Dg_q\!\restriction_{E^-(Dg_q)}$. Then there exist neighborhoods $U_1 \subset U$ and $V_1 \subset V$ and a homeomorphism $h\colon U_1 \to V_1$ such that $h \circ f = g \circ h$.*

---

[1] A direct proof of this theorem can be found in A. Katok and B. Hasselblatt, *Introduction to the Modern Theory of Dynamical Systems*, Cambridge University Press, Cambridge, 1995; here we obtain it from Theorem 9.5.8.

The following result is equivalent to Theorem 9.5.6 via a localization procedure. We obtain it as an application of the Hyperbolic Fixed-Point Theorem 9.5.4.

Theorem 9.5.8 *Suppose* $0 < \kappa_s < 1 < \kappa_u$ *and* $A \colon \mathbb{R}^n \to \mathbb{R}^n$ *is a linear map such that* $A_{\lceil \mathbb{R}^m \times \{0\}}$ *is a* κ_s-*contraction and* $A^{-1}_{\lceil \{0\} \times \mathbb{R}^{n-m}}$ *is a* $1/\kappa_u$-*contraction. Suppose* F *is a map such that* $f := F - A$ *is bounded and* $\mathrm{Lip}(f) \le \epsilon := \min\{\|A^{-1}\}^{-1}, 1 - \kappa_s, 1 - \kappa_u^{-1}\}$. *Then there is a unique homeomorphism* H *of* $\mathbb{R}^n$ *such that* $h := H - \mathrm{Id}$ *is bounded and* $H \circ A \circ H^{-1} = F$.

Proof The assumption $\mathrm{Lip}(f) \le \|A^{-1}\|^{-1}$ ensures that F is a homeomorphism because $y = F(x) \Leftrightarrow x = A^{-1}(y - f(x))$ and the right-hand side is continuous and a contraction in x; hence it determines x uniquely.

It is useful to make the proof more symmetric by introducing $G = A + g$ with g bounded and $\mathrm{Lip}(g) \le \epsilon$. The first step is to show that there is a unique continuous bounded map h of $\mathbb{R}^n$ such that

(9.5.2) $$F \circ (\mathrm{Id} + h) = (\mathrm{Id} + h) \circ G,$$

or, equivalently,

$$A \circ h \circ G^{-1} + f \circ (\mathrm{Id} + h) \circ G^{-1} + A \circ G^{-1} - \mathrm{Id} = h.$$

To this end consider the space E of bounded continuous maps from $\mathbb{R}^n$ to $\mathbb{R}^n$, which splits into the space E_s of bounded continuous maps to $\mathbb{R}^m \times \{0\}$ and the space E_u of bounded continuous maps to $\{0\} \times \mathbb{R}^{n-m}$. Define maps $\mathcal{A} \colon E \to E$ and $\mathcal{F} \colon E \to E$ by

$$\mathcal{A}(h) := A \circ h \circ G^{-1} \qquad \text{and} \qquad \mathcal{F}(h) := f \circ (\mathrm{Id} + h) \circ G^{-1} + A \circ G^{-1} - \mathrm{Id}.$$

Then $\mathcal{A}$ preserves E_s and E_u, $\mathcal{A}_{\lceil E_s}$ is a κ_s-contraction, and $\mathcal{A}^{-1}_{\lceil E_u}$ is a κ_u^{-1}-contraction. Since $\mathrm{Lip}(\mathcal{F}) \le \mathrm{Lip}(f)$, the Hyperbolic Fixed-Point Theorem 9.5.4 shows that $\mathcal{A} + \mathcal{F}$ has a unique fixed point in E. This gives the desired h.

It remains to show that H is a homeomorphism. This is where the symmetry of the preceding argument helps. It yields an $\bar{h}$ such that

$$G \circ (\mathrm{Id} + \bar{h}) = (\mathrm{Id} + \bar{h}) \circ F.$$

Combining this with (9.5.2) two ways gives the two relations

$$G \circ (\mathrm{Id} + \bar{h}) \circ (\mathrm{Id} + h) = (\mathrm{Id} + \bar{h}) \circ (\mathrm{Id} + h) \circ G$$

$$F \circ (\mathrm{Id} + h) \circ (\mathrm{Id} + \bar{h}) = (\mathrm{Id} + h) \circ (\mathrm{Id} + \bar{h}) \circ F.$$

Each of these is of the same type as (9.5.2), as it would be obtained by running through the first part of the proof with $F = G$. Since we already established uniqueness of the conjugating map, and Id conjugates F with F and G with G, we conclude

$$(\mathrm{Id} + \bar{h}) \circ (\mathrm{Id} + h) = (\mathrm{Id} + h) \circ (\mathrm{Id} + \bar{h}) = \mathrm{Id},$$

which in particular proves that $H = \mathrm{Id} + h$ is a homeomorphism. $\square$

CHAPTER 10

Hyperbolic Dynamics

This chapter synthesizes the fundamental common structure that produces the behaviors we observed earlier in the various examples of complicated dynamics. The essence of chaotic behavior in dynamics lies in a combination of stretching (to separate orbits) and folding (to create recurrence and maintain compatibility with compactness). We make this notion precise and develop some important general consequences. The facts and ideas developed here play a role in many areas in dynamics and appear throughout the remainder of this book. Proofs omitted in this chapter can be found in our book, *Introduction to the Modern Theory of Dynamical Systems* (with the same notations).

10.1 HYPERBOLIC SETS

To extract from our collection of examples the features of interest we need to deal with the invertible and noninvertible cases separately.

10.1.1 Definition
A hyperbolic set is an invariant set such that at every point x the derivative looks like it does at a hyperbolic fixed point: There are complementary subspaces E_x^u and E_x^s (expanding or "unstable" and contracting or "stable") such that $Df^{-1}(x)$ is a κ-contraction (Definition 2.2.1) on E_x^u (with image $E_{f^{-1}(x)}^u$) and $Df|_x$ is a κ-contraction on E_x^s for every x, with $\kappa < 1$ independent of x. In fact, instead of contraction one needs only eventual contraction roughly as in Definition 2.6.11, and this makes the conditions in the definition easier to verify.

Definition 10.1.1 Suppose U is an open set and f is a map defined on U. If Λ is a compact invariant set, that is, $f(\Lambda) = \Lambda$, on which f is invertible, then Λ is said to be a *hyperbolic set* if at every point $x \in \Lambda$ we have subspaces E_x^u and E_x^s such that every vector v can be written uniquely as $v = v^u + v^s$ with $v^u \in E_x^u$ and $v^s \in E_x^s$, and there are $C > 0$ and $\kappa < 1$ such that $\|Df^{-n}(x)|_{E_x^u}\| \leq C\kappa^n$ and $\|Df^n(x)|_{E_x^s}\| \leq C\kappa^n$ for every $x \in \Lambda$ and $n \in \mathbb{N}$.

The hyperbolic linear map of the torus in Section 7.1.4 is an example of this situation. The set Λ is the entire torus, and the expanding and contracting subspaces are parallel to the expanding and contracting eigenlines. The constant κ is the smaller of the two eigenvalues.

The other kind of invertible example is represented by the ternary horseshoe in Section 7.3.3 and the general linear horseshoe in Section 7.4.4. The invariant set Λ constructed there is a hyperbolic set. The expanding subspaces are horizontal lines and the contracting ones are vertical lines. The constant κ is determined by the expansion and contraction rates that one chooses for the horseshoe map. It is $1/3$ for the ternary horseshoe.

We used invertibility to define the expanding direction. For noninvertible maps this is a problem, except for those cases where one has expansion in all directions. Since our noninvertible examples are one-dimensional, this is automatic.

Definition 10.1.2 Suppose U is an open set and f is a map defined on U. A compact invariant set Λ is said to be a *hyperbolic repeller* if there is a $\kappa > 1$ such that, if $x \in \Lambda$ and v is a vector, then $\|Df^n(x)v\| \geq C\kappa^n\|v\|$.

The linear expanding maps E_m of the circle in Section 7.1.1 have this property for $\Lambda = S^1$ (and $\kappa = |m|$); so do nonlinear expanding maps (Definition 7.1.7). By compactness, the assumption $|f'(x)| > 1$ for all $x \in S^1$ implies that there is a $\kappa > 1$ such that $|f'(x)| \geq \kappa$ for all $x \in S^1$. In all of these cases one can take $C = 1$.

For $\lambda > 4$, the invariant set $\bigcap_{n \in \mathbb{N}_0} f_\lambda^{-n}([0, 1])$ in Proposition 7.4.4 of the quadratic maps in Section 2.5 and Section 7.1.2 is a hyperbolic repeller (Proposition 11.4.1). This is not as easy to verify for $4 < \lambda \leq 2 + \sqrt{5}$ as it was in the proof of Proposition 7.4.4. Also, here one cannot take $C = 1$ in Definition 10.1.2 unless one changes the metric.

Definition 10.1.3 Let Λ be a hyperbolic set for f on U. If there is an open neighborhood V of Λ such that $\Lambda = \Lambda_V^f := \bigcap_{n \in \mathbb{Z}} f^n(\overline{V})$, then Λ is said to be *locally maximal*, *isolated*, or *basic*.

This assumption is natural and pervasive, and will therefore be made most of the time henceforth, sometimes implicitly.

10.1.2 Cone Criterion

We now present an alternate characterization of hyperbolic sets whose requirements are more apparently robust and are often easier to check. A typical example is the proof of existence of the Lorenz attractor (Theorem 13.3.3).

Definition 10.1.4 The *standard horizontal γ-cone* at $p \in \mathbb{R}^n$ is defined by

$$H_p^\gamma = \{(u, v) \in T_p\mathbb{R}^n \mid \|v\| \leq \gamma\|u\|\}.$$

The *standard vertical γ-cone* at p is

$$V_p^\gamma = \{(u, v) \in T_p\mathbb{R}^n \mid \|u\| \leq \gamma\|v\|\}.$$

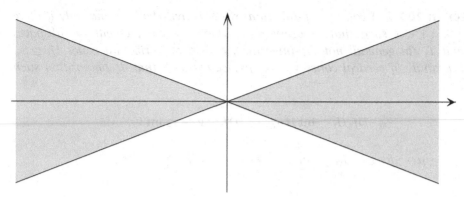

Figure 10.1.1. A horizontal cone.

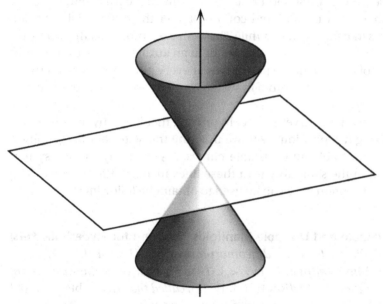

Figure 10.1.2. A vertical cone.

More generally, a *cone K* in $\mathbb{R}^n$ is defined as the image of a standard cone under an invertible linear map.

Some examples will clarify the picture. In dimension $n = 2$, all cones look alike. A horizontal cone $|x_2| \le \gamma |x_1|$ is shaded in Figure 10.1.1. The closure of its complement $|x_1| \le |x_2|/\gamma$ is a vertical cone, of which a 3-dimensional counterpart is shown in Figure 10.1.2. In dimension $n = 3$, the following is obviously a cone: Let $u = x_1, v = (x_2, x_3), \sqrt{x_2^2 + x_3^2} \le \gamma |x_1|$. But here, too, the closure of the complement of a cone is a cone; so letting $u = (x_2, x_3), v = x_1, |x_1| \le \sqrt{x_2^2 + x_3^2}/\gamma$ gives an example of a cone that does not look like those designed to hold ice cream.

By a *cone field* we mean a map that associates to every point $p \in \mathbb{R}^n$ a cone K_p at p. A diffeomorphism $f: \mathbb{R}^n \to \mathbb{R}^n$ naturally acts on cone fields by

$$(f_*K)_p := Df_{f^{-1}(p)}\left(K_{f^{-1}(p)}\right).$$

Theorem 10.1.5 *A compact f-invariant set Λ is hyperbolic if and only if there exist $\lambda < 1 < \mu$ such that at every $x \in \Lambda$ there are complementary subspaces S_x and T_x (in general, not Df-invariant), a field of horizontal cones $H_x \supset S_x$, and a family of vertical cones $V_x \supset T_x$ associated with that decomposition such that*

$$Df_x H_x \subset \text{Int } H_{f(x)}, \qquad Df_x^{-1} V_{f(x)} \subset \text{Int } V_x,$$

$$\|Df_x \xi\| \geq \mu \|\xi\| \text{ for } \xi \in H_x, \qquad \text{and} \qquad \left\| Df_x^{-1} \xi \right\| \geq \lambda^{-1} \|\xi\| \quad \text{for } \xi \in V_{f(x)}.$$

10.1.3 Stable Manifolds

Theorem 9.5.2 used the Contraction Principle to show that corresponding to the subspaces E^u and E^s that expand and contract under the differential there are curves that do the same for the map. While Theorem 9.5.2 produced these only for a hyperbolic fixed point, one can use that result in an auxiliary space to show that in case of a hyperbolic set we have such curves at every point, fixed or otherwise. Alternatively, one can modify the proof in such a way as to work along entire orbits of possibly nonfixed points.

We used these invariant curves before. In proving that linear hyperbolic maps on a torus are mixing (Proposition 7.2.9) we used the translates of the eigenlines, which are exactly the stable and unstable curves. A general hyperbolic system should be pictured in the same way with these lines made a little "wobbly". The method of proof of Theorem 9.5.2 can be used to obtain the following:

Theorem 10.1.6 (Stable and Unstable Manifolds Theorem for Hyperbolic Sets)
Let Λ be a hyperbolic set for a C^1 diffeomorphism f on U such that Df on Λ expands E^u by $\mu > 1$ and contracts E^s by $\lambda < 1$. Then for each $x \in \Lambda$ there is a pair of C^1-embedded (Definition A.2.5) disks $W^s(x)$, $W^u(x)$, called the local stable manifold *and the* local unstable manifold *of x, respectively, such that*

(1) *$E^s(x)$ is tangent to $W^s(x)$ at x, and $E^u(x)$ is tangent to $W^u(x)$ at x.*
(2) *$f(W^s(x)) \subset W^s(f(x))$, $f^{-1}(W^u(x)) \subset W^u(f^{-1}(x))$.*
(3) *For every $\delta > 0$ there exists $C(\delta)$ such that, for $n \in \mathbb{N}$,*

$$\text{dist}(f^n(x), f^n(y)) < C(\delta)(\lambda + \delta)^n \text{dist}(x, y) \qquad \text{for } y \in W^s(x),$$

$$\text{dist}(f^{-n}(x), f^{-n}(y)) < C(\delta)(\mu - \delta)^{-n} \text{dist}(x, y) \quad \text{for } y \in W^u(x).$$

(4) *There exists $\beta > 0$ and a family of neighborhoods O_x containing the ball around $x \in \Lambda$ of radius β such that*

$$W^s(x) = \{y \mid f^n(y) \in O_{f^n(x)}, \quad n \in \mathbb{N}\},$$

$$W^u(x) = \{y \mid f^{-n}(y) \in O_{f^{-n}(x)}, n \in \mathbb{N}\}.$$

This also implies that global stable and unstable manifolds

$$\widetilde{W}^s(x) = \bigcup_{n=0}^{\infty} f^{-n}(W^s(f^n(x)))$$

$$\widetilde{W}^u(x) = \bigcup_{n=0}^{\infty} f^n(W^u(f^{-n}(x)))$$

are defined independently of a particular choice of local stable and unstable manifolds and can be characterized topologically:

$$\widetilde{W}^s(x) = \{y \in U \mid \text{dist}(f^n(x), f^n(y)) \to 0, \quad n \to \infty\},$$
$$\widetilde{W}^u(x) = \{y \in U \mid \text{dist}(f^{-n}(x), f^{-n}(y)) \to 0, \quad n \to \infty\}.$$

The following useful fact is easy to show.

Proposition 10.1.7 *Denote by $W_\epsilon^s(x)$ and $W_\epsilon^u(x)$ the ϵ-balls in $\widetilde{W}^s(x)$ and $\widetilde{W}^u(x)$. Then there exists an $\epsilon > 0$ such that for any x, $y \in \Lambda$ the intersection $W_\epsilon^s(x) \cap W_\epsilon^u(y)$ consists of at most one point $[x, y]$ and there exists a $\delta > 0$ such that whenever $d(x, y) < \delta$ for some x, $y \in \Lambda$ then $W_\epsilon^s(x) \cap W_\epsilon^u(y) \neq \varnothing$.*

The argument which showed that toral automorphisms are topologically mixing also shows that, if global stable and unstable leaves are dense in a hyperbolic set, then the hyperbolic set is topologically mixing (hence chaotic by Proposition 7.2.14). The converse is true also.

Proposition 10.1.8 *If Λ is a compact locally maximal hyperbolic set for f and $f: \Lambda \to \Lambda$ is topologically mixing, then there is an $N \in \mathbb{N}$ such that for x, $y \in \Lambda$ and $n \geq N$ we have $f^n(W^u(x)) \cap W^s(y) \neq \varnothing$.*

This innocuous result has remarkable dynamical consequences, among them Theorem 10.2.8. Sensitive dependence is one of our ingredients of chaos, and Theorem 10.1.6 immediately implies it. However, it gives an even stronger property. Sensitive dependence says that nearby orbits *may* diverge, but the description of the local hyperbolic structure given by Theorem 10.1.6 shows that nearby orbits *must* diverge either in the future or in the past – the local picture near a hyperbolic point shows that other points move away either in the future, because of a nonzero unstable component, or in the past, due to a stable component, but usually both. The resulting property plays an important role.

Definition 10.1.9 A homeomorphism (correspondingly, a continuous map) $f: X \to X$ is said to be *expansive* if there exists a constant $\delta > 0$, called an *expansivity constant*, such that, if $d(f^n(x), f^n(y)) < \delta$ for all $n \in \mathbb{Z}$ (correspondingly, $n \in \mathbb{N}_0$), then $x = y$.

The maximal expansivity constant is sometimes called *the* expansivity constant for the dynamical system. By compactness, the property of being expansive does not depend on the particular choice of a metric on X defining a given topology, and hence it is an invariant of topological conjugacy. However, the expansivity constant does depend on the choice of metric.

Corollary 10.1.10 *The restriction of a diffeomorphism to a hyperbolic set is expansive.*

Proof If $\mathrm{dist}(f^n(x), f^n(y)) < \beta$ for $n \in \mathbb{Z}$, then $f^n(y) \in O_{f^n(x)}$ for $n \in \mathbb{Z}$ and $y \in W^s(x) \cap W^u(x) = \{x\}$ by Theorem 10.1.6(4). $\square$

In our examples of hyperbolic dynamical systems, topological entropy is also the exponential growth rate of periodic orbits. This is true for all hyperbolic systems, and expansivity explains half of it: It forces periodic orbits to be well separated, so these can serve as the separated sets in the definition of entropy. We return to this in Section 10.2.4.

10.1.4 The Contraction Principle

The Contraction Principle (Proposition 2.6.10) is the most basic ingredient for the general hyperbolic theory. It underlies the construction of stable and unstable manifolds whether one uses the proof in Section 9.5 or another. Their importance alone would put the Contraction Principle close to center stage. But other direct and indirect uses of the Contraction Principle pervade the hyperbolic theory. As in Chapter 9, it is applied by constructing a map that has two properties: The desired object is a fixed point of the map, and this map is a contraction.

The contraction property is obtained by using the contraction/expansion assumptions in Definition 10.1.1. The way this is done has some common features in most instances, so a useful shortcut is provided by the Hyperbolic Fixed-Point Theorem 9.5.4 whose use provides a once-and-for all device to pass from hyperbolicity to contraction. Thus numerous proofs set up a map in an auxiliary space that is hyperbolic rather than a contraction because such a map may be a little more straightforward to invent. The proof of Theorem 10.2.2 is a case in point.

10.2 ORBIT STRUCTURE AND ORBIT GROWTH

10.2.1 Density of Periodic Points, Closing Lemma

That periodic points are dense (Proposition 7.1.2, Proposition 7.1.3, Proposition 7.1.10, and Corollary 7.4.7) is not only common to our examples, but an intrinsic feature of hyperbolic dynamics.

The density of periodic points follows from the Anosov Closing Lemma, which, together with the closely related shadowing and specification theorems as well as invariant manifolds, coding, and the Contraction Principle, is part of a suite of tools that yields a highly detailed picture of the orbit structure and fundamental results about statistical behavior and structural stability.

We present the Closing Lemma with proof to give an illustration of how the Hyperbolic Fixed-Point Theorem 9.5.4 (and hence indirectly the Contraction Principle) is put to use in a concrete setting.

Definition 10.2.1 Let (X, d) be a metric space, $U \subset M$ open, and $f: U \to X$. For $a \in \mathbb{Z} \cup \{-\infty\}$ and $b \in \mathbb{Z} \cup \{\infty\}$ a sequence $\{x_n\}_{a<n<b} \subset U$ is said to be an ϵ-*orbit* or an ϵ-pseudo-orbit for f if $d(x_{n+1}, f(x_n)) < \epsilon$ whenever $a < n$ and $n+1 < b$. If, furthermore, $-\infty < a < b < \infty$ and $x_{b-1} = x_{a+1}$, then it is said to be a *periodic* ϵ-orbit.

Theorem 10.2.2 (Anosov Closing Lemma) *Let Λ be a hyperbolic set for a map f on U. Then there is a neighborhood $V \supset \Lambda$ and C, $\epsilon_0 > 0$ such that for $\epsilon < \epsilon_0$ and any periodic ϵ-orbit $(x_0, \dots, x_m) \subset V$ there is a point $y \in U$ with $f^m(y) = y$ and $\mathrm{dist}(f^k(y), x_k) < C\epsilon$ for $k = 0, \dots, m-1$.*

Remark 10.2.3 A particular case of an ϵ-periodic orbit is provided by an orbit segment $x_0, f(x_0), \dots, f^{m-1}(x_0)$ such that $\mathrm{dist}(f^m(x_0), x_0) < \epsilon$. Thus the Anosov Closing Lemma implies in particular that near any point in a hyperbolic set whose orbit nearly returns to the point there is a periodic orbit that closely follows the almost-returning segment. Another sharp way of stating when this yields density of periodic points is the following.

Definition 10.2.4 The set $\{x \mid \forall \epsilon > 0 \text{ there is a periodic } \epsilon\text{-orbit containing } x\}$ is called the *chain-recurrent set*.

The Anosov Closing Lemma shows the density of periodic points in the chain-recurrent set.

If V is an open neighborhood of Λ, then any periodic point in V is contained in Λ_V^f. If V is sufficiently small and Λ is locally maximal, then these orbits are in Λ. In that case the Closing Lemma therefore produces a periodic point $y \in \Lambda$. In particular, topological transitivity implies density of periodic points and hence that the hyperbolic set is chaotic.

Proof of Theorem 10.2.2 For each x_k there is a neighborhood V_k on which f is a small perturbation of a hyperbolic linear map given by $f_k(u, v) = (A_k u + \alpha_k(u, v), B_k v + \beta_k(u, v))$ with $\|\alpha_k\|$, $\|\beta_k\|$, $\|D\alpha_k\|$, and $\|D\beta_k\|$ bounded by $C_1\epsilon$ for all k and some $C_1 > 0$. We do not assume that the maps f_k fix the origin.

A sequence $(u_k, v_k) \in V_k$, $k = 0, \dots, m-1$, is a periodic orbit if and only if

$$(u, v) := ((u_0, v_0), (u_1, v_1), \dots, (u_{m-1}, v_{m-1}))$$

$$= (f_{m-1}(u_{m-1}, v_{m-1}), f_0(u_0, v_0), \dots, f_{m-2}(u_{m-2}, v_{m-2})) =: F(u, v).$$

Therefore we need to find a fixed point of F. We use the norm $\|(x_0, x_1, \dots, x_{m-1})\| := \max_{0 \le i \le m-1} \|x_i\|$. Represent F as "linear plus small":

$$F(u, v) = L(u, v) + S(u, v),$$

where

$$S((u_0, v_0), (u_1, v_1), \dots, (u_{m-1}, v_{m-1})) :=$$

$$((\alpha_{m-1}(u_{m-1}, v_{m-1}), \beta_{m-1}(u_{m-1}, v_{m-1})), \dots, (\alpha_{m-2}(u_{m-2}, v_{m-2}), \beta_{m-2}(u_{m-2}, v_{m-2}))),$$

$$L((u_0, v_0), (u_1, v_1), \ldots, (u_{m-1}, v_{m-1})) :=$$

$$((A_{m-1}u_{m-1}, B_{m-1}v_{m-1}), (A_0 u_0, B_0 v_0), \ldots, (A_{m-2}u_{m-2}, B_{m-2}v_{m-2})).$$

L is hyperbolic: It expands the subspace $((u_0, 0), (u_1, 0), \ldots, (u_{m-1}, 0))$ and contracts the subspace $((0, v_0), (0, v_1), \ldots, (0, v_{m-1}))$. Since $\|S(u, v) - S(u', v')\| \le C_3 \cdot \epsilon \cdot \|(u, v) - (u', v')\|$ for some $C_3 = C_3(f, \Lambda) > 0$, we can apply the Hyperbolic Fixed-Point Theorem 9.5.4 to obtain the desired closed orbit near the pseudo-orbit. $\square$

The Anosov Closing Lemma produces an abundance of periodic points, a subject that we presently develop further. But first we present related results that are stronger in several different ways.

10.2.2 The Shadowing Lemma
An immediate way to strengthen the Anosov Closing Lemma is to omit periodicity to approximate any pseudo-orbit by a genuine one. This concept plays a sufficiently prominent role to have proper name.

Definition 10.2.5 Let (X, d) be a metric space, $U \subset M$ open, and $f: U \to X$. For $a \in \mathbb{Z} \cup \{-\infty\}$ and $b \in \mathbb{Z} \cup \{\infty\}$ a pseudo-orbit $\{x_n\}_{a<n<b} \subset U$ is said to be δ-shadowed by the orbit $\mathcal{O}(x)$ of $x \in U$ if $d(x_n, f^n(x)) < \delta$ for all $a < n < b$.

Theorem 10.2.6 (Shadowing Lemma) *Let f be a diffeomorphism with a compact hyperbolic set Λ. Then Λ has a neighborhood $U(\Lambda)$ such that for any $\delta > 0$ there is an $\epsilon > 0$ such that every ϵ-orbit in $U(\Lambda)$ is δ-shadowed by an orbit of f.*

The Shadowing Lemma can be proved in a similar way to the Anosov Closing Lemma (Theorem 10.2.2) by considering a sequence of maps of $\mathbb{R}^n$ close to hyperbolic linear maps. Although it does not assert that the shadowing orbit for a periodic pseudo-orbit is periodic, one can extract this extra information by shadowing a pseudo-orbit with explicit repetition and then using expansivity. Thus, this is a genuine strengthening of the Closing Lemma.

The Shadowing Lemma is good news for numerical computation of hyperbolic sets. Sensitive dependence (and even more so expansivity) suggests that computation of a particular orbit is hopeless because roundoff deviations from a genuine orbit will be amplified exponentially by the dynamics. Shadowing reassures us that the computed pseudo-orbit closely matches *some* orbit in the hyperbolic set.

Since the intent of such numerical computation is to get a useful graphic rendering of the hyperbolic set in question, typically a strange attractor, there is a substantial issue that remains. While shadowing tells you that every plotted orbit represents a genuine orbit, it does not guarantee that these genuine orbits are at all typical. This is illustrated by the linear expanding map E_2, where the repeated doubling of binary representations in the computer will attract all computer orbits to the origin, thus missing virtually everything. So even without roundoff errors it is not clear how to make sure that one finds some typical orbits. A reassuring answer to this question in provided by the theory of SRB measures in the following, which implies that almost any initial choice gives a typical orbit.

10.2.3 Specification

One can view the Shadowing Lemma as a tool for the design of orbits. With the right choices sensitive dependence can be used to amplify the ϵ-deviations in a pseudo-orbit to macroscopic effects of a desired type, and shadowing then produces an orbit to realize these effects. From this point of view there is a beautiful refinement that allows to design orbits with remarkable specificity. In essence it says that one can take any finite collection of (finite) orbit segments whatsoever and stitch them together to occur at specified times in a real orbit, which can be chosen periodic. This fine tool can be combined with expansivity to develop most of the topological and statistical orbit structure of a hyperbolic set.

Definition 10.2.7 Let $f\colon X \to X$ be a bijection of a set X. A *specification* $S = (\tau, P)$ consists of a finite collection $\tau = \{I_1, \ldots, I_m\}$ of finite intervals $I_i = [a_i, b_i] \subset \mathbb{Z}$ and a map $P\colon T(\tau) := \bigcup_{i=1}^{m} I_i \to X$ such that, for $t_1, t_2 \in I \in \tau$, we have $f^{t_2 - t_1}(P(t_1)) = P(t_2)$. S is said to be *n-spaced* if $a_{i+1} > b_i + n$ for all $i \in \{1, \ldots, m\}$, and the minimal such n is called the *spacing* of S. We say that S *parameterizes* the collection $\{P_I \mid I \in \tau\}$ of orbit segments of f.

We let $T(S) := T(\tau)$ and $L(S) := L(\tau) := b_m - a_1$. If (X, d) is a metric space, we say that S is ϵ*-shadowed* by $x \in X$ if $d(f^n(x), P(n)) < \epsilon$ for all $n \in T(S)$.

Thus a specification is a parameterized union of orbit segments $P_{\restriction I_i}$ of f.

If (X, d) is a metric space and $f\colon X \to X$ is a homeomorphism, then f is said to have the *specification property* if for any $\epsilon > 0$ there exists an $M = M_\epsilon \in \mathbb{N}$ such that any M-spaced specification S is ϵ-shadowed by some $x \in X$ and such that, moreover, for any $q \geq M + L(S)$ there is a period-q orbit ϵ-shadowing S.

Theorem 10.2.8 (Specification Theorem) *Let Λ be a topologically mixing compact locally maximal hyperbolic set for a diffeomorphism f. Then $f_{\restriction \Lambda}$ has the specification property.*

The proof relies mainly on Proposition 10.1.8. One starts with the first orbit segment and calls its final point x. Denoting the initial point of the second segment by y, we apply Proposition 10.1.8 with sufficiently small pieces of invariant manifolds. The intersection point thus obtained defines the second approximation to the ultimate orbit segment. The required transition time depends only on the desired closeness. Being in $W^u(x)$ it is extremely close to the initial orbit segment, and being in $W^s(y)$ it also approximates the second segment well. Connecting to the third segment in the same way leaves the early points of the previous approximation almost unchanged; the total accumulation of errors is bounded by a geometric series.

Remark 10.2.9 One cannot prove this without mixing; indeed, it is easy to show that the specification property implies that $f_{\restriction \Lambda}$ is topologically mixing. This condition is not as restrictive as it seems, however. The spectral decomposition (Theorem 10.3.6) shows that it holds essentially without loss of generality.

10.2.4 Growth of Periodic Orbits

While density gives a qualitative indication of the abundance of periodic points in our examples, we also produced quantitative measures of this. We found that the number of periodic points grows exponentially (Proposition 7.1.2, Proposition 7.1.3, Proposition 7.1.10, and Corollary 7.4.7) and at a specific rate. Both of these features are common to hyperbolic dynamical systems. The specific growth rates can be obtained via coding, and they coincide with the topological entropy (Chapter 8). This means that periodic orbit growth is a significant measure of the overall dynamical complexity in hyperbolic dynamical systems.

Proposition 10.2.10 $p(f) \leq h_{top}(f)$ *for an expansive homeomorphism f of a compact metric space.*

Proof If δ_0 is the expansivity constant, then $\mathrm{Fix}(f^n)$ is (n, δ_0)-separated for all $n \in \mathbb{N}$ since if $x \neq y \in \mathrm{Fix}(f^n)$ and $\delta := \max\{d(f^i(x), f^i(y)) \mid 0 \leq i < n\}$ then $d(f^i(x), f^i(y)) \leq \delta$ for $i \in \mathbb{Z}$ and hence $\delta > \delta_0$. Thus $P_n(f) \leq N(f, \epsilon, n)$ for $\epsilon < \delta_0$, implying the claim. $\square$

Conversely, the specification property makes it possible to mimic separated sets with sets of periodic orbits to produce the reverse inequality.

Theorem 10.2.11 $p(f_{\restriction_\Lambda}) = h_{top}(f_{\restriction_\Lambda})$ *for an expansive homeomorphism f of a compact metric space with the specification property.*

Proof Any element of an (n, ϵ)-separated set E_n can be $\epsilon/2$-shadowed by a periodic point of period $n + M_{\epsilon/2}$. These points are distinct by the triangle inequality in the d_n^f metric. Thus there are at least $\mathrm{card}(E_n)$ distinct periodic points of period $n + M_{\epsilon/2}$ and consequently $P_{n+M_{\epsilon/2}}(f') \geq N(f', \epsilon, n)$, implying $p(f') \geq h_{top}(f')$. $\square$

Although locally maximal hyperbolic sets of diffeomorphisms provide the prime examples of expansive maps with specification, other important classes of transformations are covered, too; notably, transitive topological Markov chains and some more general classes of symbolic systems such as *sofic systems* are expansive with specification.

The Specification Property was not used to the fullest in this proof. Using it more diligently gives a much refined result:

Theorem 10.2.12 *Let X be a compact metric space and $f: X \to X$ an expansive homeomorphism with the specification property. Then there exist $c_1, c_2 > 0$ such that for $n \in \mathbb{N}$*

$$c_1 e^{nh_{top}(f)} \leq P_n(f) \leq c_2 e^{nh_{top}(f)}.$$

Coding gives a sharper result because the periodic orbit growth of topological Markov chains is $\lambda^n +$ smaller exponentials due to Corollary 7.3.6. (However, for continuous-time systems coding is not as powerful.)

10.2.5 The Shadowing Theorem

While the Specification Theorem improves on the Shadowing Lemma to give the most precise tool for the creation of individual orbits of a specified kind, the Shadowing Lemma can be improved in a different direction to give tight global control of the orbit structure. The distinctive feature of this Shadowing Theorem is that it provides for coherent shadowing of families of orbits that may in an evident sense be as complex as the entire orbit structure of a hyperbolic set.

Theorem 10.2.13 (Shadowing Theorem) *Let M be a Riemannian manifold, d the natural distance function, $U \subset M$ open, $f \colon U \to M$ a diffeomorphism, and $\Lambda \subset U$ a compact hyperbolic set for f.*

Then there exist a neighborhood $U(\Lambda) \supset \Lambda$ and $\epsilon_0, \delta_0 > 0$ such that for all $\delta > 0$ there is an $\epsilon > 0$ with the following property:

If $f' \colon U(\Lambda) \to M$ is a C^2 diffeomorphism ϵ_0-close to f in the C^1 topology, Y is a topological space, $g \colon Y \to Y$ is a homeomorphism, $\alpha \in C^0(Y, U(\Lambda))$, and $d_{C^0}(\alpha g, f'\alpha) := \sup_{y \in Y} d(\alpha g(y), f'\alpha(y)) < \epsilon$, then there is a $\beta \in C^0(Y, U(\Lambda))$ such that $\beta g = f'\beta$ and $d_{C^0}(\alpha, \beta) < \delta$.

Furthermore, β is locally unique: If $\bar\beta g = f'\bar\beta$ and $d_{C^0}(\alpha, \bar\beta) < \delta_0$, then $\bar\beta = \beta$.

Remark 10.2.14

(1) Local maximality of Λ is not required.

(2) To get the Shadowing Lemma take $Y = (\mathbb{Z}, \text{discrete topology})$, $f' = f$, $\epsilon_0 = 0$, $g(n) = n+1$ and replace $\alpha \in C^0(Y, U(\Lambda))$ by $\{x_n\}_{n \in \mathbb{Z}} \subset U(\Lambda)$ and "$\beta \in C^0(Y, U(\Lambda))$ such that $\beta g = f'\beta$" by $\{f^n(x)\}_{n \in \mathbb{Z}} \subset U(\Lambda)$. Then $d(x_n, f^n(x)) < \delta$ for all $n \in \mathbb{Z}$.

(3) The Closing Lemma is another particular case corresponding to $f' = f$, $Y = \mathbb{Z}/n\mathbb{Z}$, $g(k) = k+1 \pmod{n}$.

10.2.6 Stability and Classification

A smooth dynamical system is said to be C^1-structurally stable if every C^1-perturbation is topologically conjugate to it.

At first sight it is natural to believe that structural stability is only possible when there are finitely many (hyperbolic) periodic points. To each of these one can apply Proposition 9.3.1 for some definite perturbation size. The smallest of these sizes preserves all of these orbits; one can then tidy up the remainder of the picture. However, in the presence of infinitely many periodic points this does not work, because the allowed perturbation sizes may get arbitrarily small. Yet, Theorem 7.4.3 shows that expanding circle maps are structurally stable. Indeed, it was a remarkable discovery that, even though hyperbolic dynamical systems have infinitely many periodic points, the tight intertwining of the orbit structure makes it robust on the whole.

Structural stability holds for all hyperbolic dynamical systems, and it is a distinctive feature of hyperbolic dynamics that has provided one of the foremost motivations for their study. It automatically provides for a wealth of further examples of hyperbolic dynamical systems, namely, perturbations of our examples. In Section 7.4.4 it is immaterial that we assumed linearity in constructing the horseshoe. Nonlinear horseshoes exhibit exactly the same behavior. They arise naturally in many applications; this is described in Chapter 12.

There is an intimate connection between shadowing and structural stability. Structural stability (with continuous dependence of the conjugating homeomorphism on the perturbation) certainly implies that orbits of a perturbation are shadowed by unperturbed orbits. Conversely, the orbits of a perturbation of a dynamical system are ϵ-orbits for the original system. Since they are shadowed by unperturbed orbits, the correspondence that sends perturbed orbits to the unperturbed orbits shadowing them gives a candidate for a conjugacy. In this way the Shadowing Theorem can be used to prove the structural stability of hyperbolic sets.

Theorem 10.2.15 (Strong Structural Stability of Hyperbolic Sets) *Let $\Lambda \subset M$ be a hyperbolic set of the diffeomorphism $f: U \to M$. Then for any open neighborhood $V \subset U$ of Λ and every $\delta > 0$ there exists $\epsilon > 0$ such that, if $f': U \to M$ and $d_{C^1}(f_{\lceil V}, f') < \epsilon$, there is a hyperbolic set $\Lambda' = f'(\Lambda') \subset V$ for f' and a homeomorphism $h: \Lambda' \to \Lambda$ with $d_{C^0}(\mathrm{Id}, h) + d_{C^0}(\mathrm{Id}, h^{-1}) < \delta$ such that $h \circ f'_{\lceil \Lambda'} = f_{\lceil \Lambda} \circ h$. Moreover, h is unique when δ is small enough.*

Remark. The proof uses the Shadowing Theorem 10.2.13, which is stated here for C^2 maps. One can prove structural stability of C^1 maps directly or by sharpening the Shadowing Theorem.

Proof We use the Shadowing Theorem 10.2.13 three times. First take $\delta_0 < \delta$ as in the Shadowing Theorem and apply the Shadowing Theorem with $\epsilon < \delta_0/2$, $Y = \Lambda$, $\alpha = \mathrm{Id}_{\lceil \Lambda}$ the inclusion, and $g = f$ to obtain a unique $\beta: \Lambda \to U(\Lambda)$ such that $\beta \circ f = f' \circ \beta$. The cone criterion for hyperbolicity can be used to show that orbits staying sufficiently near a hyperbolic set must be hyperbolic, that is, that $\Lambda' := \beta(\Lambda)$ is hyperbolic.

To show that β is injective, apply the Shadowing Theorem the other way around: Take ϵ as before, $y = \Lambda'$, $\alpha' = \mathrm{Id}_{\lceil \Lambda'}$ the inclusion, and $g = f'$ to obtain a map h such that $h \circ f' = f \circ h$. It is important to keep in mind that we are allowed to use f' instead of f in the Shadowing Theorem if ϵ here is chosen small enough. We claim that $h \circ \beta = \mathrm{Id}$ and hence $h = \beta^{-1}$ is a homeomorphism.

Apply the uniqueness part of the Shadowing Theorem now in the "$f = f'$" case, when trivially $\alpha \circ f = f \circ \alpha$ and at the same time by the above $\bar{\beta} \circ f = f \circ \bar{\beta}$, where $\bar{\beta} := h \circ \beta$.

Since $d_{C^0}(\alpha, \bar{\beta}) = d_{C^0}(\mathrm{Id}, h \circ \beta) \leq d_{C^0}(\mathrm{Id}, \mathrm{Id} \circ \beta) + d_{C^0}(\mathrm{Id} \circ \beta, h \circ \beta) = d_{C^0}(\mathrm{Id}, \beta) + d(\mathrm{Id}, h) < \delta_0$, the uniqueness assertion of the Shadowing Theorem implies $\bar{\beta} = \alpha = \mathrm{Id}_{\lceil \Lambda}$, as claimed. $\square$

One of the outstanding achievements of dynamics in the twentieth century was to show that it is *precisely* hyperbolic dynamical systems that are C^1-structurally stable, that is, that C^1-structural stability is equivalent to hyperbolicity.[1] The

---

[1] Joel Robbin: A Structural Stability Theorem, *Annals of Mathematics* **94**, no. 2 (1971), 447–493, R. Clark Robinson, Structural Stability of C^1 Diffeomorphisms, *Journal of Differential Equations* **22**, no. 1 (1976), 28–73; Ricardo Mañé, A Proof of the C^1 Stability Conjecture, Publications Mathématiques de l'Institut des Hautes Études Scientifiques **66** (1988), 161–210.

corresponding issue for C^2-structural stability (defined analogously) is still open.

Striving for a classification is a lofty goal, but for some important collections of hyperbolic dynamical systems it has been achieved. The first instance is Theorem 7.4.3, which shows that an expanding map of the circle is equivalent to the linear expanding map with the same degree. The main example is that if the hyperbolic set is a torus, then the map on it is equivalent to a linear toral map. Which linear toral map it is can also be determined from degree-like global information. The degree of the expanding map E_m is the number m, and the analog of degree of the map F_L is the matrix L.[2]

Further examples of a classification are provided by hyperbolic attractors in low-dimensional dynamical systems.

10.3 CODING AND MIXING

10.3.1 Coding

The close relation between each of our examples and a symbolic system (Section 7.3.1, Proposition 7.4.2, Proposition 7.4.4, Proposition 7.4.6, and Corollary 7.4.10) arises from one of the primary features of hyperbolic dynamics, being modeled by a topological Markov chain up to occasional "bookkeeping" to adjust for the slight overlaps on the edges of a partition. This coding uses a Markov partition of a hyperbolic set. This is a finite cover by closed sets that overlap only on their boundaries, and with the "Markov" property that if U, V, W are among these sets and $f(U) \cap V \neq \varnothing \neq f(V) \cap W$, then $f^2(U) \cap W \neq \varnothing$. It is essentially this property that makes the symbolic system defined by possible itineraries with respect to this partition into a topological Markov chain. This Markov property arises from the geometric fact that $f(U) \cap V \neq \varnothing$ implies that "$f(U)$ goes across V".

We write Int_Λ and ∂_Λ to refer to the interior and boundary relative to Λ.

Definition 10.3.1 Let Λ be a compact locally maximal hyperbolic set, take ϵ, δ and $[x, y]$ as in Proposition 10.1.7, and let $\eta = \epsilon$. Then $R \subset \Lambda$ is said to be a *rectangle* if the diameter of R is smaller than $\eta/10$ and $[x, y] \in R$ whenever $x, y \in R$. A rectangle R is *proper* if $R = \overline{\text{Int}_\Lambda R}$. We write $W_R^i(x) := W_\eta^i(x) \cap R$ for $x \in R$, $i = u, s$, and set $\partial^s R := \{x \in R \mid x \notin \text{Int}_{\Lambda \cap W_\eta^u(x)} W_R^u(x)\}$, $\partial^u R := \{x \in R \mid x \notin \text{Int}_{\Lambda \cap W_\eta^s(x)} W_R^s(x)\}$.

A *Markov partition* is a finite cover $\mathcal{R} = \{R_0, \dots, R_{m-1}\}$ of Λ by proper rectangles such that

(1) $\text{Int } R_i \cap \text{Int } R_j = \varnothing$ for $i \neq j$;
(2) whenever $x \in \text{Int } R_i$ and $f(x) \in \text{Int } R_j$, then $W_{R_j}^u(f(x)) \subset f(W_{R_i}^u(x))$ and $f(W_{R_i}^s(x)) \subset W_{R_j}^s(f(x))$.

While the word "rectangle" is natural and appropriate, the elements of a Markov partition are rarely as simple as in our example so far. In general, and already for toral automorphisms in dimension 3 and higher, Markov partitions have complicated geometry with fractal boundaries.

---

[2] This is presented, for example, in Katok and Hasselblatt, *Introduction to the Modern Theory of Dynamical Systems*, pp. 330, 587.

It is useful to observe the following:

Lemma 10.3.2 *If R is a rectangle, then $\partial_\Lambda R = \partial^s R \cup \partial^u R$.*

Proof $x \in \text{Int}_\Lambda R \Rightarrow x \in \text{Int}_{\Lambda \cap W_\eta^u(x)}(R \cap W_\eta^u(x) \cap \Lambda) = \text{Int}_{\Lambda \cap W_\eta^u(x)} W_R^u(x)$ since R is a neighborhood of x in Λ. Thus $\partial^s R \subset \partial_\Lambda R$. Likewise, $\partial^u R \subset \partial_\Lambda R$. If $x \in (\text{Int}_{\Lambda \cap W_\eta^s(x)} W_R^s(x)) \cap (\text{Int}_{\Lambda \cap W_\eta^u(x)} W_R^u(x))$, then by continuity of $[\cdot, \cdot]$ there is a neighborhood U of x in Λ such that for all $y \in U$ we have $[x, y], [y, x] \in R$ and hence $y' := [[y, x], [x, y]] \in R \cap W_\eta^s(x) \cap W_\eta^u(y) \subset W_\eta^s(x) \cap W_\eta^u(y) \subset \{y\}$, so $x \in \text{Int}_\Lambda R$. $\square$

Theorem 10.3.3 *A compact locally maximal hyperbolic set admits Markov partitions of arbitrarily small diameter.*

Proof outline First take $\delta > 0$ small, ϵ as in Theorem 10.2.13, $\gamma < \epsilon/2$ such that $d(f(x), f(y)) < \epsilon/2$ when $d(x, y) < \gamma$, and a γ-dense set $P := \{p_0, \ldots, p_{N-1}\}$ in the hyperbolic set Λ. Then $\Omega(P) := \{\omega \in \Omega_N \mid d(f(p_{\omega_i}), p_{\omega_{i+1}}) < \epsilon\}$ is a topological Markov chain. For each ϵ-orbit from $\Omega(P)$ there is a unique $\beta(\omega) \in \Lambda$ that δ-shadows $\alpha(\omega) := \{p_{\omega_i}\}_{i \in \mathbb{Z}}$. One can show that β is surjective and continuous. We extend $[\cdot, \cdot]$ to ϵ-orbits by setting

$$[\omega, \omega']_i = \begin{cases} \omega_i & \text{for } i \geq 0, \\ \omega_i' & \text{for } i \leq 0, \end{cases}$$

for any $\omega, \omega' \in \Omega(P)$ with $\omega_0 = \omega_0'$. Then $[\cdot, \cdot]$ commutes with β, that is, $\beta([\omega, \omega']) \in W_{2\delta}^s(\beta(\omega)) \cap W_{2\delta}^u(\beta(\omega')) = \{[\beta(\omega), \beta(\omega')]\}$.

Then $R_i' := \{\beta(\omega) \mid \omega_0 = i\}$ is a rectangle since for $x = \beta(\omega)$, $y = \beta(\omega') \in R_i'$ we have $[\omega, \omega']_0 = i$ and thus $[x, y] = [\beta(\omega), \beta(\omega')] = \beta([\omega, \omega']) \in R_i'$. It is not hard to obtain (2) in Definition 10.3.1. To obtain a Markov partition we need, however, proper rectangles with pairwise disjoint interiors. To that end we modify these rectangles.

For $x \in \Lambda$ let $\mathcal{R}(x)$ be the set of rectangles from $\mathcal{R}'$ that contain x and $\mathcal{R}^*(x)$ be the set of rectangles from $\mathcal{R}'$ that intersect a rectangle from $\mathcal{R}'(x)$. Then $A := \{x \in \Lambda \mid W_\eta^s(x) \cap \partial^s R_i' = \varnothing, W_\eta^u(x) \cap \partial^u R_i' = \varnothing \text{ for all } i\}$ is open and dense. If $R_i' \cap R_j' \neq \varnothing$, then we cut R_j' into four rectangles as follows:

$$R(i, j, su) := R_i' \cap R_j',$$

$$R(i, j, 0u) := \{x \in R_j' \mid W_{R_i'}^s(x) \cap R_j' = \varnothing, \ W_{R_i'}^u(x) \cap R_j' \neq \varnothing\},$$

$$R(i, j, s0) := \{x \in R_j' \mid W_{R_i'}^s(x) \cap R_j' \neq \varnothing, \ W_{R_i'}^u(x) \cap R_j' = \varnothing\},$$

$$R(i, j, 00) := \{x \in R_j' \mid W_{R_i'}^s(x) \cap R_j' = \varnothing, \ W_{R_i'}^u(x) \cap R_j' = \varnothing\},$$

and for $x \in A$ let $R(x) := \bigcap\{\text{Int}_\Lambda R(i, j, q) \mid x \in R_i', \ R_i' \cap R_j' \neq \varnothing, \ x \in R(i, j, q), q \in \{su, 0u, s0, 00\}\}$. Then $\overline{R(x)}$ are rectangles covering $R_i' \cap A$ and the $R(x)$ are finitely many pairwise disjoint open rectangles, so

$$\mathcal{R} := \{\overline{R(x)} \mid x \in A\} =: \{R_0, \ldots, R_{m-1}\}$$

is a finite cover of Λ by proper rectangles with pairwise disjoint interiors. One can show that this is the desired Markov partition by showing (2) of Definition 10.3.1. $\square$

The desired consequence is the existence of a semiconjugacy between a compact locally maximal hyperbolic set and a topological Markov chain:

Theorem 10.3.4 *If Λ is a compact locally maximal hyperbolic set, $\mathcal{R} = \{R_1, \ldots, R_m\}$ a Markov partition of sufficiently small diameter, and*

$$
A_{ij} := \begin{cases} 1 & \text{if } R_i \cap f^{-1}(R_j) \neq \varnothing, \\ 0 & \text{otherwise,} \end{cases}
$$

then $f_{\restriction \Lambda}$ is a topological factor of the topological Markov chain (Ω_A, σ_A). The semiconjugacy $h \colon \Omega_A \to \Lambda$ is injective on $h^{-1}(\Lambda')$, where $\Lambda' := \Lambda \smallsetminus \bigcup_{i \in \mathbb{Z}} f^i(\partial^s \mathcal{R} \cup \partial^u \mathcal{R})$ and $\partial^s \mathcal{R} := \bigcup_{R \in \mathcal{R}} \partial^s R$, $\partial^s \mathcal{R} := \bigcup_{R \in \mathcal{R}} \partial^u R$.

Proof For $\omega \in \Omega_A$ define $h(\omega) = \bigcap_{i \in \mathbb{Z}} f^{-i}(R_{\omega_i})$. The main effort is to use the Markov property to show that the intersection is nonempty by checking the finite intersection property. It cannot contain more than one point by expansivity. h is continuous and it is surjective because $h(\Omega_A)$ is a compact set containing Λ'. Clearly, $h \circ \sigma_A = f \circ h$, and it is clear that every $x \in \Lambda'$ has only one preimage. $\square$

10.3.2 Topological Mixing and Spectral Decomposition

Our examples turn out to be not only topologically transitive, but also topologically mixing by Proposition 7.2.7 (linear expanding maps), Proposition 7.2.9 (toral automorphisms), Corollary 7.4.7 (horseshoes), Theorem 7.4.3 (nonlinear expanding maps), and Proposition 7.3.4 (quadratic maps via coding).

Proposition 7.3.12 says that a transitive topological Markov chain is topologically mixing. However, a topological Markov chain can be topologically transitive without having a transitive transition matrix. The issue is that there may be restrictions about the times at which images of one set overlap another. However, a topologically transitive topological Markov chain always permutes finitely many pieces with a topologically mixing return map. Here is a brief argument. Assume A is a 0-1 $m \times m$ matrix with at least one 1 in each row and each column. If $i \in \{0, \ldots, m-1\}$, then $\Omega_{A,i} = \{\omega \in \Omega_A \mid \omega_0 = i\} \neq \varnothing$. If there is an $\omega \in \Omega_A$ that contains the symbol i at least twice, then i is said to be *essential*. Two essential symbols i and j are *equivalent* if there exist $\omega, \omega' \in \Omega_A$, $k_1 < k_2$, $l_1 < l_2$ such that $\omega_{k_1} = \omega'_{l_2} = i$, $\omega_{k_2} = \omega'_{l_1} = j$. This is an equivalence relation. If σ_A has a dense positive semiorbit, then all symbols are essential and equivalent. Let N be the greatest common divisor of lengths of cycles (sequences beginning and ending at the same symbol) and identify two symbols if they are connected by a path whose length is a multiple of N. Let $\Lambda_1, \ldots, \Lambda_N = \Lambda_0$ be the equivalence classes. Show that the restriction of $(\sigma_A)^N$ to each Λ_i is topologically mixing.

Via coding hyperbolic sets decompose analogously, even without being topologically transitive. Here we indicate how to obtain the decomposition directly. In the case of quadratic maps and horseshoes, it is luck that they are topologically mixing. But for expanding maps and linear toral maps this is not an accident. It happens because there is no nontrivial partition of the circle or torus into compact sets. For the same reason any connected hyperbolic set is topologically mixing.

Before obtaining the decomposition into mixing pieces a correction is in order. A hyperbolic set might contain an attracting and a repelling fixed point together with a heteroclinic orbit (Definition 2.3.4). This is incompatible with transitivity, and we must therefore discard points that wander in this fashion.

Definition 10.3.5 A point $x \in X$ is *nonwandering* with respect to the map $f\colon X \to X$ if for any open set $U \ni x$ there is an $N > 0$ such that $f^N(U) \cap U \neq \varnothing$. The set of all nonwandering points of f is denoted by $NW(f)$.

So, in fact, the spectral decomposition decomposes the nonwandering set of a compact locally maximal hyperbolic set into finitely many components on which the appropriate iterate of f is topologically mixing.

Theorem 10.3.6 (Spectral Decomposition) *Let M be a Riemannian manifold, $U \subset M$ open, $f\colon U \to M$ a diffeomorphism, and $\Lambda \subset U$ a compact locally maximal hyperbolic set for f. Then there exist disjoint closed sets $\Lambda_1, \dots, \Lambda_m$ and a permutation σ of $\{1, \dots, m\}$ such that $NW(f_{\restriction_\Lambda}) = \bigcup_{i=1}^m \Lambda_i$, $f(\Lambda_i) = \Lambda_{\sigma(i)}$, and when $\sigma^k(i) = i$, $f^k_{\restriction_{\Lambda_i}}$ is topologically mixing.*

The proof uses an equivalence relation on $\mathrm{Per}(f_{\restriction_\Lambda})$ defined by $x \sim y$ if and only if $W^u(x) \cap W^s(y) \neq \varnothing$ and $W^s(x) \cap W^u(y) \neq \varnothing$ with both intersections transverse at at least one point. The Λ_i's are the closures of equivalence classes.

10.4 STATISTICAL PROPERTIES

Especially for hyperbolic dynamics, the interplay between topological properties and statistical ones is important. Section 7.5 and Section 7.6 clearly outline the features and concepts relevant for studying statistical properties of hyperbolic systems. These are nonunique ergodicity (Section 7.5.1), uniform distribution/ergodicity (Proposition 7.5.1, Theorem 7.5.6, and Section 7.5.4), and mixing (Proposition 7.6.6 and Proposition 7.6.7).

10.4.1 Expanding Maps; Difficulties of the Direct Approach

Nonlinear expanding maps of S^1 lend themselves as a model for some general features of statistical analysis.

Theorem 7.5.6 tells us that for the doubling map E_2 on S^1 the Birkhoff averages $\mathcal{B}_n(\varphi)(x)$ of a continuous function φ converge to $\int_{S^1} \varphi(t)\, dt$ almost everywhere. An analogous result for arbitrary expanding maps (even just those of degree 2) is clearly desirable, although these do not preserve length. We should look for weighted equidistribution as suggested in Section 7.6.4, where orbits have a skewed but coherent asymptotic distribution. Ideally, there should be a continuous function $g\colon [0, 1] \to [0, 1]$ such that the "weighted length" $\int_I g(x)\, dx$ of intervals I is invariant under the expanding map (see also Section A.3.2). There is a pretty obvious approach to defining such a function because a nonlinear expanding map of degree 2 is topologically conjugate to E_2 (Theorem 7.4.3). If h is the conjugacy, then $h \circ f = E_2 \circ h$, so we should be able to define a density g from the condition $\int_a^b g(x)\, dx = h(b) - h(a)$. This does not work because the conjugacy h looks rather like the distribution functions illustrated in Figure 7.6.1. It is far from differentiable, unlike the expression $\int_a^b g(x)\,dx$ for continuous g. We do not obtain a density function in this way. This is a subtle problem, but it is related to the elementary observation that the conjugacy cannot be differentiable if the derivative of f at the fixed point is not 2, that is, if it disagrees with that of E_2.

Although constructing an invariant density is therefore not trivial, such a density does indeed exist, and one can then inquire about uniform distribution with respect to it. For a nonlinear expanding map, especially when the degree is 2, uniform distribution can then be established (in principle) by probabilistic arguments as in the proof of Proposition 7.6.1.

This is rather involved, although as a reward one can establish much stronger statistical properties than simple uniform distribution. For the latter purpose the most appropriate tool is the Birkhoff Ergodic Theorem,[3] which produces uniform distribution from ergodicity defined in the proper framework of measure theory, which we do no touch in this book. However, checking ergodicity for an invariant measure is relatively straightforward when measure theory is in hand.

The conclusion for the present context is that expanding maps have an invariant density, and almost all points are uniformly distributed with this density.

10.4.2 Abundance of Invariant Measures

To elaborate on the failure of unique ergodicity in hyperbolic dynamical systems it is useful to describe it in terms of a purely qualitative property. This is expressed in terms of invariant measures.

Definition 10.4.1 Let X be a compact metric space and $f\colon X \to X$ continuous. An *invariant integral* for f is a nonzero real-valued linear map (linear functional) $\mathfrak{I}\colon C(X) \to \mathbb{R}$ on the space of continuous functions on X that is continuous in the topology of uniform convergence [that is, if $\phi_n \to \phi$ uniformly, then $\mathfrak{I}(\phi_n) \to \mathfrak{I}(\phi)$] and is f-invariant, that is, $\mathfrak{I}(\phi \circ f) = \mathfrak{I}(\phi)$ for any $\phi \in C(X)$. If $\mathfrak{I}(1) = 1$, then the invariant integral is also said to be an *invariant measure* or invariant probability.

A simple example occurs in Section 11.4.3.2 (as well as Section A.3.2) and is given by $\mathfrak{I}(\phi) := \int \phi\rho\,dx$, where ρ is a reasonable function. Measures that can be represented in this fashion are said to be *absolutely continuous*.

Proposition 10.4.2 *A transformation f is uniquely ergodic if and only if it has exactly one invariant measure.*

In terms of this result the failure of unique ergodicity as stated in Section 7.5.1 can be reexpressed as the presence of several invariant measures. Indeed, hyperbolic dynamical systems have an abundance of invariant measures. This is evident because there are so many periodic points (Proposition 7.1.2, Proposition 7.1.3, Proposition 7.1.10, and Corollary 7.4.7), and each periodic orbit $\mathcal{O}(x) = \{x, f(x), \ldots, f^{n-1}(x)\}$ carries an invariant measure $\mathfrak{I}$ defined by $\mathfrak{I}(\phi) = \sum_{i=0}^{n-1} \phi(f^i(x))/n$. Of course, these measures are not absolutely continuous.

---

[3] Katok and Hasselblatt, *Introduction to the Modern Theory of Dynamical Systems*, Theorem 4.1.2.

There are various ways of creating new invariant measures from these. The most basic is *convex combination*: If $\mathfrak{I}$ and $\mathfrak{J}$ are invariant measures and $a, b \geq 0$, $a + b = 1$, then $a\mathfrak{I} + b\mathfrak{J}$ is an invariant measure. In other words, the space of invariant measures is convex (Definition 2.2.13). In particular, finite convex combinations of "periodic measures" are invariant measures.

Although this is not obvious, the space of invariant measures is also compact.[4] Therefore, every sequence $(\mathfrak{I}_i)_{i \in \mathbb{N}}$ of invariant measures has a convergent subsequence. Its limit defines a new invariant measure. Applying these limit processes to the set of periodic measures yields a great wealth of invariant measures. The Specification Theorem provides a way of designing periodic orbits in a highly specific way and thereby lends itself to the construction of invariant measures with specific properties, such as *Gibbs measures*.

10.4.3 Equidistribution

In the discussion of shadowing (Section 10.2.2) it was pointed out that shadowing does not guarantee that computed orbits are typical. While the example of the doubling map is natural, it is also quite specific in that the binary rationals whose behavior is pointed out form a null set of points whose orbits are not uniformly distributed on the interval. Unless such a specific circumstance forces sampling from a null set, uniform distribution guarantees that almost every randomly chosen orbit behaves typically.

1. Absolutely Continuous Invariant Measures. The discussion of the doubling map in Section 7.6.4 produced an example of uniform asymptotic distribution with a complicated distribution function. This extended notion of equidistribution is appropriate for hyperbolic dynamical systems in general, and the appropriate framework is outlined in Section A.3. A new collection of examples arises from the quadratic family, which has an absolutely continuous invariant measure for many parameter values (Section 11.4). This results in *weighted equidistribution*, which is a little stronger than uniform asymptotic distribution because there is a reasonable density function that reflects the distribution. It is worth emphasizing again that, in combination with nonunique ergodicity (which reflects a rather heterogeneous orbit structure), this is one aspect of highly complicated asymptotic behavior. For added emphasis we list again some of the places where equidistribution made a prior appearance: Proposition 7.5.1, Theorem 7.5.6, Section 7.5.4, Proposition 7.6.6, and Proposition 7.6.7.

2. Absolutely Continuous Measures on Attractors. Chapter 13 is dedicated to (strange) attractors. Equidistribution issues are of particular importance for these. However, they have to be framed carefully. Orbits should be uniformly distributed on the attractor in some sense, and since an attractor often is a null set in phase space, this requires a significant adaptation of the notion of equidistribution.

To fix ideas we keep in mind as a model the solenoid illustrated in Figure 13.2.1. Under the map in question, points in the attractor are stretched apart while points off

---

[4] This follows from the Alaoglu Theorem in functional analysis.

the attractor move toward it. Since uniform distribution of an orbit on the attractor must mean uniform distribution conditioned on overlooking everything outside the attractor, a natural way to define it in terms similar to those above is to look at measures *on the attractor*, that is, measures that operate on continuous functions that are defined only on the attractor. Densities (for weighted equidistribution) should be defined only on the attractor, that is, we desire an absolutely continuous measure on the attractor (see Section A.3). We want it to be ergodic to obtain equidistribution.

The problem is that this means asking for a functional of the form $\mathfrak{I}(\phi) := \int \phi g\, dx$ without knowing what the integral means. After all, the whole attractor is a null set. But thinking of the solenoid one can make sense of the integral because the attractor locally is simply an accumulation of smooth curves (unstable manifolds), and integrating along any piece of such curves is unproblematic even though in $\mathbb{R}^3$ it is a null set. To make this more explicit requires a description of how to descend on a particular unstable leaf. Because these are null sets even within the attractor, one has to go through some normalization along the way. To this end take an interval I in a specific unstable leaf and suppose I_n is the union of the stable disks of radius $1/n$ with center in I, that is, a tube around I of radius $1/n$. The required property of our measure is that there is a function g on the attractor such that for any such interval and any continuous function ϕ we get $\lim_{n\to\infty} \mathfrak{I}(\phi\chi_{I_n})/\mathfrak{I}(\chi_{I_n}) = \int_I \phi g\, dx$, where dx denotes arc length on I.

This explains what we mean by an ergodic invariant measure on the attractor that is absolutely continuous on unstable manifolds, and why it is natural to look for one. At this point we know that having such a measure produces (weighted) equidistribution of almost all orbits that lie on the attractor. ("Almost all" makes sense because null sets on smooth curves make sense.)

3. Sinai–Ruelle–Bowen Measure. Such a measure is called a "Sinai–Ruelle–Bowen measure" (or SRB measure). Its description so far leaves a practical problem. Strange attractors are not described analytically. They emerge as computer pictures obtained by following a more or less random orbit. But since the attractor is a null set, chances are that one computes orbits that do not lie on the attractor. The preceding development has nothing to say about these, yet. One needs to tie the asymptotics of an orbit that approaches the attractor to the asymptotics of orbits on the attractor. The solenoid is again the perfect illustration. A point x near the attractor is in the stable leaf of a point p on the attractor because (in this case) every constant-angle slice is a stable leaf. Its motion under the map therefore decomposes into movement towards the attractor and the motion of the point p, which dictates the motion of the stable slice. This means that once the iterates of x are close enough to the attractor, they are indistinguishable from the corresponding iterates of p. If these are uniformly distributed on the attractor, then the orbit of x reflects the density of the SRB measure.

We wish to deduce that almost every nearby orbit of the solenoid map eventually looks equidistributed as described above. The set to be excluded is that of points x on stable manifolds of points p in the null set of bad points for the SRB measure. By the definition of a null set (on a curve piece from the attractor) the set of such p can be covered by countably many intervals whose lengths have an arbitrarily small sum. The union of stable slices through all points of these dangerous intervals is a countable collection of "wedges" of the solid torus that are close enough to being rectangles,[5] and the volume in $\mathbb{R}^3$ of each of these agrees with the length of the base interval on the solenoid up to the factor of area of a stable slice. Therefore, the volumes of these wedges have an arbitrarily small sum and the nontypical points in this neighborhood form a null set in $\mathbb{R}^3$.

One of the important results in the theory of hyperbolic dynamical systems is that hyperbolic attractors carry a Sinai–Ruelle–Bowen measure. As explained above, this implies that, except for a null set, every point in the basin of an attractor is uniformly distributed with the corresponding density. This gives us assurance that numerical simulations of "strange attractors" produce pictures that show the real attractor and all of it. The Sinai–Ruelle–Bowen measure is also referred to as the "physically observed measure."

10.5 NONUNIFORMLY HYPERBOLIC DYNAMICAL SYSTEMS

The theory of hyperbolic dynamics is remarkable for the strength of its conclusions about systems with vastly complicated orbit structure. The strength of the results and the comparative ease with which they can be obtained owes a lot to the uniformity of the hyperbolicity assumption used throughout.

As in our examples so far, orbit complexity in real-life systems is due in large part to a combination of stretching and folding. However, the folding is not always as neat, as we will see in the succeeding chapters. To widen the applicability of the hyperbolic theory it has been extended to include systems where each point is subject to hyperbolic action by the derivative, but the contraction and expansion rates vary between points, with no uniform separation from 1. Compared to the parent theory outlined here, the theory of nonuniformly hyperbolic dynamical systems relies far more on measure theory. Therefore it appears here only by way of the forthcoming examples. We note a few basic features.

The invariant directions E^u and E^s may be undefined on a null set and need not depend continuously on the point, nor is there a lower bound for the angle between them. Stable and unstable manifolds exist, but since the proof of this theorem operates locally and there is no uniformity in the size of neighborhood, the sizes of the leaves are not uniform, nor is there good control of the angle between them. Instead of the uniformly hyperbolic picture as for linear toral automorphisms, where along a piece of unstable curve the stable curves line up to form a uniform stack of leaves, a good mental picture of the nonuniform situation is of a "fence" consisting of segments crossing a Cantor set (but not a null set) in an unstable curve, the gaps containing similar fences with much shorter and more crooked segments, and so on.

In part because of the central role of measure theory, there is great interest in stochastic behavior. This plays a role with respect to quadratic maps and attractors. For the latter it is always of particular interest to establish the existence (and uniqueness) of a Sinai–Ruelle–Bowen measure to validate numerical pictures.

---

[5] This faith in the harmlessness of the deformation skips a serious technical point addressed only by *absolute continuity of the stable foliation.*

CHAPTER 11

Quadratic Maps

11.1 PRELIMINARIES

11.1.1 Simple and Complicated Behavior in the Quadratic Family

In Section 2.5 the quadratic family provided examples of nonmonotone interval maps that display fairly simple asymptotic behavior, which can be fully described in rather pedestrian terms. Such simple behavior occurs for $0 \leq \lambda \leq 3$. One can extend the same kind of analysis to slightly larger parameter values. In the next section we describe the first step of this extension in some detail (although without complete proof) and indicate what happens after the next significant bifurcation (cascade of *period doublings*). During this process of gradual buildup of complexity the behavior can still be classified as simple in the sense that all asymptotic regimes are periodic, albeit with a growing number of orbits of different periods. This ends, however, beyond $\lambda_\infty \approx 3.58$.

For larger λ the topological entropy of f_λ is positive and the global behavior is increasingly complex. In particular, the topological entropy grows with λ, and periodic points of periods other than powers of 2 start appearing. For $\lambda = 1 + \sqrt{8}$ an orbit of period 3 appears for the first time (Figure 11.1.1). By then periodic points of all periods coexist (Proposition 11.3.8). However, as λ increases, the degree to which there is any uniformity in the asymptotic behavior changes in a highly irregular way. Accordingly, this coexistence of a multitude of periodic orbits does not preclude simple periodic asymptotic for *most* (as opposed to all) initial conditions. For $3.8284 \approx 1 + \sqrt{8} \leq \lambda \leq 3.841499008,$[1] the iterates of a randomly chosen initial condition asymptotically display periodic behavior with period three (Figure 11.2.3). On the other hand, for other parameter values, for example, $\lambda = 4$, the asymptotic behavior is uniformly distributed (with a density), analogously to that of the linear expanding map (see Section 11.4.3).

---

[1] While this parameter has an expression in radicals, it is rather more complicated than the preferred parameters we produce later: $\lambda = 1 + \sqrt{52800 - 3900z + 285z^2 + 15\sqrt{201}z(20 + z)}/120$, where $z = \sqrt[3]{460 + 60\sqrt{201}}$. This is due to Sharon Chuba and Andrew Scherer.

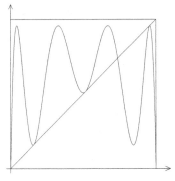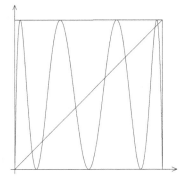

Figure 11.1.1. The maps $f^3_{1+\sqrt{8}}$ and f^3_4.

A considerable amount of research has been done on the quadratic family and on similar classes of one-dimensional maps in order to identify the possible types of behavior and to establish how various types appear in the parameter space. Some of the structural features reflect general paradigms in the theory of dynamical systems (hyperbolicity, structural stability, Markov partitions, existence of absolutely continuous invariant measures), while others are specific to one-dimensional dynamics [Sharkovsky ordering of periods (Theorem 11.3.9), kneading theory]. The research uses methods from general differentiable dynamics such as fixed-point methods based on hyperbolicity, methods specific to dimension one (Intermediate-Value Theorem, distortion estimates), and specific properties of analytic functions, or of polynomials. All of this makes the quadratic family an excellent proving ground for various approaches to dynamics and a model case for difficulties that appear even in simple-sounding problems.

As a result of a number of deep and brilliant works, the structure of the quadratic family (also sometimes called a *real quadratic family*) is comprehensively understood in its main features.

11.1.2 Attracting Periodic Orbits

Beginning with Proposition 11.2.1, we describe the first bifurcations in the quadratic family, which cause the appearance of heterogeneous asymptotics. While the detection and classification of periodic orbits is a matter of explicit computation, some simple topological facts will strengthen the insights to be obtained. Therefore, this section develops some basic global information pertinent to the existence of attracting periodic points.[2]

Recall that an m-periodic point x_0 of a continuous map f of an interval I into itself is said to be *attracting* if there is an $\epsilon > 0$ such that $|x - x_0| < \epsilon$ implies $|f^n(x) - f^n(x_0)| \to 0$ as $n \to \infty$. Equivalently, it is an attracting fixed point for f^m (Definition 2.2.22). A periodic orbit $\{x_0, f(x_0), \ldots, f^{m-1}(x_0)\}$ is said to be attracting if one (and hence all) of its points are attracting.

---

[2] The initial doublings were studied by Myrberg as part of a series of papers in the 1950s and 1960s; see, for example, Pekka Juhana Myrberg, Iteration der reellen Polynome zweiten Grades. II., *Annales Academiæ Scientiarum Fennicæ Mathematica Ser. A* **I 268** (1959).

Let $f: I \to I$ be differentiable and x_0 an m-periodic point. If $|(f^m)'(x_0)| < 1$, then x_0 is attracting (Proposition 2.2.17); if $|(f^m)'(x_0)| > 1$, then is it not. If $|(f^m)'(x_0)| = 1$, it may or may not be attracting.

Definition 11.1.1 The *basin of attraction* of an attracting periodic orbit $\mathcal{O} = \{x_0, f(x_0), \ldots, f^{m-1}(x_0)\}$ is the set of x such that $|f^n(x) - f^{n+k}(x_0)| \to 0$ as $n \to \infty$ for some k.

The basin of attraction of the orbit $\mathcal{O}$ consists of those points whose asymptotic behavior follows one of the points on the orbit. Such a point is unique since different points on the orbit cannot approach each other. Thus one can talk about the *phase* for each point in the basin of attraction.

Definition 11.1.2 The *immediate basin of attraction* of the periodic point x_0 is the maximal interval J containing x_0 such that $|f^n(x) - f^n(x_0)| \to 0$ as $n \to \infty$ for any $x \in J$. The *immediate basin of attraction* of a periodic orbit is the union of immediate basins of attraction of all points on the orbit.

Remark 11.1.3 Basins and immediate basins are open sets.

Lemma 11.1.4 *Let $I = [a, b]$ and $f: I \to I$ concave and twice differentiable such that $f(a) = f(b) = a$. Then the immediate basin of attraction of any attracting periodic orbit contains a critical point.*

Proof If $f'(a) \leq 1$, then by concavity $f'(x) < 1$ for $x > a$ and hence $f(x) < x$ for $x > a$, so I lies the immediate basin of attraction of a. In this case we are done. Henceforth we assume $f'(a) > 1$.

Let J be the immediate basin of attraction of an attracting m-periodic point x_0 and $c < d$ its endpoints. Then $f^m(J)$ is an interval by the Intermediate-Value Theorem and contains x_0. The interval $J \cup f^m(J)$ belongs to the immediate basin of attraction of x_0, so $f^m(J) \subset J$.

If, contrary to the assertion, none of the intervals $f^i(J)$, $(0 \leq i \leq m-1)$ contain the critical point of f, then f^m is monotone on J. We aim to derive a contradiction from this.

Consider first the case when one of the images of the endpoints is an endpoint of I. Since $f(b) = a$, we may assume without loss of generality that this endpoint is a. Replacing x_0 by its iterate, we may assume that a is an endpoint of J, that is, $c = a$.

Since a is a fixed point, $f'(a) > 1$ implies $(f^m)'(a) > 1$; hence $(f^m)'$ is positive and decreasing on $[a, d]$ due to the presence of the attracting point x_0. (This uses the chain rule and that f^m is monotone on $[a, d]$). Thus $0 < (f^m)'(d) < (f^m)'(x_0) < 1$, so f^m is a contraction on the interval $[x_0, d]$ and $d < b$ since $[a, d]$ does not contain the critical point of f. Thus the map f is a contraction on $[x_0, d + \epsilon]$ for sufficiently small $\epsilon > 0$, contrary to maximality of J.

Therefore none of the images of c or d is an endpoint of I. Then $f^m(J) \subset J$ implies $f^m(J) = J$ because neither c nor d is in the basin of attraction. Hence c and d are periodic points of period m or $2m$. Since the derivative of f^m (and hence

f^{2m}) has constant sign on J and f is concave, the derivatives of f^m and f^{2m} are monotone on J by the chain rule (they are products of monotone functions with constant sign). The average of $(f^m)'$ [and of $(f^{2m})'$] over J is ± 1 because $f^m(J) = J$, so the value at one endpoint is less than 1 in absolute value, and at least one of the points c or d is an attracting periodic point for f. This is impossible since its immediate basin of attraction would overlap with J, the immediate basin of x_0. $\Box$

Proposition 11.1.5 *Let $I = [a, b]$ and $f\colon I \to I$ concave and twice differentiable such that $f(a) = f(b) = a$. Then f has at most one attracting periodic orbit.*

Proof Immediate basins of attraction of different periodic orbits do not overlap. Since f has one critical point, the proposition follows from Lemma 11.1.4. $\Box$

Corollary 11.1.6 *A quadratic map f_λ has at most one attracting periodic orbit.*

One may ask what happens in the presence of a periodic point outside the basin of attraction. We will see later that the behavior on the remaining set R, which is invariant in both positive and negative direction, may still be quite complicated. This set is called *the universal repeller* (because it is the maximal repelling set; see A. Katok and B. Hasselblatt, *Introduction to the Modern Theory of Dynamical Systems*, Cambridge University Press, 1995, p. 519).

The following crucial fact holds in greater generality, but we formulate it for the quadratic family.

Theorem 11.1.7 *If a quadratic map f_λ has an attracting periodic point, then the universal repeller is a nowhere dense null set.*

Thus, both in the topological sense and in the sense of probability most points are attracted to the single periodic orbit.

The internal structure of the restriction of a quadratic map to a universal repeller is well descibed by Markov models (Definition 7.3.2, Section 7.3.7).

Theorem 11.1.8 *The restriction of a quadratic map with an attracting periodic point to the universal repeller is topologically conjugate to a one-sided topological Markov chain (Definition 7.3.2).*

Both theorems follow from *hyperbolicity* of the universal repeller (Definition 10.1.2). This is the reason for the following terminology:

Definition 11.1.9 Quadratic maps with an attracting periodic point are said to be *hyperbolic*.

Remark 11.1.10 Since the existence of an attracting periodic point is an open condition, the set of parameters for which a quadratic map is hyperbolic is open.

In light of the above discussion, the following question becomes natural and interesting.

Question How big is the set of parameters λ for which the quadratic map f_λ is hyperbolic, that is, has an attracting periodic orbit? In particular, is it dense? Is its complement a null set?

Both of the latter questions have been answered, the last negatively by Michael Jakobson in the early eighties, and the other positively by Gregorz Świątek and his collaborators in the late nineties. Their respective works represent the high points of one-dimensional dynamics in the corresponding periods. Moreover, they establish two principal paradigms of behavior in one-dimensional real dynamics, which we will refer to as "hyperbolic" and "stochastic."

Before telling the story we look at the development of behavior after the first crucial bifurcation at $\lambda = 3$. The principal feature is a succession of period doublings.

11.2 DEVELOPMENT OF SIMPLE BEHAVIOR BEYOND THE FIRST BIFURCATION

11.2.1 First Period Doubling

For the next parameter interval after $\lambda = 3$ orbits are asymptotic to a periodic orbit rather than a fixed point.

Proposition 11.2.1 *For $3 < \lambda \leq 1 + \sqrt{6}$ all orbits of $f_\lambda(x) = \lambda x(1 - x)$ on $[0, 1]$, except for 0 and x_λ and their preimages, are asymptotic to a unique periodic orbit of period 2.*

Sketch of proof Unlike in the previous case (Proposition 2.5.2) we do not prove the entire statement but simply calculate the periodic points and determine their stability. The full dynamics can be unraveled in a similar way as before, but the argument would be a bit more lengthy. To show that there is a period-2 orbit look for fixed points of f_λ^2 by solving the quartic equation

$$x = f_\lambda^2(x) = f_\lambda(f_\lambda(x)) = \lambda(\lambda x(1 - x))(1 - \lambda x(1 - x)).$$

First discard the known solutions – the fixed points – by dividing out $f_\lambda(x) - x$ from $f_\lambda^2(x) - x$:

$$\frac{f_\lambda^2(x) - x}{f_\lambda(x) - x} = \frac{\lambda^2 x(1 - x)(1 - \lambda x(1 - x)) - x}{\lambda x(1 - x) - x} = -(\lambda x)^2 + (\lambda + 1)\lambda x - (\lambda + 1).$$

This is a quadratic function of λx with roots

$$\lambda x = \frac{-(\lambda + 1) \pm \sqrt{(\lambda + 1)^2 - 4(\lambda + 1)}}{-2},$$

so there are two points of period 2:

(11.2.1)
$$x_{\lambda,\pm} = \frac{\lambda + 1 \pm \sqrt{(\lambda + 1)(\lambda - 3)}}{2\lambda}.$$

Let us note a few features of this explicit description. For $\lambda < 3$ there are no real solutions and $x_{3,\pm} = 2/3 = x_3$. Therefore, as λ increases beyond 3, these two periodic points appear by splitting off from the nonzero fixed point $x_\lambda = (\lambda - 1)/\lambda$. The distance from x_λ is

$$(11.2.2) \quad x_{\lambda,\pm} - x_\lambda = \frac{\lambda + 1 \pm \sqrt{(\lambda + 1)(\lambda - 3)} - 2(\lambda - 1)}{2\lambda}$$

$$= \frac{3 - \lambda \pm \sqrt{(\lambda + 1)(\lambda - 3)}}{2\lambda} \approx \frac{3 - \lambda}{6} \pm \frac{\sqrt{4(\lambda - 3)}}{6} \approx \pm \frac{\sqrt{\lambda - 3}}{3}.$$

This is the reason for the parabolic curves in the bifurcation diagram (Figure 11.2.3.)

This qualitative change in the orbit structure is a bifurcation different from those in Section 2.3.2. In this case, the change is from having two fixed points and all other points asymptotic to one of them to having two fixed points as well as a period-2 orbit.

If $\lambda > 3$, then $|f'_\lambda(x_\lambda)| = \lambda|1 - 2x_\lambda| = |2 - \lambda| > 1$, so the nonzero fixed point becomes repelling beyond exactly the parameter value at which it spawns the periodic points. This particular kind of transition is referred to as a *period doubling bifurcation*. An intuitive way of understanding the necessity of such transitions is to note that for $\lambda < 3$ all points near the fixed point are attracted to it (while switching sides in each iteration). For λ slightly larger than 3, nearby points are repelled, whereas faraway points still try to approach the fixed point. Consequently, there must be a point on either side of the fixed point to separate these two behaviors. These two points form the attracting periodic orbit.

For $\lambda < 1 + \sqrt{6}$, the periodic orbit is attracting. To see this, calculate the derivative of f_λ at $x_{\lambda,\pm}$. Using $f'_\lambda(x) = \lambda(1 - 2x)$ with $x_{\lambda,\pm}$ given by (11.2.1) gives

$$f'_\lambda(x_{\lambda,\pm}) = \lambda \left(1 - \frac{\lambda + 1 \pm \sqrt{(\lambda + 1)(\lambda - 3)}}{\lambda}\right) = -1 \pm \sqrt{(\lambda + 1)(\lambda - 3)},$$

and $(f^2_\lambda)'(x_{\lambda,\pm})$ is the product of these two numbers: $(f^2_\lambda)'(x_{\lambda,\pm}) = 1 - (\lambda + 1)(\lambda - 3)$. This is 1 when $\lambda = 3$. It decreases to -1 as λ increases to the positive solution $\lambda = 1 + \sqrt{6}$ of $(\lambda + 1)(\lambda - 3) = 2$. Therefore, the period-2 orbit is attracting for $3 < \lambda < 1 + \sqrt{6}$.

For $\lambda = 1 + \sqrt{6}$, all nonfixed orbits still converge to the period-2 orbit $x_{1+\sqrt{6},\pm} = (\sqrt{2} + \sqrt{3} \pm 1)/(\sqrt{2} + 2\sqrt{3})$, but now with subexponential speed; see Figure 11.2.1 or Figure 11.2.2. $\square$

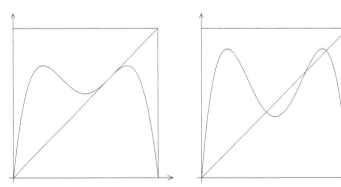

Figure 11.2.1. The maps f^2_3 and $f^2_{1+\sqrt{6}}$.

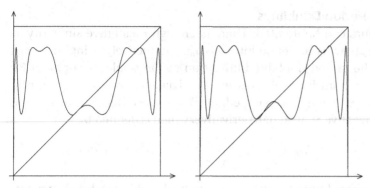

Figure 11.2.2. The maps $f^4_{1+\sqrt{6}}$ and $f^4_{1.05+\sqrt{6}}$.

We describe the basins of attraction for $3 < \lambda \leq 1 + \sqrt{6}$. Other than itself, the unstable fixed point x_λ has one preimage,

$$x_{\lambda,1} := 1 - x_\lambda = 1/\lambda,$$

which in turn has two preimages,

$$x^{\pm}_{\lambda,2} := 1/2 \pm \sqrt{\lambda^2 - 4}/2\lambda.$$

Direct inspection using the order of the points shows that the immediate basins of attraction of $x_{\lambda,\pm}$ are

(11.2.3) $\qquad (x_\lambda, x^+_{\lambda,2}) \qquad$ and $\qquad I_1(\lambda) := (x_{\lambda,1}, x_\lambda),$

correspondingly.

Further preimages of the unstable fixed point x_λ form two sequences that converge to the endpoints 0 and 1, respectively. Those preimages separate the basin of attraction of the stable orbit of period 2 into countably many disjoint intervals. For the map f^2_λ, the basin of attraction of each of its two stable fixed points consists of countably many disjoint intervals, which alternate on both sides of the immediate basins of attraction.

The central observation that makes it possible to carry this analysis further is that, if we restrict the map f^2_λ for $3 < \lambda < 1 + \sqrt{6}$ to the immediate basin of attraction of either of its two attracting fixed points, we obtain a map that looks qualitatively like the original map f_λ on the whole interval $[0, 1]$ for $\lambda < 3$. It has a unique critical point, a repelling fixed point at one end, the preimage of that point on the other end, and an attracting fixed point with negative derivative.

As the parameter increases, the attracting fixed point of this new map becomes repelling in a period doubling bifurcation. Beyond that bifurcation one can restrict $f^4_\lambda = (f^2_\lambda)^2$ to the interval $I_2(\lambda)$ defined as the interval between the preimage of the unstable fixed point and the point itself [this corresponds to the interval $I_1(\lambda)$ for $f^2_\lambda \restriction_{I_1(\lambda)}$]. This can be continued indefinitely.

This is the first germ of the idea of *renormalization*.

11.2.2 Cascade of Period Doublings

Indeed, the procedure can be iterated. There is enough qualitative similarity as described above to allow us to define intervals $I_n(\lambda)$ inductively using the maps $f_\lambda^{2^{n-1}}$ and observe the period-doubling bifurcation of the stable fixed points for the latter map on such an interval. This is not trivial, but it is not specific to the quadratic family, either. The key ingredient that makes this renormalization possible is to have negative Schwarzian derivative, which is defined by

$$(11.2.4) \qquad\qquad Sf := \frac{f'''}{f'} + \frac{3}{2}\left(\frac{f''}{f'}\right)^2.$$

This property is preserved under composition. Imposing this condition prevents iterates from having almost flat pieces away from critical points. Analytically it is used to control distortion under the map.

Before discussing this general aspect of the picture we describe results in this specific case. The next theorem gives a qualitative picture of the orbit structure during this cascade of period doublings.

Theorem 11.2.2[3] *There is a monotone sequence of parameter values* $\lambda_1 = 3, \lambda_2 = 1 + \sqrt{6}, \lambda_3, \ldots,$ *such that for* $\lambda_n < \lambda \le \lambda_{n+1}$ *the quadratic map* f_λ *has one attracting periodic orbit* $\mathcal{O}_n(\lambda)$ *of period* 2^n, *two repelling fixed points* 0 *and* x_λ, *and one repelling periodic orbit* $\mathcal{O}_k(\lambda)$ *of period* 2^k *for each* $k = 1, 2, \ldots, n-1$. *The basin of attraction of the orbit* $\mathcal{O}(\lambda)$ *is dense and consists of all points other than these periodic orbits and their preimages. At* $\lambda = \lambda_n$, *the orbit* $\mathcal{O}_n(\lambda)$ *undergoes a period-doubling bifurcation.*

During the period doubling the points of the nascent attracting 2^{n+1}-periodic orbit appear in pairs around the points of the period-2^n orbit, so the order of points on various periodic orbits is easy to describe: 11, 1212, 14241424, 1848284818482848, and so on.

It is curious that the interesting parameter values that we have encountered so far can be written in a consistent form: For $\lambda = 1 + \sqrt{0}$ a nonzero fixed point appears, at $\lambda = 1 + \sqrt{1}$ its "orientation" changes (the derivative becomes negative), at $\lambda = 1 + \sqrt{4}$ the period-2 orbit appears, and at $\lambda = 1 + \sqrt{6}$ the period-4 orbit appears. At $\lambda = 1 + \sqrt{8}$ we first have a period-3 orbit and, indeed, all periods (Proposition 11.3.8). Finally, for $\lambda = 1 + \sqrt{9}$ we obtain the maximal complexity (Section 7.1.2). Furthermore, the derivative at the stable period-2 orbit changes sign when $(f_\lambda^2)'(x_{\lambda,\pm}) = 1 - (\lambda + 1)(\lambda - 3)$ is zero, that is, when $(\lambda + 1)(\lambda - 3) = 1$ or $\lambda = 1 + \sqrt{5}$. These special parameters $1 + \sqrt{n}$ for $n = 0, 1, 4, 5, 6, 8, 9$ are marked by vertical dotted lines in Figure 11.2.3. Unstable points have dashed lines. Also, additional points of a given period may appear later. An interesting example is the appearance of a second orbit of period 4 for $\lambda = 1 + \sqrt{4 + \sqrt[3]{108}} = 1 + \sqrt{4 + 3\sqrt[3]{4}}$, indicated by a small tick mark on the x-axis in Figure 11.2.3.[4] To verify the appearance of a period-3

---

[3] Welington de Melo and Sebastian van Strien, *One-Dimensional Dynamics*, Springer-Verlag, Berlin, 1993.

[4] This is due to An Nguyen.

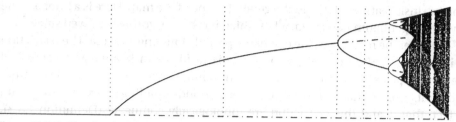

Figure 11.2.3. The fixed and periodic points as a function of λ.

orbit at $\lambda = 1 + \sqrt{8}$ use that if $g_\alpha(x) := \alpha - x^2$ and $h_\lambda = \lambda(x - \frac{1}{2})$, then $h_\lambda(f_\lambda(x)) = g_\alpha(h_\lambda(x))$, where $\alpha = (\lambda^2/4) - (\lambda/2)$ (Exercise 2.5.2 (p. 61)). For $\lambda = 1 + \sqrt{n}$ this gives $\alpha = (n - 1)/4$, so the issue is the appearance of a period-3 orbit in $g_\alpha(x) = \alpha - x^2$ for $\alpha = 7/4$. This happens because $g^3_{7/4}(x) - x$ has five roots, of which three are also roots of $(g^3_{7/4})'(x) - 1$: Verify that $64 \cdot (g^3_{7/4}(x) - x) = (g_{7/4}(x) - x)(1 - 18x - 4x^2 + 8x^3)^2$ and $64 \cdot ((g^3_{7/4})'(x) - 1) = -(8 - 3x - 22x^2 + 4x^3 + 8x^4)(1 - 18x - 4x^2 + 8x^3)$.

The mere *appearance* of infinitely many period doublings can be derived in much greater generality than for the quadratic family (see Section 11.3.2). Further develoment of this general approach is based on a general device known as the *kneading theory*.[5] Monotonicity, that is, the absence of relapses from period 2^k back to 2^{k-1} as the parameter increases, is a subtle fact that in general uses negative Schwarzian derivative (11.2.4). So is the fact that the whole cascade appears only once and no two bifurcations occur at the same time. While those features are not restricted to the quadratic family, they require much more specific structure of the family in question.

11.3 ONSET OF COMPLEXITY

11.3.1 Feigenbaum Universality

The period-doubling cascade was known in the early 1960s. At a 1975 conference, Steven Smale suggested that there is something interesting about how the doublings accumulate on an eventual parameter. Mitchell Feigenbaum then found numerically that in the quadratic family there is a regular pattern to the rate at which successive period-doubling bifurcations occur.[6] Specifically, the distance between successive bifurcations decreases eventually in a geometric progression. As the numerics suggested, and later proof has shown, $\delta := \lim_{n\to\infty}(\lambda_n - \lambda_{n-1})/(\lambda_{n+1} - \lambda_n)$ exists. In fact, Feigenbaum achieved great precision, finding $\delta \approx 4.6992016090$.

This is an interesting observation to begin with. More interestingly, period-doubling cascades occur for all nearby one-parameter families of unimodal maps and the sequence of bifurcation parameters produces an eventually geometric progression as in the quadratic case. What is truly astonishing is that this occurs *with the same limiting ratio*. That is, the specific numerical value obtained for the quadratic family has nothing to do with the particular structure of those maps but

---

[5] A description and references can be found in Katok and Hasselblatt, *Introduction to the Modern Theory of Dynamical Systems,* Cambridge University Press, Cambridge, 1995.

[6] Mitchell J. Feigenbaum, Quantitative Universality for a Class of Nonlinear Transformations, *Journal of Statistical Physics* **19**, no. 1 (1978), 25–52.

is an intrinsic feature related to the general shape of the map. Here is a brief outline of what Feigenbaum perceived, all of which has been rigorously proved since.[7]

Consider the map f_{λ_∞} at the end of the period-doubling cascade. The restriction of its square to the interval $[1 - x_{\lambda_\infty}, x_{\lambda_\infty}]$ is (aside from being upside-down) of the same type as f_{λ_∞} in that it fixes one endpoint, which is the image of the other endpoint, is unimodal, and is at the end of a period-doubling cascade. It turns out, in fact, that f and this restriction are topologically conjugate (Definition 7.3.3), that is, they agree up to a continuous change of variable. Feigenbaum wondered whether they are more closely related, namely, by a linear change of variable. Although they are not, he found something as useful. To describe this with more ease, change variables so as to make the quadratic family take the form $f(x) = \alpha - x^2$ on a symmetric interval (Exercise 2.5.2). Then there is one positive fixed point and 0 is the critical point. Feigenbaum found that there is a unique even analytic map g that is linearly related to a restriction of its square. This means that there is a unique real number α (roughly -2.5) such that $\alpha g^2(x/\alpha) = g(x)$. One can obtain g approximately by solving numerically for the coefficients in its power series. Indeed, $g(x) \approx 1 - 1.52763x^2 + 0.10481x^4 - 0.0267057x^6 + \cdots$.

How would one get at this distinguished self-similar map? It is the unique fixed point of the renormalization operator defined by

$$\mathcal{R}f(x) = \alpha f^2(x\alpha)$$

for x between the positive fixed point x_f of f and $-x_f$. Here f is supposed to be an even function with $f(0) = 1$ and $\alpha = 1/f^2(0) = 1/f(1)$. (We won't worry about the fact that this is an infinite-dimensional space of functions.)

Studying this operator leads to insights about the rate at which period-doubling bifurcations appear. The self-similar map g is a hyperbolic fixed point of the operator $\mathcal{R}$. Moreover, the differential of $\mathcal{R}$ at g has a one-dimensional eigenspace with eigenvalue $\delta \approx 4.69920166$, and there is a complementary invariant subspace on which it is contracting. As is always the case with a hyperbolic fixed point (Section 9.5), the contracting space for the differential of $\mathcal{R}$ corresponds to a surface consisting of functions positively asymptotic to g under iteration of $\mathcal{R}$, the stable manifold. It can be described intrinsically as the set of maps for which the orbit of the critical point is ordered in the same way as for g. In fact, S is the limit of surfaces S_n that consist of functions for which the critical point is 2^n-periodic. Note that $\mathcal{R}(S_{n+1}) = S_n$. This implies that the distance from S_n to S is approximately $1/\delta^n$ for large n, because δ is the rate at which the S_n are "pushed away" from S under $\mathcal{R}$. (See Figure 11.3.1.)

The method described in Section 9.5 suggests how to get at a standard one-parameter family of functions that contains g: For any one-parameter family of functions that crosses S close to g define $\tilde{\mathcal{R}}(f_\lambda) := \mathcal{R}(f_{\lambda/\delta})$. This "normalizes away" the stretching under $\mathcal{R}$ by δ in the one-dimensional direction. As a result there is a fixed point of this adapted operator, which gives an $\mathcal{R}$-invariant one-parameter family that includes g.

---

[7] Oscar Lanford III, A Computer-Assisted Proof of the Feigenbaum Conjectures, *Bulletin of the American Mathematical Society* (N.S.) **6**, no. 3, (1982), 427–434.

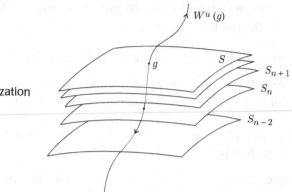

Figure 11.3.1. The renormalization operator.

The point of this last observation is that distances between successive period doubling bifurcations in a one-parameter family can be compared via such a renormalization. Since we have just described the limiting behavior as universal, any period doubling cascade happens with asymptotic ratio $1/\delta$ between successive doublings.

11.3.2 Appearance of Powers of 2

The verification of the assertions in the preceding discussion requires formidable analysis. To show that some more basic features are even more universal requires no more than the Intermediate-Value Theorem. We now show that the order in which powers of 2 appear in the quadratic family is the only order possible for any family.

Definition 11.3.1 Consider a continuous map $f \colon I \to I$ of an interval I. We say that an interval $J \subset I$ *covers* (or f-*covers*) $K \subset I$ (under f) if $K \subset f(J)$, and we denote this situation by $J \to K$. If we consider a finite collection of subintervals of I, then the graph with these as vertices and arrows determined by the f-covering relation is called the associated *Markov graph*.

If $J \to K$, then the covering takes place in the obvious way:

Lemma 11.3.2 *If J, K are intervals, K is closed, and $J \to K$, then there exists a closed interval $L \subset J$ such that $f(L) = K$.*

Proof Write $K = [a, b]$ and $c := \max f^{-1}(\{a\})$. Take $L = [c, d]$ with $d := \min((c, \infty) \cap f^{-1}(\{b\}))$, if this is defined. Otherwise, $L = [c', d']$ with $c' := \max((-\infty, c) \cap f^{-1}(\{b\}))$ and $d' := \min((c', \infty) \cap f^{-1}(\{a\}))$ is as desired. $\square$

Thus if $J \to K$, then there are several intervals $L_1, \ldots, L_k \subset J$ with pairwise disjoint interiors such that $f(L_i) = K$. Sometimes we write this as $J \rightrightarrows K$ with k arrows, if k is the maximal number of such subintervals L_i. These L_i are called *full components* associated to $J \to K$. Note that the preimage of K in J may contain infinitely many intervals, even though there are only finitely many full components by compactness.

The next two lemmas provide a connection between this covering relation and periodic points.

Lemma 11.3.3 *If $J \to J$, then f has a fixed point $x \in J$.*

Proof If $J = [a, b]$, then $J \to J$ implies that there are $c, d \in J$ with $f(c) = a \leq c$ and $f(d) = b \geq d$. By the Intermediate-Value Theorem, $f(x) - x$ has a zero in J. $\square$

Lemma 11.3.4 *If $I_0 \to I_1 \to I_2 \to \cdots \to I_n$, then $\bigcap_{i=0}^{n} f^{-i}(I_i)$ contains an interval Δ_n such that $f^n(\Delta_n) = I_n$.*

Proof Let $\Delta_0 = I_n$ and recursively take a full component $\Delta_i \subset I_{n-i}$ associated to $I_{n-i} \to \Delta_{i-1}$. $\square$

By virtue of Lemma 11.3.3 this has the

Corollary 11.3.5 *If $I_0 \to I_1 \to I_2 \to \cdots \to I_{n-1} \to I_0$, then there exists $x \in \text{Fix } f^n$ such that $f^i(x) \in I_i$ for $0 \leq i < n$.*

Lemma 11.3.6 (Barton–Burns) *If f has a nonfixed periodic point, then it has a nonfixed point of period 2.*

Proof We show that if $x \neq f^2(x)$ has period p, then f has a nonfixed point of period less than p.

Consider the intervals $I_1, \ldots, I_{p-1}$ whose endpoints are adjacent iterates of x. For each $i \in \{1, \ldots, p-1\}$ there is a $j \neq i$ such that $I_i \to I_j$ since the endpoints are not 2-periodic. Thus, there is a nontrivial loop in the Markov graph that visits $k \leq p-1$ intervals. By Corollary 11.3.5 there is a nonfixed k-periodic point. $\square$

This implies that powers of 2 appear in increasing order:

Theorem 11.3.7 *If $f : I \to I$ is a continuous map of a closed interval with a periodic point of period 2^n, then f has periodic points of all periods 2^m for $m \leq n$.*

Proof If $m = 0$, use Lemma 11.3.3; otherwise, apply Lemma 11.3.6 to $f, f^2, \ldots, f^{2^{n-2}}$. $\square$

11.3.3 Further Period-Forcing Relations

There is a nice complementary result that is also of historical interest, because the paper in which it was published introduced the term "chaos" to a wide audience.[8]

---

[8] The paper was Tien-Yien Li and James A. Yorke, Period Three Implies Chaos, *American Mathematical Monthly* **82**, no. 10 (1975), 985–992. Earlier and more precise uses of chaos, but not necessarily deterministic chaos, appeared in Norbert Wiener, The Homogeneous Chaos, *American Journal of Mathematics* **60**, no. 4 (1938), 897–936, and Norbert Wiener and Aurel Wintner, The Discrete Chaos, *American Journal of Mathematics* **65**, no. 2 (1943), 279–298. Scientists were more likely to see Robert M. May, Biological Populations with Non-Overlapping Generations: Stable Point, Stable Cycles and Chaos, *Science* **186** (1974), 645–647, or Simple Mathematical Models with Very Complicated Dynamics, *Nature* **261** (1976), 459–467, who emphatically credits Yorke for the minting of the term.

Proposition 11.3.8 *If* $f: I \to I$ *is a continuous map of a closed interval with a periodic point of period 3, then* f *has periodic points of all periods.*

Proof Label the points of the period-3 orbit by $\{x_1 < x_2 < x_3\}$. Consider the intervals $I_1 = [x_1, x_2]$ and $I_2 = [x_2, x_3]$. Suppose $f(x_2) = x_3$. Then $f^2(x_2) = x_1$ and hence I_2 f-covers both I_1 and I_2, and I_1 covers I_2. If $f(x_2) = x_1$, we relabel I_1 and I_2 to get the same conclusion. Thus the Markov graph associated to I_1, I_2 contains the graph

(11.3.1)
$$I_1 \rightleftarrows I_2,$$
$$\circlearrowleft$$

and for any $n \in \mathbb{N}$ we have a loop $I_1 \to I_2 \to I_2 \to \cdots \to I_2 \to I_1$ (with $n-1$ occurrences of I_2) which, by Corollary 11.3.5, gives a periodic point of period n which evidently cannot have any smaller period. $\square$

This result is a special case of a much more general one, which was discovered earlier:

Theorem 11.3.9 (Sharkovsky Theorem)[9] *If* $I \subset \mathbb{R}$ *is a closed interval,* $f: I \to I$ *continuous with a periodic point of prime period* p, *and* $q \lhd p$, *where "*$\lhd$*" is defined by*

$$1 \lhd 2 \lhd 2^2 \lhd 2^3 \lhd \cdots \lhd 2^m \lhd \cdots \lhd 2^k(2n-1) \lhd \cdots \lhd 2^k \cdot 3 \lhd \cdots \lhd 2 \cdot 3 \lhd \cdots \lhd$$

$$2n-1 \lhd \cdots \lhd 9 \lhd 7 \lhd 5 \lhd 3,$$

then f *has a periodic point of prime period* q.

The two special cases (Theorem 11.3.7 and Proposition 11.3.8) contain the essential ingredients for the proof of the Sharkovsky Theorem. One can develop this theme further: While the presence of any particular period other than 3 does not imply that all other periods exist, one can show, for example, that if $f: [0, 1] \to [0, 1]$ has a periodic orbit $\{x_1 < x_2 < x_3 < x_4\}$ such that $f(x_i) = x_{i+1}$ for $i < 4$ and $f(x_4) = x_1$, then f has periodic points of all periods. This leads to the study of which "patterns" of periodic orbits force the existence of other periodic orbits.

11.3.4 The Feigenbaum–Misiurewicz Attractor

The map f_{λ_∞} at the end of the period doubling cascade should be expected to have a great deal of self-similarity in some way. The discussion of Feigenbaum universality suggests that it is in some essential way invariant under renormalization. We now establish in an explicit way some important self-similarity of the dynamics.

We will do this in two ways. First we briefly describe where the periodic points of f_{λ_∞} are located relative to each other. Then we describe the dynamics on an important invariant set that contains no periodic points.

---

[9] Alexander N. Sharkovskiĭ, Coexistence of Cycles of a Continuous Map of the Line into Itself, *Ukrainskiĭ Matematicheskiĭ Zhurnal* **16**, no. 1 (1964), 61–71; English translation: *International Journal of Bifurcation and Chaos in Applied Sciences and Engineering* **5**, no. 5 (1995), 1263–1273; On Cycles and Structure of a Continuous Map, *Ukrainskiĭ Matematicheskiĭ Zhurnal* **17**, no. 3 (1965), 104–111.

1. Combinatorics of Periodic Points. The combinatorial pattern of periodic points was already noted after Theorem 11.2.2: 11, 1212, 14241424, 1848284818482848, and so on. Moreover, one can determine the dynamical ordering of each of these orbits. For example, the period-4 orbit naturally has a left half and a right half, and these two halves are interchanged by f_{λ_∞} because they are "dragged along" with the two points of the period-2 orbit. Likewise, the left and right halves of any of the period-2^n points are interchanged by f_{λ_∞}. Self-similarity begins to show when we describe, for example, how the left half of the period-8 orbit is mapped to itself under $f_{\lambda_\infty}^2$, or how it is mapped to the right half by f_{λ_∞}. Because the two "packets" associated with the two left points of the period-4 orbit are "dragged along" with those points, the period-8 orbit is mapped in packets of two. Specifically, if we label the points $x_1, \ldots, x_8$ according to their ordering in the interval, then $f_{\lambda_\infty}(\{x_1, x_2\}) = \{x_i, x_{i+1}\}$ with $i = 5$ or $i = 7$. Whether it is 5 or 7 is determined from the period-4 orbit.

Therefore the period-2^n orbits for various n are all properly intertwined with each orbit mapped in "packets" that can be tracked recursively according to the combinatorics of the orbit of half the period. This is one way in which self-similarity appears.

2. The Feigenbaum–Misiurewicz Attractor. The self-similar dynamics in f_{λ_∞} is abstractly given in Problem 4.1.15 and can be described in more familiar terms as follows.

Definition 11.3.10 Let Ω_2^R be the space of one-sided 0-1-sequences as defined in Section 7.3.4. The map $A\colon \Omega_2^R \to \Omega_2^R$ given by

$$(A\omega)_i = \begin{cases} 1 - \omega_i & \text{if } \omega_i = 1 \text{ for all } j < i, \\ \omega_i & \text{otherwise} \end{cases}$$

of the space Ω_2^R is called the *dyadic adding machine*.

Theorem 11.3.11 *The map f_{λ_∞} has a closed invariant set that consists of isolated repelling periodic orbits of periods 2^n for $n = 0, 1, 2 \ldots$ (two for $n = 0$ and one orbit for each other period) and a Cantor set S the dynamics on which is topologically conjugate to the dyadic adding machine.*

Furthermore, S is exactly the set of accumulation points for the set of periodic points. S is also the ω-limit set (Definition 4.3.18) of any point that is not eventually periodic (that is, a preimage of a periodic point).

Sketch of proof An equivalent description of the preceding combinatorial analysis of periodic points is that of the intervals defined by a period-2^{n+1} orbit none containing a period-2^n point f_{λ_∞}-covers one containing a period-2^{n-1} point. We use this to construct S.

First consider the interval I whose endpoints are the period-2 points and let S^0 be a full component of $I \to I$. There are two intervals whose endpoints are on the period-4 orbit and that contain a period-2 point, I_1 on the left and I_2 on the right. We let S^1 be the union of one full component each of $I_1 \to I_2 \to I_1$ and $I_2 \to I_1 \to I_2$. Then S^1 has positive distance from the fixed point. S^2 is likewise obtained as a union of full components corresponding to the length-four loops associated with the four intervals obtained similarly from the period-8 orbit. Thus S^2 is separated from the points with periods 1 and 2. Likewise we obtain sets S^n consisting of 2^{n-1} intervals separated from the points of period up to 2^{n-1}. Next we let S be the boundary of $\bigcup_{m=1}^{\infty} \bigcap_{n=m}^{\infty} S^n$, that is, S is the collection of nonperiodic points in the closure of the set of periodic points. Denote by y_n the right endpoint of the leftmost interval in S^n and let $S^n = \{x \in S \mid x \leq y_n\}$.

Define $h \colon S \to \Omega_2^R$ by $h(S^n) = \{\omega \in \Omega_2^R \mid \omega_1 = \cdots = \omega_n = 1\}$ and $h \circ f_{\lambda_\infty} = A \circ h$. Then h is continuous and surjective, that is, the adding machine A is a factor of $f_{\lambda_\infty}\restriction_S$. In this sense the dynamics of A on Ω_2^R is contained in that of f_{λ_∞} on S. Furthermore, the size of the intervals in the construction above tends to 0, and this implies that h is injective. $\square$

In light of Theorem 11.3.11, it is natural to pay attention to the intrinsic dynamics of the dyadic adding machine.

Proposition 11.3.12 *The dyadic adding machine is uniquely ergodic.*

Proof The phase space of the dyadic adding machine A has a nested structure. At the nth step there are 2^n disjoint Cantor subsets called the *cylinders of rank n*, which are both closed and open and which are cyclically interchanged by the dynamics. Every function constant on each cylinder of a given rank is continuous, and every continuous function is the uniform limit of functions constant on cylinders of certain ranks. For a function ϕ that is constant on cylinders of rank n one has

$$\text{const.} = \frac{1}{2^n} \sum_{i=0}^{2^n-1} \phi \circ A^i = \lim_{N \to \infty} \frac{1}{N} \sum_{i=0}^{N} \phi \circ A^i.$$

Uniform approximation of an arbitrary continuous function by functions constant on cylinders yields uniform convergence of averages. $\square$

Thus the adding machine has some common features with irrational rotations of the circle. In some sense it is even simpler (small pieces return back precisely, not approximately, as do small intervals on the circle), but its unique ergodicity is less "perfect" because even iterates are not uniquely ergodic.

The following fact shows that in a sense the dyadic adding machine is the only model of relatively simple nonperiodic recurrent behavior for interval maps.[10]

Theorem 11.3.13 *Suppose f is a continuous self-map of an interval with zero topological entropy and S is a closed f-invariant set without periodic points and with a dense orbit (that is, topologically transitive; Definition 4.1.3). Then the restriction of f to S is topologically conjugate to the dyadic adding machine.*

---

[10] The proof is contained in the proof of Theorem 15.4.2 of Katok and Hasselblatt, *Introduction to the Modern Theory of Dynamical Systems*, Cambridge University Press, Cambridge, 1995.

11.4 HYPERBOLIC AND STOCHASTIC BEHAVIOR

We now change our point of view and study the dynamics of quadratic maps not by way of the combinatorics of orbits, but by their stability or instability. This returns us to the questions posed in Section 11.1.2.

11.4.1 Hyperbolic Cantor Repeller

Proposition 7.4.4 holds for all $\lambda > 4$; the assumption $\lambda > 2 + \sqrt{5}$ was made earlier to make the proof feasible in our setting. For the remaining values the argument is more involved and uses negativity of the Schwarzian derivative (11.2.4). Therefore one has, in fact,

Proposition 11.4.1 *There is a homeomorphism $h\colon \Omega_2^R \to \Lambda := \bigcap_{n \in \mathbb{N}_0} f^{-n}([0, 1])$ such that $h \circ \sigma^R = f \circ h$, where $f\colon \mathbb{R} \to \mathbb{R}$, $x \to \lambda x (1 - x)$ with $\lambda > 4$.*

In fact, the proof of Proposition 7.4.4 shows that the set Λ is a hyperbolic repeller (Definition 10.1.2).

This is a helpful model to keep in mind when discussing the possible behaviors of quadratic maps. For parameters $\lambda < \lambda_\infty$, the asymptotics are straightforward: Orbits are attracted to the 2^n-periodic orbit with the largest n available. In the previous section the dynamics of f_{λ_∞} was described in terms of periodic points (all repelling) and a self-similar invariant set S.

It remains to study λ between λ_∞ and 4. This requires sophisticated analysis well beyond the scope of this book, but the results of this analysis can be described well enough. We begin by describing one of two main types of behaviors.

The first step gives a general outline of increasing complexity

Theorem 11.4.2 [11] *The topological entropy of the map f_λ is nondecreasing, is zero for $\lambda \le \lambda_\infty$, and is positive for $\lambda > \lambda_\infty$.*

This implies[12]

Corollary 11.4.3 *For $\lambda > \lambda_\infty$, the map f_λ has infinitely many periodic orbits whose periods are not powers of 2.*

11.4.2 Periodic Attractor and Markov Repeller

Recall from Section 11.1.2 that a hyperbolic quadratic map is one for which all recurrence takes place on periodic orbits and an invariant Cantor set that may be empty and is a null set (Definition 7.5.3). There is only one attracting periodic orbit, and for $\lambda < 4$ any hyperbolic map has such an orbit. Every orbit not in the Cantor set is positively asymptotic to this orbit. The Cantor set is a hyperbolic repeller (Definition 10.1.2) and it is nonempty for $\lambda > \lambda_\infty$.

---

[11] Welington de Melo and Sebastian van Strien, *One-Dimensional Dynamics*, Springer-Verlag, Berlin, 1993.

[12] Positive topological entropy implies the existence of a (one-dimensional) horseshoe for an iterate, which in turn implies the corollary; see Katok and Hasselblatt: *Introduction to the Modern Theory of Dynamical Systems*, Corollary 15.2.4.

In this hyperbolic situation a straightforward numerical exploration would show only the attracting periodic orbit. Although the dynamics on the Cantor set is complicated, a computer will miss it because it is a null set. Even if the initial value for a numerical computation were in the Cantor set, a generic roundoff error would eject the computed orbit from it and produce an orbit that tends to a periodic one.

Theorem 11.4.4 (Graczyk–Świątek) [13] *The set of $\lambda \in [0, 4]$ for which f_λ is hyperbolic is open and dense.*

The difficult part of this theorem is density, since openness follows almost directly from the definition of hyperbolicity. This statement illustrates the need for mathematical insight over numerical exploration, because this set of parameters does not look prominent in the bifurcation diagram (Figure 11.2.3), where only a few windows (light vertical strips) are visible. The reason is twofold. Except for the smallest periods, any one of these windows is fantastically narrow. And orbits of rather large period not only contain enough points to blend into the shaded environment, but it also takes a computer many more iterations to get close to the orbit. Nevertheless, the bifurcation diagram suggests that there is another prominent set of parameters.

11.4.3 Stochastic Behavior

There can be complicated behavior of a type rather different from the kinds just discussed, namely, behavior where complex dynamics coexists with equidistribution on the interval. The overall complexity of the picture is mind-boggling; however many phenomena appear in rather exceptional situations. Still, there is one type of nonuniformly hyperbolic behavior that is particularly important both because it appears for a nonnull set of parameter values and because of its intrinsic structure.

1. The Tent Map. In order to introduce this kind of behavior consider first the model situation where $g: [0, 1] \to [0, 1]$ is the "tent" map

$$g(x) = \begin{cases} 2x, & 0 \le x \le 1/2, \\ 2 - 2x, & 1/2 \le x \le 1. \end{cases}$$

It is clear (see Section 7.1.1) that this map preserves measure in the following sense: For any interval $I \subset [0, 1]$, the measure of $g^{-1}(I)$ agrees with the measure of I. (Here the measure of a disjoint union of intervals is defined to be the sum of their lengths, as in Definition 4.2.2.) Except for the up–down reversal on the right half of the interval, the tent map looks exactly like the expanding map. Indeed, one can trace through the arguments that prove equidistribution of orbits for the doubling map (Theorem 7.5.6) to obtain equidistribution of orbits for the tent map. This engenders a fair amount of dynamical complexity, but this complexity is necessarily

[13] Jacek Graczyk and Grzegorz Świątek, Generic Hyperbolicity in the Logistic Family, *Annals of Mathematics* (2) **146**, no. 1 (1997), 1–52.

distributed over the entire interval, in contrast to the hyperbolic behavior discussed above.

2. Chebyshev–von Neumann–Ulam Map. The preceding observation is relevant to the quadratic family because the tent map g is topologically conjugate to the quadratic map $f_4\colon x \mapsto 4x(1-x)$ via $h(x) = \sin^2(\pi x/2)$. The map f_4 represents the *second Chebyshev polynomial*. Pafnutij L. Chebyshev was aware of the conjugacy but did not consider its dynamical implications, unlike von John von Neumann and Stanislav Ulam. This is the reason that f_4 is sometimes called the *von Neumann–Ulam map*.

The expression for f_4 is easily the simplest single algebraic formula (without absolute values, reductions modulo 1, or other mathematical trickery) that produces a map with genuine "stochastic" behavior. The conjugacy with the tent map is simply a matter of a trigonometric identity. This does not mean that f_4 preserves measure in the same sense, but rather that, due to the mild distortion under the coordinate change h, there is a positive "density" function $\rho\colon [0, 1] \to \mathbb{R}$, given in this case by $\pi\rho(x) = 1/\sqrt{x(1-x)}$, such that f_4 preserves the measure obtained via Definition 4.2.2 by taking "lengths" of intervals to be $l_\rho(I) := \int_I \rho(x)\,dx$. This is the first nontrivial example of what we refer to as having *weighted uniform distribution* or *stochasticity* (see Section A.3). It implies that with respect to information about being in the left or right half of the interval we have (via the tent map, which is materially the same as E_2) the same statistical complexity as the coin toss in Section 7.5.1. On one hand, this gives various forms of the law of large numbers. On the other hand, this means also that there are points that have no long-term average at all, as described in Section 7.5.1.[14]

For purposes of classifying quadratic maps we allow the density to be nonnegative, and we do not require the set of zeros to be a null set.

3. Mechanisms of Stochasticity. The presence of the critical point $1/2$ where the derivative of a quadratic map is zero is an obvious obstruction to local expansion, which is needed for hyperbolic repellers and stochastic behavior. In the hyperbolic cases, this point is simply absorbed by the basin of attraction and is thus harmless for the repeller, which is disjoint from it and its preimages. For the Chebyshev–von Neumann–Ulam map, the critical point is mapped to zero, which is an expanding fixed point. Thus points close to the critical point (even if recurrent) have enough time to recover the local expansion before coming back into the area with small derivatives. This mechanism can be increasingly generalized and modified to produce ever larger sets of parameters with stochastic behavior.

4. Ruelle Maps. The most straighforward generalization of the Chebyshev–von Neumann–Ulam situation appears when *an iterate of the critical point is a repelling periodic point*. The set of such parameters is countable and the invariant density is piecewise smooth.

---

[14] This was also observed by Edward Lorenz, who used the quadratic map as a metaphor for the possibilities of weather and concluded that it is not automatic that a "climate" exists, where climate is understood to mean long-term average of weather data.

5. Misiurewicz Maps. What makes the previous situation work is that the critical point is not recurrent. Misiurewicz used this observation to construct the first uncountable (albeit still null) set of parameter values producing stochastic behavior. Misiurewicz's condition is that *the critical point is not recurrent*, that is, the ω-limit set of $1/2$ does not contain $1/2$. In this case the ω-limit set of the critical point is a hyperbolic repeller, and there is still an invariant density, which is smooth on a countable set of intervals but may be discontinuous on a Cantor set. The basic reasoning for stochasticity in the Misiurewicz situation is still similar to the Chebyshev–von Neumann–Ulam case: Any recurrent point that comes close to the critical one has enough time to recover expansion while wandering near the repelling ω-limit set.

6. Jakobson– and Collet–Eckmann Maps. To produce a nonnull set of parameters with stochastic behavior one has to consider situations when the critical point returns arbitrarily close to itself. The model for stochasticity in such a situation was described by Collet and Eckmann, but it was Jakobson in his landmark 1980 work[15] who showed that stochastic behavior appears on nonnull set and is in fact "prevalent" among parameter values close to 4 and to the Ruelle and Misiurewicz values.

The Jakobson–Collet–Eckmann mechanism allows the critical point to return close to itself but sufficiently unfrequently so that, *between* the returns, points sufficiently close to return (but not so close as to have the derivative almost annihilated at the previous round) recover enough expansion. A sophisticated inductive procedure shows that for enough points expansion prevails on all time scales over occasional contractions caused by returns.

The most important observation, due to Jakobson, is the *parameter exclusion method*, which controls parameter values for which a too-close accidental return of the critical point produces attracting periodic orbits and thus locks the contraction forever. The heart of the method is an estimate showing that, when this happens, after many iterates the speed of moving the corresponding iterate of the critical point is so high that the "dangerous zone" where the locking may appear at the given return is passed very quickly.

In the Jakobson–Collet–Eckmann case the invariant density tends to be highly discontinuous but the main qualitative features seen in the tent map still prevail.

---

[15] Michael V. Jakobson, Absolutely Continuous Invariant Measures for One-Parameter Families of One-Dimensional Maps, *Communications in Mathematical Physics* **81**, no. 1 (1981), 39–88.

CHAPTER 12

Homoclinic Tangles

Our study of complicated dynamics in the course developed mostly by looking at examples. One reason is that this is an effective way to develop important concepts in a natural fashion. But another reason is that those examples almost fully represent the range of phenomena responsible for chaotic behavior. We now return to the horseshoe and explain why this seemingly particular example is an important mechanism that gives rise to chaotic dynamics for some orbits. Specifically, we show how it appears in real systems and that it does so often, and we describe how it has been used as an important tool in deciding fundamental questions in dynamics. We first present the principal mechanism that produces horseshoes and then give an account of the various ways this scenario arises in real problems.

12.1 NONLINEAR HORSESHOES

For the discussion of the horseshoe in Section 7.4.4 it was convenient to assume linearity, but as we mentioned in Section 10.2.6, it is not essential. We now introduce nonlinear horseshoes.

It is easy and useful to define horseshoes in arbitrary dimension, but there are several reasons to restrict ourselves to the planar situation. It makes it easier to picture the arguments, and, accordingly, in the development of the theory of dynamical systems the planar case played the leading role. Finally, it is in dimension two that the full topological entropy of a dynamical system can be accounted for by looking only at horseshoes in it.

In the definition we use the coordinate projections $\pi_1 \colon (x, y) \mapsto x$ and $\pi_2 \colon (x, y) \mapsto y$ on $\mathbb{R}^2$.

Definition 12.1.1 If $U \subset \mathbb{R}^2$ is open, then a rectangle $\Delta = D_1 \times D_2 \subset U \subset \mathbb{R}^2$ (where D_1 and D_2 are intervals) is said to be a *horseshoe* for a diffeomorphism $f \colon U \to \mathbb{R}^2$ if $\Delta \cap f(\Delta)$ contains at least two connected components Δ_0 and Δ_1

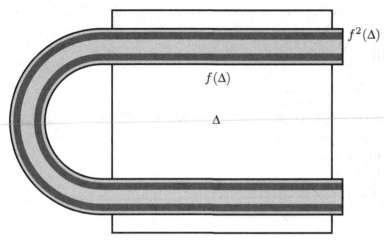

Figure 12.1.1. The horseshoe.

such that, if we write $\Delta_i = f(\Delta_i')$ for $i = 1, 2$ and $\Delta' = \Delta_0 \cup \Delta_1$, then

(1) $\pi_2(\Delta_i') = D_2$,

(2) $\pi_1\big|_{f(\Delta_i' \cap (D_1 \times \pi_2(z)))}$ is a bijection onto D_1 for any $z \in \Delta_i'$,

(3) $\pi_2(\Delta') \subset \operatorname{int} D_2$, $\pi_1(f^{-1}(\Delta')) \subset \operatorname{int} D_1$,

(4) $D(f\big|_{f^{-1}(\Delta')})$ preserves and expands a horizontal cone family on $f^{-1}(\Delta')$,

(5) $D(f^{-1}\big|_{\Delta'})$ preserves and expands a vertical cone family on Δ' (Definition 10.1.4).

In Figure 12.1.1, Δ_0 and Δ_1 are the two lightly shaded horizontal rectangles that make up $\Delta \cap f(\Delta)$. Their preimages are vertical rectangles that go all the way through Δ vertically. This is the content of the first item. The second requirement says that these rectangles themselves must go entirely across Δ horizontally, and without too much wiggling. The third requirement is that these strips stay away from the top and bottom part of the boundary, and their preimages stay away from the sides. In other words, there is room to spare in either direction. This is useful for stability under perturbation. The last two conditions are horizontal expansion and vertical contraction combined with the requirement that almost horizontal lines remain almost horizontal, and almost vertical lines remain almost vertical.

This definition makes relatively mild and robust requirements: A horseshoe looks like the one in Figure 12.1.1 but is allowed to be a little deformed and have variable expansion and contraction. In Section 12.3 we will see how such horseshoes arise naturally. By structural stability (Section 10.2.6) the dynamics on a linear horseshoe and a small perturbation are identical up to a homeomorphic coordinate change that moves no point by much. This implies in particular that our results about the orbit structure of linear horseshoes also hold for their perturbations (coding in Proposition 7.4.6 and periodic orbit growth and density as well as transitivity in Corollary 7.4.7).

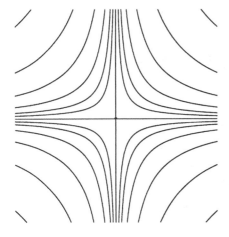

Figure 12.2.1. Hyperbolic fixed point.

12.2 HOMOCLINIC POINTS

We now describe a scenario and some of its outstanding features that lead to a description of why horseshoes are so ubiquitous.

As the starting point we take the linear hyperbolic map $A: \mathbb{R}^2 \to \mathbb{R}^2$, $A(x, y) = (2x, y/2)$. (See Figure 12.2.1.) The y-axis consists of points asymptotic to the origin in positive time, and the x-axis consists of points asymptotic to the origin in negative time, while all other points move along hyperbolas $xy =$ const.

Now we consider a nonlinear diffeomorphism f that is close to A in a neighborhood of the origin. By the Stable Manifold Theorem (Theorem 9.5.2) there is a *stable* curve (replacing the y-axis for A) of points asymptotic to 0 and an *unstable* curve of points negatively asymptotic to 0. This description implies that both curves are invariant. Furthermore, neither of these curves can self-intersect. If the stable curve did, the images of the resulting loop would accumulate at 0, and that would mean that near 0 the set of points asymptotic to 0 is not a curve, contrary to Theorem 9.5.2.

Suppose that far from the origin the nonlinearities are such that these curves bend enough to intersect at a point p and form a nonzero angle (see Figure 12.2.2). Such a point p is called a *transverse homoclinic point* for the fixed point 0. It is a homoclinic point as in Definition 2.3.4, and this situation corresponds to breaking

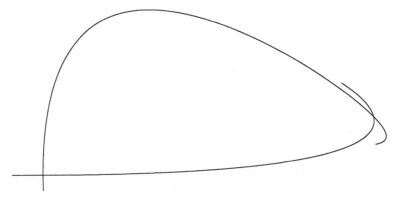

Figure 12.2.2. Transverse homoclinic point.

up a homoclinic loop such as in Section 6.2.2. (Contrast Figure 6.2.2 and Figure 4.3.3 with Figure 12.2.2.)

It is fair to ask how we know that this can be accomplished. One way of looking at it is to alter the description slightly by starting from A in a neighborhood of 0 only and then extending the definition in such a way that the images/preimages of the horizontal and vertical segments through 0 follow the desired pattern. On the other hand, here is an explicit map of the plane (that preserves area) with the desired property:

$$\begin{pmatrix} x \\ y \end{pmatrix} \mapsto \begin{pmatrix} 3(x + (x - y)^2) \\ \frac{1}{3}(y + (x - y)^2) \end{pmatrix}.$$

This *Cremona map*[1] is invertible; indeed, the inverse is equally simple:

$$\begin{pmatrix} x \\ y \end{pmatrix} \mapsto \begin{pmatrix} \frac{x}{3} - (3y - \frac{x}{3})^2 \\ 3y - (3y - \frac{x}{3})^2 \end{pmatrix}.$$

Finally, as we remark at the end, this very picture appears naturally in all kinds of real-world problems, although often in a slightly different way: In a simplified problem the two curves bend to form a smooth loop in the first quadrant, and when one perturbs this problem to the real one, a nonzero intersection angle results. Deformations of an elliptic billiard table would be a case in point, and perturbations of the time-1 map for the mathematical pendulum (Section 6.2.2) can achieve the same end.

The complications that this causes begin to become apparent along the orbit $\mathcal{O}(p)$ of the homoclinic point p. Since p lies on the stable and on the unstable curve, so does every point of the orbit. Therefore, all of these points are also homoclinic points. Since f is a diffeomorphism, the intersection angles are always nonzero. These points accumulate on 0 along both curves.

Any two adjacent points of $\mathcal{O}(p)$ define a loop with a stable and an unstable boundary piece. As these loops accumulate on 0 (from above) in positive time they are compressed vertically and stretched horizontally. Since the unstable curve cannot self-intersect, these ever longer lobes follow the unstable curve ever more closely, as in Figure 12.2.3. In negative time, a complementary accumulation occurs with vertical stretching. This gives the complete picture in Figure 12.2.3. Note from the picture that this gives rise to "second-generation" homoclinic points, and recursively to ever higher complexity.

Heteroclinic tangles arising from a transverse heteroclinic point are similarly complex and occur frequently as well; for example, they appear in regions of instability of twist maps (Corollary 14.3.3). However, transverse heteroclinic points may appear without engendering much recurrence. Such is the case for the gradient flow of the height function on a torus (bagel-shaped surface) that is standing up almost but not quite vertical.[2] In this case no appreciable complexity appears. Tangles appear in the case of heteroclinic *loops*. Because of the fundamental similarity, we do not treat heteroclinic tangles here.

---

[1] This map was brought to our attention by Alex Dragt.
[2] See Katok and Hasselblatt, *Introduction to the Modern Theory of Dynamical Systems*, Section 1.6.

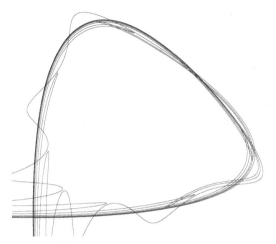

Figure 12.2.3. Tangles.

12.3 THE APPEARANCE OF HORSESHOES

We now show that the stretching and folding of the invariant curves produce the stretching and folding of a rectangle into a horseshoe.

Theorem 12.3.1 (Birkhoff–Smale) *Let $U \subset \mathbb{R}^2$ be a neighborhood of 0 and $f: U \to \mathbb{R}^2$ an embedding (Definition A.2.5) for which 0 is a hyperbolic fixed point with a transverse homoclinic point p. Then in an arbitrarily small neighborhood of 0 there exists a horseshoe for some iterate of f.*

Remark 12.3.2 This is a remarkable result because it tells us that any hyperbolic fixed point can be affected by "faraway" circumstances in such a way that the dynamics nearby has the full complexity of the horseshoe.

Proof Figure 12.3.1 tells the whole story. If one takes a small rectangle in the first quadrant with two sides just beyond the invariant curves, then the stretching and folding of the tangles pull the rectangle along and cause an overlap after several iterations. The picture shows that the geometry is right, and the iterations of f force the required horizontal stretch and vertical contraction.

To make the notations simpler we assume that near zero the invariant curves coincide with the coordinate axes (this can be arranged by a coordinate change). Then take a rectangle $\Delta = D_1 \times D_2$, where D_1 and D_2 are small intervals containing 0 in their interior. In Figure 12.3.1 we chose them to have the left endpoint of D_1 and the lower endpoint of D_2 especially close to 0.

There is a preimage $p' = f^{-n}(p)$ in the interior of D_1, and the nearby portion of a stable curve through q' is almost vertical. The pieces of unstable loops nearby are all close to horizontal. On one hand, for any $n \in \mathbb{N}$, the origin is contained in one component Δ_0 of $f^n(\Delta) \cap \Delta$. On the other hand, taking n large enough there will be a second component Δ_1 satisfying the first three conditions in Definition 12.1.1.

That the expansion and contraction conditions in Definition 12.1.1 are satisfied is clearly no problem, because most of the applications of f that are needed to produce the right geometry are near 0 and therefore stretch and contract almost

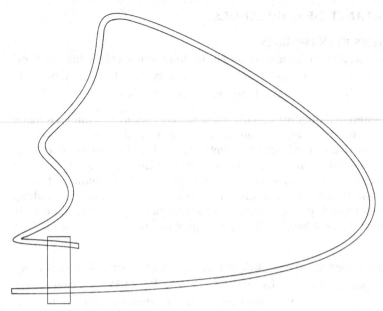

Figure 12.3.1. Obtaining a horseshoe.

precisely like the linear part. If needed, one can shrink D_1 and D_2 slightly to increase the number of iterates of f. All additional iterates contribute uniform distortion. □

In realistic situations rather few iterations (on the order of 10) are needed, but already these few produce extreme stretching and contraction. Indeed, in Figure 12.3.1 one sees the enormous stretch directly, and an exact rendition would show the elongated rectangle as a single curve due to a corresponding rate of contraction in the stable direction.

Figure 12.3.2 shows a better-proportioned horseshoe in the Cremona map that we obtained by searching in a neighborhood of 0.

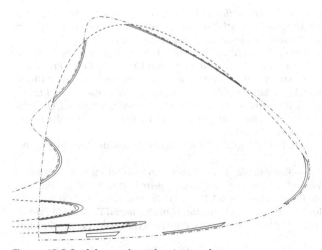

Figure 12.3.2. A horseshoe due to tangles.

12.4 THE IMPORTANCE OF HORSESHOES

12.4.1 From Tangles to Horseshoes

Homoclinic tangles were first observed by Henri Jules Poincaré in his work on the three-body problem, and it gave the first indication that there may be truly complex dynamics in the solar system. Later he described this situation as follows:

> When we try to represent the figure formed by these two curves and their infinitely many intersections, each corresponding to a doubly asymptotic solution, these intersections form a type of trellis, tissue, or grid with infinitely fine mesh. Neither of the two curves must ever cut across itself again, but it must bend back upon itself in a very complex manner in order to cut across all of the meshes in the grid an infinite number of times.
>
> The complexity of this figure is striking, and I shall not even try to draw it. Nothing is more suitable for providing us with an idea of the complex nature of the three-body problem, and of all the problems of dynamics in general, where there is no uniform integral.[3]

Several decades later, George David Birkhoff proved that in this situation there are many periodic points near the fixed point. During World War II, Mary Lucy Cartwright and John Edensor Littlewood built upon the work of Poincaré in their analysis of radar circuitry tuned to parameters outside the normal operating range, which exhibited an erratic blinking of a control light in a phenomenon known as relaxation oscillations.[4] Also in the 1940s, Norman Levinson analyzed the van der Pol equation (relevant for vacuum tubes) and found infinitely many periodic solutions (combined with structural stability).[5] Around 1960 he brought this to the attention of Steven Smale, who extracted from this work the geometric picture of the horseshoe as shown in Figure 12.1.1 and proved Theorem 12.3.1 in the 1960s.[6] Although horseshoes occur in higher dimension, this entire history played out in two dimensions.

It should be emphasized that the horseshoes obtained from tangles are null sets, so the presence of complexity in such a dynamical system may be confined to a set of the kind we are usually prepared to neglect. In other words, tangles provide no guarantee that complicated orbits will be ubiquitous in a given system.

---

[3] In Chapter 33, §396, of Jules Henri Poincaré, *Les méthodes nouvelles de la mécanique céleste*, Paris, 1892–1899; English translation: *New Methods of Celestial Mechanics*, edited by Daniel Goroff, History of Modern Physics and Astronomy **13**, American Institute of Physics, New York, 1993.

[4] Mary Lucy Cartwright, Forced Oscillations in Nonlinear Systems, in *Contributions to the Theory of Nonlinear Oscillations*, Annals of Mathematics Studies **20**, Princeton University Press, Princeton, NJ, 1950, pp. 149–241; Mary Lucy Cartwright and John Edensor Littlewood, On Non-Linear Differential Equations of the Second Order. I. The Equation $\ddot{y} - k(1 - y^2)y + y = b\lambda k \cos(\lambda t + a)$, k Large, *Journal of the London Mathematical Society* **20** (1945), 180–189; John Edensor Littlewood, On Non-Linear Differential Equations of the Second Order. IV. The General Equation $\ddot{y} + kf(y)\dot{y} + g(y) = bkp(\phi)$, $\phi = t + \alpha$, *Acta Mathematica* **98** (1957), 1–110.

[5] Norman Levinson, A Second Order Differential Equation with Singular Solutions, *Annals of Mathematics* (2) **50** (1949), 127–153

[6] Steven Smale, A Structurally Stable Differentiable Homeomorphism with an Infinite Number of Periodic Points, in *Qualitative Methods in the Theory of Non-Linear Vibrations* (*Proc. Internat. Sympos. Non-Linear Vibrations, Vol. II, 1961*), Izdat. Akad. Nauk Ukrain. SSR, Kiev, 1963, pp. 365–366; Diffeomorphisms with Many Periodic Points, in *Differential and Combinatorial Topology* (*A Symposium in Honor of Marston Morse*), edited by Stewart S. Cairns, Princeton University Press, Princeton, NJ, 1965, pp. 63–80.

Their importance lies in providing the possibility of such complexity as well as in affecting the behavior of other orbits for long albeit finite time.

12.4.2 The Ubiquity of Horseshoes in Applications

Horseshoes are present in many applications. Numerous dynamical systems that arise directly from scientific questions have a transverse homoclinic point and hence contain a horseshoe. Furthermore, given any dynamical system with homoclinic tangles, all sufficiently small perturbations have homoclinic tangles as well, and hence also contain a horseshoe. This certainly implies physical importance because it means that once the differences between the model and reality are made small enough, a horseshoe in the model corresponds to a horseshoe in reality. This is the fundamental importance of all persistent or (structurally) stable phenomena. Aside from being relevant to perturbations of specific examples, the persistence of transverse homoclinic points also shows that among all dynamical systems those with homoclinic tangles are fairly common.

We emphasize once more that in every such situation this directly gives strong conclusions about the orbit complexity of the dynamical systems at hand. For example, it has positive topological entropy and exponential growth of periodic points with their period, and it contains a topological Markov chain as a subsystem. Of course, the complexity due to a horseshoe may be rather localized and might only be sustained on a null set. However, the mere possibility of exponential orbit complexity is interesting in itself (and firmly rules out that the system is integrable). Moreover, even though the horseshoe is a null set, some other orbits will be entrained with orbits in the horseshoe for long enough to acquire a fair sensitivity to initial conditions and other features of complex behavior for finite time intervals. In the quadratic family positive topological entropy is attributable to (one-dimensional) horseshoes, but it may be confined to a null set by periodic orbits that attract all other points. At the other extreme is the hyperbolic toral automorphism, where horseshoes abound, but any countable collection of these only amounts to a null set, while exponential orbit complexity fills the torus entirely.

Here are several concrete occurrences of tangles. Homoclinic tangles, and hence horseshoes, are generically present in twist maps (Definition 14.2.1), which arise in the study of various dynamical problems such as billiards, celestial mechanics, and the design of particle accelerators. The *Størmer problem* of the motion of charged particles (from solar wind) in the earth's magnetic field also exhibits homoclinic tangles.[7] This is known as the cause for "insolubility of the Størmer problem." There have also been models of transport phenomena in fluid mechanics that involve *heteroclinic* tangles, which arise from mutual intersections of the stable and unstable curves of two hyperbolic points. In these models the plume of lobes away from the tangled region (similar to what one would see beyond the left edge of Figure 12.2.3) looks strikingly similar to the

---

[7] Alex J. Dragt and John M. Finn, Insolubility of Trapped Particle Motion in a Magnetic Dipole Field, *J. Geophys. Res.* **81** (1976), 2327–2340.

turbulent wake photographed in experiments.[8] This is a prime example where the entrainment of transient orbits with those on the horseshoe produces significant effects.

One important instance where a horseshoe was used as an ingredient to settle a fundamental mathematical and physical question is provided by celestial mechanics. Until the 1960s astronomers universally assumed that without external influences there could not be capture in an n-body problem such as our solar system. Specifically, it was taken for granted that there is a symmetry between long-term behavior in positive and negative time. A body could not have an unbounded negative-time orbit without also having an unbounded positive-time orbit. In other words, the solar system could not "capture" another celestial body (if all bodies are treated as point masses). Conversely, this would mean that if we trust that the earth has always been in the solar system, then it cannot possibly be ejected from it without special external influences. In studying the three-body problem Alekseev dispelled this belief by producing a horseshoe and showing that its presence created the possibility of capture.[9] This is an example where the mere possibility of complexity is decisive, even if it is confined to a null set, that is, even if its consequences may have probability zero.

A more recent application of tangles to force orbit complexity via a horseshoe is a result of Knieper and Weiss that the geodesic flow on a ellipsoid (which is completely integrable and hence has very little orbit complexity) acquires homoclinic tangles (hence a horseshoe, positive topological entropy, exponential growth of closed geodesics, etc.) for some arbitrarily small deformations of the ellipsoid.[10] Here the distinction between complexity on a null set and pervasive complexity has raised questions that remain open: Can one construct such perturbations that have not only positive topological entropy, that is, the maximal possible orbit complexity is characterized by exponential behavior, but also positive *metric* or *measure-theoretic* entropy, which measures the *average* orbit complexity?

12.4.3 Planar Chaos Comes from Horseshoes

We close with a most striking result. If a planar dynamical system has positive topological entropy, then all of this entropy is due to horseshoes.[11]

[8] Vered Rom-Kedar, A. Leonard and Stephen Wiggins: An Analytical Study of Transport, Mixing and Chaos in an Unsteady Vortical Flow, *Journal of Fluid Mechanics* **214** (1990), 347–394.

[9] Vladimir Mihkaĭlovich Alekseev, On the Possibility of Capture in the Three-Body Problem with a Negative Value for the Total Energy Constant, *Akademiya Nauk SSSR i Moskovskoe Matematicheskoe Obshchestvo. Uspekhi Matematicheskikh Nauk* **24** no. 1 (1969), (145), 185–186.

[10] Gerhard Knieper and Howard Weiss: A Surface with Positive Curvature and Positive Topological Entropy, *Journal of Differential Geometry* **39** no. 2 (1994), 229–249; see also Gabriel P. Paternain, Real Analytic Convex Surfaces with Positive Topological Entropy and Rigid Body Dynamics, *Manuscripta Mathematica* **78**, no. 4 (1993), 397–402, and Victor J. Donnay, *Transverse Homoclinic Connections for Geodesic Flows, Hamiltonian Dynamical Systems (Cincinnati, OH, 1992)*, IMA Volues in Mathematics and its Applications **63**, Springer, New York, 1995, pp. 115–125.

[11] A. Katok, Nonuniform Hyperbolicity and Structure of Smooth Dynamical Systems, *Proceedings of the International Congress of Mathematicians, Warszawa* **2**, (1983), 1245–1254. See also the Supplement in Katok and Hasselblatt, *Introduction to the Modern Theory of Dynamical Systems*, Cambridge University Press, Cambridge, 1995.

Theorem 12.4.1 *For a surface diffeomorphism the topological entropy is arbitrarily well approximated by restrictions to a horseshoe.*

In this sense horseshoes are the *only* needed mechanism of chaos in the plane. This is one way in which the dynamics of exponentially complicated maps is exceptionally well understood in two dimensions.

This fact helps control the behavior of entropy under perturbations.

Theorem 12.4.2 *Entropy as a function on the space of surface diffeomorphisms is lower semicontinuous in the C^1-topology (Section A.2.2).*

Proof An equivalent restatement is the following: For any surface diffeomorphism f and any $\epsilon > 0$ there is a $\delta > 0$ such that if $d(f, g) < \delta$ (in the C^1-metric) then $h_{\text{top}}(g) \geq h_{\text{top}}(f) - \epsilon$.

To prove this pick any such f. If $h_{\text{top}}(f) = 0$, there is nothing to show. Otherwise, there is a horseshoe Λ of f with $h_{\text{top}}(f_{\restriction_\Lambda}) \geq h_{\text{top}}(f) - \epsilon$. Take $\delta > 0$ such that (using structural stability of horseshoes) any g with $d(f, g) < \delta$ contains a horsesoe Λ' that is topologically conjugate to Λ. Its entropy then agrees with that of Λ, so $h_{\text{top}}(g) \geq h_{\text{top}}(g_{\restriction_{\Lambda'}}) = h_{\text{top}}(f_{\restriction_\Lambda}) \geq h_{\text{top}}(f) - \epsilon$. $\square$

An analogous result also holds for continuous maps of an interval,[12] but Exercise 8.3.7 shows that it can fail in other situations. Upper semicontinuity can be obtained without dimension restrictions, but it is a far more delicate matter in terms of the smoothness required: It holds for C^∞ maps,[13] but the proof relies on a subtle estimate[14] that fails maps of finite differentiability. For these there is only some control over the degree to which upper semicontinuity fails.

12.5 DETECTING HOMOCLINIC TANGLES: THE POINCARÉ–MELNIKOV METHOD

The preceding sections have shown that transverse homoclinic points and the ensuing tangles are of great interest for mathematical and scientific applications. Therefore it is also important to have a method by which the presence of transverse homoclinic points can be detected. Numerical calculations alone may not give unambiguous results if the angle at the homoclinic point is small, because then the entire web of tangles is compressed into a thin neighborhood of the local stable and unstable leaf of the hyperbolic point, and computer pictures may not detect it reliably. In addition, numerical pictures do not constitute proof of the presence of this phenomenon.

To explain the Poincaré–Melnikov method consider a flow φ^t in the plane defined by a differential equation $\dot{x} = f(x)$ that has a hyperbolic saddle at 0 and a homoclinic loop $\Gamma := \{\varphi^t(q) \mid t \in \mathbb{R}\} \cup \{0\}$. The Poincaré–Melnikov method

---

[12] Michał Misiurewicz, Horseshoes for Continuous Mappings of an Interval, in *Dynamical Systems (Bressanone, 1978)*, Liguori, Naples, 1980, pp. 125–135.

[13] Sheldon Newhouse, Continuity Properties of Entropy, *Annals of Mathematics* (2) **129** (1989), 215–235.

[14] Yosef Yomdin, Volume Growth and Entropy, *Israel Journal of Mathematics* **57** (1987), 285–300.

detects the appearance of a transverse homoclinic point under perturbation.[15] Thus, perturb the flow to flows φ_ϵ^t defined by $\dot{x} = f(x) + \epsilon g_t(x)$, where we assume that g_t is periodic in t and, for simplicity, that $g_t(0) = 0$. Concrete examples are deformations of an elliptic billiard and the work of Knieper and Weiss mentioned previously.

We need to study the stable curve W_ϵ^s of 0 for φ_ϵ^t and the corresponding unstable curve W_ϵ^u. Orbits on these are of the form $\varphi^t(q) + \epsilon q^s(t) + O(\epsilon^2)$ $(t \geq 0)$ and $\varphi^t(q) + \epsilon q^u(t) + O(\epsilon^2)$ $(t \leq 0)$, respectively, where

$$\dot{q}^s(t) = Df(\varphi^t(q))q^s(t) + g_t(\varphi^t(q)) \qquad \text{for } t \geq 0$$

$$\dot{q}^u(t) = Df(\varphi^t(q))q^u(t) + g_t(\varphi^t(q)) \qquad \text{for } t \leq 0$$

(the linearized flow). Up to order ϵ^2 the separation $d(q)$ between W_ϵ^s and W_ϵ^u near q is given by the projection to the normal Jf of $\epsilon(q^u(0) - q^s(0))$, so $d(q) = \epsilon(q^u(0) - q^s(0))Jf(q)/\|f(q)\|$. Since Q^u and Q^s can be determined from the linearized flow, so can this difference. The *Melnikov function* is defined on Γ by

$$M(q) = \int_{-\infty}^{\infty} g_t(\varphi^t(q))Jf(\varphi^t(q))\,dt.$$

Note that this integral is computed using only information about the unperturbed flow and the perturbation term itself.

Theorem 12.5.1 *If the Melnikov function has any simple zero along Γ, then for all sufficiently small $\epsilon \neq 0$ there is a transverse homoclinic point for φ_ϵ^t. If $M \neq 0$ except at 0, then $W_\epsilon^s \cap W_\epsilon^u = \{0\}$.*

In some important situations the splitting is exponentially small. This causes difficulties that were noticed 20 years ago[16] and have been worked on since.[17]

12.6 HOMOCLINIC TANGENCIES

In a one-parameter family f_ϵ of diffeomorphisms of $\mathbb{R}^2$ with homoclinic tangles associated to a hyperbolic fixed point p it may happen that for $\epsilon = 0$ a "tongue" of the stable tangle is tangent to the unstable leaf at a point q. This is called a *homoclinic tangency*. It might be that q is the "primary" intersection and there are no tangles at all for either negative ϵ or for positive ϵ. Or q may be a higher-order intersection arising from a transverse homoclinic point that persists for all small ϵ. Either way, this situation is a mechanism for the production of additional dynamical complexity. Note that in investigating this situation we are studying hyperbolicity

---

[15] V. K. Melnikov, On the Stability of the Center for Time-Periodic Perturbations, *Trudy Moskovskogo Matematičeskogo Obščestva* **12** (1963), 3–52; *Transactions of the Moscow Mathematical Society* **12** 1–57.

[16] Jan A. Sanders, *A Note on the Validity of Melnikov's Method*, Report 139, Wiskundig Seminarium, Vrije Universiteit Amsterdam.

[17] See, for example, Vasily G. Gelfreich, A Proof of the Exponentially Small Transversality of the Separatrices for the Standard Map, *Communications in Mathematical Physics* **201**, no. 1 (1999), 155–216.

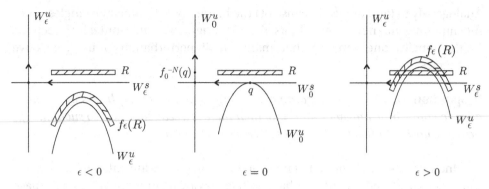

Figure 12.6.1. Homoclinic tangencies.

but at the same time leave the uniformly hyperbolic context of the preceding sections, because a transient homoclinic tangency produces a global bifurcation, which is incompatible with the structural stability of hyperbolic sets. Thus we are pushing techniques further, but in doing so we rely strongly on the underlying hyperbolicity.

We suppose that the homoclinic tangency is of generic type. Specifically, we assume that there is a local coordinate system near the tangency point q in which $W^s(p)$ is represented as the x-axis and $W^u(p)$ is given by $y = \epsilon - x^2$.

The source of additional dynamical complexity is the appearance of extra horseshoes.

Proposition 12.6.1 *If a one-parameter family f_ϵ of diffeomorphisms of $\mathbb{R}^2$ is volume-contracting at p (that is, $\|Df_\epsilon(p)\| < 1$) and has a generic homoclinic tangency for $\epsilon = 0$, then for $\epsilon > 0$ an extra horseshoe appears.*

This effect is evident from Figure 12.6.1, which shows a tangent bifurcation with the requisite rectangle and its image included.

By producing more horseshoes this phenomenon adds further dynamical complexity of the same type as that already present in any web of tangles. However, there are also effects that are of an altogether different nature. These appear just before the bifurcation:

Proposition 12.6.2 (Newhouse Phenomenon) [18] *If a one-parameter family f_ϵ of diffeomorphisms of $\mathbb{R}^2$ is volume-contracting at p (that is, $\|Df_\epsilon(p)\| < 1$) and has a generic homoclinic tangency for $\epsilon = 0$, then for a residual set of small $\epsilon < 0$ there are infinitely many attracting periodic orbits (sinks).*

A residual set is the intersection of a countable collection of open and dense sets (this implies density by Lemma A.1.15).

At this point it is rather clear that every generic homoclinic tangency is the source of dynamical complexity, but also of significant change in the dynamics.

---

[18] Sheldon Newhouse, Diffeomorphisms with Infinitely Many Sinks, *Topology* **13** (1974), 9–18.

Analogously to (and indeed because of) the hierarchy of homoclinic tangles, a homoclinic tangency never comes by itself. By looking at the higher-order intersections in the tangles one can see that many small perturbations must also have tangencies:

Proposition 12.6.3 *If a one-parameter family f_ϵ of diffeomorphisms of $\mathbb{R}^2$ has a generic homoclinic tangency for $\epsilon = 0$, then there is a sequence $\epsilon_n \to 0$ such that f_{ϵ_n} has a homoclinic tangency for all $n \in \mathbb{N}$ at points $q_n \to q$.*

One should keep in mind that most of these tangencies are obtained from high-order "tongues", and accordingly have effects that are highly concentrated in space and are associated with high iterates of the diffeomorphism. Accordingly, complexity is added on small spatial scales and large time scales. If one considers the fact that the creation of a horseshoe increases entropy, then it is clear that continuity of entropy (which follows from Theorem 12.4.2 and the upper semicontinuity result mentioned after it) requires most of these changes to be of this type in order to limit the accumulation of orbit complexity. Thus in this complex sequence of homoclinic tangency bifurcations all manner of tiny horseshoes for high iterates appear to make their contributions to orbit complexity and entropy. Yet, for each value of the parameter the bulk of the orbit complexity can be understood in terms of the same generic properties we developed for all uniformly hyperbolic dynamical systems.

CHAPTER 13

Strange Attractors

Strange attractors are a popular subject in dynamics. They are attractors with a complicated geometric structure, in particular attractors that are not simple curves or surfaces. Before looking at strange attractors it is appropriate to look at some geometrically simple ones. From there we will get to strange attractors via an important explicit model situation.

Hyperbolicity of some kind is a characteristic feature of strange attractors, but the study of the most popular examples is difficult because the hyperbolicity of those attractors is of a weaker form than the uniform hyperbolicity discussed in Chapter 10. There is enough of it to produce a great deal of complexity, but not enough to apply the tools from Chapter 10 directly. It is often similar to the nonuniformly hyperbolic behavior that appears in stochastic quadratic maps (the Jakobson–Collet–Eckmann case; see Section 11.4.3.6), and, in fact, those are used as both models and as the basis for perturbative constructions for some popular strange attractors. However, we consider only the Lorenz attractor, where the difficulties are of a different sort and may be described as "uniform hyperbolicity with singularities." The proof of existence of the Lorenz attractor is one of the most spectacular examples of computer-assisted proofs in continuous mathematics.

13.1 FAMILIAR ATTRACTORS

Not all attractors are strangers to us at this point. The simplest ones are attracting fixed points, which were formally introduced in Definition 2.2.22. Here is an equivalent definition of an attracting fixed point that is more conducive to generalization.

Definition 13.1.1 A fixed point p of a map $f: X \to X$ of a metric space is said to be an attracting fixed point if there is a neighborhood U of p such that $f(U) \subset U$ and $\bigcap_{n \in \mathbb{N}} f^n(U) = \{p\}$.

The other familiar attractor is a limit cycle, as seen in Section 2.4.3. We defined it as a periodic point p that has a neighborhood whose every point is positively asymptotic to $\mathcal{O}(p)$. Analogously to Definition 13.1.1, we can give the following definition.

Definition 13.1.2 A periodic point p is said to be an attracting periodic point (or limit cycle) of a flow ϕ^t if there is a neighborhood U of $\mathcal{O}(p)$ such that $\phi^t(U) \subset U$ for $t \geq 0$ and $\bigcap_{t \geq 0} \phi^t(U) = \mathcal{O}(p)$.

The attractive feature of these definitions is that they do not need the notion of "asymptotic to," which is a little more subtle when the target of the asymptotics is more than a point.

Suppose p is an attracting periodic point of a flow ϕ^t and consider the map $f(x) = \phi^1(x)$. Then the orbit of p under the flow is a set that attracts points under repeated application of f. To make this precise we modify Definition 13.1.2. In fact, we can replace $\mathcal{O}(p)$ by any set there to determine whether that set is attracting.

Definition 13.1.3 Suppose $f \colon X \to X$ is a map. A compact set $A \subset X$ is said to be an *attractor* for f if there is a neighborhood U of A such that $f(U) \subset U$ and $\bigcap_{n \in \mathbb{N}} f^n(U) = A$. We usually require A to have no proper subsets with the same property.

The requirement that A is "smallest", that is, has no proper attracting subsets, makes a difference in the following example situation. Figure 13.1.1 shows a dynamical system for which the set $A = \{(x, y) \in \mathbb{R}^2 \mid x \in [0, 1]\}$ has most properties required by Definition 13.1.3, but it contains the attractors $(-1, 0)$ and $(1, 0)$. Therefore we do not call A an attractor in this case.

A slightly more substantial but still simple example can be constructed as follows. Consider a circle map f_1 with an attracting fixed point, for example as in Figure 4.3.3. Suppose this point is 0. If f_2 is any map of $\mathbb{T}^2$, such as a translation or a hyperbolic automorphism, then the map $f \colon \mathbb{T}^3 \to \mathbb{T}^3$, $f(x, y, z) = (f_1(x), f_2(y, z))$ has an attracting 2-torus $\{0\} \times \mathbb{T}^2$.

So attracting fixed points, limit cycles, and attracting periodic orbits are examples of attractors, as are such attracting surfaces. However, there are plenty of other examples. Those that do not have a comparably simple structure gave rise to the name "strange" attractor.

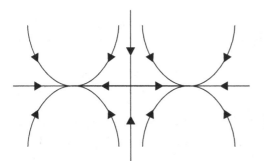

Figure 13.1.1. Not an attractor.

13.2 THE SOLENOID

Strange attractors abound in systems arising from applications of dynamics. Those attractors can be hard to analyze effectively, and therefore it is useful to look at a model situation first.

We describe one from a class of examples presented by Smale in the 1960s. It is called the *Smale attractor* or *solenoid*. The construction can be visualized as similar to doubling over a rubber band.

Consider the solid torus $M := S^1 \times D^2$, where D^2 is the unit disk in $\mathbb{R}^2$. This looks like a bagel. On it we define coordinates (φ, x, y) such that $\varphi \in S^1$ and $(x, y) \in D^2$, that is, $x^2 + y^2 \le 1$. Using these coordinates we define the map by doubling up and shrinking the thickness by five.

Proposition 13.2.1 *The map*

$$f: M \to M := S^1 \times D^2, \qquad f(\varphi, x, y) = (2\varphi, \tfrac{1}{5}x + \tfrac{1}{2}\cos\varphi, \tfrac{1}{5}y + \tfrac{1}{2}\sin\varphi)$$

is well defined and injective.

Proof The map is well defined, that is, $f(M) \subset M$, because

$$(\tfrac{1}{5}x + \tfrac{1}{2}\cos\varphi)^2 + (\tfrac{1}{5}y + \tfrac{1}{2}\sin\varphi)^2 = \tfrac{1}{25}(x^2 + y^2) + \tfrac{1}{5}(x\cos\varphi + y\sin\varphi)$$

$$+ \tfrac{1}{4}(\cos^2\varphi + \sin^2\varphi) \le \tfrac{1}{25} + \tfrac{2}{5} + \tfrac{1}{4} < 1.$$

Thus, in fact, $f(M)$ is contained in the interior of M.

That f is injective is not surprising, because we shrink the thickness a lot. Suppose $f(\varphi_1, x_1, y_1) = f(\varphi_2, x_2, y_2)$. Then

$$2\varphi_1 = 2\varphi_2 \pmod{2\pi},$$

$$\tfrac{1}{5}x_1 + \tfrac{1}{2}\cos\varphi_1 = \tfrac{1}{5}x_2 + \tfrac{1}{2}\cos\varphi_2,$$

$$\tfrac{1}{5}y_1 + \tfrac{1}{2}\sin\varphi_1 = \tfrac{1}{5}y_2 + \tfrac{1}{2}\sin\varphi_2.$$

If $\varphi_1 = \varphi_2$, the trigonometric terms cancel, so $x_1 = x_2$ and $y_1 = y_2$. If $\varphi_1 = \varphi_2 + \pi$, then

$$\tfrac{1}{5}x_1 + \tfrac{1}{2}\cos\varphi_1 = \tfrac{1}{5}x_2 - \tfrac{1}{2}\cos\varphi_1,$$

$$\tfrac{1}{5}y_1 + \tfrac{1}{2}\sin\varphi_1 = \tfrac{1}{5}y_2 - \tfrac{1}{2}\sin\varphi_1,$$

or

$$\tfrac{1}{5}(x_2 - x_1) = \cos\varphi_1 \qquad \text{and} \qquad \tfrac{1}{5}(y_2 - y_1) = \sin\varphi_1,$$

which implies that

$$(x_2 - x_1)^2 + (y_2 - y_1)^2 = 25.$$

Since the left-hand side is at most 8, this is impossible. $\square$

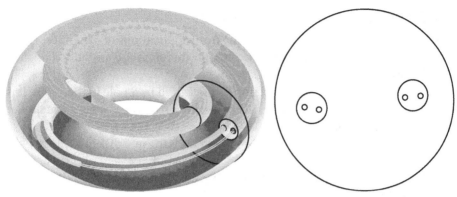

Figure 13.2.1. The Smale attractor and a cross section.

A much milder contraction than by a factor of 5 will do. The same argument, but using $x \cos \varphi + y \sin \varphi \leq x + y \leq \sqrt{2}$, shows that

$$f: M \to M := S^1 \times D^2, \qquad f(\varphi, x, y) = (2\varphi, \tfrac{1}{3}x + \tfrac{1}{2}\cos\varphi, \tfrac{1}{3}y + \tfrac{1}{2}\sin\varphi)$$

is well defined and injective.

The image $f(M)$ intersects any cross section $C = \{\theta\} \times D^2$ of M in two disjoint disks of radius $1/5$ as shown in Figure 13.2.1 and on the book cover. Indeed, $C \cap f(M) = f(C_1) \cup f(C_2)$, where C_1 and C_2 are cross sections. Under iteration this picture repeats on smaller scales.

Clearly $f^2(M) \subset f(M)$, but moreover $C \cap f^2(M) = f(C_1 \cap f(M)) \cup f(C_2 \cap f(M))$, where C_1 and C_2 are as before. Thus $C \cap f^2(M)$ consists of four little disks, two each in $f(C_1)$ and $f(C_2)$ (Figure 13.2.1), and $f^2(M)$ winds around M four times. Pictorially we have doubled up the rubber band a second time.

Recursively, $C \cap f^{l+1}(M)$ consists of 2^{l+1} disks, two each in the disks of $C \cap f^l(M)$.

What happens here is geometrically similar to what happens in the taffy machines that one often sees in seaside towns. Those machines constantly stretch and fold the taffy (which consists of molasses or sugar), and this stretching and refolding produces a remarkably aligned stringy structure – it is definitely not a curve or a surface. There are other situations where a similar technique is employed to produce particular materials. One is the production of Japanese swords, which are made by repeatedly folding over and flattening a piece of steel to produce a highly aligned molecular structure. Twenty foldings of a piece of metal less than 1 cm thick give layers of thickness less than 2^{-20} cm $< 10^{-6}$ cm $= 10$ nm, about the size of an atom. This stretching and folding is also rather like the construction of the horseshoe in Section 7.4.4.

Both of these real-life examples differ from the corresponding model situation in that our models lose some volume, as it were. In the Smale attractor, the fivefold shrinking in two directions combined with only a doubling in the remaining direction reduces the total volume by more than 10 in each iteration. In the horseshoe we lose volume by discarding parts that fall outside the rectangle. Nevertheless, the taffy and sword examples are real-life illustrations of the stretching and folding that are jointly responsible for the complex dynamics we saw in Chapter 7 and for similar complexity in the Smale attractor. And losing volume is an essential feature of having an attractor, as one can see from Definition 13.1.3.

The Smale attractor is a *hyperbolic* attractor: The stretching is uniform (by a factor 2 everywhere) because we arranged for constant expansion in the angular coordinate. This makes the dynamics maximally chaotic by producing the full complexity developed in Chapter 10. It also happens to make the global attractor relatively easy to study with the methods from Chapter 10.

A direct relation to the inverse limit construction (Definition 7.1.12) is the following:

Proposition 13.2.2 *The Smale attractor is the inverse limit of the circle doubling E_2.*

Because hyperbolic attractors are so tractable, the term "strange" attractor is often reserved for attractors with a comparable degree of complexity that do not have a uniform lower bound on the expansion rate. We next consider a famous example of that sort.

13.3 THE LORENZ ATTRACTOR

In 1961, the meteorologist Edward Norton Lorenz studied a weather model with his new electronic computer, based on rules that expressed the relationships between temperature and pressure, pressure and wind speed, and so forth. At one time he wanted to extend a prior run further in time, and instead of starting at the beginning typed in a number well into that run. But instead of duplicating the old run, the new one diverged rapidly from it. He realized that instead of the six decimal places used in the computation he had typed in only the three digits from the printout, and the small difference of a few thousandths quickly accumulated to macroscopic errors.

By 1963 Lorenz had radically simplified a convection model used in atmospheric science to the differential equations

$$\dot{x} = \sigma(y - x)$$

(13.3.1)
$$\dot{y} = rx - y - xz$$

$$\dot{z} = xy - bz,$$

where $\sigma, r, b > 0$. (He quickly singled out the particular values $\sigma = 10$, $b = 8/3$, $r = 28$ as warranting careful study.) The same equations have since appeared in various other models. They happen to describe a dynamo (the precursor of modern generators), and the direction reversals they exhibit are thought to potentially explain the occasional reversals of the earth's magnetic field in geological times. A chaotic waterwheel provides a mechanical realization.[1]

The qualitative behavior of solutions to these differential equations depends rather strongly on the values of the parameters. Nowadays one can easily verify this numerically (to the extent that one trusts numerical simulations to give a complete and accurate picture), but standard techniques from ordinary differential equations enabled Lorenz to carry the analysis of these differential equations rather far.

---

[1] Steven Strogatz, *Nonlinear Dynamics and Chaos*, Addison-Wesley, 1994, p. 302.

13.3.1 The Case $r < 1$

The simplest behavior occurs for $r < 1$:

Proposition 13.3.1 *For $r < 1$, all solutions of the Lorenz equations (13.3.1) tend to the origin as $t \to \infty$.*

Proof First, for $r < 1$ the origin is the only fixed point: From (13.3.1) one sees that $\dot{x} = 0$ implies $x = y$ and $0 = \dot{z} = xy - bz = x^2 - bz$ then gives $z \geq 0$, hence $r - 1 - z < 0$. Together with $0 = \dot{y} = rx - y - xz = x(r - 1 - z)$ this gives $x = 0$, hence $y = 0$ and (from $\dot{z} = 0$) $z = 0$.

Although it is not necessary for this proof we note that the linearization

$$
(13.3.2) \qquad \begin{pmatrix} \dot{x} \\ \dot{y} \\ \dot{z} \end{pmatrix} = \begin{pmatrix} \sigma(x - y) \\ rx - y \\ -bz \end{pmatrix} = \begin{pmatrix} -\sigma & \sigma & 0 \\ r & -1 & 0 \\ 0 & 0 & -b \end{pmatrix} \begin{pmatrix} x \\ y \\ z \end{pmatrix}
$$

of (13.3.1) has only negative eigenvalues when $r < 1$. This means that the origin is an attractor.

To show that it is a global attractor we employ a *Lyapunov function*. This is a positive function that decreases along orbits. Inventing one for a given system is an art form. Here we use $L(x, y, z) := x^2/\sigma + y^2 + z^2$, whose level sets are ellipsoids centered at the origin. Then L decreases along each orbit. By the chain rule,

$$
\begin{aligned}
\dot{L} &= \frac{d}{dt}(x^2/\sigma + y^2 + z^2) = 2x\dot{x}/\sigma + 2y\dot{y} + 2z\dot{z} \\
&= 2x(y - x) + 2y(rx - y - xz) + 2z(xy - bz) \\
&= 2(r + 1)xy - 2x^2 - 2y^2 - 2bz^2 \\
&= -2\left(x - \frac{r+1}{2}y\right)^2 - 2\left(1 - \left(\frac{r+1}{2}\right)^2\right)y^2 - 2bz^2,
\end{aligned}
$$

which is never positive and is zero only if each term is zero, which implies that $y = z = 0$ (from the last two terms) and then $x = 0$ from the first term. Thus, L decreases along each orbit, which means that orbits constantly move to smaller and smaller ellipsoids, which means that they get closer to the origin.

That L decreases along each orbit does not in itself imply that it decreases to 0. But this must be so because for each orbit we must have $\dot{L} \to 0$ (since L is bounded below), which from the above implies $(x, y, z) \to 0$. Thus, 0 is a *global attractor*. $\square$

Looking back at the above arguments one can see that for $r = 1$ the qualitative behavior is largely the same. The analysis of the Lyapunov function becomes more elaborate because the term $(1 - (r + 1/2)^2)y^2$ disappears, but the complications are still well within the range of standard techniques in ordinary differential equations.

13.3.2 The Case $r > 1$

When $r > 1$, two other equilibria appear. $\dot{x} = 0$ still implies $x = y$. From (13.3.1) one sees that $0 = \dot{y} = rx - y - xz = x(r - 1 - z)$ gives either $x = 0$ (hence $y = 0$ and then

$z = 0$ because $0 = \dot{z} = xy - bz = -bz$) or $z = r - 1$. The first case gives the known equilibrium at the origin. The case $z = r - 1$ combines with $0 = \dot{z} = xy - bz = x^2 - bz$ to give $x^2 - b(r-1) = 0$, that is, $x = \pm\sqrt{b(r-1)}$. This gives two new equilibria, $(x_0, y_0, z_0) = (\sqrt{b(r-1)}, \sqrt{b(r-1)}, r-1)$ and $(-\sqrt{b(r-1)}, -\sqrt{b(r-1)}, r-1)$, that tend to 0 as $r \to 1^+$. Thus, $r = 1$ is a bifurcation value. [That the rate of approximation is of square-root type corresponds to what we observed in the period-doubling bifurcations of the quadratic family in (11.2.2) and Figure 11.2.3.]

Proposition 13.3.2 *If $\sigma > b + 1$ and $1 < r < \sigma((\sigma + b + 3)/(\sigma - b - 1))$, then the equilibria $(\pm\sqrt{b(r-1)}, \pm\sqrt{b(r-1)}, r - 1)$ are stable.*

Outline of proof To linearize (13.3.1) at these equilibria set $\xi := x - x_0$, $\eta := y - y_0$ and $\mu := z - z_0$. Then

$$\dot{\xi} = \sigma(x_0 + \xi - y_0 - \eta) \qquad\qquad = [\sigma(x_0 - y_0)] + \sigma(\xi - \eta)$$

$$\dot{\eta} = r(x_0 + \xi) - y_0 - \eta - (x_0 + \xi)(z_0 + \mu) = [rx_0 - y_0 - x_0 z_0] + r\xi - \eta$$

$$\qquad\qquad\qquad\qquad - (x_0\mu + z_0\xi + \xi\mu)$$

$$\dot{\mu} = (x_0 + \xi)(y_0 + \eta) - b(z_0 + \mu) \qquad = [x_0 y_0 - bz_0] + x_0\eta + y_0\xi + \xi\eta - b\mu.$$

The bracketed terms are zero. Dropping terms that are not linear in ξ, η, μ and using $x_0 = y_0$ and $z_0 = r - 1$ leaves

$$\begin{pmatrix} \dot{\xi} \\ \dot{\eta} \\ \dot{\mu} \end{pmatrix} = \begin{pmatrix} \sigma(\xi - \eta) \\ r\xi - \eta - x_0\mu - z_0\xi \\ x_0\eta + y_0\xi - b\mu \end{pmatrix} = \begin{pmatrix} \sigma(\xi - \eta) \\ \xi - \eta - x_0\mu \\ x_0\xi + x_0\eta - b\mu \end{pmatrix} = \begin{pmatrix} \sigma & -\sigma & 0 \\ 1 & -1 & -x_0 \\ x_0 & x_0 & -b \end{pmatrix} \begin{pmatrix} \xi \\ \eta \\ \mu \end{pmatrix}.$$

The characteristic polynomial of the last matrix is

$$\lambda^3 + (b + 1 - \sigma)\lambda^2 + \lambda b(\sigma + r) - 2\sigma b(r - 1).$$

Check that the solutions have a negative real part when $1 < r < \sigma((\sigma + b + 3)/(\sigma - b - 1))$. $\square$

For this parameter range there are also two unstable periodic solutions that merge into the equilibria, $(\sqrt{b(r-1)}, \sqrt{b(r-1)}, r - 1)$ and $(-\sqrt{b(r-1)}, -\sqrt{b(r-1)}, r - 1)$, respectively, when $r = \sigma(\sigma + b + 3/\sigma - b - 1)$.

13.3.3 The Case $r > \sigma(\sigma + b + 3/\sigma - b - 1)$

This is where things get interesting. Already for $r > 1$ the origin ceases to be stable because the characteristic polynomial of the linearization matrix in (13.3.2) is $(b + \lambda)[\lambda^2 + \lambda(\sigma + 1) + \sigma(1 - r)]$ with roots $-b < 0$ and $(1/2)[-\sigma - 1 \pm \sqrt{(\sigma + 1)^2 - 4\sigma(1 - r)}]$, both of which are negative only when $r < 1$. There are two further unstable fixed points, but the unstable periodic solutions have disappeared. Lorenz gave a plausible argument to the effect that there could be no stable periodic solutions. At the same time it is not so hard to observe that solutions cannot grow too much and are indeed bounded.

At this point it becomes believable that the dynamics must be complicated. The orbits are trapped in a compact region devoid of stable equilibria and, apparently, limit cycles. Compactness forces them to accumulate somewhere. Lorenz decided to concentrate on particular values of the parameters

$$\sigma = 10, \qquad b = 8/3, \qquad r = 28 > 10\frac{47}{19} = 10\frac{10 + (8/3) + 3}{10 - (8/3) - 1} = \sigma\frac{\sigma + b + 3}{\sigma - b - 1}$$

and resorted to a computed picture. What Figure 13.3.1 shows has been known as the Lorenz attractor for several decades. For almost four decades formidable technical difficulties stood in the way of proving that there is indeed an attractor. It is only now, as we are writing this book, that there is finally proof of the existence of an attractor.

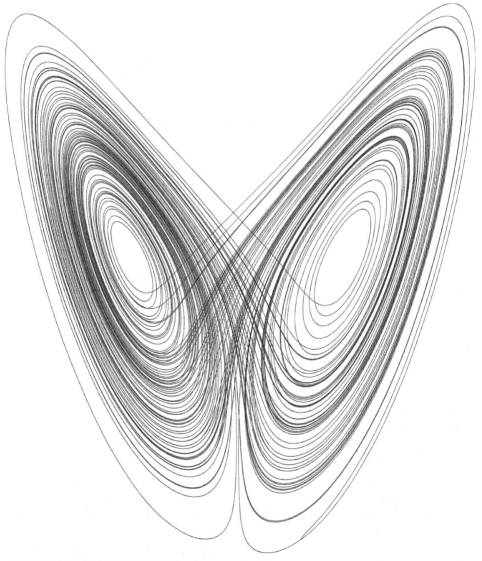

Figure 13.3.1. The Lorenz attractor.

13.3.4 The Tucker Theorem

The proof that for the "classical" parameter values there actually is an attractor in the Lorenz system is based on a sophisticated blend of mathematical theory and rigorous computation. Warwick Tucker developed an algorithm for computing rigorous solutions to a large class of ordinary differential equations based on a partitioning process and the use of interval arithmetic with directed rounding.

This gives important conclusions that go beyond the existence of an attractor.

Theorem 13.3.3 *The Lorenz equations for* $\sigma = 10, b = 8/3, r = 28$ *have an attractor. Moreover, the attractor is robust, that is, it persists under small parameter changes, and it carries a unique "SRB measure" (see Section 10.4.3.3).*

This means, essentially, that the attractor is actually the object that numerical computations show. More precisely, this means that there is an invariant measure (Section 10.4.3.3) defined on the attractor (this means that it assigns zero to continuous functions that vanish on the attractor) with respect to which almost every orbit that starts in a neighborhood of the attractor is uniformly distributed: The time averages of any continuous function coincide with the *space average* determined by the invariant measure. This is therefore called the "physically observed" density. These two added conclusions show that there is not only some attractor, but it is both physically meaningful (persistent) and observed.

We describe the proof in some detail.

Proof outline[2] The first step of the proof is standard and had been taken as a starting point of prior attempts long before. Judiciously choose a rectangle S in the plane $z = r - 1$ and consider the return map R of the Lorenz flow to this rectangle. Specifically, the rectangle includes $(0, 0, r - 1)$, and the equilibria $(\sqrt{b(r - 1)}, \sqrt{b(r - 1)}, r - 1)$ and $(-\sqrt{b(r - 1)}, -\sqrt{b(r - 1)}, r - 1)$ are the midpoints of two sides. The Lorenz flow swirls around these points, crossing the rectangle from above. Aside from a line Γ through $(0, 0, r - 1)$ parallel to the sides through the equilibria, the return map R is defined on S. Points on Γ are attracted to the origin. This is why the return map is not defined on Γ, and moreover it is the reason that numerical attacks failed to resolve the issue of existence of the attractor: Since return times get arbitrarily large near Γ, one cannot control the errors in numerical solutions of the Lorenz equations.

A known feature of the return map to S is that it maps $S \setminus \Gamma$ to two triangular regions in S, each of which crosses Γ as shown in Figure 13.3.2. The tips of these triangles correspond to Γ. The curve Γ therefore provides an opportunity for orbits to separate without having large derivative of the return map. This is the "singularity" that we alluded to at the beginning of the chapter.

Having established this much, the standard line of reasoning imposed some benign assumptions on more specific aspects of the way in which R maps $S \setminus \Gamma$ into

[2] Warwick Tucker, The Lorenz Attractor Exists, *Comptes Rendus des Séances de l'Académie des Sciences. Série I. Mathématique* **328**, no. 12 (1999), 1197–1202; A Rigorous ODE Solver and Smale's 4th Problem, *Foundations of Computational Mathematics* **2**, no. 1 (2002), 53–117.

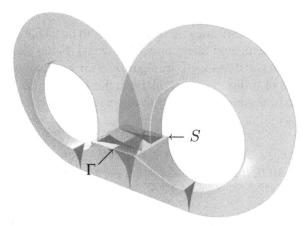

Figure 13.3.2. Section for the Lorenz map.

S. These were called the "geometric Lorenz attractor," and allowed to establish the existence of an attractor. However, it was not clear whether these assumptions hold, that is, whether the attractor thus obtained is the Lorenz attractor. This is a good moment to mention that the amount of nonhyperbolicity is reasonably benign here. While a uniformly hyperbolic attractor such as the solenoid is structurally stable, this is not the case with the Lorenz or geometric Lorenz attractor because of the singularity at the origin. However, the failure of structural stability is moderate. There is more than one conjugacy class nearby, but the conjugacy classes form only a two-parameter family. This is significantly more benign than the general case, in which there may be infinitely many moduli of conjugacy, or, worse, the conjugacy classes may be highly irregular.

Tucker's approach is a good example of how to analyze specific strange attractors. He combines analytic and numerical methods, using numerical methods for the bulk of the attractor and analytic methods to deal with the singularity that confounds all numerical approaches. In this way he establishes the following three facts.

First, there is a compact set $N \subset S$ such that $N \smallsetminus \Gamma$ is strictly R-invariant, that is, $R(N \smallsetminus \Gamma) \subset \text{Int}(N)$. This establishes the existence of an attractor $\mathcal{L}$ for the Lorenz flow, which intersects S in $\Lambda := \bigcap_{n=0}^{\infty} R^n(N)$.

In order to make sure that the attractor is as nontrivial as expected, two more facts are developed. There is an R-invariant cone field on N; that is, if $C(x)$ is the sector with $10°$ opening centered around a curve that approximates Λ, then $DR(x)C(x) \subset C(R(x))$ (see Definition 10.1.4). Finally, vectors inside cones are eventually expanded under repeated application of DR. Specifically, there is some $C > 0$ and $\lambda > \sqrt[29]{2}$ such that, if v is a vector in $C(x)$ for some $x \in N$, then $\|DR^n(x)v\| \geq C\lambda^n\|v\|$ for all $n \in \mathbb{N}$. These two facts about the cone field correspond to the properties of a cone field centered around the angular direction in the solenoid, where hyperbolicity of the map implies these properties. Here these two facts prove hyperbolic behavior (by Theorem 10.1.5). Tucker noticed, by the way, that the positive constant C in the cone-expansion property cannot be 1, that is, vectors in those cones could be contracted in the first few steps. This was not expected.

The proof of these properties combines two complementary approaches. One of these is normal form theory. In a small cube, centered at the origin (the equilibrium) there is an almost linear analytic coordinate change that transforms the Lorenz equations to the form $\dot{v} = Av + F(v)$, where A is diagonal and F is of order 20 in $\|v\|$. Specifically, on a cube of size r we have $\|F\| \leq 7 \cdot 10^{-9} r^{20}/(1 - 3r)$. This means that on the small cube the new coordinate representation is phenomenally close to linear. One the same cube, the nonlinear part of the coordinate change is bounded by $r^2/2$.

The coordinate change as well as the remarkable estimates are obtained by a method of Poincaré. Basically, one writes down the change of variables equation with (undetermined) Taylor series for the new coordinate representation and the coordinate change. If one desires to eliminate nonlinear terms up to order 20 from the normal form, then the Taylor coefficients of the coordinate change are given explicitly in terms of those of the Lorenz flow, appearing with coefficients of the type $1/(n_1\lambda_1 + n_2\lambda_2 + n_3\lambda_3)$, where the λ_i's are the eigenvalues of the derivative of the Lorenz flow at 0 and $n_i \in \mathbb{N}$. For large n_i there are explicit upper bounds on these fractions, and numerical work can check the remaining (finite but enormous) number of cases.

The second ingredient is rigorous numerical orbit computation. This is done using interval arithmetic (in each coordinate). After the first integration step the numerical value is bracketed by error terms: Upper bounds for the error of the method of integration compounded with roundoff error are known. Adding and subtracting these from the numerical value gives the endpoints of an interval containing the true value. Then these endpoints are integrated one further step, and the smaller result is diminished while the larger is augmented by the error bound. This is repeated at every step, ultimately leading to an interval (or parallelepiped) that is guaranteed to contain the true answer. Explicit values for the partial derivatives of the Lorenz flow follow directly from the Lorenz equations and can be evolved in time by the same method. This is used for the cone estimates.

This method encounters a problem that confounded all prior attempts to establish the reality of the Lorenz attractor: In computing orbits that approach the origin, the return times become arbitrarily large and errors compound out of control. This is where the normal form provides the decisive advantage. For an orbit entering a fixed box around 0 it provides precise control over how that orbit exits the little box.

Thus orbits are tracked by a combined method: While outside the box, around the origin they are integrated numerically with interval arithmetic. When an orbit enters the box one directly transits through the box using the transfer map provided by the normal form. $\square$

CHAPTER 14

Variational Methods, Twist Maps, and Closed Geodesics

In this chapter we leave the area of hyperbolic dynamics, which dominated the Panorama so far, and we describe some applications of the variational approach to dynamics that are among the most impressive and at the same time easily visualizable. The variational approach was first mentioned in Chapter 6 in connection with the Lagrangian formalism and then in the specific context of billiards. We begin with the study of twist maps, which include billiards in convex domains as a special case and provide an ideal setting for the variational approach due to the relatively simple structure of the phase space. Aubry–Mather theory establishes the existence of a full range of motions with the features of orbits of one-dimensional maps whose rotation numbers are compatible with the amount of twist in the system. These motions can be viewed as "traces" or "ghosts" of invariant curves. Thus we obtain both the remnants of "order," and the presence of chaos, since in the absence of genuine invariant curves the interaction between periodic orbits and Cantor-like sets produces not only homoclinic and heteroclinic tangles but also complex behavior of an even more baffling kind and puts a rigorous description of typical behavior way beyond the reach of current methods.

We continue with results that involve fewer assumptions on dynamics and deeper use of topology and end up with the description of one of the most impressive applications of dynamics to geometry: the existence of infinitely many closed geodesics for every Riemannian metric on the two-dimensional sphere.

14.1 THE VARIATIONAL METHOD AND BIRKHOFF PERIODIC ORBITS FOR BILLIARDS

14.1.1 Periodic States and Action Functional

In Section 6.4.5 we constructed two special period-2 orbits for a convex billiard by finding critical points of a functional related to the generating function and defined on a space of "potential orbits." In that particular simple case "potential orbits" were *period-2 states*, that is, pairs of points on the boundary with the functional being the length of the chord connecting the two points. This kind of approach is called the variational method, and in this section it is used to find many periodic orbits of a specific nature, in particular, orbits of any desired period. Following the original approach of G. D. Birkhoff, we now describe the basic results in the case of billiards, where geometric pictures help to visualize the situation nicely. In the next section we present a more general context that includes billiards as a particular case, explain the results about the existence of periodic points on a more technical level, and develop this approach beyond the periodic case.

Let $p, q \in \mathbb{Z}$ be relatively prime. Without loss of generality we assume $q > 0$ and $1 \leq p \leq q - 1$. For a convex billiard we look for specific periodic orbits of period q that wind around the table p times in such a manner that each orbit moves exactly p steps in the positive (counterclockwise) direction. In other words, the billiard map restricted to such an orbit behaves in the same way as the rotation $R_{p/q}$ by the angle $2\pi p/q$. Such orbits are called *Birkhoff periodic orbits of type* (p, q).

We now outline the construction of *at least two different* Birkhoff periodic orbits of type (p, q) for any $q > 0$ and $1 \leq p \leq q - 1$. Altogether this will produce infinitely many different periodic orbits with arbitrarily long periods. The space $C_{p,q}$ of potential orbits will naturally be the space of (p, q) periodic states. It can be visualized as q-gons (in general, self-intersecting) inscribed into the billiard table with a marked vertex, and with edges between vertices p apart. The marked vertex x_0 corresponds to the origin in counting; then in the cyclic order $x_1, \ldots, x_{q-1}$ will be the other vertices. Extend this sequence periodically; in other words, if $0 \leq k \leq q - 1$ and $l \in \mathbb{Z}$, define $x_{k+lq} = x_k$. The functional $A_{p,q}$ for the variational problem is the total length or perimeter of the polygon described above, namely, connecting x_0 with x_p, then with x_{2p}, and so on, until $x_{qp} = x_0$. Now we use the representation of the billiard map via the generating function (the negative of the distance) given by (6.4.1). Let $s_0, s_1, \ldots, s_{q-1}$ be the values of the length parameter corresponding to $x_0, x_p, \ldots, x_{p(q-1)}$. Then

$$A_{p,q}(x_0, x_1, \ldots, x_{q-1}) = -(H(s_0, s_1) + H(s_1, s_2) + \cdots + H(s_{q-1}, s_0)).$$

The negative of this functional is often called the *action functional*. Equation (6.4.2) shows that three successive vertices form an orbit segment if and only if the partial derivative of $A_{p,q}$ with respect to the position of the middle vertex vanishes; hence, critical points of the functional $A_{p,q}$ on the space $C_{p,q}$ are exactly the configurations corresponding to Birkhoff periodic orbits of type (p, q).

14.1.2 Existence of Two Birkhoff Periodic Orbits

It remains to show that there are at least two critical points for the perimeter functional $A_{p,q}$ (see Figure 14.1.1). In the case of period-2 orbits, which corresponds to $p = 1$, $q = 2$, these two critical points correspond to the diameter and the width (see Section 6.4.5). The first orbit is obtained from a configuration corresponding to the maximal value of the functional $A_{p,q}$. Since the space $C_{p,q}$ is not compact, an argument is needed to show that such a maximal value is attained. This is done in a predictable fashion: The space $C_{p,q}$ is extended to its natural closure in the space of q-tuples of points, namely, one adds configurations where the order is not exact: Several successive points may coincide. This space is compact, the functional $A_{p,q}$ naturally extends to it, and it reaches its maximal value. Now it is enough to show

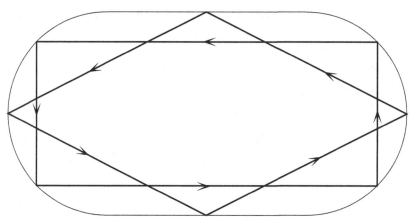

Figure 14.1.1. Two Birkhoff orbits.

that the maximal length cannot be achieved at an added degenerate configuration. This is almost immediate in the case of period-3 orbits and also can be done in an elementary fashion for the period-4 case, but for longer orbits some care is required and we outline the argument in the next section, when the problem is treated in greater generality. The idea is that for any degenerate configuration there is a perturbation within the extended space that makes the polygon longer and, in fact, makes it "less degenerate," that is, with fewer vertices coinciding. See the proof of Theorem 14.2.5 below for details.

Having found a maximum for the perimeter functional $A_{p,q}$ we immediately notice that the same configuration gives rise to q different maxima by moving the marked point along the orbit. This observation is the key to the construction of the second Birkhoff periodic orbit of type (p, q), which is based on the *minimax* or *mountain pass principle*. The name suggests a visualization of the argument: To traverse the ridge between two peaks in a mountain chain with minimal loss of altitude one has to pass through a saddle or a mountain pass. By changing the sign of the functional represented in this picture by the altitude (and thus reverting to the original generating function in the construction of the functional) one gets a version for less adventurous mountaineers: To pass from one mountain valley to another with minimal altitude gain one has to traverse a pass.

For billiards, the mountain pass argument works like this. Let $x = (x_0, x_1, \ldots, x_{q-1}) \in C_{p,q}$ be a configuration at which $A_{p,q}$ reaches its maximum. Consider smooth paths $x(t) = (x_0(t), x_1(t), \ldots, x_{q-1}(t))$, $0 \leq t \leq 1$ in $C_{p,q}$ that connect x with $x' = (x_1, \ldots, x_{q-1}, x_0)$, and such that $x_i(t)$ remains between x_i and x_{i+1} for $i = 0, \ldots, q - 1$. On such a path the functional $A_{p,q}$ is either constant [and then trivially every configuration $x(t)$ generates a different Birkhoff periodic orbit of type (p, q)] or, much more likely, attains a minimal value, which is strictly less than the common value at x and x'. A simple differentiation again shows that if such a value is maximal over all possible paths of the above-described type, then it has to correspond to a critical point of the functional $A_{p,q}$ (the mountain pass; see Figure 14.1.2). The remaining issue is again the existence of such a path where the minimum attains its maximal possible value. This is by far the most subtle part of

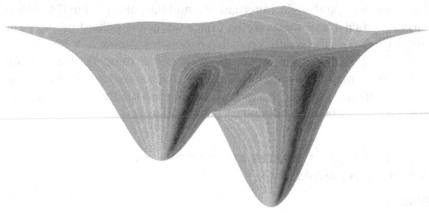

Figure 14.1.2. A mountain pass.

the whole argument, although intuitively it looks quite convincing that it must be more advantageous to move all vertices of the maximal configuration away rather than freeze some of them. (See Figure 14.1.3.)

14.1.3 Lift to the Strip

Our alternative description of the billiard system is to parametrize the boundary not by a simple closed curve, that is, not by a parameter in S^1, but rather by a periodic curve with period 1. Then, of course, every "physical" boundary point corresponds to infinitely many parameter values, all of them integer translates of each other. These are called lifts of this point, and all points $(x, y) \in S := \mathbb{R} \times (-1, 1)$ that project to a given point $(s, r) \in C$ are called lifts of (s, r). This corresponds exactly to an "unrolling" of the boundary circle that reverses the process described in Section 2.6.2 (and there is a lift construction corresponding to Proposition 4.3.1). The billiard map can still be unambiguously described in this model: Given a parameter $x \in \mathbb{R}$ and an angle y, find the corresponding ray in the table. It

Figure 14.1.3. The minimax orbit.

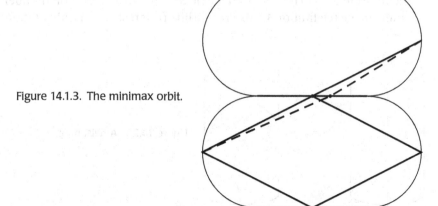

determines a new point and angle. [Alternatively, find the point (s, y) in C to which (x, y) projects modulo 1 and then take its image under the billiard map ϕ.] For the new point we choose the smallest possible parameter value $x' > x$. [Or, for the resulting point (s', y'), choose a point (x', y') where $x' = s'$ (mod 1).] This way we obtain a continuous map (as is seen by letting $y \to 0$ for a fixed s). From this new map $\Phi: S \to S$, which is periodic in s and which we shall call the *lift* of ϕ, we can easily recover ϕ itself by taking all boundary data modulo 1.

14.2 BIRKHOFF PERIODIC ORBITS AND AUBRY–MATHER THEORY FOR TWIST MAPS

14.2.1 Twist Maps

Any cylinder map can be lifted to the strip $\mathbb{R} \times (-1, 1)$ in a fashion completely similar to that described above for the billiard map. To distinguish between a map and its lift we always denote the cyclic coordinate of the cylinder by s and the first coordinate on the strip by x.

Definition 14.2.1 A diffeomorphism $\phi: C \to C$ of the open cylinder $C = S^1 \times (-1, 1)$ is said to be a *twist map* if

  (1) it preserves orientation and preserves boundary components in the sense that there exists an $\epsilon > 0$ such that, if $(x, y) \in S^1 \times (-1, \epsilon - 1)$, then $\phi(x, y) \in S^1 \times (-1, 0)$, and
  (2) $(\partial/\partial y)\Phi_1(x, y) > 0$, where $\Phi = (\Phi_1, \Phi_2)$ is a lift of ϕ to $S = \mathbb{R} \times (0, 1)$. See Figure 14.2.1.
  (3) The map ϕ extends to a homeomorphism $\bar{\phi}$ (not necessarily smooth) of the closed cylinder $S^1 \times [-1, 1]$.

ϕ is said to be a *differentiable twist* if, in fact, for $\epsilon > 0$ there is a $\delta > 0$ such that $(\partial/\partial y)\Phi_1(x, y) > \delta$ on $C_\epsilon := S^1 \times [\epsilon - 1, 1 - \epsilon]$.

The last condition in the definition is not essential. However, it simplifies some of the considerations like the definition of the generating function below. Furthermore, it helps to quantify "the amount of twist" present in a twist map. The restriction of the homeomorphism $\bar{\phi}$ to the "bottom" circle $S^1 \times \{-1\}$ has a rotation number defined up to an integer (Definition 4.3.6). Fixing a lift of this restriction with rotation number

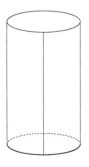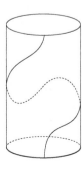

Figure 14.2.1. A twist map.

ρ_- defines a unique lift of $\bar{\phi}$; the restriction of this lift to the "top" circle $S^1 \times \{1\}$ has a uniquely defined rotation number ρ_+. The interval $[\rho_-, \rho_+]$ changes if the original lift is chosen differently, but only by an integer translation. We call this interval *the twist interval* of the twist map ϕ. This way the notion of the twist map can be defined for any homeomorphism of the closed cylinder that preserves boundary components.

Basic geometry, (6.4.4), and Proposition 6.4.2 imply

Proposition 14.2.2 *The billiard map* $\phi: C \to C$ *of the open cylinder* $C = S^1 \times (-1, 1)$ *is an area-preserving differentiable twist map with the additional property that a lift* Φ *satisfies* $\Phi_1(x, y) \xrightarrow[y \to -1]{} x$ *and* $\Phi_1(x, y) \xrightarrow[y \to 1]{} x + 1$. *Thus the twist interval for any billiard map is* $[0, 1]$.

14.2.2 Generating Function for a Twist Map

Area-preserving twist maps share most essential features with billiard maps. We begin by showing that every area-preserving differentiable twist map can be represented via a smooth generating function in the form (6.4.3). To avoid dealing with possible overlaps of domains in the cylinder and counting areas with multiplicities, we describe the generating function for the lift.

Let $\Phi(x, y) = (x', y')$. Fixing x and x', consider the "triangle" formed by the vertical segment with the coordinate x', the image under Φ of the vertical segment with the coordinate x, and the segment of the bottom horizontal connecting the bottom of the latter curve to that of the former. Let $H(x, x')$ be the area of this domain. (See Figure 14.2.2.) Then

$$\frac{\partial}{\partial s'} H(x, x') = y'.$$

Applying Φ^{-1} and using preservation of area gives

$$\frac{\partial}{\partial s} H(x, x') = -y.$$

While the definition of the function does not require the differentiable twist condition, this is needed to ensure existence of the second derivatives $\partial^2 H/\partial s^2$ and $\partial^2 H/\partial s'^2$. On the other hand, the twist condition guarantees that the mixed partial derivative $\partial^2 H/\partial s \partial s' = -\partial y/\partial s'$ exists and is nonpositive. It is negative for a differentiable twist map.

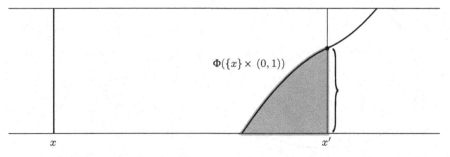

Figure 14.2.2. The generating function.

The generating function is obviously uniquely defined up to an additive constant. An alternative way of constructing it is as follows. By the twist condition y and y' are uniquely defined by the values of x and x'. The differentiable twist condition implies that if $y(x, x')$ and $y'(x, x')$ are defined for a pair of values (x, x'), then they are defined and differentiable in a neighborhood of that pair. In order to find H *locally* one has to check the well-known exactness condition $\partial y / \partial x' = \partial y' / \partial x$. This requires a small calculation, which shows that exactness is equivalent to area-preservation. Thus the generating function H is defined locally up to an additive constant, and it can be extended to all admissible pairs (x, x') by gluing the local definitions and adjusting constants. One expects that $H(x + 1, x' + 1) - H(x, x')$ is a constant, and the fact that the map Φ preserves a strip implies that this constant vanishes.

The principal qualitative properties known for billiard maps extend to area-preserving differentiable twist maps. One obvious advantage of using this notion is that it covers many other important cases such as periodically forced oscillators, neighborhoods of most elliptic points in general area-preserving maps, small perturbations of Hamiltonian systems with two degrees of freedom, and outer billiards. An outer billiard map is defined on points outside a convex curve by drawing a tangent line and moving to the opposite point on it (equal distance to the point of tangency). See Figure 14.2.3. From the intrinsic point of view, the difference between considering, say, billiard maps directly or treating them as twist maps is similar to the difference between the Lagrangian and Hamiltonian formalisms in classical mechanics and, in fact, constitutes a particular case of the discrete-time version of that duality. In general, the Hamiltonian approach makes the dynamical (in the sense used in this book) nature of the problem more apparent by considering dynamical systems in the phase space without making a particular distinction between positions and momenta. The Lagrangian approach, which separates the configuration space and divides phase space coordinates into positions and momenta (or velocities), is sometimes useful because it provides a nice geometric visualization of methods and results.

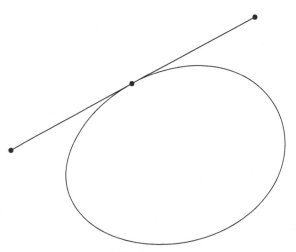

Figure 14.2.3. Outer billiard.

14.2.3 Birkhoff Periodic Orbits

Birkhoff periodic orbits appeared in the previous section for billiards. Now we discuss them in the context of twist maps.

Definition 14.2.3 Given a twist map ϕ and its lift Φ, a point $w \in C$ is said to be a *Birkhoff periodic point of type* (p, q) and its orbit a *Birkhoff periodic orbit of type* (p, q) if for a lift $z \in S$ of w there exists a sequence $((x_n, y_n))_{n \in \mathbb{Z}}$ in S such that

(1) $(x_0, y_0) = z$,
(2) $x_{n+1} > x_n \ (n \in \mathbb{N})$,
(3) $(x_{n+q}, y_{n+q}) = (x_n + 1, y_n)$,
(4) $(x_{n+p}, y_{n+p}) = \Phi(x_n, y_n)$.

Remark 14.2.4 The sequence (x_n, y_n) does not parametrize the orbit according to the "dynamical ordering" induced by passing from (x, y) to $\Phi(x, y)$, but rather in the "geometric ordering" of its projection to S^1 (see Section 14.1.1). In fact, this order coincides with the ordering of iterates of the rational rotation $R_{p/q}$ on the circle. Moreover, the projection of a Birkhoff periodic orbit of type (p, q) to the circle is a finite set, and the map induced by projecting Φ can be extended piecewise linearly to a homeomorphism of the circle.

Now we give the more technical discussion of the existence of the first (action-minimizing) Birkhoff periodic orbit promised in Section 14.1.1, including the explanation of why the minimum is achieved inside the space of (p, q)-states.

Theorem 14.2.5 *Let $\phi: S \to S$ be a differentiable twist map. If $p, q \in \mathbb{N}$ are relatively prime and p/q lies in the twist interval of ϕ, then there is a Birkhoff periodic orbit of type (p, q) for ϕ.*

Proof Take a lift Φ of ϕ such that p/q lies within the twist interval for that lift. Denote the restrictions of Φ to the "bottom" $\mathbb{R} \times \{-1\}$ and the "top" $\mathbb{R} \times \{-1\}$ by Φ_- and Φ_+, correspondingly. We obtain the Birkhoff periodic orbit by finding the sequence of its x-coordinates as a global minimum of an appropriate action defined on a space of sequences of points in $\mathbb{R}$. As reasonable candidates for x-coordinates of an orbit consider the following space Σ. First, let $\tilde{\Sigma}$ be the set of nondecreasing sequences $(x_n)_{n \in \mathbb{Z}}$ of real numbers such that

(14.2.1) $$x_{n+q} = x_n + 1$$

and

(14.2.2) $$\Phi(x_n \times [\epsilon - 1, 1 - \epsilon]) \cap (x_{n+p} \times [\epsilon - 1, 1 - \epsilon]) \neq \varnothing,$$

where $\epsilon > 0$ is as follows: Since p/q belongs to the twist interval of Φ, there exists $\delta \in (0, 1)$ such that $x_{k+1} \leq \Phi_-(x_k) + \delta \ (k = 0, 1, \ldots, q - 1)$ implies $x_q < x_0 + p$ and similarly, if $x_{k+1} \geq \Phi_+(x_k) - \delta \ (k = 0, 1, \ldots, q - 1)$, then $x_q > x_0 + p$. Take $\epsilon > 0$ such that

$$\bigcup_{i=0}^{q-1} \Phi^i(\mathbb{R} \times ((-1, \epsilon - 1] \cup [1 - \epsilon, 1))) \subset \mathbb{R} \times ((-1, \delta - 1] \cup [1 - \delta, 1)).$$

We call these sequences *ordered states of type* (p, q).

Thus any orbit on S whose x-coordinates satisfy (14.2.1) and (14.2.2) has y-coordinates in $(\epsilon - 1, 1 - \epsilon)$. Define an equivalence relation $\sim$ on $\tilde{\Sigma}$ by $x \sim x'$ if $x_i - x_i' = k$ for all i and some fixed $k \in \mathbb{Z}$. Let $\Sigma := \tilde{\Sigma}/\sim$ be the set of equivalence classes.

Condition (14.2.1) is periodicity, and, since p/q belongs to the twist interval of Φ, condition (14.2.2) ensures that there is a point (x_n, y_n) with $\Phi(x_n, y_n) = (x_{n+p}, y_{n+p})$ for some y_{n+p}. A sequence satisfying (14.2.1) and (14.2.2) is usually not the x-projection of an orbit, but we will find a sequence that is, and the corresponding orbit is the desired Birkhoff periodic orbit of type (p, q).

Each sequence has only q "independent variables" $x_0, \ldots, x_{q-1}$, say, by (14.2.1), that is, $\tilde{\Sigma}$ is naturally embedded in $\mathbb{R}^q$. Condition (14.2.2) applied inductively shows that $\{x_n - x_0\}_{n=0}^{q-1}$ is bounded for any $x \in \tilde{\Sigma}$, so Σ is a closed and bounded, hence compact, subset of $\mathbb{R}^q/\mathbb{Z} \sim \mathbb{R}^{q-1} \times S^1$.

Define the *action functional*

$$L(s) := \sum_{n=0}^{q-1} H(x_n, x_{n+p})$$

on Σ, where H is the generating function. Since p and q are relatively prime, it follows from (14.2.1) that $L(s) = \sum_{n=0}^{q-1} H(x_j, x_{j+np})$ for any $j \in \mathbb{Z}$. Since L is invariant under the integer shift, it is defined on the compact set Σ and hence attains its maximum and minimum, but it could be on the boundary. We show that the minimum corresponds to a Birkhoff periodic orbit of type (p, q) and deduce that it is also not on the boundary.

Consider any sequence $x \in \Sigma$. It is not constant by (14.2.1). So for any $m \in \mathbb{Z}$ there are $n \in \mathbb{Z}$ and $k \geq 0$ such that $n \leq m \leq n + k$ and $x_{n-1} < x_n = \cdots = x_{n+k} < x_{n+k+1}$. (If $k > 0$, then x is a boundary point of $\tilde{\Sigma}$.) Define $h_1(x, x')$ and $h_2(x, x')$ by

$$(14.2.3) \qquad \Phi(x, h_2(x, x')) = (x', h_1(x, x')).$$

Since x is nondecreasing, the twist condition (Definition 14.2.1.(2)) implies

$$\epsilon - 1 \leq h_1(x_{n+k-p}, x_{n+k}) \leq \cdots \leq h_1(x_{n-p}, x_n) \leq 1 - \epsilon,$$

$$\epsilon - 1 \leq h_2(x_n, x_{n+p}) \leq \cdots \leq h_2(x_{n+k}, x_{n+k+p}) \leq 1 - \epsilon,$$

so either

$$(14.2.4) \qquad h_2(x_n, x_{n+p}) < h_1(x_{n-p}, x_n)$$

or

$$(14.2.5) \qquad h_1(x_{n+k-p}, x_{n+k}) < h_2(x_{n+k}, x_{n+k+p})$$

or

$$(14.2.6) \qquad h_1(x_{n+l-p}, x_{n+l}) = h_2(x_{n+l}, x_{n+l+p}) \text{ for } l \in \{0, \ldots, k\}.$$

For case (14.2.4) note that considering $s = x_n$ as an independent variable and keeping all other x_i fixed we have

$$\frac{d}{ds}|_{s=x_n} L(x) = \frac{d}{ds}|_{s=x_n} \sum_{i=0}^{q-1} H(x_i, x_{i+p}) = \frac{d}{ds}|_{s=x_n} (H(x_{n-p}, s) + H(s, x_{n+p}))$$

$$= h_1(x_{n-p}, x_n) - h_2(x_n, x_{n+p}) > 0,$$

and that by (14.2.4) we can decrease x_n – and hence $L(x)$ – slightly, without leaving Σ, so x is not a minimum. For case (14.2.5) we find similarly that setting $s = x_{n+k}$

we get $d/ds|_{s=x_{n+k}} L(x) < 0$, so by (14.2.5) we can increase x_{n+k} slightly – and hence decrease $L(x)$ slightly – without leaving Σ, so s is not a minimum. Thus if $x = (x_m)_{m \in \mathbb{Z}}$ is a minimum, then for all $m \in \mathbb{Z}$ the analysis above leads to (14.2.6). Therefore

$$(14.2.7) \qquad h_1(x_{m-p}, x_m) = h_2(x_m, x_{m+p}) \qquad \text{for all } m \in \mathbb{Z}.$$

Setting $(s_n, y_n) = (x_n, h_1 x_{n-p}, x_n)$ now yields a periodic orbit.

Now $y_n \in (\epsilon - 1, 1 - \epsilon)$ for *all* $n \in \mathbb{Z}$, since having $y_n \leq \epsilon - 1$ for any $n \in \mathbb{Z}$ implies $y_n < \delta$ for all $n \in \mathbb{Z}$, which is incompatible with (14.2.1) and (14.2.2) by the choice of δ. Thus, to show that (x_n, y_n) is a Birkhoff periodic orbit of type (p, q) and s is not on the boundary of Σ, it suffices to show that $s_n = x_n$ is strictly increasing.

Suppose $s_n = s_{n+1}$. By choosing a different n, if necessary, we may assume that either $s_{n-1} < s_n$ or $s_{n+1} < s_{n+2}$ (since s is not constant). Then, since s is nondecreasing, the twist condition and (14.2.7) yield $y_{n+1} = h_1(s_{n-p+1}, s_{n+1}) \leq h_1(s_{n-p}, s_{n+1}) \leq h_1(s_{n-p}, s_n) = y_n = h_2(s_n, s_{n+p}) \leq h_2(s_n, s_{n+p+1}) = y_{n+1}$ with at least one strict inequality, which is absurd.

Thus we have found a Birkhoff periodic orbit of type (p, q) such that the sequence of its x-coordinates is a global minimum of L in the interior of Σ. $\square$

The construction of the second (minimax) Birkhoff periodic orbit uses a version of the mountain pass principle applied to the same functional restricted to the space of states that stay between the state defining a maximal Birkhoff periodic orbit and its shift. For the discussion of nonperiodic orbits in the rest of this chapter only the existence of one Birkhoff periodic orbit is essential.

14.2.4 Ordered Orbits
In this section we prove that any order-preserving orbit of a twist map forms part of the graph of a Lipschitz function whose Lipschitz constant can be taken bounded on any closed annulus in S. As in Section 14.2.1, we frequently work with lifts.

Definition 14.2.6 (Compare Definition 14.2.3.) Consider a twist diffeomorphism $\phi: C \to C$. An orbit segment (or orbit) $\{(x_m, y_m), \ldots, (x_n, y_n)\}$ of ϕ with $-\infty \leq m < n \leq \infty$, which may be infinite in one or both directions, is said to be *ordered* or *order-preserving* if $x_i \neq x_j$ when $i \neq j$ and $(i, j) \neq (n, m)$ and ϕ preserves the cyclic ordering of the x-coordinates; that is, if x_i, x_j, x_k, where $i, j, k < n$, are positively ordered (with respect to a chosen orientation on S^1), then $x_{i+1}, x_{j+1}, x_{k+1}$ are ordered in the same way

Lemma 14.2.7 *Let* $\Phi: \mathbb{R} \times (-1, 1) \to \mathbb{R} \times (-1, 1)$ *be the lift of a twist diffeomorphism* $\phi: C \to C$ *(not necessarily area-preserving). If* $(x_i, y_i) = F^i(x_0, y_0)$ *and* $(x_i', y_i') = F^i(x_0', y_0')$ *and* $x_i' > x_i$ *for* $i = -1, 0, 1$, *then there exists* $M \in \mathbb{R}$ *such that* $|y_0' - y_0| < M|x_0' - x_0|$. *$M$ can be chosen uniformly on any closed annulus in* C.

Proof Suppose first that $y_0' < y_0$. If $(\bar{x}, \bar{y}) = \Phi(x_0', y_0)$, then the twist condition yields

$$\bar{x} > x_1' + c(y_0 - y_0'),$$

where c is bounded away from zero on any closed annulus in C. On the other hand, differentiability of ϕ implies that there is a constant L (bounded on compact annuli in C) such that

$$x_1' > x_1 > \tilde{x} - L(x_0' - x_0).$$

Taking $M = Lc^{-1}$ we obtain the claim. If $y_0' > y_0$, repeat the same argument with ϕ^{-1} instead of ϕ. $\square$

Corollary 14.2.8 *Consider an area-preserving twist map $\phi: C \to C$ and an order-preserving orbit segment $\{(x_m, y_m), \ldots, (x_n, y_n)\}$ with $-\infty \le m < n \le \infty$ of ϕ that is contained in a closed annulus in C. Then $|y_i - y_j| < M|x_i - x_j|$ for all i, j such that $m < i, j < n$.*

Proof Apply Lemma 14.2.7 to the triples $(i - 1, i, i + 1)$ and $(j - 1, j, j + 1)$. $\square$

This corollary shows that the closure E of an ordered orbit is contained in the graph of a Lipschitz function $\varphi: S^1 \to (-1, 1)$. Note that $\phi_{\restriction E}$ projects to a homeomorphism of the projection of E to S^1, which we can also extend linearly into the gaps of that set to obtain a circle homeomorphism. We can thus define the *rotation number* of an ordered orbit to be the rotation number of this induced circle homeomorphism. So the intrinsic dynamics of ordered orbits of a twist map of the annulus is essentially one-dimensional. Now we will see that for every rotation number from the twist interval of a twist map some such one-dimensional dynamics is represented the twist map.

14.2.5 Aubry–Mather Sets

Our next goal is to show that every irrational number from the twist interval is the rotation number of an ordered orbit. We will furthermore see that such orbits are not isolated like Birkhoff periodic orbits; there are many of them for each rotation number. The construction of such orbits can be carried out by a rather sophisticated version of the variational approach applied to an appropriate infinite-dimensional space. This is a complicated but powerful method, whose further development yields lots of additional information both about ordered orbits and about more complicated types of orbits. However, it is quite remarkable that fairly simple continuity arguments yield ordered periodic orbits with irrational rotation numbers as limits of Birkhoff periodic orbits. Thus the results of the rest of this section (with the exception of Theorem 14.2.15) do not use preservation of area directly but only the existence of Birkhoff periodic orbits (which can be shown under weaker assumptions).

Definition 14.2.9 Let $\phi: C \to C$ be a twist map. A closed invariant set $E \subset C$ is called an *ordered set* if it projects one-to-one to a subset of the circle and ϕ preserves the cyclic order on E. An *Aubry–Mather set* is a minimal ordered invariant set projecting one-to-one on a Cantor set of S^1.

Any orbit in an ordered set is an ordered orbit. The complement of the projection of an Aubry–Mather set is the union of countably many intervals on the circle. We call those intervals the *gaps* of the Aubry–Mather set. The endpoints of each interval are the projections of points on the Aubry–Mather set which we shall also call *endpoints*. Corollary 14.2.8 immediately yields:

Corollary 14.2.10 *Let* $\phi\colon C \to C$ *be a twist diffeomorphism and A an Aubry–Mather set for* ϕ. *Then there is a Lipschitz-continuous function* $\varphi\colon S^1 \to (-1, 1)$ *whose graph contains A.*

Proof Corollary 14.2.8 gives us such a function defined in the projection of A to S^1. Extending it linearly through the gaps of that Cantor set gives a function with the same Lipschitz constant. $\square$

We define the *rotation number* of an Aubry–Mather set or an invariant circle to be the rotation number of any of its orbits as defined at the end of Section 14.2.4. We can now prove one of the central results in the theory of twist maps.

Theorem 14.2.11 *Let* $\phi\colon C \to C$ *be an area-preserving differentiable twist map. For any irrational number* α *from the twist interval of* ϕ *there exists an Aubry–Mather set A with rotation number* α *or an invariant circle graph(φ), where* φ *is a Lipschitz function, with rotation number* α.

Proof Let p_n/q_n be a sequence of rationals in lowest terms that approximates α. Apply Theorem 14.2.5 and take any sequence w_n of Birkhoff periodic orbits of type (p_n, q_n). According to Corollary 14.2.8, we can construct a Lipschitz function $\varphi_n\colon S^1 \to (-1, 1)$ whose graph contains w_n. By an argument similar to the one that yielded (14.2.2), we observe that all of these orbits are contained in a closed annulus in C, so the Lipschitz constant can be chosen independently of n. Using precompactness of this equicontinuous family of functions (the Arzelá–Ascoli Theorem) we may without loss of generality assume that these functions converge to a Lipschitz function φ. The graph of φ may not be ϕ-invariant, but it always contains a closed ϕ-invariant set A, which is obtained as follows. The domain of φ_n contains the projection of the Birkhoff periodic orbit of type (p_n, q_n) to S^1. These Birkhoff periodic orbits of type (p_n, q_n) are closed ϕ-invariant subsets of C, and thus in the topology of the Hausdorff metric (see Definition A.1.28) they have an accumulation point $A \subset C$. The set A obviously belongs to the graph of φ and is ϕ-invariant by Lemma A.1.27. Furthermore, ϕ preserves the cyclic ordering of A (since this is true for the Birkhoff periodic orbit w_n and is a closed property). If we denote by ϕ_n the extensions to S^1 of the projections of ϕ from the Birkhoff periodic orbits of type (p_n, q_n) to S^1, and by ϕ_α the extension of the projection of $\phi_{\restriction A}$ to S^1, then $\phi_n \to \phi_\alpha$ uniformly. Thus by continuity of the rotation number in the C^0-topology (Proposition 4.4.5) the rotation number of A is α. Consider now the minimal set of ϕ_α. By the dichotomy of Proposition 4.3.19 it is either the whole circle or it is an invariant Cantor set. In the latter case the image of this Cantor set under $\mathrm{Id} \times \varphi$ is then an Aubry–Mather set with rotation number α. $\square$

Remark 14.2.12 The Aubry–Mather set obtained in Theorem 14.2.11 may be a subset of an invariant circle for ϕ. However, if both the map and the invariant circle are C^2, then the restriction of the map to the circle is a C^2 diffeomorphism of the circle, and by the Denjoy Theorem (see Section 4.4.3) it is topologically transitive. Thus for an Aubry–Mather set to lie on an invariant circle either the map, or the circle, or both should fail to be C^2. Michael Herman found a remarkable construction by which he managed to embed Denjoy-type examples of nontransitive $C^{2-\epsilon}$ ($\epsilon > 0$) circle diffeomorphisms into $C^{3-\epsilon}$ area-preserving differentiable twist maps, thus gaining an extra derivative over obvious constructions. It is not known however whether a C^3 diffeomorphism can have an invariant circle with an Aubry–Mather set.

The Hausdorff limit of the Birkhoff periodic orbits of type (p_n, q_n) may be larger than an Aubry–Mather set, although it is always an order-preserving set. If it is not a minimal set, then it contains a set of orbits that are homoclinic to the Aubry–Mather set. By taking Hausdorff limits of the minimax Birkhoff periodic orbits and using some careful variational estimates one can show that such orbits always exist.

Replacing the Birkhoff periodic orbits w_n in the preceding arguments by arbitrary invariant ordered sets that converge in the Hausdorff metric, we obtain the following:

Proposition 14.2.13 *The rotation number of an ordered invariant set is continuous in the topology of the Hausdorff metric (see Definition 4.1.25).*

This, in turn, implies

Corollary 14.2.14 *The rotation number of ordered orbits is a continuous function of the initial condition.*

Proof Let $x_n \to x$ be a convergent sequence of points with ordered orbits. Without loss of generality we may assume that the rotation numbers α_n of the orbits of the x_n converge. Consider the collection of orbits of the x_n. By compactness of the topology of the Hausdorff metric (see Lemma A.1.26) it contains a subsequence that converges to an ordered set that contains the orbit closure of x. Thus, by Proposition 14.2.13, the limit of the rotation numbers of the orbit of x_n is the rotation number of the orbit of x. $\square$

We can now show that for any irrational number there is at most one invariant circle with that rotation number.

Theorem 14.2.15 *Let $\phi\colon C \to C$ be an area-preserving twist map and α an irrational number in the twist interval. Then ϕ has at most one invariant circle of the form graph(φ) with rotation number α. If there is such an invariant circle, then ϕ has no Aubry–Mather sets with rotation number α outside this circle, and hence has at most one such Aubry–Mather set.*

Remark 14.2.16 It is, in fact, possible for a twist map to have several invariant circles with the same rational rotation number. This is the case in the elliptic billiard (see Figure 6.3.9), where two branches of heteroclinic loops form a pair of invariant circles with rotation number $1/2$. A time-t map (for small t) of the mathematical pendulum (Section 6.2.2) exhibits a similar phenomenon for rotation number zero.

Lemma 14.2.17 *Suppose ϕ has an invariant circle R [of the form graph(φ)] with rotation number α. Then every order-preserving orbit whose closure is disjoint from R has rotation number different from α.*

Proof The circle R divides the annulus C into an upper and a lower component. Suppose x is a point in the upper component of $C \smallsetminus R$ whose orbit is order-preserving and bounded away from R. Then ϕ restricted to the orbit of x projects to a map of a subset E of the circle S^1. We want to extend it to a map ϕ_2 of S^1, which is strictly ahead (in the sense of Definition 4.4.6) of the map ϕ_1 induced by $\phi_{\restriction R}$, that is, $\phi_1 \prec \phi_2$. This relation holds on E already, so we need only take care to extend carefully from E. Extending to the closure of E does not change the strict inequalities, since we have the twist condition and the assumption that the orbit of x is bounded away from R. To define ϕ_2 on the intervals complementary to $\bar{E}$ denote the endpoints of such an interval by x_1 and x_2 and let $\delta := \min\{\phi_2(x_1) - \phi_1(x_1), \phi_2(x_2) - \phi_1(x_2)\}$. Set $\phi_2(tx_1 + (1-t)x_2) = \max(t\phi_2(x_1) + (1-t)\phi_2(x_2), \delta + \phi_1(tx_1 + (1-t)x_2))$. Then ϕ_2 is monotone and $\phi_1 \prec \phi_2$. Consequently, Proposition 4.4.9 implies that the rotation number of ϕ_2 is greater than α. Likewise, there cannot be an order-preserving orbit of rotation number α in the lower component of $C \smallsetminus R$. $\square$

Proof of Theorem 14.2.15 Suppose there are two invariant circles with rotation number α. Their intersection is invariant, so if at least one of them is transitive, then they are disjoint, which is impossible by the lemma. Otherwise the intersection contains a common Aubry–Mather set A and the two circles form the graphs of two distinct functions φ_1 and φ_2, which coincide on the projection of A. The graphs of both $\max(\varphi_1, \varphi_2)$ and $\min(\varphi_1, \varphi_2)$ are invariant, and hence so is the area between these graphs. But the latter area has to have infinitely many connected components, since it projects into the nonrecurrent complementary intervals to the projection of the Aubry–Mather set. Thus we obtain an open disk with pairwise disjoint images, which is impossible by area-preservation (see the Poincaré Recurrence Theorem 6.1.6). Here we use irrationality of the rotation number, without which there could be finitely many components that are permuted by ϕ.

The lemma also yields the impossibility of having an Aubry–Mather set of rotation number α outside an invariant circle of rotation number α. $\square$

Remark 14.2.18 In the absence of an invariant circle with rotation number α there may be many Aubry–Mather sets with that rotation number. In fact, often there are multiparameter families of such sets.[1]

14.2.6 Homoclinic and Heteroclinic Orbits

We now turn the process around and approximate a rational number by irrational ones and consider the limits of the corresponding Aubry–Mather sets in order to construct nonperiodic orbits with rational rotation number.

---

[1] John Mather, *More Denjoy Minimal Sets for Area Preserving Diffeomorphisms, Commentarii Mathematici Helvetici* **60**, no. 4 (1985), 508–557.

Proposition 14.2.19 *Let $\phi\colon C \to C$ be an area-preserving twist map and p/q a rational number in the twist interval. Then there exists an order-preserving closed ϕ-invariant set with rotation number p/q that is either an invariant circle consisting of periodic orbits or contains nonperiodic points. Moreover in the latter case the two endpoints of each complementary interval are nonperiodic.*

Proof Let $(\alpha_n)_{n\in\mathbb{N}}$ be a sequence of irrational numbers in the twist interval that approximates p/q. Consider the corresponding invariant minimal order-preserving sets A_n with rotation number α_n. Without loss of generality we may assume that the A_n's converge to a set A in the topology of the Hausdorff metric as $n \to \infty$. A is clearly ϕ-invariant and ordered. If infinitely many of the A_n's are circles, then A is also a circle and by continuity of the rotation number the restriction of ϕ to this circle has rotation number p/q. By the classification of circle maps with rational rotation number (see Proposition 4.3.12) we are done in this case. So we may assume that all A_n's are Aubry–Mather sets. To understand the dynamics of A we consider the gaps, that is, the intervals in S^1 complementary to the projection of A to S^1. Each of these gaps $G \subset S^1$ has a well-defined length $l(G)$, and we want to show that the two endpoints of such a gap are not periodic.

A gap G of A is the limit of the corresponding gaps G_n of A_n in the Hausdorff metric. Denote by ϕ_n an extension to a circle homeomorphism of the projection of $\phi_{\lceil A_n}$ to S^1 and by ϕ_0 the same extension corresponding to $\phi_{\lceil A}$. Since ϕ_n has irrational rotation number, the images of the gap G_n under the iterates of ϕ_n are pairwise disjoint, so $\sum_{m\in\mathbb{N}} l(f_n^m(G_n)) \leq 1$. If both endpoints of G are periodic, then the gap G is periodic, that is, $\sum_{n\in\mathbb{N}} l(\phi_0(G))$ diverges. But $l(\phi_n^m G_n) \to l(\phi_0^m G)$ for all $m \in \mathbb{N}$, which gives a contradiction. Thus one of the endpoints of G is nonperiodic.

The other endpoint of the gap G must then also be nonperiodic, since otherwise $\phi_0^q(G)$ is a gap that intersects G nontrivially without coinciding with G. $\square$

We can thus describe the structure of such an invariant set in the generic case when it contains only finitely many periodic orbits:

Corollary 14.2.20 *If a closed order-preserving ϕ-invariant set A with rational rotation number p/q contains only finitely many periodic orbits, then there is a complete set of heteroclinic connections in the following way: If $\gamma_1, \ldots, \gamma_s$ denote the periodic orbits in A, ordered according to the induced cyclic ordering of the circle, then there are heteroclinic orbits $\sigma_1, \ldots, \sigma_n$ such that either*

$$\gamma_1 = \omega(\sigma_s) = \alpha(\sigma_1),$$

$$\gamma_2 = \omega(\sigma_1) = \alpha(\sigma_2),$$

$$\vdots$$

$$\gamma_s = \omega(\sigma_{s-1}) = \alpha(\sigma_s)$$

or that the same situation holds with α and ω interchanged. Here α and ω denote the α- and ω-limits sets of an orbit (see Definition 4.3.18). If $s = 1$, the orbit σ_1 is, of course, a homoclinic one.

14.3 INVARIANT CIRCLES AND REGIONS OF INSTABILITY

14.3.1 Global Structure of Invariant Circles

In the previous section we encountered invariant curves for a twist map that appear as limits of Birkhoff periodic orbits and hence are graphs of Lipschitz maps $S^1 \to [-1, 1]$. The existence of such a circle is an alternative to the existence of a Cantor-like Aubry–Mather set. While in general these two possibilies are not mutually exclusive, they often are, namely, if the circle has dense orbits. Of course, the basic difference between an invariant circle and an Aubry–Mather set is that the former separates the phase space. Since the boundary components are preserved, any orbit that starts on one side of an invariant circle stays there forever. Thus the existence of even a single invariant circle provides substantial information about the behavior of *all* orbits. It is natural to ask whether there are invariant sets other than Lipschitz graphs that separate the phase space. Such sets may appear around certain periodic orbits, as we already saw in the case of the elliptic billiard and the pairs of invariant curves surrounding the stable period-2 orbits, which correspond to hyperbolas as caustics (Section 6.3.5.3). However, if we only consider sets that separate the cylinder in such a way that the two boundary components lie in different pieces, then the following classical result of Birkhoff shows that this only happens due to the presence of invariant curves that are Lipschitz graphs.

Theorem 14.3.1 [2] *If U is an open invariant set of a differentiable twist map ϕ that contains a neighborhood of the "bottom" $S^1 \times \{-1\}$ and has connected boundary, then the boundary of U is the graph of a Lipschitz function.*

The dynamics of a twist map on the union of invariant circles is reasonably well understood in terms of the dynamics of circle maps (Section 4.3). Thus one needs to understand what happens outside of the union of invariant circles.

First consider a simple example. For the billiard in an ellipse there is exactly one invariant circle with any rotation number other than $1/2$; such circles correspond to confocal ellipses as caustics. For the rotation number $1/2$ there are two invariant circles corresponding to the orbits passing through the foci: In the phase space picture (Figure 6.3.9) they are formed by the upper and lower branches of the separatrices of the hyperbolic orbit of period 2 (the larger axis or diameter). The rest are orbits circling around the elliptic orbit of period 2 (the smaller axis), corresponding to the hyperbolas as caustics. Such a picture is only possible because an invariant circle with rational rotation number does not have to be unique. By Theorem 14.2.15 there is at most one invariant circle with a given irrational rotation number, so such invariant circles are ordered by rotation numbers. Each circle lies completely within one component of the complement to any other circle. Furthermore, the limit of a sequence of invariant circles is an invariant circle and the rotation number is continuous on the set of invariant circles.

---

[2] Katok and Hasselblatt, *Introduction to the Modern Theory of Dynamical Systems*, Theorem 13.2.13.

14.3.2 Regions of Instability

Thus, each complementary interval [a, b] to the set of rotation numbers of invariant circles generates a unique region whose boundary components are disjoint invariant circles with rotation numbers a and b, correspondingly. Such a region is called a *region of instability*.

We showed that there are billiard maps with no caustics (Theorem 6.4.7) and hence without invariant curves. For such maps the whole cylinder is a single region of instability, and hence for *every* irrational number between 0 and 1 there is a nowhere-dense Aubry–Mather set.

A simple coordinate change allows us to consider the restriction of the twist map to a region of instability as a twist map whose twist interval is the interval between the rotation number of the two boundary invariant circles. Thus, for any rational (correspondingly irrational) number from that interval there are necessarily Birkhoff periodic orbits (correspondingly Aubry–Mather sets) inside the region of instability. There are no "barriers" in the form of invariant circles that prevent orbits from wandering around the region, in particular bouncing between the boundary components.

1. Difficulties Beyond Comprehension. The dynamics in the regions of instability is complicated. Many specific orbits can be found either for all cases or in "typical" situations, and there are plausible conjectures about the behavior of most orbits. ("Most" may mean either orbits covering an open dense set or a complement to a null set.) It seems that a rigorous analysis of typical orbits is beyond the reach of currently available or conceivable methods even under the most optimistic projections. The difficulty of this problem likely exceeds that of some famous problems with huge price tags (such as the Poincaré conjecture in three-dimensional topology) and no substantial progress is to be expected during the twenty-first century.

2. Entropy and Horseshoes. Disallowing exponential growth puts severe restrictions on the dynamics of twist maps.

Theorem 14.3.2 [3] *A twist map with zero topological entropy has invariant circles for any rotation number in the twist interval; in particular, there are no regions of instability.*

Using Theorem 12.4.1 we obtain

Corollary 14.3.3 *For any C^2 twist diffeomorphism and any region of instability there is a horseshoe and hence a hyperbolic periodic point with a transverse homoclinic point in that region.*

Thus all complexity compatible with area preservation and discussed in Chapter 12 appears within any region of instability.

---

[3] Sigurd B. Angenent, A Remark on The Topological Entropy and Invariant Circles of an Area Preserving Twistmap, in *Twist Mappings and Their Applications*, Springer–Verlag, New York, 1992, pp. 1–5.

3. Special Orbits by Variational Methods. Further developments of the variational approach for twist maps involve critical points of properly constructed action functionals on carefully and often ingeniously constructed spaces of states satisfying conditions that preclude such simple solutions as ordered orbits. This approach has been developed to great depth by John Mather.

It yields orbits that travel from one boundary component of a region of instability to the other in either direction, that is, orbits homoclinic and heteroclinic to the boundary components, and orbits that oscillate between the boundary components in a prescribed fashion. Furthermore, there are heteroclinic orbits for pairs of Aubry–Mather sets with different rotation numbers and more complicated orbits that wander among different collections of such sets in a prescribed fashion. And all of this wealth only covers a set that is expected to be always thin both metrically (a null set) and topologically (nowhere dense).

4. Complexity in Typical Situations. There are dynamical restrictions on the behavior of Birkhoff periodic orbits. For example, the maximal orbits cannot be elliptic, that is, have a pair of complex conjugate eigenvalues. Such orbits, if nondegenerate, are hyperbolic, typically with heteroclinic tangles between different points on the orbit. Thus the horseshoe structure that always appears due to positivity of entropy in this case shows up in a rather specific form (Theorem 12.3.1).

The minimax Birkhoff periodic orbits are often elliptic. It is a widely held illusion, especially among scientists and engineers dealing with models that produce twist or similar maps, that barring a degeneracy (a double eigenvalue one) these orbits are always elliptic. This is not the case. The minimax orbits may be hyperbolic but with negative eigenvalues, as in the famous example of the billiard in the "stadium" shown in Figure 14.1.3.[4] Still, ellipticity of the minimax orbits is a common phenomenon, for example, it happens for small perturbations of integrable twist maps $f(x, y) = (x + g(y), y)$. Elliptic periodic orbits typically produce islands of relative stability due to the fact that around such an orbit the period map in properly chosen coordinates becomes a twist and in fact has invariant curves surrounding the orbits (imagine what happens in the areas of instability for *that* little twist map and then try to iterate the picture in your imagination!). These islands are thus excluded at least from the global complexity play since, for example, all of the wealth of variationally construced orbits as well as heteroclinic tangles described above lie outside the islands.

5. The Impossible Problem of Peaceful Coexistence. Thus the orbit picture for a "typical" area-preserving twist map in a region of instability emerges. There are elliptic periodic points surrounded by islands of relative stability from which orbits cannot escape. There are also hyperbolic points with homoclinic and heteroclinic tangles as well as other orbits that exhibit hyperbolic behavior in various ways. The set of questions referred to above as difficult beyond comprehension concern

---

[4] Katok and Hasselblatt, *Introduction to the Modern Theory of Dynamical Systems*, Section 9.2.

the prevalence of either type of behavior and the mechanisms of their forced coexistence. Here is a sample.

(1) Is the union of elliptic islands ever (or typically) dense?
(2) Is the complement to the union of elliptic islands ever (or typically) a null set?
(3) Does the closure of the stable manifold of a hyperbolic Birkhoff periodic orbit ever (or typically, or always under natural nondegeneracy assumptions) contain an open set?
(4) Is the union of closures from the previous item ever (or typically, or always) a null set?
(5) Is the Kolmogorov entropy of a C^2 area-preserving twist map ever (or typically) positive?

14.4 PERIODIC POINTS FOR MAPS OF THE CYLINDER

14.4.1 Cylinder Maps with Weak Twist Conditions

Part of the theory outlined in the previous section can be generalized to maps with much weaker twist-like properties. This part concerns the existence of infinitely many periodic points of growing periods. A classical result in this direction is much older than the theory of twist maps. It was formulated and proved in some particular cases by Poincaré shortly before his death. Birkhoff, who realized the importance of the results in many problems of classical mechanics and geometry, gave a rigorous proof.

Remember that the notion of twist interval is defined in Section 14.2.1 for any homeomorphism of a closed cylinder that preserves boundary components. We assume the latter condition throughout this section without restating it.

Theorem 14.4.1 (Poincaré's Last Geometric Theorem)[5] *Let f be an area-preserving homeomophism of a closed cylinder whose twist interval contains zero in the interior. Then f has a fixed point inside the cylinder*

Applying this theorem to a properly chosen extension of an iterate of the map one obtains a statement about the existence of periodic orbits, which, while not necessarily ordered, still possess some features of Birkhoff orbits.

Corollary 14.4.2 *Let p and q be relatively prime integers, q > 0, and f be an area-preserving homeomophism of a closed cylinder whose twist interval contains the number p/q in the interior. Then f has a periodic point (s, y) of period q such that for some lift F of f and any lift (x, y) of the point (s, y) one has $F^q(x, y) = (x + p, q)$.*

Of course in the absence of any control inside the cylinder the limit process described in Section 14.2 does not work, and hence there is no natural extension of this result to irrational rotation numbers.

---

[5] George David Birkhoff, *Dynamical Systems*, American Mathematical Society Colloquium Publications **9**.

In the 1980s, John Franks generalized the last result using a much weaker property than rotation numbers of the boundary components being on the different sides of p/q. In fact, his condition is the weakest twist-like condition imaginable. While Franks also assumes a weaker recurrence property than preservation of area, we retain the latter assumption.

Theorem 14.4.3 *Let p and q be relatively prime integers, $q > 0$, f be an area-preserving homeomophism of a closed cylinder, and F be a lift of f. Assume that there are points u and v on the universal cover such that*

$$(14.4.1) \qquad \lim_{n \to \infty} \frac{F^n(u) - u}{n} \le \frac{p}{q} \le \overline{\lim_{n \to \infty}} \frac{F^n(v) - v}{n}.$$

Then f has a periodic point of period q such that any lift w of that point satisfies $F^q(w) = w + (0, p)$.

The solution to Problem 4.3.8 uses a condition that is stronger than (14.4.1) but of a similar type.

If u, v are points on the universal cover and

$$(14.4.2) \qquad \lim_{n \to \infty} \frac{F^n(u) - u}{n} < \overline{\lim_{n \to \infty}} \frac{F^n(v) - v}{n},$$

then there are infinitely many different rational numbers satisfying (14.4.1), and hence f has infinitely many different periodic orbits.

14.4.2 Periodic Points Without Twist

Thus, having only finitely many (or no) periodic points implies that (14.4.2) does not hold. This is equivalent to all points in the annulus having the same rotation number, that is,

$$\lim_{n \to \infty} \frac{F^n(v) - v}{n}$$

exists for every point v on the universal cover and is independent of v. (In fact, independence of v follows from existence for all v.) We may then call this the rotation number of f. If it is irrational, there are no periodic points at all. If it is rational, then there is at least one periodic orbit by Theorem 14.4.3 applied to a boundary circle, but Franks made a remarkable improvement on the last conclusion:

Theorem 14.4.4 *An area-preserving homeomorphism of a closed annulus with rational rotation number has infinitely many periodic points in the interior.*

Therefore, having only finitely many periodic points forces irrational rotation number and hence no periodic points at all, which leads to the final conclusion of the Franks theory.

Theorem 14.4.5[6] *If a homeomorphism of a closed annulus that preserves area and boundary components has at least one (fixed or) periodic point, then it has infinitely many periodic points in the interior.*

This is the main dynamical ingredient in the proof of existence of infinitely many closed geodesics on any 2-sphere, which is discussed in the next section.

In Theorem 14.4.4 it is not guaranteed that there are points of arbitrarily high period, since the identity map has only fixed points. As it turns out, this is the only exception: Any other map has periodic points whose minimal period is arbitrarily high.[7]

Proof outline of Theorem 14.4.4 First, assume that there are only finitely many periodic points. Taking an iterate one can make all of these points fixed and the rotation number equal to zero. Deleting those points, one gets a noncompact surface S of finite genus and an area-preserving homeomorphism of it without fixed points. Using an important tool from low-dimensional topology, the classification (due to William Thurston) of surface homeomorphisms up to homotopy, one can make sure that, by passing to another iterate, the homeomorphism can be made homotopic to the identity, that is, that it can be continuously deformed (within the surface) to the identity map. Now lift the homeomorphism (with the fixed points deleted) to the universal cover of S (which looks quite different from the strip). One can make sense of the rotation number in this setting, and it turns out to be zero. This allows us to produce a perturbation of the map that has a periodic point with zero rotation number in the interior. This contradicts a deep result by Michael Handel, which states that in the absence of fixed points any periodic point must have a nonzero rotation number. □

14.5 GEODESICS ON THE SPHERE

Section 1.3.3 presented the problem of finding closed geodesics on a (deformed) sphere. Put differently, one can consider the motion of particles confined to (the surface of) a dented sphere and moving without any external forces (Section 6.2.8). The specific question is whether there are infinitely many different ways of flying around in a periodic motion. That this is the case has been believed for a long time, but it was only proved in full generality relatively recently. The proof involves a unique blend of techniques from differential geometry, variational calculus, low-dimensional topology, and dynamical systems. While we cannot present it here, we can give an idea of how several dynamical ingredients come into play.

We begin with a quick historical account of the geodesics problem. The existence of at least one closed geodesic is not particularly specific to the sphere (it works for any compact Riemannian manifold). It is based on another version of the mountain pass argument. In the case of a sphere it works like this. Consider smooth one-parameter families of closed curves "anchored" at a particular point and

---

[6] John Franks, Geodesics on S^2 and Periodic Points of Annulus Homeomorphisms, *Inventiones Mathematicae* **108**, no. 2 (1992), 403–418.

[7] Patrice Le Calvez, private communication; unpublished.

covering the whole sphere (that is, a smooth surjective map from $[0, 1] \times [0, 1]$ to S^2 that maps the boundary of the square to a single point). In each such family there is a longest curve. Obviously the length of such a longest curve is bounded away from zero independently of the family. If there is a family for which the length of such a longest curve attains its minimal value over all families, then it follows from the variational description of geodesics that the longest curve is a closed geodesic. The existence of such a family is based on a general argument that shows that the lower bound over all families is the same as over a certain collection of families that can be made compact in a topology in which the maximal-length functional is still continuous.

A somewhat more subtle fact is the existence of a *simple* closed geodesic, that is, one that does not intersect itself. It can still be produced via a variational argument by finding a critical point of a length functional in a proper space of simple closed curves.

A much deeper and more specific result also proved by variational methods is due to Lazar A. Lyusternik and Lev G. Shnirelman.

Theorem 14.5.1 [8] *There are always at least three different simple closed geodesics on the two-dimensional sphere.*

The seminal 1930 paper was available in the West only in severely abridged translation, and other proofs were later published by several mathematicians.

The Lyusternik–Shnirelman Theorem uses a fairly crude but quite remarkable invariant of topological spaces called the *Lyusternik–Shnirelman category*. This is simply the minimal number of subsets into which the space can be decomposed and which are *contractible*, that is, they can be continuously deformed (within themselves) into a point. The connection with variational problems is that for spaces (even infinite-dimensional ones) where differentiation can be defined and hence the notion of a critical point for a function makes sense, the Lyusternik–Shnirelman category gives a lower bound for the number of critical points of *any* differentiable function.

The space to which this criterion is applied in the geodesic problem is constructed from a space of (parametrized) simple smooth curves. The functional is related to the length, and the critical point condition guarantees that the curve is a closed geodesic parametrized by the length parameter. The topological part of the proof is to show that the Lyusternik–Shnirelman category of this space is three; this is naturally independent of the metric.

[8] Lazar A. Lyusternik, and Lev G. Shnirelman, Sur le problème de trois géodesiques fermées sur les surfaces de genre 0, *Comptes Rendus des Séances de l'Académie des Sciences. Série I. Mathématique* **189** (1929), 269–271; Topological Methods in Variational Problems, *Proceedings of the Institute of Mathematics and Mechanics* (1930); Topological Methods in Variational Problems and Their Application to the Differential Geometry of Surfaces, *Akademiya Nauk SSSR i Moskovskoe Matematicheskoe Obshchestvo. Uspekhi Matematicheskikh Nauk* **2**, no. 1(17), (1947), 166–217. The only article available in English is, Lazar A. Lyusternik, The Topology of the Calculus of Variations in the Large, *Trudy Mat. Inst. Steklov* **19** (1947); translation: Translations of Mathematical Monographs **16**, American Mathematical Society, 1966.

The Lyusternik–Shnirelman result is optimal in the sense that there are metrics with no more than three simple closed geodesics. A triaxial ellipsoid is an example.

Even before the work of Lyusternik and Shnirelman, Birkhoff suggested an approach to finding infinitely many closed geodesics. It uses the existence of a simple closed geodesic, which Birkhoff did not know to hold in full generality, although he found the minimax argument mentioned above, which gives one closed geodesic. By the Jordan Curve Theorem a simple closed geodesic divides the sphere into two connected components, which we will call the northern and southern hemispheres. The geodesic itself will be called the equator. Consider the set of all unit tangent vectors to the sphere with foot point on the equator and pointing into the northern hemisphere (either choice of which hemisphere is "northern" will do). The set S of these vectors is parametrized by a circle (the equator) times an interval [angles in $(0, \pi)$] and is hence an open cylinder. Every one of these vectors determines a unique geodesic. If this geodesic, which initially enters the northern hemisphere, leaves the northern hemisphere and arrives at the equator again from the south, then it determines a new vector of the same type. This defines a map from a subset R of S to S. Periodic points of this map correspond to closed geodesics, and different periodic orbits produce different closed geodesics. If $R = S$, the question arises about extending the section map to the boundary of the open cylinder S. If such an extension is possible, then the obvious symmetry shows that the maps on two boundary components are inverse to each other; hence its twist interval contains zero. Now there is a dichotomy. One possibility is that the interval has positive length, and then Theorem 14.4.1 produces infinitely many periodic orbits and hence infinitely many closed geodesics (this conclusion was one of the chief reasons for Birkhoff's interest in Poincaré's Last Geometric Theorem). Alternatively, the rotation numbers on the boundary components are equal to zero. This is where the matter stood for about 60 years.

The results of Franks, Theorem 14.4.3 and Theorem 14.4.4, finish the proof of the existence of infinitely many different closed geodesics in the case when the section map is defined in the open cylinder S and is extendable by continuity to the closed cylinder.

The remaining case was resolved by Victor Bangert, based partly on his earlier joint work with Wilhelm Klingenberg. There are two cases to consider. If the map is defined in the open cylinder but does not extend to the boundary, then it follows that along the original simple closed geodesic there are no "conjugate points". Roughly speaking, nearby geodesics diverge. This implies a special structure of the length functional, which makes it amenable to a variational treatment that produces infinitely many closed geodesics. If the section map is not defined, then Bangert shows that there is another simple closed geodesic without conjugate points and the problem reduces to the previous case.

The final outcome of these considerations is the following remarkable result.

Theorem 14.5.2 *On any smooth sphere there are infinitely many closed geodesics.*

CHAPTER 15

Dynamics, Number Theory, and Diophantine Approximation

Some questions in the theory of numbers that go back to the work of Dirichlet, Jacobi, Kronecker, and Weyl can be viewed retrospectively as one of the principal sources of modern dynamics. A fruitful interaction has developed in both directions: Dynamical methods often provide new and sometimes unexpected insights into problems in analytic number theory, and, on the other hand, algebraic number theory provides tools for studying some model dynamical systems to greater depth than more general analytic, topological, and geometric methods allow.

The contents of the first four sections of this chapter are classical and are meant to demonstrate the usefulness of the dynamical approach to the problems of uniform distribution and Diophantine approximation of numbers. Along the way the prominent role of hyperbolic geometry becomes apparent. The last section gives a brief account of one of the top achievements of the dynamical approach to analytic number theory: the proof of Oppenheim's conjecture about small values of quadratic forms in three variables.

15.1 UNIFORM DISTRIBUTION OF THE FRACTIONAL PARTS OF POLYNOMIALS

In this section we describe how some general dynamical arguments establish the uniform distribution results (unique ergodicity; see Section 4.1.4) for special dynamical systems needed to solve a number-theoretic problem discussed in the Introduction (Section 1.3.5) and its generalizations. Unique ergodicity (Definition 4.1.18) is essential for establishing uniform distribution for specific sequences because a given sequence is associated with a particular orbit of the dynamical system. Convergence outside of a null set is not sufficient, since that particular orbit may turn out to be in the exceptional null set.

15.1.1 Quadratic Polynomials and Affine Maps of the 2-Torus

We return to the problem of the distribution of the last digits before the decimal point for the sequence $x_n = n^2 \sqrt{2}$ first introduced in Section 1.3.5. There is nothing special about $\sqrt{2}$ aside from irrationality, so we consider the sequence αn^2 for any irrational α. To deal with the question about the last digits we need to consider the first digits of the sequence of the fractional parts of $\alpha n^2/10$. Unlike some other sequences discussed earlier in this book, such a sequence does not appear as the sequence of successive iterates of a one-dimensional map and cannot be modified naturally into such a sequence. There is however a method that bypasses this difficulty. One constructs a dynamical system in higher dimension and interprets the sequence in question as a sequence of coordinates (or, more generally, values of a function) for the successive iterates of a particular initial condition. For our question, the proper dynamical system is the following *affine* map of the 2-torus

$$A_{\alpha/5}(x, y) = (x + \alpha/5, y + x) \pmod 1.$$

This map is "integrable" in the sense that there is a closed formula for its iterates:

$$A^n_{\alpha/5}(x, y) = (x + n\alpha/5, y + nx + n(n-1)\alpha/10) \pmod 1.$$

The sequence of x-coordinates is familiar from Section 4.1.1, but the sequence of y-coordinates includes the quadratic term we need. To eliminate the unwanted linear term one picks $x = \alpha/10$ and $y = 0$.

Following the arguments from Section 4.2.2, we see that in order to establish the uniform distribution of the last digits for the sequence αn^2 it suffices to show that for any decimal interval $\Delta_k = [k/10, (k+1)/10]$, $k = 0, 1 \ldots, 9$ the Birkhoff averages of the characteristic function χ_{Δ_k},

$$\frac{1}{n} \sum_{i=0}^{n} \chi_{\Delta_k} \circ A^i_{\alpha/5},$$

uniformly converge to $1/10$. Again, there is nothing special about these characteristic functions aside from the fact that they are Riemann integrable (Section 4.1.5).

There are two principal methods for establishing uniform convergence of Birkhoff averages for all Riemann-integrable functions for these maps. The first method was introduced by Hermann Weyl in 1916. It is a refinement of the Kronecker–Weyl method we used for rotations of the circle in Section 4.1.6 and for translations of the torus in Section 5.1.6. It involves estimating the Birkhoff averages for the characters $\exp 2\pi i(kx + ly)$, $(k, l \in \mathbb{Z})$ by directly using a more sophisticated calculation than simply summing the geometric progression as in Section 4.1.6. The expressions in the Weyl calculations are a particular case of *trigonometric sums*, which play a prominent role in analytic number theory. After that the Weierstraß Theorem establishes uniform convergence for all continuous functions (unique ergodicity; see Definition 4.1.18) and the standard argument as in Section 4.1.6 extends it to Riemann-integrable functions. The principal strength of this method is that for certain classes of functions it allows us to establish the *speed* of convergence of Birkhoff averages.

15.1.2 Ergodicity and Unique Ergodicity

We discuss a more qualitative approach, which was pioneered by Hillel Furstenberg around 1960. While it does not provide any estimates on the speed of convergence for Birkhoff averages, it does not depend on clever calculations with trigonometric sums and is hence more broadly applicable. The key point is that both unique ergodicity (uniform convergence of averages for continuous functions) and the weaker property of convergence of averages outside of a null set (see Section 7.5.3) are equivalent to certain purely qualitative properties.

In the case of unique ergodicity, the corresponding qualitative property is the *uniqueness of an invariant measure* or integral (Definition 10.4.1, and Proposition 10.4.2). In the case of convergence of averages outside of a null set, the corresponding qualitative property is called *ergodicity*. In the case of a volume-preserving transformation f, ergodicity means that every "reasonable" f-invariant set is either a null set or the complement of a null set, or, equivalently, that every "reasonable" f-invariant function is constant outside of a null set. The rigorous definition of "reasonable" in either case requires the notion of measurability, which is not discussed in this book beyond the concept of a null set.

The second method of proving the uniform distribution for polynomials is based on a general criterion of unique ergodicity for a class of maps called group extensions, which we formulate here in a special case.

Proposition 15.1.1 *Let f be a continuous volume-preserving uniquely ergodic map of the torus $\mathbb{T}^k$ and $\phi: \mathbb{T}^k \to S^1$ continuous. If the map $f_\phi: \mathbb{T}^{k+1} \to \mathbb{T}^{k+1}$ defined by*

$$f_\phi(x, y) = (f(x), y + \phi(x))$$

is ergodic, then it is uniquely ergodic.

Outline of proof We will use the equivalence of unique ergodicity to the uniqueness of an invariant integral, as well as certain properties of invariant integrals that follow from *ergodic decomposition*.

Since the map f_ϕ is ergodic (with respect to the usual volume measure on the torus $\mathbb{T}^{k+1}$), every orbit outside of a null set is uniformly distributed. But, since the volume is invariant with respect to any translation, in particular with respect to a "vertical" translation $(x, y) \to (x, y + \beta)$, and f_ϕ commutes with vertical translations, it follows that the set A of uniformly distributed orbits is also invariant under vertical translations and hence consists of complete circles $x = $ const. We say that such a set is *saturated*. The set A is not a null set since its complement is a null set (the union of two null sets is a null set, and the whole torus is not a null set).

Now suppose f_ϕ is not uniquely ergodic. This means that there is at least one more invariant integral different from the one generated by the standard volume and hence (by ergodic decomposition) there is another asymptotic uniform distribution. The set of points with this new asymptotic uniform distribution is disjoint from the set of points uniformly distributed with respect to the standard volume and hence belongs to a null set of circles $x = $ const. Projecting this invariant integral to the x-coordinate, one obtains an invariant integral for f, which by assumption is unique and hence coincides with the standard one. This gives us a contradiction: The set A of all points that are uniformly distributed with respect to the usual volume measure on the torus $\mathbb{T}^{k+1}$ is a null set with respect to the second invariant integral, but the latter projects to the standard integral on $\mathbb{T}^k$, and hence the set A, which is saturated, projects to a null set. $\square$

15.1.3 Uniform Distribution of Squares
With these ingredients in hand the proof of equidistribution of squares is not very hard.

Proposition 15.1.2 *The map* $A_\beta(x, y) = (x + \beta, y + x) \pmod 1$ *of* $\mathbb{T}^2$ *is uniquely ergodic.*

Proof This map is of the form described in Proposition 15.1.1, where $f = R_\beta$. Thus it suffices to prove ergodicity, that is, that any "reasonable" invariant function is a constant outside a null set. It is sufficient to consider L^2 functions, for which the Fourier series expansion is uniquely and unambigously defined, although it may not converge at some points. Checking the absence of invariant nonconstant functions boils down to writing equations for the Fourier cofficients of such a function that imply that all coefficients except one vanish. Specifically, suppose g is an invariant L^2-function. Then in terms of Fourier series

$$\sum_{i,j} a_{i+j\,j} e^{2\pi ix} e^{2\pi jx} e^{2\pi jy} = \sum_{i,j} a_{ij} e^{2\pi ix} e^{2\pi jy} = g(x, y)$$

$$= g(A_\beta(x, y)) = g(x + \beta, y + x) = \sum_{i,j} a_{ij} e^{2\pi i(x+\beta)} e^{2\pi j(y+x)},$$

so $a_{i+j\,j} = a_{ij} e^{2\pi i\beta}$ for all $i, j \in \mathbb{Z}$. This implies that $|a_{i+kj\,j}| = |a_{ij}|$ for all $i, j, k \in \mathbb{Z}$. The Fourier coefficients of an L^2-function are square-summable, so $\lim_{l\to\infty} a_{lj} = 0$ and $a_{ij} = 0$ for $i, j \in \mathbb{Z} \setminus \{0\}$ (using $e^{2\pi i\beta} \neq 1$ for $i \neq 0$ since $\beta \notin \mathbb{Q}$). If $i = 0$, then $a_{i+j\,j} = a_{ij} e^{2\pi i\beta}$ gives $a_{0j} = a_{jj} = 0$ for $j \neq 0$. If $j = 0$, then $a_{i+j\,j} = a_{ij} e^{2\pi i\beta}$ gives $a_{i0} = a_{i0} e^{2\pi i\beta}$, which implies $a_{i0} = 0$ for $i \neq 0$. Thus $a_{ij} = 0$ unless $i = j = 0$. $\square$

Corollary 15.1.3 *If* $\alpha \notin \mathbb{Q}$, *then the sequence of last digits of* αn^2 *is uniformly distributed.*

Proof Since $A^n_{\alpha/5}(x, y) = (x + n\alpha/5, y + nx + n(n-1)\alpha/10) \pmod 1$ and $A_{\alpha/5}$ is uniquely ergodic, every orbit is uniformly distributed. For $x = \alpha/10$ and $y = 0$, this implies that $y + nx + n(n-1)\alpha/10 = \alpha n^2/10$ is uniformly distributed on the unit interval and in particular so are the first digits after the decimal point of this sequence. $\square$

15.1.4 Uniform Distribution of Polynomials
The method described above can be applied inductively to prove the following.

Proposition 15.1.4 *If* α *is irrational, then the map* $A_{k,\alpha} \colon \mathbb{T}^{k+1} \to \mathbb{T}^{k+1}$ *defined by*

$$A_{k,\alpha}(x_0, x_1, \ldots, x_k) = (x_0 + \alpha, x_1 + x_0, \ldots, x_k + x_{k-1}) \pmod 1$$

is uniquely ergodic.

For any polynomial P of degree $k + 1$ with irrational leading coefficient one can find α and initial conditions $x_0, x_1, \ldots, x_k$ such that the last coordinate of $A^n_{k,\alpha}(x_0, x_1, \ldots, x_k)$ is $P(n)$. This is done by first finding a closed formula for the iterates $A^n_{k,\alpha}$ and then solving a system of linear equations. Taking into account the periodicity of the fractional parts of polynomials with *rational* coefficients, one obtains the equidistribution theorem for polynomials.

Theorem 15.1.5 *The fractional parts of the values $P(n)$ for a polynomial with at least one irrational coefficient are uniformly distributed.*

Since the last k digits before the decimal point of a real number become the first k digits after the decimal point upon division by 10^k, this theorem implies

Corollary 15.1.6 *If P is a polynomial with at least one irrational coefficient, then the last k digits before the decimal point of the numbers $(P(n))_{n\in\mathbb{N}}$ are uniformly distributed.*

15.2 CONTINUED FRACTIONS AND RATIONAL APPROXIMATION

As we saw in Proposition 6.1.12, all orbits of a circle rotation R_α are (uniformly) recurrent. Considering the orbit of 0, one sees that this means that αn gets arbitrarily close to integers for some large $n \in \mathbb{N}$, or that $|\alpha n - m|$ can be made arbitrarily small by choosing $n, m \in \mathbb{N}$ appropriately.

15.2.1 Best Approximation

To refine this in a quantitative way, one can ask how large n, m have to be to get $|\alpha n - m| < \epsilon$. Put slightly differently, given bounds on n, minimize $\min_{m\in\mathbb{Z}} |\alpha n - m|$. For a rational α, this is a finite problem that is of no interest for us in the present context. We henceforth assume $\alpha \notin \mathbb{Q}$, and also $0 < \alpha < 1$ for the moment.

Definition 15.2.1 A rational number p/q is said to be a *best rational approximation* or simply a *best approximation* of the irrational number α if $q > 0$ and $|q\alpha - p| \le |n\alpha - m|$ whenever $n, m \in \mathbb{Z}$ and $|n| \le q$.

Clearly this implies that p and q are relatively prime. It follows from the topological transitivity of an irrational rotation that there are infinitely many different best approximations for any given irrational number.

Now we approach the question of finding best approximations by tracking patiently how close αn gets to integers (and what those integers are). Specifically, we take note of those $n \in \mathbb{N}$ for which αn is closer to $\mathbb{Z}$ than any αi for $i < n$. In doing so we try to discern a recursive pattern. From a dynamical point of view, we are looking at how closely the orbit of 0 under the iterates of the rotation R_α returns back to 0 and registering the closest returns.

Denote the first such integer by a_1. It is determined by $a_1\alpha < 1 < (a_1 + 1)\alpha$, that is, by being closer to $\mathbb{Z}$ than α. This means that $a_1 = \lfloor 1/\alpha \rfloor$, where $\lfloor \cdot \rfloor$ denotes integer part. In terms of approximating α by a rational number, this corresponds to $a_1\alpha \approx 1$ or $\alpha \approx 1/a_1$. For the next step it helps to imagine that this is a good approximation, that is, that the error $\delta := 1 - a_1\alpha < \alpha$ is small.

The next close approximation of an integer occurs with the first $n \in \mathbb{N}$ for which $n\alpha$ is within less than δ of an integer. Imagining δ to be small, note that in the a_1

steps after the previous encounter the numbers $i\alpha$ for $a_1 < i \le 2a_1$ lie well between 1 and 2, and that at the end of this run we have $2a_1\alpha = 2(1 - \delta) = 2 - 2\delta$; so we have fallen behind by another δ, as it were.

Let a_2 be the last integer n for which $n\delta < \alpha$, that is, $a_2 = \lfloor \alpha/\delta \rfloor$. Then $0 < \delta_1 := \alpha - a_2\delta < \delta$ and $a_1 a_2 \alpha$ has "fallen behind integers" by close to α (by $\alpha - \delta_1$, to be exact). Thus, $(a_1 a_2 + 1)\alpha$ is closer to an integer (specifically, a_2) than all previous $i\alpha$'s, because $(a_1 a_2 + 1)\alpha - a_2 = \alpha - a_2(1 - a_1\alpha) = \alpha - a_2\delta = \delta_1 < \delta$. Note that the corresponding rational approximation is $(a_1 a_2 + 1)\alpha \approx a_2$ or

$$\alpha \approx \frac{a_2}{a_1 a_2 + 1} = \frac{1}{a_1 + \dfrac{1}{a_2}}.$$

To discern a pattern in the choice of a_2 note that $\alpha_2 := \delta/\alpha = (1 - a_1\alpha)/\alpha = (1/\alpha) - a_1 = \{1/\alpha\}$, where $\{\cdot\}$ now denotes fractional part, and therefore $a_2 = \lfloor 1/\alpha_2 \rfloor$.

15.2.2 Continued Fraction Representation

The preceding scheme gives an algorithm for the best rational approximation of irrational numbers and for best representation of any real number by a continued fraction.

Theorem 15.2.2 *Given* $\alpha \in \mathbb{R} \setminus \mathbb{Q}$, *define* $(a_i)_{i=0}^\infty$ *and* $(\alpha_i)_{i=1}^\infty$ *recursively by*

$$a_0 := \lfloor \alpha \rfloor, \qquad \alpha_1 := \{\alpha\}, \qquad a_i := \left\lfloor \frac{1}{\alpha_i} \right\rfloor, \qquad \alpha_{i+1} := \left\{ \frac{1}{\alpha_i} \right\}$$

and set

$$a_0 + \cfrac{1}{a_1 + \cfrac{1}{\ddots + \cfrac{1}{a_n}}} =: \frac{p_n}{q_n}$$

in lowest terms with $q_n > 0$. *(These are called the* convergents.*)*
   Then

$$\alpha = a_0 + \cfrac{1}{a_1 + \cfrac{1}{a_2 + \ldots}} := \lim_{n \to \infty} a_0 + \cfrac{1}{a_1 + \cfrac{1}{\ddots + \cfrac{1}{a_n}}}.$$

If $\alpha \in \mathbb{Q}$, *then the above recursion terminates with* $\alpha_{i+1} = 0$ *for some* i, *and*

$$\alpha = a_0 + \cfrac{1}{a_1 + \cfrac{1}{\ddots + \cfrac{1}{a_i}}}.$$

Both sequences p_n *and* q_n *satisfy the two-step recursion*

$$x_{n+1} = x_{n-1} + a_{n+1}x_n,$$

with initial conditions $p_0 = a_0$, $p_1 = a_0 a_1 + 1$, $q_0 = 1$, $q_1 = a_0$; $p_{n+1} q_n - p_n q_{n+1} = (-1)^n$; and, finally, every best rational approximation of an irrational number is a convergent and every convergent is a best approximation.

Remark 15.2.3 A little thought shows that the continued fraction representation of a number is unique. In comparison with the decimal expansion of reals there are a few things to note. Decimal expansion is not always unique. Continued fractions use a sequence of integers with no a priori bound, whereas decimal expansion uses single digits. Rationals and irrationals can be distinguished using the decimal expansion (periodic versus not), but this distinction can only be made either way if the entire expansion is in hand. By contrast, a number is rational if and only if the continued fraction expansion terminates. If this is the case, it is apparent from a finite-level continued fraction expansion.

Proof that these a_i's work Given the form of the continued-fraction expansion, it is easy to check that the a_i's are determined by the procedure described: If

$$\alpha = a_0 + \cfrac{1}{a_1 + \cfrac{1}{a_2 + \dots}},$$

then $\lfloor \alpha \rfloor = a_0$ and with $\alpha_1 := \alpha - a_0 = \{\alpha\}$ we get

$$\frac{1}{\alpha_1} = a_1 + \cfrac{1}{a_2 + \cfrac{1}{a_3 + \dots}},$$

so $a_1 = \lfloor 1/\alpha_1 \rfloor$ and

$$\alpha_2 := \left\{ \frac{1}{\alpha_1} \right\} = \frac{1}{\alpha_1} - a_1 = \cfrac{1}{a_2 + \cfrac{1}{a_3 + \dots}};$$

hence $a_2 = \lfloor 1/\alpha_2 \rfloor$, and so on. $\square$

Remark 15.2.4 Note that the p_n/q_n lie alternately above and below α. This can be seen from the discussion of closest approximation, or from the alternating effect obtained by adding a term to a finite continued-fraction approximation.

Example 15.2.5 The two-step recursion for p_n and q_n implies that the Fibonacci numbers $(b_n)_{n \in \mathbb{N}_0} = (1, 1, 2, 3, 5, 8, \dots)$ produce the convergents $1, 2, 3/2, 5/3, 8/5, \dots$ of the golden mean

$$\cfrac{1}{1 + \cfrac{1}{1 + \cfrac{1}{1 + \dots}}} = \frac{\sqrt{5} - 1}{2}.$$

Example 15.2.6 Consider the simple continued fraction

$$x = \cfrac{1}{2 + \cfrac{1}{2 + \cfrac{1}{2 + \ldots}}},$$

which is the positive root of the quadratic equation $1/x = x + 2$. Thus $x^2 + 2x - 1 + 0$ and $x = \sqrt{2} - 1$. The previous example can be treated similarly.

These examples suggest a fruitful generalization. Assume that the continued fraction expansion of a number is eventually periodic, that is, it becomes periodic after finitely many terms. Then one can express it as a root of a quadratic equation with rational coefficients. As it turns out, the converse is also true. Thus a real number has eventually periodic continued-fraction expansion if and only if the number is a *quadratic irrationality*, that is, it has the form $a + \sqrt{b}$, where a and b are rational numbers.

15.2.3 Speed of Approximation and Dynamics

There is another natural way to measure the quality of a rational approximation p/q of an irrational number α, namely, by comparing the absolute value of the difference $\alpha - p/q$ with a given function of the denominator q. Naturally, the convergents come to the fore in considerations of this kind.

Proposition 15.2.7 *If $\alpha \notin \mathbb{Q}$ and $n \in \mathbb{N}$, then*

$$\left| \alpha - \frac{p_n}{q_n} \right| < \frac{1}{a_{n+1}q_n^2} \le \frac{1}{q_n^2}.$$

Proof The convergents oscillate around α, so Theorem 15.2.2 gives

$$(15.2.1) \quad \left| \alpha - \frac{p_n}{q_n} \right| < \left| \frac{p_{n+1}}{q_{n+1}} - \frac{p_n}{q_n} \right| = \frac{|p_{n+1}q_n - p_n q_{n+1}|}{q_n q_{n+1}} = \frac{1}{q_n q_{n+1}} \le \frac{1}{a_{n+1}q_n^2}. \quad \square$$

One can also show that

$$(15.2.2) \quad \left| \alpha - \frac{p_n}{q_n} \right| > \frac{1}{(a_{n+1} + 2)q_n^2},$$

which gives a tight connection between the size of a_n and the actual error. A slightly more delicate argument refines Proposition 15.2.7 to give $|\alpha - p_n/q_n| < 1/2q_n^2$ infinitely often; the estimate never fails for two consecutive n's. This is fairly sharp in that for relatively prime p and q the inequality

$$|\alpha - p/q| < 1/2q^2$$

holds only if p/q is a convergent for α. The sharpest possible result is that $|\alpha - p/q| \leq 1/\sqrt{5}q^2$ for infinitely many (p, q). The fraction for the golden mean

$$\cfrac{1}{1 + \cfrac{1}{1 + \cfrac{1}{1 + \ldots}}}$$

realizes this bound.

Thus, the inverse square speed of approximation is the fastest speed that is achieved universally, and in order to find the fastest approximations for a given number one has to look at the convergents and only at them. It is clear from (15.2.1) and (15.2.2) that the growth properties of the sequence a_n essentially determine the speed of approximation of α by rationals.

Now we look at how a particularly fast approximation may appear. It is conceivable, for example, that the first error δ is microscopic compared to α. This means, of course, that $\alpha \approx 1/a_1$ is an enormously good approximation (with error δ/a_1). This would appear to be a desirable situation. One effect is that $a_2 = \lfloor \alpha/\delta \rfloor$ is quite large. Thus, the size of the terms a_i in the continued fraction expansion is a measure of the quality of approximation at the previous step. (Indeed, this is directly related to the next q_n being substantially larger than the current one, that is, the approximation being good means that it takes a long time to improve on it.) A remarkable concrete example of this phenomenon is the excellent rational approximation of π by $p_3/q_3 = 355/113$, which was known in ancient China. This is correct to six decimal places because a_4 turns out to be unexpectedly large so early in the game: It is 292. Accordingly, q_4 jumps to the rather large value of 33102.

Such fast approximation affects the dynamics of the circle rotation by α. Reconsider our first example by projecting it to the circle $\mathbb{R}/\mathbb{Z}$, taking $a_1 = 5$ and $\delta \approx 10^{-6}\alpha$. We are tracking the orbit of 0 and find the first five iterates evenly spaced around the circle (up to δ mismatch). The first 500,000 iterates lie within $1/10$ of the first 5, so there are 5 intervals of length $1/10$, none of which contains any of the first half-million iterates. While this does not contradict uniform distribution (Proposition 4.1.7), it takes an enormous number of iterates to balance the distribution, even on a crude scale. The possibility that such close encounters may happen at any number of levels in the rational approximation scheme implies that relatively rapid approximation can affect the dynamics in undesirable ways. It gives rise to *Liouvillian phenomena*. One aspect of such effects is that circle rotations by a rapidly approximable angle tend to be far less robust in some sense than those with moderate continued-fraction denominators.

15.2.4 Metric Theory of Diophantine Approximations

The approximation questions addressed so far are of the type that inquires about the possible rates of approximation by convergents or about the approximation rate of a particular irrational. Metric number theory addresses a related question of converse type: Given a particular rate of approximation, how many numbers exhibit this rate? This turns out to be a question important for dynamics.

If $r \colon \mathbb{N} \to (0, \infty)$ is a nonincreasing "rate" function, then $\alpha \in \mathbb{R}$ is said to be r-approximable if $|q\alpha - p| \leq r(|q|)$ for infinitely many $(p, q) \in \mathbb{Z} \times \mathbb{N}$.

The inverse power scale is of particular importance both for number theory and for many applications of rational approximations.

Definition 15.2.8 The *Diophantine condition with exponent $\beta \geq 0$ and constant $C > 0$* for a number $\alpha \in \mathbb{R}$ is that

$$|\alpha - (p/q)| \geq C/q^{2+\beta}$$

for all $p, q \in \mathbb{Z}, q \neq 0$, that is, α is not $C/x^{1+\beta}$-approximable for some $C > 0$. Denote the set of all such numbers by $D_{\beta,C}$. The number α is said to be *Diophantine* if it satisfies a Diophantine condition for some $\beta, C > 0$, and is said to be *Liouvillian* otherwise; that is, if there are sequences $p_n, q_n \in \mathbb{Z}$ such that

$$(15.2.3) \qquad\qquad \left| \alpha - \frac{p_n}{q_n} \right| = o(q_n^{-\gamma}) \qquad \text{for all } \gamma > 0.$$

Each $D_{\beta,C}$ is closed and nowhere dense, but in a different sense most numbers satisfy the Diophantine condition with any positive exponent β and a constant depending on β: The complement of $\bigcap_{\beta>0} \bigcup_C D_{\beta,C}$ is a null set.

The definitive results concerning the size of sets allowing approximation with a particaular speed are due to Alexander Ya. Khinchine.[1] If the series $\sum_{n=1}^{\infty} r(n)$ converges, then the set of r-approximable numbers is a null set. The converse, which is more difficult, is also true: If the series diverges, then only a null set of numbers fails to be r-approximable.

15.3 THE GAUSS MAP

Khinchine's theorems are proved by a standard but subtle method from real analysis: looking directly at the structure of the sets of numbers that allow (or do not allow) approximation with a particular speed, and constructing appropriate coverings. This method is quite efficient in establishing the best rate of approximation for various classes of numbers. A more thorough investigation of rational approximation would study how often a particularly fast or not-so-fast "best" approximation takes place. In other words, one would like to know the distribution of the partials a_n for various classes of numbers, in particular, some asymptotic statistical properties of such distributions.

15.3.1 Coding of Continued Fraction Expansion

Recall that the distribution of digits in a decimal expansion (or, more generally, expansion in base *m*) is related to the asymptotic behavior of the linear expanding map E_{10} (or E_m). Analogously, the continued-fraction expansion is naturally associated with iterations of the following map.

---

[1] Alexander Ya. Khinchine, Einige Sätze über Kettenbrüche, mit Anwendungen auf die Theorie der diophantischen Approximationen, *Mathematische Annalen* **92** (1924), 115–125.

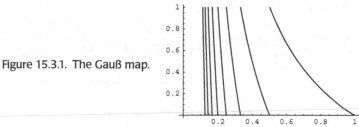

Figure 15.3.1. The Gauß map.

Definition 15.3.1 The map $G\colon [0, 1] \to [0, 1)$ defined by $x \mapsto \{1/x\}$ (fractional part) is called the *Gauß map* (see Figure 15.3.1).

This map has jump discontinuities at $1/i$ for $i \in \mathbb{N}$ and a discontinuity without a right-sided limit at 0. Projecting to the circle $S^1 = \mathbb{R}/\mathbb{Z}$ would fix the discontinuities away from zero, but as a dynamical system this map is most naturally defined on the interval.

If $x \in (0, 1]$, the first term a_1 in the continued fraction expansion of x is i if and only if $1/(i + 1) < x \le 1/i$. Similarly, the second term a_2 is i if and only if $1/(i + 1) < G(x) \le 1/i$, and so on. Thus the distribution of the partials a_n in the continued-fraction expansion for x coincides with the distribution of the iterates $G^n(x)$ between the half-open intervals $(1/(i + 1), 1/i]$, $i \in \mathbb{N}$.

The Gauß map is discontinuous, and the discontinuity at zero is of a serious sort, but it is almost expanding away from the discontinuities. In fact, G^2 is expanding because the minimum of the derivative is greater than one. Thus it is not entirely surprising that the Gauß map has some of the properties related to hyperbolic (in particular, expanding) behavior that we observed in earlier chapters. It is far from being uniquely ergodic: It has a great variety of invariant integrals and asymptotic distributions, periodic orbits are dense (they are quadratic irrationalities of a particular kind), and a great variety of orbit behavior is possible.

15.3.2 Uniform Distribution
The central fact about the Gauß map relevant for our discussion is the existence of a weighted uniform distribution given by a density, much like many quadratic maps (Section 11.4.3.2), or expanding or tent maps (Theorem 7.5.6 and Section 11.4.3, respectively).

Proposition 15.3.2 *The Gauß map preserves the measure defined by the density* $1/(1 + x)$.

Proof The measure of an interval $[a, b]$ is

$$\int_a^b \frac{1}{1 + x}\, dx = \log(b + 1) - \log(a + 1).$$

The preimage is obtained by noting that $a = \{1/x\}$ if and only if $1/x = a + n$ for some $n \in \mathbb{N}$, and likewise for b. Therefore the measure of $G^{-1}([a, b])$ is a sum of weighted lengths as follows:

$$\sum_{n\in\mathbb{N}}\int_{\frac{1}{b+n}}^{\frac{1}{a+n}}\frac{1}{1+x}\,dx = \sum_{n\in\mathbb{N}}\log\left(1+\frac{1}{a+n}\right)-\log\left(1+\frac{1}{b+n}\right)$$

$$= \sum_{n\in\mathbb{N}}\log(a+n+1)-\log(a+n)-\log(b+n+1)+\log(b+n)$$

$$= \log(b+1)-\log(a+1)$$

(telescoping sum). These agree, as required. $\square$

Indeed, with moderate further effort one can show

Proposition 15.3.3 *The Gauß map is ergodic.*

Corollary 15.3.4 *The set of $\alpha \in [0,1]$ with bounded continued-fraction coefficients is a null set.*

Proof Since $a_i = \lfloor 1/G^{i-1}(\alpha)\rfloor$, this is the set of points whose orbits under the Gauß map are bounded away from zero and are therefore not uniformly distributed. Therefore this is a null set. $\square$

Proposition 15.3.5 *For almost every number (see Definition 7.5.3) the continued-fraction coefficients are distributed as follows: The number $n \in \mathbb{N}$ occurs as a continued-fraction coefficient with asymptotic frequency*

(15.3.1)
$$\frac{\log\left(1+\frac{1}{n}\right)-\log\left(1+\frac{1}{n+1}\right)}{\log 2}.$$

Sketch of proof The Gauß map is ergodic with respect to the measure defined by the density $1/(1+x)$. Hence the orbits of the Gauß map are uniformly distributed with respect to the density $1/(1+x)$, so the probability that $a_i = \lfloor 1/G^{i-1}(\alpha)\rfloor = n$ is the probability that

$$\frac{1}{n+1} < G^{i-1}(\alpha) \le \frac{1}{n},$$

which is $\int_{1/(n+1)}^{1/n}(1/1+x)\,dx/\log 2$, where $\log 2 = \int_0^1 (1/1+x)\,dx$ is the normalization constant. This integral gives (15.3.1). $\square$

15.3.3 Gauß Map and Inducing for Circle Rotations

Now we will give a different interpretation of the Gauß map which is an instance of a very fruitful approach to dynamics. Sometimes dynamical systems from a certain class can be considered as elements of a certain space with a natural ("global") dynamics in it. In other words, there is a canonically defined operation within the class, and the asymptotic behavior of an orbit with respect to this operation often reveals essential properties of the system corresponding to the initial condition. Often the global dynamics involved goes under the revealing name of "renormalization"; rather than giving a speculative explanation of this term in general we will illustrate this approach by our example at hand.

Let us identify the real number α, $0 < \alpha < 1$ with the rotation R_α. By cutting the circle at 0 we can represent the rotation as the following map of the half-open interval $[0, 1)$:

$$R_\alpha(x) = \begin{cases} x + \alpha & \text{if } 0 \le x < 1 - \alpha, \\ x + \alpha - 1 & \text{if } 1 - \alpha \le x < 1. \end{cases}$$

Geometrically this can be visualized as the interchange of two intervals $[0, 1 - \alpha)$ and $[1 - \alpha, 1)$. Now consider the *first return* map on the interval $[0, \alpha)$. An immediate inspection shows that this map amounts to the exchange of two intervals $[0, \beta)$ and $[\beta, \alpha)$, where $\beta = \{n\alpha\}$ and the integer n is defined by the inequality $(n-1)\alpha < 1 \le n\alpha$, that is, $n = [1/\alpha] + 1$ and hence $\beta = [1/\alpha]\alpha + \alpha - 1$.

Now consider this exchange as a representation of a rotation as above. Of course for that we need to normalize the interval $[0, \alpha)$ to length one. This makes the induced map the representation of the rotation by $1 - (\beta/\alpha) = 1 - [1/\alpha] + 1 + (1/\alpha) = \{1/\alpha\} = G(\alpha)$.

Thus iterations of the Gauß map correspond to the process of taking the first return map successively. Since the operation of taking the first return on decreasing intervals is obviously transitive, we can view the iterations as taking the first return on specially chosen smaller and smaller intervals (the lengths of these intervals can be easily computed and are closely related with the continued-fraction expansion) and "renormalizing" those back to full size. This procedure reveals the microscopic behavior of the original rotation, and thus the dynamics of the Gauß map tells us about the properties of inducing.

For example, if α is rational, then a certain iterate $G(\alpha)$ is equal to 0, after which the process stops: The induced map is the identity, which is simply another way of saying that the original rotation is periodic.

If α is a quadratic irrationality, the corresponding orbit of the Gauß map is eventually periodic, and hence the inducing process exhibits *self-similarity*: Beginning from a finite step the inducing process becomes periodic, so in a sense the microscopic structure at any level is the same as the macroscopic one.

Finally, for a typical α (outside of a null set), the orbit of the Gauß map is uniformly distributed with respect to the density $1/(1 + x)$. Hence the induced rotations are distributed according to the same density, and the macroscopic structure at different levels changes in a fairly random way.

We have already encountered a renormalization procedure of a different kind in Section 11.3.1. The main difference is that there the auxiliary dynamics took place in a neighborhood of a hyperbolic fixed point and hence did not exhibit much recurrence. Indeed, renormalization of an individual map would make it more similar to a map with an attractor of the Feigenbaum–Misiurewicz type.

15.4 HOMOGENEOUS DYNAMICS, GEOMETRY, AND NUMBER THEORY

A natural class of dynamical systems comes from the following general algebraic construction. Let G be a locally compact group, $H \subset G$ a closed subgroup, and $M = H\backslash G$ the left homogeneous space, which in this case has the natural locally compact topology. Sometimes for noncompact G and H the homogeneous space may happen to be compact similarly to the abelian case of $G = \mathbb{R}^n$, $H = \mathbb{Z}^n$. Since left and right multiplications in a group commute, the action of G on itself by right translations projects to M. The restriction of this action to a subgroup Λ of G is called a *(right) homogeneous action*; if Λ is a one-parameter subgroup, its action is called a *(right) homogeneous flow*. Naturally, by considering right homogeneous spaces and left actions one defines left homogeneous actions, not surprisingly, certain homogeneous actions play a central role in the interaction between number theory, geometry, and dynamics.

15.4.1 The Modular Surface, Geodesic, and Horocycle Flows

We begin with a description of a famous homogeneous action closely related to continued fractions and rational approximation. From the point of view of hyperbolic geometry, the action is the geodesic flow on a particular surface of constant negative curvature. Accordingly we start from some geometric preparations.

1. The Upper Half-Plane. Consider the upper half of the complex plane defined by points having positive imaginary part: $\mathbb{H} := \{(x + iy) \in \mathbb{C} \mid y > 0\}$. We define a non-euclidean distance on it by agreeing that the length of a vector v at a point $(x, y) \in \mathbb{H}$ is $\|v\|/y$. Equivalently, the length of a curve in $\mathbb{H}$ is $\int (1/y)(dx + dy)$, the integral taken along the curve. This defines the *hyperbolic metric*.

The geodesics are easy to describe: Every geodesic is either a vertical line $\{x + iy \mid y > 0\}$ or a semicircle whose endpoints are real, that is, lie on the x-axis.

The geodesic flow is defined as usual on the space $S\mathbb{H}$ of unit tangent vectors to $\mathbb{H}$. Unlike the sphere or torus, whose geodesic flows we encountered earlier (in Section 1.3.3 and Section 5.2.2), (and similarly to the euclidean plane), this space is rather "big". It is noncompact and has infinite volume. There is however a big difference between the behavior of geodesics in Euclidean and non-Euclidean geometry. To obtain a smaller space we will later perform a construction similar to that in Section 2.6.4, where the torus is obtained by identifying points in $\mathbb{R}^2$ if one is an integer translate of the other.

2. Möbius Transformations. The group $SL(2, \mathbb{R})$ acts isometrically and transitively on $\mathbb{H}$ by fractional-linear transformations, often called *Möbius transformations*:

$$f_g(z) = \frac{az + b}{cz + d}, \qquad z \in \mathbb{H}, \qquad g = \begin{pmatrix} a & b \\ c & d \end{pmatrix} \in SL(2, \mathbb{R}),$$

with kernel $\mathbb{Z}_2 = \{\pm \mathrm{Id}\} \subset SL(2, \mathbb{R})$. These transformations preserve the structure developed so far: They preserve the hyperbolic metric (that is, lengths and angles) and (thus) send geodesics to geodesics. The factor group $G = PSL(2, \mathbb{R}) = SL(2, \mathbb{R})/\mathbb{Z}_2 \simeq SO(1, 2)^0$ acts effectively on $\mathbb{H}$, and the isotropy subgroup of the point $z_0 = i \in \mathbb{H}$ is $C = PSO(2) = SO(2)/\mathbb{Z}_2$; hence $\mathbb{H}$ can be identified with G/C.

The differential of the action of $PSL_2(\mathbb{R})$ on $\mathbb{H}$ defines a transitive and free action on $S\mathbb{H}$, so the latter can be identified with G by fixing any unit tangent vector to $\mathbb{H}$ as the identity element in the group. It is convenient to pick the upward-looking vertical vector at the point i for that purpose.

3. Fuchsian Groups and Surfaces of Constant Negative Curvature. Any smooth surface M of constant Gaussian curvature -1 is of the form $M = \Gamma\backslash\mathbb{H} = \Gamma\backslash G/C$ for some discrete subgroup Γ of G (such groups are called *Fuchsian*) having no torsion. The unit tangent bundle SM is therefore the *left* homogeneous space $\Gamma\backslash S\mathbb{H} = \Gamma\backslash G$. The volume form for the Riemannian metric on M induces a G-invariant measure on SM, which is called the *Liouville measure*, and clearly coincides with the appropriately normalized Haar measure on $\Gamma\backslash G$. In particular, a surface $M = \Gamma\backslash\mathbb{H}$ is of finite area if and only if Γ is a lattice in G (i.e., $SM = \Gamma\backslash G$ is of finite volume).

4. The Modular Surface. Now we consider a special Fuchsian group that is a close analog of the lattice of integer translations in a Euclidean space. It is the *modular group* $PSL(2, \mathbb{Z})$, which consists of transformations

(15.4.1) $$z \mapsto \frac{az + b}{cz + d} \quad \text{with} \quad a, b, c, d \in \mathbb{Z}, \quad \text{and} \quad ad - bc = 1.$$

In Section 2.6.4 the torus was obtained from gluing together opposite edges of the fundamental domain $[0, 1] \times [0, 1]$. In the present situation a convenient fundamental domain is

$$\{x + iy \mid |x| \leq 1/2,\ x^2 + y^2 \geq 1\},$$

which is bounded by the half-lines $\{(1/2) + iy \mid y \geq \sqrt{3}/2\}$ and $\{-(1/2) + iy \mid y \geq \sqrt{3}/2\}$ as well as the connecting arc of the unit circle $\{x + iy \mid -1 \leq 2x \leq 1 \text{ and } y = \sqrt{1 - x^2}\}$. In the case of the torus, one identifies the two vertical sides of the fundamental square because the integer translation by $(1, 0)$ sends the left side to the right side. Analogously, the vertical half-lines in bounding the hyperbolic fundamental domain are identified by the same translation, which arises from $a = b = d = 1$, $c = 0$ in (15.4.1). Furthermore, the halves of the circle arc are identified by the map $z \mapsto -1/z$ [obtained from $a = -b = c = 1, d = 0$ in (15.4.1)], which acts like a mirror symmetry on that arc. One can picture the identification as rolling up the domain around the imaginary axis like a newspaper to identify the lines, and then "zipping" the bottom half-arcs together to close it up. The result is topologically a half-infinite cylinder closed at the bottom. Geometrically, however, outside of the bottom it looks more like a pseudosphere: The length of the section decreases exponentially along the geodesics represented by vertical lines. These geodesics are *parallel* in the non-Euclidean sense: The distance between any two of them decreases exponentially.

The geodesic flow on this *modular surface* can be described by tracking a geodesic in the region until it encounters the boundary. Then it emerges back into the domain from the corresponding boundary point. For the vertical segments this would be the "opposite" point, and the geodesic continues in the same direction. Encounters with the circle arc result in a jump from $x + iy$ to $-x + iy$ that preserves the angle with the boundary. (See Figure 15.4.1.)

The geodesic flow has a section whose return map is closely related to the Gauß map (Definition 15.3.1). It is not surprising then that properties of rational approximation are closely related to properties of the geodesic flow on the modular surface.

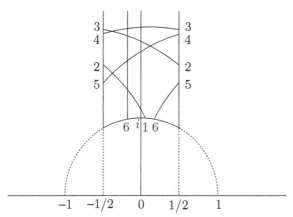

Figure 15.4.1. A geodesic ray on the modular surface.

The identification between $S\mathbb{H}$ and $PSL(2,\mathbb{R})$ described in Section 15.4.1.2 identifies the geodesic flow with the right homogeneous flow on $PSL(2,\mathbb{R})$ generated by the subgroup of diagonal matrices. We briefly mention dynamical properties of the geodesic flow in Section 15.4.4.

5. The Horocycle Flow. There is another homogeneous flow on $PSL(2,\mathbb{Z})\backslash PSL(2,\mathbb{R})$ (or, equivalently, on the unit tangent bundle to the modular surface) that is even more central for number-theoretic considerations than the geodesic flow and is closely related to the latter. This is the right homogeneous flow defined by the one-parameter group

$$\begin{pmatrix} 1 & t \\ 0 & 1 \end{pmatrix}, \quad t \in \mathbb{R}$$

and called the *the horocycle flow*[2] on the modular surface. The orbits of the horocycle flow are actually unstable manifolds for the geodesic flow. There are fascinating connections between questions about the horocycle flow and the Riemann Hypothesis in number theory.

For our considerations it is important to point out that the dynamics of the horocycle flow is very different from that of the geodesic flow (see Section 15.4.4). There is a one-parameter family of closed orbits represented by horizontal segments in the fundamental domain with vertical tangent vectors looking upward as well as their images under the geodesic flow. However, every other orbit is dense and, in fact, uniformly distributed with respect to the Liouville measure. Thus the picture is closer to unique ergodicity than to the complicated behavior represented by hyperbolicity.[3]

---

[2] The name came from *horocycles* or limit cycles, which are the limits of the circles in the hyperbolic plane when the center moves to infinity along a geodesic. Horocycles are represented by horizontal lines and circles tangent to the real line; geometrically, the horocycle flow is represented by the motion of a unit vector along one of the two horocycles perpendicular to it.

[3] For a Fuchsian group Γ such that $\Gamma\backslash PSL(2,\mathbb{R})$ is *compact*, the corresponding horocycle flow is in fact uniquely ergodic.

15.4.2 The Space of Lattices

A natural generalization of the construction of the modular surface that plays an important role in applications of homogeneous dynamics to number theory is represented by homogeneous spaces $\Omega_k := SL(k, \mathbb{Z})\backslash SL(k, \mathbb{R})$. The standard right action of $SL(k, \mathbb{R})$ on $\mathbb{R}^k$ keeps the lattice $\mathbb{Z}^k \subset \mathbb{R}^k$ invariant, and, given any $g \in G$, the lattice $g\mathbb{Z}^k$ is *unimodular*, that is, its fundamental set is of unit volume. On the other hand, any unimodular lattice Λ in $\mathbb{R}^k$ is of the form $\mathbb{Z}^k g$ for some $g \in G$. Hence the space Ω_k is identified with the space of all unimodular lattices in $\mathbb{R}^k$, and clearly this identification sends the right action of $SL(k, \mathbb{R})$ on $SL(k, \mathbb{Z})\backslash SL(k, \mathbb{R})$ to the linear action on lattices: For a lattice $\Lambda \subset \mathbb{R}^k$, one has $g\Lambda = \{gx : x \in \Lambda\}$.

Thus, in particular, the unit tangent bundle to the modular surface is naturally identified with the space of all lattices in the Euclidean plane, thus providing a fundamental link between Euclidean and non-Euclidean geometry. Similarly to the case of a modular surface, the homogeneous spaces Ω_k are not compact but have finite volume invariant under the action of G by left translation. An infinite "cusp" in the modular surface described above that can be easily visualized is replaced by a much more complicated asymptotic geometry.

The main link between number theory and the theory of homogeneous actions is the following the *Mahler criterion*, which gives a certain grasp of this asymptotic geometry:

Theorem 15.4.1 *A sequence of lattices $g_i SL(k, \mathbb{Z})$ goes to infinity in $\Omega_k \iff$ there exists a sequence $\{x_i \in \mathbb{Z}^k \setminus \{0\}\}$ such that $g_i(x_i) \to 0$ as $i \to \infty$. Equivalently, fix a norm on $\mathbb{R}^k$ and define the function δ on Ω_k by*

$$(15.4.2) \qquad\qquad \delta(\Lambda) := \max_{x \in \Lambda \setminus \{0\}} \|x\|.$$

Then a subset K of Ω_k is bounded if and only if the restriction of δ on K is bounded away from zero.

The right homogeneous actions of $SL(k, \mathbb{R})$ and its subgroups on the space Ω_k of lattices in $\mathbb{R}^k$ are very important for applications of homogeneous dynamics to number theory.

15.4.3 Indefinite Quadratic Forms on Lattices

Now we consider a model number-theoretic situation in which these actions arise, and in the next section we will briefly describe one of the most spectacular successes in the application of dynamical methods to classical problems in number theory. Let $Q(x)$ be a homogeneous polynomial in k variables (for example, a quadratic form), and suppose that one wants to study nonzero integer vectors x such that the value $Q(x)$ is small. Denote by H the stabilizer of Q in $SL(k, \mathbb{R})$ [that is, $H = \{g \in SL(k, \mathbb{R}) : Q(gx) = Q(x)$ for all $x \in \mathbb{R}^k\}$]. Then one can state the following elementary

Lemma 15.4.2 *There exists a sequence of nonzero integer vectors x_n such that $Q(x_n) \to 0$ if and only if there exists a sequence $h_n \in H$ such that $\delta(h_n\mathbb{Z}^k) \to 0$.*

The latter condition, in view of Mahler's criterion, amounts to the orbit $H\mathbb{Z}^k$ being unbounded in Ω_k. This gives a special number-theoretic importance to studying the long-term behavior of various trajectories on Ω_k.

We consider real quadratic forms in k variables and their values at integer points. Naturally, if such a form Q is positive or negative definite, the set $Q(\mathbb{Z}^k \setminus \{0\})$ has empty intersection with some neighborhood of zero. Now take an indefinite form and call it *rational* if it is a multiple of a form with rational coefficients and *irrational* otherwise. For $k = 2, 3, 4$ it is easy to construct rational forms that do not attain small values at nonzero integers;[4] therefore, a natural assumption to make is that the form is irrational.

15.4.4 Quadratic Forms in Two Variables

The case $k = 2$ for indefinite forms is rather special. We restrict our considerations to the special but representative case of the forms $Q_\lambda(x_1, x_2) := x_1^2 - \lambda x_2^2$, where $\lambda > 0$. Notice that $x_1^2 - \lambda x_2^2 = (x_1 - \sqrt{\lambda}x_2)(x_1 + \sqrt{\lambda}x_2)$, and, assuming that x_1 and x_2 are integers of the same sign (without loss of generality they then can be assumed to be positive), the second factor must be large. Hence, in order for the product to be small, the first factor should be small compared to $(\max(x_1, x_2))^{-1}$. In other words, there should be sequences of positive integers p_n and q_n, $n \in N$ such that

$$\lim_{n \to \infty} q_n^2 |\sqrt{\lambda} - p_n/q_n| = 0.$$

By (15.2.1) and (15.2.2), this condition is equivalent to the unboundedness of the coefficients a_n in the continued-fraction expansion of the number $\sqrt{\lambda}$. If these coefficients are eventually periodic, then as we know $\sqrt{\lambda}$ is a quadratic irrationality and so is λ. But there are uncountably many other numbers (although still a null set) of *bounded type*, that is, with bounded continued-function expansion.

Note that dynamically such numbers are characterized as having orbits under the Gauß map bounded away from zero or, equivalently, having compact closure in the half-open interval $(0, 1]$. The stabilizer of the form Q_λ in $SL(2, \mathbb{R})$ is a one-parameter hyperbolic subgroup that is thus conjugate to the diagonal subgroup, and its action is conjugate to the geodesic flow on the modular surface. This flow possesses all of the features of hyperbolic behavior discussed in Chapter 10 and is very similar to those of the Gauß map. In particular typical orbits are uniformly distributed and hence unbounded, while there is still a rich collection of bounded orbits. The Mahler criterion concerns the boundedness or unboundedness of the orbit with a particular initial condition (the standard lattice $\mathbb{Z}^2$), but a proper coordinate change reduces this to the corresponding question about the geodesic flow (or the left homogeneous action of the diagonal subgroup) with a variable collection of initial conditions. In fact, if one considers arbitrary indefinite quadratic forms rather than the special forms Q_λ, this set of initial conditions becomes complete. As before, the correspondence can be described

---

[4] However, by Meyer's Theorem (see J. W. S. Cassels, *An Introduction to Diophantine Approximation*, Cambridge Tracts in Mathematics and Mathematical Physics **45**, Cambridge University Press, New York, 1957) if Q is a nondegenerate indefinite rational quadratic form in $k \geq 5$ variables, then Q represents zero over $\mathbb{Z}$ nontrivially, i.e., there exist a nonzero integer vector x such that $Q(x) = 0$.

explicitly. Thus one can characterize quadratic forms that take arbitrarily small positive values at the integer vectors as those that, under this correspondence, are mapped to unit tangent vectors with unbounded geodesics on the modular surface.

15.5 QUADRATIC FORMS IN THREE VARIABLES

The situation is quite different in higher dimensions.

15.5.1 Oppenheim Conjecture

In 1986, Gregory A. Margulis[5] proved the following result, which resolved then a 60-year-old conjecture due to A. Oppenheim.

Theorem 15.5.1 *Let Q be a real indefinite nondegenerate irrational quadratic form in $k \geq 3$ variables.[6] Then, given any $\epsilon > 0$, there exists an integer vector $x \in \mathbb{Z}^k \setminus \{0\}$ such that $|Q(x)| < \epsilon$.*

For $k \geq 21$, this result was proved by analytic number theory methods in the 1950s.[7] In particular, it has been known that the validity of the conjecture for some k_0 implies its validity for all $k \geq k_0$; in other words, Theorem 15.5.1 reduces to the case $k = 3$.

The key observation was made implicitly in the 1950s by Cassels and Swinnerton-Dyer and later explicitly by Raghunathan (which motivated the latter's famous conjecture). It asserts that Theorem 15.5.1 is equivalent to a certain statement about the dynamics of a particular homogeneous action.

15.5.2 Dynamical Approach and the Margulis Theorem

The statement in question, which was eventually proved by Margulis, is the following:

Theorem 15.5.2 *Let Q be a real indefinite nondegenerate quadratic form in three variables, and let H_Q be the stabilizer of Q in $SL(3, \mathbb{R})$. Then any orbit $H_Q \Lambda$, Λ a lattice in $\mathbb{R}^3$, of the right homogeneous action of H_Q in Ω_3 with compact closure is compact.*

Outline of the reduction of Theorem 15.5.1 to Theorem 15.5.2 One uses Lemma 15.4.2 and certain considerations from the theory of lattices in Lie groups. Indeed, suppose for some $\epsilon > 0$ one has $\inf_{x \in \mathbb{Z}^3 \setminus \{0\}} |Q(x)| \geq \epsilon$. Then by Lemma 15.4.2 the orbit $H_Q \mathbb{Z}^3$ is unbounded in the space Ω_3 of unimodular lattices in $\mathbb{R}^3$; hence it is compact in view of Theorem 15.5.2. But since this orbit can be identified with $H_Q / H_Q \cap SL(3, \mathbb{Z})$, this shows that $H_Q \cap SL(3, \mathbb{Z})$ is a cocompact lattice (a discrete subgroup with a compact homogeneous space) in H_Q; hence it is Zariski-dense by Borel's Density

---

[5] G.A. Margulis, Formes quadratriques indéfinies et flots unipotents sur les espaces homogènes. *C. R. Acad. Sci. Paris Sér. I Math.* **304**(10) (1987), 249–253; G.A. Margulis, Indefinite quadratic forms and unipotent flows on homogeneous spaces, in *Dynamical Systems and Ergodic Theory (Warsaw, 1986)*, PWN, Warsaw, 1989, pp. 399–409.

[6] The original conjecture of Oppenheim assumed $k \geq 5$; later it was extended to $k \geq 3$ by Davenport.

[7] For the history of the problem see G. A. Margulis, Oppenheim conjecture, in *Fields Medallists' Lectures*, World Science Publishing, River Edge, NJ, 1997, pp. 272–327.

Theorem. The latter is not hard to show to be equivalent to H_Q being defined over $\mathbb{Q}$, which, in turn, is equivalent to Q being proportional to a form with rational coefficients.[8] □

15.5.3 Rigidity of Dynamics for Unipotent Actions

Margulis' result was a part of a broad development to understand the dynamical and statistical properties of a special class of homogeneous actions of which the horocycle flows described in Section 15.4.1.5 represent the simplest examples. Similarly to the previous section, one considers right homogeneous actions on the left homogeneous space $\Gamma \backslash G$, where G is a connected Lie group and Γ is a *lattice*, that is, a discrete subgroup of G such that the homogeneous space has finite volume invariant under the right action of G. The latter condition is obviously satisfied if the homogeneous space is compact, but in many cases, including the most interesting ones (such as Ω_k), it is not.

The defining local feature of the horocycle flow is its *unipotence*. This means that in a proper (moving) coordinate system the derivatives of the transformations look like linear maps with eigenvalues one only (in the particular case of the horocycle flow the maps in question have a single Jordan block of size three). It was shown in the 1970s by Furstenberg, Dani, and others that horocycle flows possess certain rigidity properties, beginning from unique ergodicity in the cocompact case. This led Raghunathan in the late 1970s to a general conjecture that any orbit closure for a homogeneous action that is generated by unipotent elements is algebraic in a natural sense, that is, it is a projection of a single coset of a closed subgroup.

Initial progress was reached for horocycle flows and more general unipotent actions that are closely associated to certain hyperbolic or partially hyperbolic actions; such unipotent subgroups bear the general name of horospherical. Since an intertwined hyperbolic action provides a renormalization for a horospherical action, this connection is of great assistance in the study of orbit closures and invariant measures. In the early 1980s Marina Ratner went much beyond simple unique ergodicity by establishing an impressive panoply of rigidity properties for horocycle flows, their products, and the like. In 1986, Margulis established Raghunathan's conjecture for a certain class of nonhorospherical unipotent subgroups that included the subgroups generating H_Q (see below). Finally, in 1990, Ratner established the Raghunathan conjecture in full generality[9] as well as the corresponding statement about invariant measures.[10] The Ratner theorems and the developments that followed have a great number of implications for number theory, geometry, and even group theory that are still being actively pursued.

Outline of proof Theorem 15.5.2 By a unimodular coordinate change any indefinite quadratic form in three variables can be brought to the form $\pm Q_0$, where

$$Q_0(x_1, x_2, x_3) = 2x_1x_2 + x^3.$$

---

[8] A detailed argument can be found in Alexander N. Starkov, *Dynamical Systems on Homogeneous Spaces*. American Mathematical Society, Providence, RI, 2000.

[9] Marina Ratner, Raghunathan's Topological Conjecture and Distributions of Unipotent Flows. *Duke Math. J.*, 63(1)(1991), 235–280.

[10] Marina Ratner, On Raghunathan's Measure Conjecture. *Ann. of Math. (2)* **134**(3) (1991), 545–607.

Thus the the group H_Q is isomorphic to the group H_{Q_0}, which contains a unipotent one-parameter subgroup

$$\begin{pmatrix} 1 & t & 0 \\ 0 & 1 & 0 \\ 0 & 0 & 1 \end{pmatrix}, \qquad t \in \mathbb{R},$$

and in fact can be easily shown to be generated by unipotent subgroups. The closure of an H_Q orbit in Ω_3 must, by Margulis' result or the general theorem of Ratner, be a homogeneous space of a proper closed subgroup H of $SL(3, \mathbb{Z})$, where obviously $H_Q \subset H$. Since there are no intermediate subgroups between H_{Q_0} and $SL(3, \mathbb{R})$ [and hence between H_{Q_0} and $SL(3, \mathbb{R})$], we have $H = H_Q$; hence the action is transitive on the orbit closure and the orbit is compact. $\square$

Reading

CONCURRENT READING

Several popular books complement this one well. An excellent treatment with a focus on applications is given by

Steven H. Strogatz, *Nonlinear Dynamics and Chaos*, Addison–Wesley, Reading, MA, 1994.

The following two books take a different approach to increasing complexity by going up in dimension from 1 to 2 to higher dimension:

Robert Devaney, *An Introduction to Chaotic Dynamical Systems*, Addison–Wesley, Reading, MA, 1989.
Robert Devaney, *A First Course in Chaotic Dynamical Systems*, Addison-Wesley, Reading, MA, 1992.

These also provide easy access to complex dynamics (Julia sets, etc.).

FURTHER READING

Numerous books provide an advanced treatment of dynamical systems suitable for continuing an education in dynamics. Naturally, these rely on a broader and deeper mathematical background, for which we have suggestions below.
The most natural continuation from here would be to use our book

Anatole Katok and Boris Hasselblatt, *Introduction to the Modern Theory of Dynamical Systems*, Cambridge University Press, New York, 1995.

This is a self-contained comprehensive exposition of dynamical systems.

The geometric theory of dynamical systems is treated excellently by

Clark Robinson, *Dynamical Systems, Stability, Symbolic Dynamics, and Chaos*, second edition, CRC Press, Boca Raton, FL, 1999.

We recommend several books for their presentation of important areas in dynamics. For continuous-time systems, mechanics, and applications it is useful to be well versed in the theory of differential equations. The classical treatment of differential equations from the point of view of dynamical systems is given by

Vladimir I. Arnold, *Geometrical Methods in the Theory of Ordinary Differential Equations*, Springer-Verlag, New York–Berlin, 1983.

We also recommend his text on mechanics, which emphasizes the geometric, global, and structural perspectives:

Vladimir I. Arnold, *Mathematical Methods of Classical Mechanics*, Springer-Verlag, New York, 1989.

Low-dimensional dynamics is presented in a definitive way by

Welington de Melo and Sebastian van Strien, *One-Dimensional Dynamics*, Springer-Verlag, Berlin, 1993.

A good and well-motivated account of ergodic theory is due to

Karl Petersen, *Ergodic Theory*, Cambridge University Press, Cambridge, 1983, 1989.

Symbolic dynamics and interesting applications are well-presented by

Douglas Lind and Brian Marcus, *An Introduction to Symbolic Dynamics and Coding*, Cambridge University Press, Cambridge, 1995.

There are two books that are of particular interest to readers of the Panorama who seek a broad and up-to-date high-level overview of dynamics. The first of these resulted from a research institute in Seattle in the summer of 1999 that featured lecture series and expository talks aimed at students and experts alike. The resulting volume consists of lecture notes, surveys, and original papers, and will be of great interest:

Anatole Katok, Rafael de la Llave, Yakov Pesin and Howard Weiss (eds.), *Smooth Ergodic Theory and Its Applications*, Proceedings of Symposia in Pure Mathematics **69**, Summer Research Institute, Seattle, WA, 1999, American Mathematical Society, Providence, RI, 2001

The second book consists of surveys that together cover a vast spectrum of research areas in dynamics in a coherent way. It is part of a series of handbooks on dynamical systems that will have at least four volumes:

Boris Hasselblatt and Anatole Katok (eds.), *Handbook of Dynamical Systems,* vol. 1A, Elsevier, Amsterdam, 2002, vol. 1B, Elsevier, Amsterdam, to appear.

BACKGROUND READING

Dynamical systems builds on a mathematical background in several broad areas. A reader may want to consult an appropriate textbook while learning from this book, and further study may be needed to follow our suggestions for further reading. A concise summary of a wide variety of subjects useful for dynamical systems is given in the appendix to our book, *Introduction to the Modern Theory of Dynamical Systems.*
   Three useful books for reading at the level used here are

Jerrold E. Marsden and Michael J. Hoffman, *Elementary Classical Analysis,* W. H. Freeman, New York, 1993.
Charles C. Pugh, *Real Mathematical Analysis,* Springer-Verlag, New York, 2002.
Walter Rudin, *Principles of Mathematical Analysis,* McGraw-Hill, New York–Auckland–Düsseldorf, 1976.

Among many introductory books on ordinary differential equations

Vladimir I. Arnold, *Ordinary Differential Equations,* Springer-Verlag, Berlin, 1992, is suitable as a preparation for the study of dynamical systems.

The study of statistical properties (ergodic theory) requires a background in measure theory and functional analysis. Two classic texts also include some other material of use for dynamics:

Halsey L. Royden, *Real Analysis,* Macmillan, New York, 1988,
Walter Rudin, *Real and Complex Analysis,* McGraw-Hill, New York, 1987.

An even more comprehensive source for measure theory is

Paul R. Halmos, *Measure Theory,* Springer-Verlag, New York, 1974.

Smooth dynamics also requires a background in topology and geometry. We suggest the books

John W. Milnor, *Topology from the Differentiable Viewpoint,* Princeton University Press, Princeton, NJ, 1997.
James R. Munkres, *Elementary Differential Topology,* Annals of Mathematics Studies **54**, Princeton University Press, Princeton, NJ, 1966.

APPENDIX

A.1 METRIC SPACES

Some interesting dynamical systems do not naturally "live" in Euclidean space, and there are occasions where the study of a dynamical system benefits from considerations in an auxiliary space. Therefore we use metric spaces in some generality.

A.1.1 Definitions

Definition A.1.1 If X is a set, then $d\colon X \times X \to \mathbb{R}$ is said to be a *metric* or *distance function* if

 (1) $d(x, y) = d(y, x)$ *(symmetry)*,
 (2) $d(x, y) = 0 \Leftrightarrow x = y$ *(positivity)*,
 (3) $d(x, y) + d(y, z) \geq d(x, z)$ *(triangle inequality)*.

If d is a metric, then (X, d) is said to be a *metric space*.

Remark A.1.2 Putting $z = x$ in (3) and using (1) and (2) shows that $d(x, y) \geq 0$.

Remark A.1.3 A subset of a metric space is itself a metric space by using the metric of the space (this is then called the *induced metric*).

The following notions generalize familiar concepts from Euclidean space.

Definition A.1.4 The set $B(x, r) := \{y \in X \mid d(x, y) < r\}$ is called the *(open) r-ball* around x. A set $A \subset X$ is said to be *bounded* if it is contained in a ball.

A set $O \subset X$ is said to be *open* if for every $x \in O$ there exists $r > 0$ such that $B(x, r) \subset O$. (This immediately implies that any union of open sets is open.) The *interior* of a set S is the union Int S of all open sets contained in it. Equivalently, it is the set of $x \in S$ such that $B(x, r) \subset S$ for some $r > 0$. If $x \in X$ and O is an open set containing x, then O is said to be a *neighborhood* of x. A point $x \in X$ is called a *boundary point* of $S \subset X$ if for every neighborhood U of x we have $U \cap S \neq \emptyset$ and $U \setminus S \neq \emptyset$. The *boundary* of S is the set ∂A of its boundary points.

For $A \subset X$ the set $\bar{A} := \{x \in X \mid B(x, r) \cap A \neq \emptyset$ for all $r > 0\}$ is called the *closure* of A. A is said to be *closed* if $\bar{A} = A$. A set $A \subset X$ is said to be *dense* if $\bar{A} = X$ and ϵ-*dense* if $X \subset \bigcup\{B(x, \epsilon) \mid x \in A\}$. A set is said to be *nowhere dense* if its closure has empty interior (that is, contains no nonempty open set). This is true for finite sets but fails for $\mathbb{Q}$ and intervals. A sequence $(x_n)_{n \in \mathbb{N}}$ in X is said to *converge* to $x \in X$ if for all $\epsilon > 0$ there exists an $N \in \mathbb{N}$ such that for every $n \geq N$ we have $d(x_n, x) < \epsilon$.

It is easy to see that a set is closed if and only if its complement is open. (Therefore, any intersection of closed sets is closed.) Another way to define a closed set is via accumulation points:

Definition A.1.5 An *accumulation point* of a set A is a point x for which every ball $B(x, \epsilon)$ intersects $A \setminus \{x\}$. The set of accumulation points of A is called the *derived set* of A and denoted by A'. A set is closed if $A' \subset A$ and the closure $\bar{A}$ of a set A is $\bar{A} = A \cup A'$. A set A is said to be *perfect* if $A' = A$, that is, there are no points missing (all accumulation points are there) nor any extraneous (isolated) ones.

Note that $x \in A'$ if and only if there is a sequence of points in A that does not include x but converges to x.

Example A.1.6 Perfect sets are closed. $\mathbb{R}$ is perfect, as are $[0, 1]$, closed balls in $\mathbb{R}^n$, S^1, and the middle-third Cantor set (see Section A.1.7). But $\mathbb{Z}$ or finite subsets of $\mathbb{R}^n$ are not (they have no accumulation points) and nor are the rationals $\mathbb{Q}$ (they have irrational accumulation points).

On the real line finite sets are nowhere dense, but this fails for $\mathbb{Q}$ and intervals. The ternary Cantor set is nowhere dense, because it is closed and has empty interior (contains no interval).

Here is an interesting, pertinent special case of Theorem A.1.38:

Proposition A.1.7 *All sets in $\mathbb{R}$ that are bounded, perfect, and nowhere dense are homeomorphic to the ternary Cantor set.*

Definition A.1.8 A metric space X is said to be *connected* if it contains no two disjoint nonempty open sets. A *totally disconnected* space is a space X where for every two points $x_1, x_2 \in X$ there exist disjoint open sets $O_1, O_2 \subset X$ containing x_1, x_2, respectively, whose union is X.

$\mathbb{R}$ or any interval of $\mathbb{R}$, as well as $\mathbb{R}^n$ and open balls in $\mathbb{R}^n$, or the circle in $\mathbb{R}^2$ are connected. Examples of totally disconnected spaces are provided by finite subsets of $\mathbb{R}$ with at least two elements as well as the rationals and, in fact, any countable subset of $\mathbb{R}$. The ternary Cantor set is an uncountable totally disconnected set.

A.1.2 Completeness

One important property sets apart the real number system from that of rational numbers. This property is called *completeness*, and it reflects the fact that the real line "has no holes," like the rationals do. There are several equivalent ways of expressing this property precisely, and different versions may be useful in different circumstances.

(1) If a nondecreasing sequence of real numbers is bounded above, then it is convergent.
(2) If a subset of $\mathbb{R}$ has an upper bound, then it has a smallest upper bound.
(3) A Cauchy sequence of real numbers converges.

A Cauchy sequence is a sequence $(a_n)_{n \in \mathbb{N}}$ such that for any $\epsilon > 0$ there exists an $n \in \mathbb{N}$ such that $|a_n - a_m| < \epsilon$ for any $n, m \geq N$.

The first two versions of completeness refer to the ordering of the real numbers (by using the notions of upper bound and nondecreasing). The last one does not, and it is used to define completeness of metric spaces.

Definition A.1.9 A sequence $(x_i)_{i \in \mathbb{N}}$ is said to be a *Cauchy sequence* if for all $\epsilon > 0$ there exists an $N \in \mathbb{N}$ such that $d(x_i, x_j) < \epsilon$ whenever $i, j \geq N$. A metric space X is said to be *complete* if every Cauchy sequence converges.

Example A.1.10 For example, $\mathbb{R}$ is complete, whereas an open interval is not, when one uses the usual metric $d(x, y) = |x - y|$ (the endpoints are "missing"). If, however, we define a metric on the open interval $(-\pi/2, \pi/2)$ by $d_*(x, y) = |\tan x - \tan y|$, then this unusual metric space is indeed complete. The endpoints are no longer perceived as "missing" because sequences that look like they converge to an endpoint are not Cauchy sequences with respect to this metric since it stretches distances near the endpoints.

Remark A.1.11 This is an example of the *pullback* of a metric. If (Y, d) is a metric space and $h \colon X \to Y$ is an injective map, then $d_*(x, y) := d(h(x), h(y))$ defines a metric on X. Here we took $X = (-\pi/2, \pi/2)$, $Y = \mathbb{R}$, and $h = \tan$.

Lemma A.1.12 *A closed subset Y of a complete metric space X is itself a complete metric space.*

Proof A Cauchy sequence in Y is a Cauchy sequence in X and hence converges to some $x \in X$. Then $x \in Y$ because Y is closed. $\square$

An important example is the space of continuous functions (Definition A.1.16).

Theorem A.1.13 *The space*

$$\mathcal{C}([0, 1], \mathbb{R}^n) := \{f \colon [0, 1] \to \mathbb{R}^n \mid f \text{ is continuous}\}$$

is a complete metric space with the metric induced by the norm $\|f\| := \max_{x \in [0,1]} \|f(x)\|$ (see Section A.1.5).

Proof Suppose $(f_n)_{n\in\mathbb{N}}$ is a Cauchy sequence in $\mathcal{C}([0,1],\mathbb{R}^n)$. Then it is easy to see that $(f_n(x))_{n\in\mathbb{N}}$ is a Cauchy sequence in $\mathbb{R}^n$ for all $x \in [0,1]$. Therefore, $f(x) := \lim_{n\to\infty} f_n(x)$ is well defined by completeness of $\mathbb{R}^n$. To prove $f_n \to f$ uniformly fix any $\epsilon > 0$ and find $N \in \mathbb{N}$ such that $\|f_k - f_l\| < \epsilon/2$ whenever $k,l \geq N$. Now fix $k \geq N$. For any $x \in [0,1]$ there is an N_x such that $l \geq N_x \Rightarrow \|f_l(x) - f(x)\| < \epsilon/2$. Taking $l \geq N$ gives $\|f_k(x) - f(x)\| \leq \|f_k(x) - f_l(x)\| + \|f_l(x) - f(x)\| < \epsilon$. This proves the claim because k was chosen independently of x. $\square$

Likewise one proves completeness of the space of bounded sequences.

Theorem A.1.14 *The space l^∞ of bounded sequences $(x_n)_{n\in\mathbb{N}_0}$ with the sup-norm $\|(x_n)_{n\in\mathbb{N}_0}\|_\infty := \sup_{n\in\mathbb{N}_0} |x_n|$ is complete.*

Proof The proof is the same, except that the domain is $\mathbb{N}$ rather than $[0,1]$. (Boundedness is assumed to make the norm well defined; for continuous functions on $[0,1]$ it is automatic.) $\square$

Lemma A.1.15 (Baire Category Theorem). *In a complete metric space any intersection of countably many open dense sets is dense.*

Proof If $\{O_i\}_{i\in\mathbb{N}}$ are open and dense in X and $\varnothing \neq B_0 \subset X$ is open, then inductively choose a ball B_{i+1} of radius at most ϵ/i such that $\bar{B}_{i+1} \subset O_{i+1} \cap B_i$. The centers form a Cauchy sequence and hence converge by completeness. Thus $\varnothing \neq \bigcap_i \bar{B}_i \subset B_0 \cap \bigcap_i O_i$. $\square$

A.1.3 Continuity

Definition A.1.16 Let (X,d), (Y,d') be metric spaces. A map $f: X \to Y$ is said to be an *isometry* if $d'(f(x), f(y)) = d(x,y)$ for all $x, y \in X$. It is said to be *continuous* at $x \in X$ if for every $\epsilon > 0$ there exists a $\delta > 0$ such that $f(B(x,\delta)) \subset B(f(x),\epsilon)$ or, equivalently, if $d(x,y) < \delta$ implies $d'(f(x), f(y)) < \epsilon$. f is said to be continuous if f is continuous at x for every $x \in X$. An equivalent characterization is that the preimage of each open set is open. f is said to be *uniformly continuous* if the choice of δ does not depend on x, that is, for all $\epsilon > 0$ there is a $\delta > 0$ such that for all $x, y \in X$ with $d(x,y) < \delta$ we have $d'(f(x), f(y)) < \epsilon$. f is said to be an *open map* if it maps open sets to open sets.

A continuous bijection (one-to-one and onto map) with continuous inverse is said to be a *homeomorphism*. A map $f: X \to Y$ is said to be *Lipschitz-continuous* (or Lipschitz) with Lipschitz constant C, or C-Lipschitz, if $d'(f(x), f(y)) \leq Cd(x,y)$. A map is said to be a *contraction* (or, more specifically, a λ-*contraction*) if it is Lipschitz-continuous with Lipschitz constant $\lambda < 1$.

Continuity does not imply that the image of an open set is open. For example, the map x^2 sends $(-1,1)$ or $\mathbb{R}$ to sets that are not open.

There are various ways in which two metrics can be similar, or equivalent. The easiest way to describe these is to view the process of changing metrics as taking the identity map on X as a map between two different metric spaces.

Definition A.1.17 We say that two metrics are *isometric* if the identity establishes an isometry between them. Two metrics are said to be *uniformly equivalent* (sometimes just equivalent) if the identity and its inverse are Lipschitz maps between the two metric spaces. Finally, two metrics are said to be *homeomorphic* (sometimes also equivalent) if the identity is a homeomorphism between them.

A.1.4 Compactness

An important class of metric spaces is that of *compact* ones:

Definition A.1.18 A metric space (X, d) is said to be *compact* if any open cover of X has a finite subcover; that is, if, whenever $\{O_i \mid i \in I\}$ is a collection of open sets of X indexed by I such that $X \subset \bigcup_{i \in I} O_i$, there is a finite subcollection $\{O_{i_1}, O_{i_2}, \ldots, O_{i_n}\}$ such that $X \subset \bigcup_{l=1}^{n} O_{i_l}$.

Proposition A.1.19 *Compact sets are closed and bounded.*

Proof Suppose X is a metric space and $C \subset X$ is compact. If $x \notin C$, then the sets $O_n := \{y \in X \mid d(x, y) > 1/n\}$ form an open cover of $X \smallsetminus \{x\}$ and hence of C. There is a finite subcover $\mathcal{O}$ of $\{O_n\}_{n \in \mathbb{N}}$. Let $n_0 := \max\{n \in \mathbb{N} \mid O_n \in \mathcal{O}\}$. Then $d(x, y) > 1/n_0$ for all $y \in C$, so $x \notin \bar{C}$. This proves $\bar{C} \subset C$.

C is bounded because the open cover $\{B(x, r) \mid r > 0\}$ has a finite subcover. $\square$

The Heine–Borel Theorem tells us that in Euclidean space a set is compact if and only if it is closed and bounded. In some important metric spaces, closed bounded sets may fail to be compact, however, and this definition of compactness describes the property that is useful in a general metric space. Indeed, this definition uses the metric only to the extent that it involves open sets.

If a metric is given, compactness is equivalent to being both complete and totally bounded:

Definition A.1.20 A metric space is said to be totally bounded if for any $r > 0$ there is a finite set C such that the r-balls with center in C cover the space.

Proposition A.1.21 *Compact sets are totally bounded.*

Proof If C is compact and $r > 0$, then $\{B(x, r) \mid x \in C\}$ has a finite subcover. $\square$

Proposition A.1.22 *If (X, d) and (Y, d') are metric spaces, X is compact, and $f \colon X \to Y$ is a continuous map, then f is uniformly continuous and $f(X) \subset Y$ is compact; hence it is closed and bounded. If $Y = \mathbb{R}$, this shows that f attains its minimum and maximum.*

Among the most used facts about compact spaces is this last observation that a continuous real-valued function on a compact set attains its minimum and maximum.

Proof For every $\epsilon > 0$ there is a $\delta = \delta(x, \epsilon) > 0$ such that $d'(f(x), f(y)) < \epsilon/2$ whenever $d(x, y) < \delta$. The balls $B(x, \delta(x, \epsilon)/2)$ cover X, so by compactness of X there is a finite subcover by balls $B(x_i, \delta(x_i, \epsilon)/2)$. Let $\delta_0 = (1/2) \min\{\delta(x_i, \epsilon)\}$.

If $x, y \in X$ with $d(x, y) < \delta_0$, then $d(x, x_i) < \delta_0 < \delta(x_i, \epsilon)$ for some x_i and, by the triangle inequality, $d(y, x_i) \leq d(x, x_i) + d(x, y) < \delta_0 + \delta_0 \leq \delta(x_i, \epsilon)$. These two facts imply $d'(f(x), f(y)) \leq d'(f(x), f(x_i)) + d'(f(y), f(x_i)) < \epsilon/2 + \epsilon/2 = \epsilon$.

To see that $f(X) \subset Y$ is compact, consider any open cover $f(X) \subset \bigcup_{i \in I} O_i$ of $f(X)$. Then the sets $f^{-1}(O_i) = \{x \mid f(x) \in O_i\}$ cover X, and hence there is a finite subcover $X \subset \bigcup_{l=1}^{n} f^{-1}(O_{i_l})$. But then $f(X) \subset \bigcup_{l=1}^{n} O_{i_l}$. $\square$

Proposition A.1.23 *Suppose $\{C_i \mid i \in I\}$ is a collection of compact sets in a metric space X such that $\bigcap_{l=1}^{n} C_{i_l} \neq \varnothing$ for any finite subcollection $\{C_{i_l} \mid 1 \leq l \leq n\}$. Then $\bigcap_{i \in I} C_i \neq \varnothing$.*

Proof We prove the contrapositive: Suppose $\{C_i \mid i \in I\}$ is a collection of compact sets with $\bigcap_{i \in I} C_i = \varnothing$. Let $O_i = C_1 \smallsetminus C_i$ for $i \in I$. Then $\bigcap_{i \in I} C_i = \varnothing$ implies that $\bigcup_{i \in I} O_i = C_1$, that is, the O_i form an open cover of the compact set C_1. Thus there is a finite subcover $\bigcup_{l=1}^{n} O_{i_l} = C_1$. This means that $\bigcap_{l=1}^{n} C_{i_l} = \varnothing$. $\square$

Proposition A.1.24

(1) *A closed subset of a compact set is compact.*
(2) *The intersection of compact sets is compact.*
(3) *A continuous bijection between compact spaces is a homeomorphism.*
(4) *A sequence in a compact set has a convergent subsequence.*

Proof (1) Suppose $C \in X$ is a closed subset of a compact space and $\bigcup_{i \in I} O_i$ is an open cover of C. If $O = X \smallsetminus C$, then $X = O \cup C \subset O \cup \bigcup_{i \in I} O_i$ is an open cover of X and hence has a finite subcover $O \cup \bigcup_{l=1}^{n} O_{i_l}$. Since $O \cap C = \varnothing$, we get a finite subcover $\bigcup_{l=1}^{n} O_{i_l}$ of C.

(2) The intersection of compact sets is an intersection of closed subsets and hence a closed subset of any of these compact sets. Therefore it is compact by (1).

(3) We need to show that the image of an open set is open. Using bijectivity, note that the complement of the image of an open set O is the image of the complement O^c of O. O^c is a closed subset of a compact space, hence it is compact, and thus its image is compact, and hence closed. Its complement, the image of O, is then open, as required.

(4) Given a sequence $(a_n)_{n \in \mathbb{N}}$, let $A_n := \{a_i \mid i \geq n\}$ for $n \in \mathbb{N}$. Then the closures $\overline{A_n}$ satisfy the hypotheses of Proposition A.1.23 and there exists an $a_0 \in \bigcap_{n \in \mathbb{N}} \overline{A_n}$. This means that for every $k \in \mathbb{N}$ there exists an $n_k > n_{k-1}$ such that $a_{n_k} \in B(a_0, 1/k)$, that is, $a_{n_k} \to a_0$. $\square$

An interesting example of a metric space is given by the Hausdorff metric:

Definition A.1.25 If (X, d) is a compact metric space and $K(X)$ denotes the collection of closed subsets of X, then the *Hausdorff metric* d_H on $K(X)$ is defined by

$$d_H(A, B) := \sup_{a \in A} d(a, B) + \sup_{b \in B} d(b, A),$$

where $d(x, Y) := \inf_{y \in Y} d(x, y)$ for $Y \subset X$.

Notice that d_H is symmetric by construction and is zero if and only if the two sets coincide (here we use that these sets are closed, and hence compact, so the "sup" are actually "max"). Checking the triangle inequality requires a little extra work. To show that $d_H(A, B) \leq d_H(A, C) + d_H(C, B)$, note that $d(a, b) \leq d(a, c) + d(c, b)$ for $a \in A$, $b \in B$, $c \in C$, so taking the infimum over b we get $d(a, B) \leq d(a, c) + d(c, B)$ for $a \in A$, $c \in C$. Therefore, $d(a, B) \leq d(a, C) + \sup_{c \in C} d(c, B)$ and $\sup_{a \in A} d(a, B) \leq \sup_{a \in A} d(a, C) + \sup_{c \in C} d(c, B)$. Likewise, one gets $\sup_{b \in B} d(b, A) \leq \sup_{b \in B} d(b, C) + \sup_{c \in C} d(c, A)$. Adding the last two inequalities gives the triangle inequality.

Lemma A.1.26 *The Hausdorff metric on the closed subsets of a compact metric space defines a compact topology.*

Proof We need to verify total boundedness and completeness. Pick a finite $\epsilon/2$-net N. Any closed set $A \subset X$ is covered by a union of ϵ-balls centered at points of N, and the closure of the union of these has Hausdorff distance at most ϵ from A. Since there are only finitely many such sets, we have shown that this metric is totally bounded. To show that it is complete, consider a Cauchy sequence (with respect to the Hausdorff metric) of closed sets $A_n \subset X$. If we let $A := \bigcap_{k \in \mathbb{N}} \overline{\bigcup_{n \geq k} A_n}$, then one can easily check that $d(A_n, A) \to 0$. $\square$

Any homeomorphism of a compact metric space X induces a natural homeomorphism of the collection of closed subsets of X with the Hausdorff metric, so we have the following:

Lemma A.1.27 *The set of closed invariant sets of a homeomorphism f of a compact metric space is a closed set with respect to the Hausdorff metric.*

Proof This is just the set of fixed points of the induced homeomorphism; hence it is closed. $\square$

Definition A.1.28 A metric space (X, d) is said to be *locally compact* if for every x and every neighborhood O of x there is a compact set K in O that contains x. It is said to be *separable* if it contains a countable dense subset (such as the rationals in $\mathbb{R}$).

A.1.5 Norms Define Metrics in $\mathbb{R}^n$

There is a particular class of metrics in the Euclidean space $\mathbb{R}^n$ that are invariant under translations.

Definition A.1.29 A function N on a linear space is said to be a *norm* if

(1) $N(\lambda x) = |\lambda| N(x)$ for $\lambda \in \mathbb{R}$ *(homogeneity)*,
(2) $N(x) \geq 0$ and $N(x) = 0 \Leftrightarrow x = 0$ *(positivity)*,
(3) $N(x + y) \leq N(x) + N(y)$ *(convexity)*.

A linear space with a norm is said to be a normed linear space.

Any norm determines a metric by setting the distance function $d(x, y) = N(x - y)$. For the metric thus defined, positivity follows from the positivity of the norm, symmetry follows from homogeneity for $\lambda = -1$, and triangle inequality follows from convexity. For such a metric the translations $T_v \colon x \to x + v$ are isometries by definition. Furthermore, the central symmetry $x \to -x$ is an isometry, and any homothety $x \to \lambda x$ multiplies all distances by $|\lambda|$ (we call the last property homogeneity of the metric).

Example A.1.30 The maximum distance on $\mathbb{R}^n$ is given by

(A.1.1) $$d(x, y) = \max_{1 \leq i \leq n} |x_i - y_i|.$$

Of course, the standard Euclidean metric is of that kind (it is also invariant under rotations, which we do not require), as is the maximum metric (A.1.1).

Example A.1.31 The linear space $C([0,1])$ of continuous functions on $[0,1]$ is a linear space and carries the norm $\| f \| := \max\{| f(x)| \mid x \in [0, 1]\}$.

The following proposition is the main reason why norms are useful devices in dynamics.

Proposition A.1.32 *All metrics in $\mathbb{R}^n$ determined by norms are uniformly equivalent.*

Proof First, since the property of uniform equivalence is transitive, it is sufficient to show that any metric determined by a norm is uniformly equivalent to the standard Euclidean metric.

Second, since translations are isometries, it is suffient to consider distances from the origin, that is, we can work with the norms directly.

Third, by homogeneity it is sufficient to consider norms of vectors whose Euclidean norm is equal to one, that is, the points on the unit sphere.

But then the other norm is a convex, and hence continuous, function with respect to Euclidean distance, so by compactness of the sphere it is bounded from above. It also achieves its minimum on the unit sphere. The minimum cannot be zero because this would imply the existence of a nonzero vector with zero norm. Thus the ratio of the norms is bounded between two positive constants. $\square$

A.1.6 Product Spaces

The construction of the torus as a product of circles illustrates the usefulness of considering products of metric spaces in general. To define the product of two metric spaces (X, d_X) and (Y, d_Y) we need to define a metric on the cartesian product $X \times Y$, such as

$$d_{X \times Y}((x_1, y_1), (x_2, y_2)) := \sqrt{(d_X(x_1, x_2))^2 + (d_Y(y_1, y_2))^2}.$$

That this defines a metric is checked in the same way as checking that the Euclidean norm on $\mathbb{R}^2$ defines a metric.

There are other choices of equivalent metrics on the product. Two evident ones are

$$d'_{X \times Y}((x_1, y_1), (x_2, y_2)) := d_X(x_1, x_2) + d_Y(y_1, y_2)$$

and

$$d''_{X \times Y}((x_1, y_1), (x_2, y_2)) := \max(d_X(x_1, x_2), d_Y(y_1, y_2)).$$

Showing that these metrics are pairwise uniformly equivalent is done in the same way as showing that the Euclidean norm, the norm $\|(x, y)\|_1 := |x| + |y|$, and the maximum norm $\|(x, y)\|_\infty := \max(|x|, |y|)$ define pairwise equivalent metrics (Proposition A.1.32). Indeed, this follows from it.

For products of finitely many spaces (X_i, d_{X_i}) $(i = 1, \ldots, n)$ one can define several uniformly equivalent metrics on the product as follows: Fix a norm $\| \cdot \|$ on $\mathbb{R}^n$, and for any two points $(x_1, x_2, \ldots, x_n)$, and $(x'_1, x'_2, \ldots, x'_n)$ define their distance to be the norm of the vector in $\mathbb{R}^n$ whose entries are $d_{X_i}(x_i, x'_i)$. That the resulting metrics are uniformly equivalent follows from the uniform equivalence of any two norms on $\mathbb{R}^n$ (Proposition A.1.32).

We also encounter products of infinitely many metric spaces (or, usually, a product of infinitely many copies of the same metric space). In an infinite cartesian product of a set X every element is specified by its components; that is, if the copies of the set X are indexed by a label i that ranges over an index set I, then an individual element of the product set is specified by assigning to each value of i an element of X, the ith coordinate. This leads to the formal definition of the infinite product $\prod_{i \in I} X =: X^I$ as the set of all functions from I to X.

Unlike in the case of finite products, we have to choose our product metric carefully. Not only do we have to keep in mind questions of convergence, but different choices may give metrics that are not equivalent, even up to homeomorphism. To define a product metric assume that I is countable. In case $I = \mathbb{N}$ and if the metric on X is bounded, that is, $d(x, y) \leq 1$, say, for all $x, y \in X$, we can define several homeomorphic metrics by setting

(A.1.2)
$$d_\lambda(x, y) := \sum_{i=1}^{\infty} \frac{d(x_i, y_i)}{\lambda^{|i|}}.$$

This converges for any $\lambda > 1$ by comparison with the corresponding geometric series.

If $I = \mathbb{Z}$, we make the same definition with summation over $\mathbb{Z}$ [this is the reason for writing $|i|$ in (A.1.2)].

Theorem A.1.33 (**Tychonoff**). *The product of compact spaces is compact.*

As a particular case we can perform this construction with $X = [0, 1]$, the unit interval. The product thus obtained is called the *Hilbert cube*. This is a new way to think of the collection of all sequences whose entries are in the unit interval.

A.1.7 Sequence Spaces

Generalizing from the ternary Cantor set introduced in Section 2.7.1 we now define a more general class of metric spaces of which there are many important examples.

Definition A.1.34 A *Cantor set* is a metric space homeomorphic to the middle-third Cantor set.

A natural and important example is the space Ω_2^R of sequences $\omega = (\omega_i)_{i=0}^\infty$ whose entries are either 0 or 1. This set is the product $\{0, 1\}^{\mathbb{N}_0}$ of countably many copies of the set $\{0, 1\}$ of two elements, so it is natural to endow it with a product metric. Up to multiplication by a constant there is only one metric on $\{0, 1\}$, which we define by setting $d(0, 1) = 1$. Referring to (A.1.2), we can endow Ω_2^R with the product metric

$$d(\omega, \omega') := \sum_{i=0}^\infty \frac{d(\omega_i, \omega_i')}{3^{i+1}}.$$

Proposition A.1.35 *The space* $\Omega_2^R = \{0, 1\}^{\mathbb{N}_0}$ *equipped with the product metric* $d(\omega, \omega') := \sum_{i=0}^\infty d(\omega_i, \omega_i') 3^{-(i+1)}$ *is a Cantor set.*

To prove this we need a homeomorphism between the ternary Cantor set C and Ω_2^R:

Lemma A.1.36 *The one-to-one correspondence between the ternary Cantor set C and Ω_2^R defined by mapping each point $x = 0.\alpha_1\alpha_2\alpha_3 \cdots = \sum_{i=1}^\infty (\alpha_i/3^i) \in C$ ($\alpha_i \neq 1$) to the sequence $f(x) := \{\alpha_i/2\}_{i=0}^\infty$ is a homeomorphism.*

Proof If $x = 0.\alpha_0\alpha_1\alpha_2 \cdots = \sum_{i=0}^\infty (\alpha_i/3^{i+1})$ ($\alpha_i \neq 1$) and $y = 0.\beta_0\beta_1\beta_2 \cdots = \sum_{i=0}^\infty (\beta_i/3^{i+1})$ ($\beta_i \neq 1$) in C, then

$$d(x, y) = |x - y| = \left| \sum_{i=0}^\infty \frac{\alpha_i}{3^{i+1}} - \sum_{i=0}^\infty \frac{\beta_i}{3^{i+1}} \right|$$

$$= \left| \sum_{i=0}^\infty \frac{\alpha_i - \beta_i}{3^{i+1}} \right| \leq \sum_{i=0}^\infty \frac{|\alpha_i - \beta_i|}{3^{i+1}} = 2d(f(x), f(y)).$$

Now let $\alpha = f(x)$, $\beta = f(y)$. Then $d(f^{-1}(\alpha), f^{-1}(\beta)) = d(x, y) \leq 2d(\alpha, \beta)$, so f^{-1} is Lipschitz-continuous with Lipschitz constant 2.

If $\omega, \omega' \in \Omega_2^R$ are two sequences with $d(\omega, \omega') \geq 3^{-n}$, then $\omega_i \neq \omega_i'$ for some $i \leq n$, because otherwise

$$d(\omega, \omega') \leq \sum_{i=n+1}^\infty 3^{-i-1} = \frac{3^{-n-2}}{1 - \frac{1}{3}} = 3^{-n-1}/2 < 3^{-n}.$$

Consequently, $f^{-1}(\omega)$ and $f^{-1}(\omega')$ differ in the ith digit for some $i \leq n$. This implies $d(f^{-1}(\omega), f^{-1}(\omega')) \geq 3^{-(n+1)}$ because the two points are in different pieces of C_{n+1}. Taking $x = f^{-1}(\omega)$, $x' = f^{-1}(\omega')$, we get $d(x, x') < 3^{-(n+1)} \Rightarrow d(f(x), f(y)) < 3^{-n}$. This shows that f is Lipschitz-continuous as well. $\square$

We have shown in particular that Ω_2^R is compact and totally disconnected. Let us note in addition that every sequence in Ω_2^R can be approximated arbitrarily well by different sequences in Ω_2^R by changing only very remote entries. Thus every point of Ω_2^R is an accumulation point and Ω_2^R is a perfect set.

Proposition A.1.37 *Cantor sets are compact, totally disconnected, and perfect.*

It is not hard to see that the space $\Omega_2 = \{0, 1\}^{\mathbb{Z}}$ with a product metric is in turn homeomorphic to Ω_2^R, and therefore it is also a Cantor set. To that end let

$$\alpha: \mathbb{Z} \to \mathbb{N}_0, \qquad n \mapsto \begin{cases} 2n & \text{if } n \geq 0 \\ 1 - 2n & \text{if } n < 0 \end{cases}$$

and $f: \Omega_2^R \to \Omega_2$, $\omega \mapsto \omega \circ a = (\ldots \omega_3 \omega_1 \omega_0 \omega_2 \omega_4 \ldots)$. Endowing Ω_2 and Ω_2^R with any two of the product metrics (A.1.2) makes f a homeomorphism because two sequences α, α' are close if and only if they agree on a large stretch of initial entries. Then the resulting sequences $\omega = f(\alpha)$ and $\omega' = f(\alpha')$ agree on a long stretch of entries around the 0th entry and hence are also close. Thus f is a continuous bijection between compact spaces and therefore a homeomorphism by Proposition A.1.24. (It is as easy to see directly that f^{-1} is continuous.)

A.1.8 General Properties of Cantor Sets

Theorem A.1.38 *Every perfect totally disconnected compact metric space is a Cantor set.*

We have seen that sequence spaces are perfect and compact; it is easy to see in general that they are totally disconnected: If $\alpha \neq \beta$ are sequences, then $\alpha_i \neq \beta_i$ for some index i. The set of sequences ω with $\omega_i = \alpha_i$ is open, and likewise the set of sequences with $\omega_i = \beta_i$. But these sets are disjoint and their union is the entire space.

Corollary A.1.39 *Every nonempty, perfect, bounded, nowhere dense set on the line is a Cantor set.*

Proof A perfect bounded set on the line is compact by the Heine–Borel Theorem (a closed bounded subset of $\mathbb{R}^n$ is compact). Being perfect, it also contains more than one point. If it is not totally disconnected, then it has a connected component with more than one point and hence contains a nontrivial interval, contrary to being nowhere dense. $\square$

A.1.9 Dyadic Integers

Define the following metric d_2 on the group $\mathbb{Z}$ of all integers: $d(n, n) = 0$ and $d_2(m, n) = \|m - n\|_2$ for $n \neq M$, where

$$\|n\|_2 = 2^{-k} \qquad \text{if } n = 2^k l \text{ with an odd number } l.$$

The completion of $\mathbb{Z}$ with respect to that metric is called the group of *dyadic integers* and is usually denoted by $\mathbb{Z}_2$. It is a compact topological group.

A.2 DIFFERENTIABILITY

A.2.1 The Derivative

A map is said to be differentiable if it admits a good linear approximation. We require that for each point there is a linear transformation that differs from the map by an error that is smaller than linear as a function of the distance to the reference point:

Definition A.2.1 Let V, W be normed linear spaces, $U \subset V$ open, and $x \in U$. A map $f: U \to W$ is said to be differentiable at x if there is a linear map $A: V \to W$ such that

$$\lim_{h \to 0} \frac{\| f(x + h) - f(x) - Ah \|}{\|h\|} = 0.$$

In this case A is said to be the derivative of f at x, and we write $Df(x) := A$.

If a map $f: \mathbb{R}^n \to \mathbb{R}^m$ is differentiable at x, then $Df(x)$ is the matrix of partial derivatives at x (see Section 2.2.4.1), but the existence of all partial derivatives does not imply differentiability.

A.2.2 The C^r-Topology

The sequence of functions $f_n(x) := \sin(nx)/n$ converges to 0 uniformly, but the sequence of derivatives does not. Therefore, if one wants to ensure convergence of derivatives of a sequence of functions, one must impose it explicitly. The C^1-topology is an elegant way to formulate this. On the space of bounded functions with bounded derivative we define the metric

$$d(f, g) := \max(\sup_x d(f(x), g(x)), \sup_x d(Df(x), Dg(x))).$$

Then $d(f_n, g) \to 0$ means that $f_n \to g$ and $Df_n \to Dg$ uniformly. Likewise, the C^r-topology is defined by the metric

$$d(f, g) := \max_{0 \le i \le r} \sup_x d(D^i f(x) D^i g(x)).$$

Theorem A.2.2 A space of bounded continuous functions with values in a complete space, endowed with the metric of uniform convergence, is a complete space. Likewise, any space of bounded functions with bounded derivative (and values in a complete space) is complete with the C^1-topology. An analogous statement holds for the C^r-topology.

This generalization of Theorem A.1.13 and Theorem A.1.14 is an important reason for using these topologies.

A.2.3 The Mean Value Theorem and the Taylor Remainder

The Mean Value Theorem is a basic and central result in differential calculus. It connects the derivative with the behavior of a function on an interval.

Theorem A.2.3 *If $f: [a, b] \to \mathbb{R}$ is continuous and f is differentiable on (a, b), then there is a point $x \in (a, b)$ such that $f(b) - f(a) = (b - a) f'(x)$.*

Proof $g(t) := t(f(b) - f(a)) - f(t)(b - a)$ is continuous on $[a, b]$ and differentiable on (a, b) and $g(a) = af(b) - bf(a) = g(b)$. If g is constant, then we are done. Otherwise, g has an extremum $g(x)$ at some $x \in (a, b)$ by continuity. g is differentiable at x, hence $0 = g'(x) = f(b) - f(a) - f'(x)(b - a)$. $\square$

A more sophisticated version of this result is used to establish the validity of a Taylor expansion.

Theorem A.2.4 *If $f: (a, b) \to \mathbb{R}$ has $k + 1$ derivatives and $x_0 \in (a, b)$, then for every $x \in (a, b)$ there exists a c between x and x_0 such that*

$$f(x) = \sum_{i=0}^{k} \frac{f^{(i)}(x_0)}{i!}(x - x_0)^i + \frac{f^{(k+1)}(x_0)}{(k+1)!}(x - x_0)^{k+1},$$

where $f^{(i)}$ denotes the ith derivative.

Proof Let $f_k(x) := \sum_{i=0}^{k} f^{(i)}(x_0)(x - x_0)^i/i!$, $z := (f(x) - f_k(x))/(x - x_0)^{k+1}$, and $g(t) := f(t) - f_k(t) - z(t - x_0)^{k+1}$ on $[a, b]$.

We will show that $g^{(k+1)}(c) = 0$ for some c between x and x_0. Since $g^{(k+1)}(t) = f^{(k+1)}(t) - (k+1)!z$, this implies $f^{(k+1)}(c) = (k+1)!z$, as required.

We use that $g^{(i)}(x_0) = 0$ for $0 \le i \le k$ since $f^{(i)}(x_0) = f_k^{(i)}(x_0)$ by definition. Combined with $g(x) = 0$ (by choice of z), this gives a c_1 between x and x_0 such that $g'(c_1) = 0$. Combining this with $g'(x_0) = 0$ gives a c_2 between c_1 and x_0 such that $g''(c_2) = 0$. Repeating k times gives the desired c. $\square$

A.2.4 Diffeomorphisms and Embeddings

The inverse of an invertible differentiable map need not be differentiable; x^3 is an example. Since having a differentiable inverse is useful, such maps have a name: A differentiable map with differentiable inverse is said to be a diffeomorphism.

For our purposes it is useful to extend this notion to maps that are not surjective (onto). We want to allow maps such as $(x, y) \mapsto (x, y, x^2 + y^2)$ from the unit disk to $\mathbb{R}^3$, but we wish to exclude $t \mapsto (t, \pi t)$ (mod 1) from $\mathbb{R}$ to the torus (see Section 2.6.4), because its "inverse" is not continuous.

Definition A.2.5 Suppose $U \subset \mathbb{R}^n$. A map $f: U \to \mathbb{R}^m$ is said to be an *embedding* if f is differentiable, its derivative has rank n at every point, and $f: U \to f(U)$ is a homeomorphism.

In this definition one can replace either or both Euclidean spaces by a torus, a cylinder, or a sphere of the corresponding dimension.

A.3 RIEMANN INTEGRATION IN METRIC SPACES

The notion of integration with respect to a "measure" appears many times throughout the book.

A.3.1 The Riemann Integral

The basic notion is that of Riemann integration with respect to length, area, or volume in space, and its subsets and related spaces such as spheres, cylinders, and tori. An important question is, What functions are Riemann integrable? In the standard definition through upper and lower sums boundedness is obviously necessary. Similarly, the function must be *compactly supported*, that is, it should vanish outside of a compact set. This is no restriction if the ambient space itself is compact, such as a closed interval, a rectangle, a sphere, or a torus. Under these assumptions every continuous function is integrable. Some of the most important functions that appear in connection with integration are discontinuous, though. First, there are characteristic functions of intervals, rectangles, and other "nice" sets where the integral is equal to the length, area, or volume, depending on the dimension.

It turns out that there is a powerful necessary and sufficient condition for integrability.

Theorem A.3.1 (Lebesgue). *A function defined in a bounded domain of Euclidean space, or on a sphere, a torus, or a similar compact differentiable manifold is Riemann integrable if and only if it is bounded and the set of its discontinuity points is a null set.*

The main idea of the proof is to connect the countably many sets in the definition of null set (Definition 7.5.3) with the finitely many rectangles in the upper and lower sums. The method is to note that the set of points near which f varies by at least ϵ can be covered by countably many rectangles whose volumes sum to arbitrarily little, and then compactness leads to a finite subcover that can be made part of a legitimate partition.[1]

Applying this criterion to the characteristic function of a compact set A (whose discontinuity points are exactly the boundary points of A) we immediately obtain

Corollary A.3.2 *For a compact set A the length, area, or volume is defined if and only if the boundary of A is a null set.*

This of course immediately extends to sets with compact closure.

A.3.2 Weighted Integration

A natural extension of Riemann integration is weighted integration with respect to a nonnegative density ρ, which can be reduced to the standard case by simply multiplying the integrand by ρ. Naturally, in order for this procedure to work, the function ρ must be Riemann integrable itself. This is also sufficient due to the following fact.

Proposition A.3.3 *The sum, product, and uniform limit of Riemann-integrable functions is Riemann integrable. Moreover, the integral behaves naturally with*

---

[1] See Jerrold E. Marsden and Michael J. Hoffman, *Elementary Classical Analysis*, W. H. Freeman, New York, 1993.

*respect to linear combinations and limits, that is, the integral of a linear com-
bination (or limit) is the corresponding linear combination (or limit) of the
integrals.*

The notion of integral can be extended beyond the above setting. Most texts
on calculus and elementary real analysis deal with situations (which often appear
in the real world) when either the domain of the function is not compact [as
for $f(x) = 1/(1 + x^2) + x^2$ on the real line], or the function is unbounded [as for
$f(x) = \log x$ on the interval $(0, 1)$], or both. In such cases natural approximation
procedures often give a notion of integral called an *improper integral*.

A.3.3 The Riemann–Stieltjes Integral on the Line

Bernoulli measures (Section 7.6.4) constitute a situation where integration is
naturally defined but cannot be reduced to any of the situations described above.
Integration with respect to any nonsymmetric Bernoulli measure cannot be
reduced to Riemann integration with respect to a density since a null set may
have positive Bernoulli measure (this is referred to as *singularity* of the measure).
Still, integration with respect to a Bernoulli measure can be defined following the
familiar procudure of upper and lower Riemann sums associated to a partition of
$[0, 1]$ into small intervals and taking the joint limit when the length of the longest
partition element converges to zero.

A general construction of this kind is called the *Riemann–Stieltjes integral*. It is
defined for functions on an interval $I \subset \mathbb{R}$. It depends on a *distribution function*
F on I. This is a function that is monotone, bounded from above and below,
and continuous from the left; that is, for an increasing sequence $x_n \in I$ one has
$\lim_{n \to \infty} F(x_n) = F(\lim_{n \to \infty} x_n)$.

We first consider the case of continuous F; this is not a serious restriction
in dynamical considerations. Define the *measure* of an interval $[a, b] \subset I$ as
$F(b) - F(a)$. Using this measure instead of length, define upper and lower Riemann
sums for a function with respect to a finite partition of I. In this case there is no
distinction between closed, open, or half-open intervals.

In general, that is, if there are discontinuity points, one considers first intervals
whose endpoints are continuity points of F and defines the measure the same
way. To avoid ambiguities in the definiton of Riemann sums, one has to restrict to
partitions where the division points are continuity points of F. This is because the
discontinuity points are not null sets: The measure of any such point x is given by
the jump of the distribution function at x. The proof that any continuous function
is integrable (that is, upper and lower Riemann sums have the same limit for any
sequence of partitions for which the maximal length of elements goes to zero) is
effectively identical to the proof for the standard Riemann integral.

The distribution function construction provides the most general treatment of
integration in one dimension; it can in fact be extended to the case of the whole
line or a half-line. In particular, it takes care of the situation of a "reasonable"
unbounded density funtion, which appears for example as an invariant measure
for some quadratic maps such as those discussed in Section 11.4.3.

A.3.4 Integrals as Positive Functionals

However, in our discussions of hyperbolic systems and strange attractors we encounter situations beyond dimension one where an asymptotic distribution exists but cannot be reduced to integration with respect to a density. Besides, some of the natural systems considered in this book such as symbolic systems act on spaces that are very different from Euclidean spaces. In the rest of this section we describe a general framework for integration in metric spaces, which in particular addresses all of these situations. In fact, what we describe is going from an integral defined for continuous function to the measure for "nice" sets and back, as was done in the case of uniform distribution on the circle in Section 4.1.4 and Section 4.1.6.

Definition A.3.4 Let X be a compact metric space. A *Riesz integral* is a nonzero linear functional $\mathcal{I}$ on the space $C(X)$ of continuous real-valued functions on X that is continuous in the uniform topology and nonnegative, that is, $\mathcal{I}(f) \geq 0$ if $f \geq 0$.

The weighted Riemann integral $\mathcal{I}(f) = \int \rho f \, dx$ as well as any Riemann–Stieltjes integral satisfy these conditions.

A Riesz integral is defined on the characteristic function χ_A only if χ_A is continuous, that is, if A is simultaneously open and closed. While this is the case with cylinder sets in sequence spaces, it is impossible for connected spaces. We extend the definition of the Riesz integral to some characteristic functions and many other functions by approximation.

Definition A.3.5 For a function $f\colon X \to \mathbb{R}$ define the *upper integral* as

$$\mathcal{I}^+(f) := \inf\{\mathcal{I}(g) \mid g \in C(X), \ g \geq f\},$$

and similarly the *lower integral* is

$$\mathcal{I}^-(f) := \sup\{\mathcal{I}(g) \mid g \in C(X), \ g \leq f\}.$$

The function f is *integrable* if $\mathcal{I}^+(f) = \mathcal{I}^-(f)$. In this case, this common value is denoted by $\mathcal{I}(f)$.

Obviously a linear combination of integrable functions is integrable. It is slightly less trivial but still not hard to see that the product of two integrable functions is integrable and the uniform limit of integrable functions is integrable. The following proposition shows the abundance of integrable functions among characteristic functions.

Proposition A.3.6 *If $x \in X$ then for all but at most countably many values of r, the characteristic functions of both the closed and open r-ball around x are integrable.*

Definition A.3.7 A set whose characteristic function is integrable is said to be *(Riemann) measurable with respect to $\mathcal{I}$. $\mathcal{I}(\chi_A)$ is called the *measure* of A.

Proposition A.3.8 *Any finite union or finite intersection of measurable sets is measurable.*

Remark A.3.9 There is an obvious weakness in this definition that is not fully apparent in the case of standard Riemann integration. Namely, some fairly "nice" sets may not be measurable. The simplest example arises for the δ-*measure*: $\delta_{x_0}(f) = f(x_0)$ for a given point $x_0 \in X$. This is obviously a Riesz integral, and obviously it is "concentrated" at the point x_0, since it vanishes at any function that vanishes at that point. However, the one-point set $\{x_0\}$ is not measurable according to our definition. The way out of this problem is to extend the notions of integrability and measurability. This leads to the theory of *Lebesgue* integration.

A.3.5 Partitions and Riemann Sums

Now we show how a Riesz integral can be reconstructed using a procedure similar to the construction of the standard Riemann integral through partitions into rectangles and Riemann sums.

Definition A.3.10 Given a Riesz integral $\mathfrak{I}$, a *measurable partition* of X is a decomposition of X into the union of finitely many measurable sets. The *size* of a measurable partition is the supremum of distances between points in the same element.

Proposition A.3.11 *For any ϵ there is a measurable partition of size less than ϵ.*

Proof By Proposition A.3.6, for every point there exists a ball around that point of radius less than $\epsilon/3$ whose characteristic function is measurable. Take a finite cover $B_1, \ldots, B_n$ by such balls and let $C_k = B_k \setminus \bigcup_{i=1}^{k-1} B_i$, $k = 1, \ldots, n$. The sets $C_1, \ldots, C_n$ form a partition with the desired property. $\square$

Given a bounded function f on X and a measurable partition $\xi = (C_1, \ldots, C_n)$ of X, we define the *upper and lower Riemann sums* as

$$U(f, \xi) := \sum_{i=1}^{n} \text{measure}(C_i) \sup\{f(x) \mid x \in C_i\}$$

and

$$L(f, \xi) := \sum_{i=1}^{n} \text{measure}(C_i) \inf\{f(x) \mid x \in C_i\},$$

correspondingly.

Theorem A.3.12 *If f is integrable and ξ_m is a sequence of partitions whose size goes to zero, then*

$$\lim_{m \to \infty} U(f, \xi_m) = \lim_{m \to \infty} L(f, \xi_m) = \mathfrak{I}(f).$$

Proof First consider a continuous function f and let $\epsilon > 0$. Since f is uniformly continuous, there exists $N \in \mathbb{N}$ such that for any $m > N$ and for any element C

of the partition ξ_m, $\sup\{f(x) \mid x \in C\} - \inf\{f(x) \mid x \in C\} < \epsilon$. Since measure is additive, this implies that $U(f, \xi_m) - L(f, \xi_m) < \epsilon$.

For an arbitrary-Riemann integrable function f there are continuous functions f^+ and f_- such that $f^- \le f \le f_+$ and $(f^+) - (f_-) < \epsilon$. Applying the previous argument to these functions gives the statement. $\square$

A.3.6 The General Riemann Integral

Finally we show that a Riesz integral arises from a measure defined on a sufficient collection of sets through the construction of Riemann sums. In analogy with the classical Riemann-integral construction we call such sets "rectangles". We assume that a collection of rectangles is fixed together with a measure defined for each rectangle. We consider a compact metric space X.

Sufficiency: The whole space is a rectangle.

Refinement: Given any $\epsilon > 0$, a rectangle can be partitioned into finitely many rectangles, each of which fits into an ϵ-ball.

Intersection: The intersection of two rectangles is a rectangle.

Additivity: If a rectangle R is partitioned into rectangles R_i $(i = 1, \ldots, k)$ then $\text{measure}(R) = \sum_{i=1}^{k} \text{measure}(R_i)$.

Suppose f is a bounded real-valued function defined on a bounded rectangle A. For any partition $\mathcal{R} = \{R_1, \ldots, R_k\}$ (that is, $A = \bigcup_{i=1}^{k} R_i$) into rectangles, define the upper sum to be

$$U(f, \mathcal{R}) := \sum_{i=1}^{k} \text{measure}(R_i) \sup\{f(x) \mid x \in R_i\}$$

and the lower sum

$$L(f, \mathcal{R}) := \sum_{i=1}^{k} \text{measure}(R_i) \inf\{f(x) \mid x \in R_i\}.$$

Lemma A.3.13 *If $\mathcal{R}$ and $\mathcal{R}'$ are partitions, then $L(f, \mathcal{R}') \le U(f, \mathcal{R})$.*

Proof If $\mathcal{R}' = \mathcal{R}$, this is obvious. To reduce to this case use the *common refinement* $\bar{\mathcal{R}} := \{R \cap R' \mid R \in \mathcal{R}, \ R' \in \mathcal{R}'\}$. This is a partition by rectangles, and it is not hard to check that

$$L(f, \mathcal{R}') \le L(f, \bar{\mathcal{R}}) \le U(f, \bar{\mathcal{R}}) \le U(f, \mathcal{R})$$

using additivity. $\square$

This lemma implies that

$$\overline{\int}_A f := \inf_{\mathcal{R}} U(f, \mathcal{R}) \qquad \text{and} \qquad \underline{\int}_A f := \sup_{\mathcal{R}} L(f, \mathcal{R})$$

are well defined and finite, and $\overline{\int}_A f \ge \underline{\int}_A f$.

Definition A.3.14 A function f defined on a rectangle A is said to be Riemann integrable over A if $\overline{\int}_A f = \underline{\int}_A f$. In this case, $\int_A f := \underline{\int}_A f$ is called the Riemann integral of f over A.

Using the refinement property above, an argument very similar to the proof of Theorem A.3.12 shows that continuous functions are Riemann integrable and that the Riemann integral thus defined on $C(X)$ is a Riesz integral as in Definition A.3.4.

Hints and Answers

Exercise 1.2.5 $kT = -\log 2$, where log is the natural logarithm.

Exercise 1.3.3 Seven. The sixth step is quite close to Heron's initial guess.

Exercise 1.3.8 Exercise 1.1.5 solves $\cos x = x$, Exercise 1.1.8 finds $\sqrt{5}$, and Exercise 1.1.9 solves $\sin x = x$.

Exercise 1.3.23 Since the sequence of the last two digits has period 20, it suffices to multiply 8 by 2^{20} until 008 reappears. It may help to truncate in between.

Exercise 2.2.6 Use graphical computing (Remark 2.3.3 and Figure 2.3.1) in the middle image of Figure 2.3.2.

Exercise 2.2.7 The last two.

Exercise 2.2.11 Use the triangle inequality to reduce the problem to continuity at the zero matrix, then apply Exercise 2.2.9.

Problem 2.2.12 Consider an annulus centered at the origin with a narrow slit (a region that looks like the letter C) and use polar coordinates to contract the radial as well as the angular component.

Problem 2.2.13 Show that the minimum of $d(f(x), x)$ exists and must occur at a fixed point.

Problem 2.2.14 $f(x) = x + e^{-x}$.

Exercise 2.3.4 Define the map on the complementary intervals to E in such a way that every point moves to the right.

Problem 2.3.5 Do the construction on the complementary intervals similarly to the previous problem, but with extra care: First, make the function infinitely differentiable at each such interval with all derivatives of the difference with the identity vanishing at both ends; second, control the derivatives as the intervals get smaller. In other words, make the function representing the deviation of your map from the identity very "flat" near the set E.

Exercise 2.4.2 Separate variables and integrate to get $s = k - 1$.

Exercise 2.4.3 This example must violate the Lipschitz condition on f. Take $\dot{x} = \sqrt{x}$ and $x(t) = 4t^2$.

Exercise 2.4.4 $\dot{x} = x^2$.

Problem 2.4.7 The return map to a section may look like $x \to x - x^3$ near zero.

Exercise 2.5.1 Prove the convexity of f_λ^2 on that interval to bound the derivative by the derivative at x_λ.

Exercise 2.6.1 If $y \in B(x, r)$, show that $B(y, r - d(x, y)) \subset B(x, r)$.

Exercise 2.6.8 See Lemma A.1.12.

Problem 2.6.9 By compactness, $d(x, f(x))$ attains its minimum at some point R_0. Use the assumption to show that the minimum is zero, as well as the uniqueness of the fixed point. *Convergence*: For $x \in X$, the sequence $(f^n(x))_{n \in \mathbb{N}}$ has an accumulation point x'. Show that $f(x')$ is also an accumulation point and that $x' \neq x_0$ contradicts the fact that $d(f^n(x), x_0)$ is decreasing in n.

Problem 2.7.5 Verify that the map $0.\alpha_1\alpha_2\alpha_3 \ldots \mapsto (0.\alpha_1\alpha_3 \ldots, 0.\alpha_2\alpha_4 \ldots)$ is a homeomorphism, where all expansions are ternary and all $\alpha_i \in \{0, 2\}$.

Problem 2.7.6 Consult Section 4.4.1.

Problem 2.7.7 Construct a homeomorphism of the unit interval onto itself that maps C' onto C by matching complementary intervals preserving their order and picking each time a longest one of the available intervals.

Exercise 3.1.4 The solutions are of the form $x\lambda^n + yn\lambda^n$.

Exercise 3.1.5 Denote the desired number by a_n. The shape of the tiles forces $a_n = a_{n-1} + 2a_{n-2}$.

Exercise 3.2.3 Rewrite the equation in polar coordinates and separate variables.

Exercise 3.3.2 Use Proposition A.1.32.

Exercise 3.3.3 Use Proposition 3.3.3 and Proposition A.1.32.

Exercise 3.3.5 Consider the absolute values of the eigenvalues. Their sum is at least 2.7, so if none exceeds 1, then the product is at least 0.7.

Problem 3.3.6 Use Jordan normal form.

Problem 3.3.7 Consider the Jordan blocks first.

Exercise 4.1.5 Show that the change of the time difference between sunrise and moonrise from one day to the next is constant; hence the evolution of this difference represents an orbit of a rotation. Conclude that this rotation must be irrational.

Exercise 4.1.6 Use $\mathbb{Z}$ as the space.

Exercise 4.1.7 Consider as the space $\{-1, 1\} \cup \{\frac{1}{n} - 1/n \in \mathbb{N}\} \cup \{1 - \frac{1}{n}/n \in \mathbb{N}\}$ and modify the previous solution.

Exercise 4.1.9 The time averages converge to the value at the fixed point.

Exercise 4.1.11 292; see Proposition 15.2.7.

Problem 4.1.14 The time averages converge to the value at 0.

Exercise 4.2.5 The angle is equal to $1/\gamma$.

Exercise 4.2.10 Distinguish between the rational and irrational cases. In the former case there are finitely many very bright points and in the latter a circle.

Exercise 4.3.1 If a is an integer of half of an integer.

Exercise 4.3.4 F is a lift of an orientation-preserving homeomorphism. Since $F(0) = 0$, $\rho(F) = 0$.

Exercise 4.3.5 Notice that the arguments for homeomorphisms used only continuity and monotonicity.

Exercise 4.3.6 It changes sign.

Exercise 4.3.7 Use Lemma 4.3.7.

Problem 4.3.8 If there is a point for which the corresponding $\overline{\lim}$ and $\underline{\lim}$ disagree, choose a rational number in between and locate a corresponding periodic point.

Problem 4.3.9 Consider the intervals between adjacent fixed points and obtain a contradiction if all points are stable or semistable.

Problem 4.4.5 Yes: If $\{O_1, O_2\}$ is a disjoint open cover of $A_{p/q}$, show that we may assume $\overline{O}_1 \cap \overline{O}_2 = \varnothing$. Use a compactness argument to obtain a contradiction.

Exercise 5.1.2 Either at least one is a perfect square or their ratio is a ratio of perfect squares. Prove this by case distinctions on the number of nonzero coefficients.

Exercise 5.1.5 Use the Chinese Remainder Theorem.

Exercise 5.1.7 If all elements of Γ are linearly dependent over $\mathbb{R}$, that is, Γ lies in a line, the statement follows from Exercise 4.2.8 by a coordinate change. Otherwise, Γ contains two linearly independent vectors. Consider the factor of Γ by the lattice generated by these vectors. This is a closed subgroup of $\mathbb{T}^2$. Classify such subgroups using the previous exercise.

Problem 5.1.9 $\mathbb{R}^k \times \mathbb{Z}^l$, where $0 \le k + l \le n$ and $k < n$.

Problem 5.1.11 $\mathbb{Z}_2$ has a Cantor-like structure. At the nth level there are 2^n sets (two elements are in the same set if their difference is a multiple of 2). Uniform distribution means that the asymptotic frequencies of visits to those sets are all equal.

Exercise 5.2.1 Eight, if the initial direction is not parallel to one of the faces.

Exercise 5.2.2 Consider the group S of eight elements generated by reflections in three faces of the cube I passing through the origin. The orbit of the unit cube under this group covers the cube C of double size centered at the origin. Use the group S for a partial unfolding. Then any billiard orbit inside I unfolds into a parallel motion in the torus obtained from S by the identification of the pairs of opposite sides.

Exercise 6.1.3 Yes.

Exercise 6.1.9 Apply the Baire Category Theorem to find the recurrent points in a closed ball contained in the domain.

Problem 6.2.6 One can use a more complicated version of the coordinate calculation in (6.2.4) or use the fact that the potential is unchanged when the coordinate system is translated (this helps with the center of gravity) or rotated (for angular momentum).

Problem 6.2.7 The motion relative to the center of gravity looks like two independent central force problems, so the orbits are ellipses.

Problem 6.2.9 This is outlined in V. I. Arnold, *Mathematical Methods of Classical Mechanics*, Springer, Berlin, 1980.

Exercise 6.3.3 Consider separately the orbits that hit the inner circle and those that do not. Both parts split into invariant circles.

Exercise 6.3.4 Consider the unfolding generated by reflections in the coordinate axes.

Problem 6.3.6 Find a combination of the squares of the three integrals (two components of velocity plus angular momentum with respect to the origin) for free particle motion that is invariant under collisions.

Exercise 6.4.3 They are formed by the corresponding numbers of equal elliptic arcs.

Exercise 6.4.4 $H(S, S_2) + \cdots + H(S_n, S_1)$.

Exercise 6.4.5 Perturb a short arc of the circle.

Problem 6.4.6 Such curves are said to have constant width. They can be obtained by rotating a segment of fixed length around a point that moves along the segment.

Problem 6.4.9 Use a modified version of the string construction.

Exercise 7.1.2 It can be written as p/q, with m and q relatively prime.

Exercise 7.1.4 Consider lifts, interpolate linearly ("straight-line deformation"), and project.

Exercise 7.1.5 If 0 is an attracting fixed point, then
$$x_0 := \sup\{x \in [0, 1] : f(y) \leq y \text{ for } y \in [0, x]\}$$
is an extra fixed point.

Exercise 7.1.9 See the hint to Exercise 7.1.5.

Exercise 7.2.1 1/4.

Exercise 7.2.7 Consider two cases: (i) both eigenvalues of absolute value > 1, and (ii) one eigenvalue of absolute value less than one. In the case (i), the map is expanding in an appropriate norm and the argument goes as in Proposition 7.2.7; in the case (ii), the argument is as in Proposition 7.2.9.

Exercise 7.3.4
$$\begin{pmatrix} 1 & 1 & 0 & 0 & 0 \\ 1 & 1 & 1 & 1 & 0 \\ 0 & 0 & 0 & 1 & 0 \\ 0 & 0 & 1 & 0 & 1 \\ 0 & 0 & 1 & 0 & 1 \end{pmatrix}.$$

Exercise 7.3.8 1.

Exercise 7.3.11 Consider the second preimages of zero.

Exercise 7.3.13 $\frac{1+\sqrt{5}}{2}$.

Problem 7.3.14 Every factor is given by a map E_n.

Problem 7.3.15 It is achieved for

$$A = \begin{pmatrix} 0 & 1 & 0 \\ 1 & 0 & 1 \\ 1 & 0 & 0 \end{pmatrix}.$$

Problem 7.4.8 Diagonalize the matrix, extend the eigenlines until they intersect sufficiently, many times, iterate the partition thus obtained, and take connected components. It is instructive to consider $\begin{pmatrix} 13 & 8 \\ 8 & 5 \end{pmatrix} = \begin{pmatrix} 2 & 1 \\ 1 & 1 \end{pmatrix}^3$ as an example.

Exercise 7.5.2 The counterpart of the crucial formula includes expressions containing factorials. The total number of terms grows cubically, rather than quadratically. Find the biggest "bad" terms as before and estimate factorials using the Stirling formula. The bound decreases exponentially, so the sum of any polynomially growing number of terms decreases exponentially, similarly to (7.5.3).

Exercise 7.6.1 For the one-sided shift, the semiconjugacy to E_2 is invertible iff a null set, so the proof remains the same. For the two-sided shift, it suffices to consider cylinders with positive indices, which is equivalent to considering a one-sided shift.

Exercise 7.6.3 If ε is small enough, the sum (7.6.1) for p and the analogous sum for q correspond to disjoint sets.

Exercise 8.1.2 1.

Exercise 8.1.5 Use the fact that cylinders are balls to construct a minimal cover, and the fact that d''_λ is an ultrametric, that is, that any triangle has at least two equal sides, to show that the covers by cylinders are optimal.

Exercise 8.1.8 Proceed as for the ternary Cantor set. For box dimension 0, take out the middle intervals of relative length $1 - (1/2^n)$, and for box dimension 1 use $1/2^n$.

Exercise 8.1.9 $\frac{\log \frac{1+\sqrt{5}}{2}}{\log 2}$.

Exercise 8.2.1 0.

Exercise 8.2.2 $\log 2$.

Exercise 8.2.3 $h_{\text{top}}(f) = 0$. Use semiconjugacy with an irrational rotation and show that wandering intervals add only a bounded number of (n, ϵ)-separated points for each $\epsilon > 0$.

Exercise 8.2.4 $h_{\text{top}}(f) = 0$. The growth of (n, ϵ)-separated sets is quadratic in n.

Exercise 8.3.3 Consider, for example, the *Hilbert cube*, the direct product of countably many copies of the unit interval indexed by integers. It is a compact space that may be given a metric. Then the shift map on H is topologically transitive and has infinite topological entropy.

A simpler but not topologically transitive example is a disjoint union of shift spaces Ω_N, $N = 2, 3 \ldots$ each with the metric d_2 scaled by 2^{-N} with a point p added to make the space compact, so that for $x \in \Omega_N$ the distance between p and x is equal to $10/2^N$.

Problem 8.3.8 For integer t the statement follows directly from Proposition 8.3.6 and Proposition 8.2.9(3). The latter statement immediately implies equality for rational t. For irrational t use rational approximation and the argument from the proof of Proposition 8.3.6 to obtain inequalities in both directions.

Solutions

Exercise 1.2.3 See Example 2.2.9.

Exercise 1.2.4 Call the sum x_n and write it out, repeating it on the line below, but "shifted" to the right. Adding corresponding terms gives $2x_n = x_n - 1 + a_{n+1} + a_n$, so $x_n = a_{n+2} - 1$.

Exercise 1.2.6 See Proposition 2.5.1.

Exercise 1.2.19 $1/3$. Show that $a_n + 2a_{n+1}$ is independent of n.

Exercise 1.3.1 $(1, 4) \mapsto (5/2, 8/5) \mapsto (41/20, 80/41) \mapsto ((41^2 + 40^2)/41 \cdot 40, 41 \cdot 160/(41^2 + 40^2))$.

Exercise 1.3.6 See Section 2.2.8.

Exercise 1.3.9 $(n + 10)^2 = n^2 + 20n + 100 = n^2$ (mod 10).

Exercise 1.3.10 $(10 - n)^2 = 100 - 10n + n^2 \equiv n^2$ (mod 10).

Exercise 1.3.11 $a_{10q-n} - a_n = (10q - n)^2 p/q - n^2 p/q = 10(10pq - 2np)$ is a multiple of 10.

Exercise 1.3.19 See Proposition 2.2.27.

Exercise 1.3.20 See Proposition 2.2.27.

Exercise 1.3.24 $2^{n+50} + 2^n = 2^n(2^{50} + 1) = 2^n(\ldots 625)$ is a multiple of 125 and of 8 for $n \geq 3$, and hence of 1000.

Exercise 2.1.1 $y_{i+1} = x_{i+1} - (b/1 - k) = kx_i + b - (b/1 - k) = k(y_i + (b/1 - k)) + b - (b/1 - k) = ky_i$.

Exercise 2.2.1 With the radian setting we are evaluating $\sin x$ repeatedly; since $|\sin x| = \sin |x| < |x|$, the sequence $a_n := \sin^n |x|$ is decreasing. Since it is bounded by 0, it converges (Section A.1.2), necessarily to a fixed point [see (2.3.1)], which must be 0. Since the derivative of $\sin x$ is 1 at 0, this map is not a contraction and moreover the ratio of successive terms tends to 1; hence the convergence is not exponential.

If we use degrees, then we are evaluating $\sin(\pi x/180)$, which is a contraction by Proposition 2.2.3, and the ratio of successive terms (quickly) increases to $a := \pi/180$. To get a factor of 10^{-10} therefore requires some $-10/\lg a < 6$ steps.

Exercise 2.2.2 The function $\sqrt{x}$ is a contraction on $[a, \infty)$ for any $a > 1/4$ with fixed point 1, so this is the limit of any such sequence. Since the derivative, $1/2\sqrt{x}$, is $1/2$ at $x = 1$, the difference from 1 about halves at every step; so after some k steps the difference is too small for the calculator, if we started from a moderate number (for a number of size roughly 2^l it takes $k + l$ steps).

Exercise 2.2.3 Since we cannot use the Contraction Principle on $(0, 1]$, we reduce to the previous case by taking reciprocals: If the initial value is x, then $\sqrt[2^n]{x} = 1/\sqrt[2^n]{1/x}$ is about as close to 1 as $\sqrt[2^n]{1/x}$. Thus we have exponential convergence and an initial value roughly 2^{-l} settles to within k binary digits in $k + l$ iterations.

Exercise 2.2.4 $|x^2 - y^2| = |x - y||x + y| \leq \lambda|x - y|$ for $x, y \in [-\lambda/2, \lambda/2]$.

Exercise 2.2.5 Each summer the population triples and then all lemmings die that were alive the previous summer. Thus, $b_{n+1} = 3b_n - b_{n-1}$. Dividing by b_n gives

$$a_n := \frac{b_{n+1}}{b_n} = \frac{3b_n - b_{n-1}}{b_n} = 3 - \frac{1}{a_{n-1}} =: g(a_{n-1}).$$

Now $g(2) = 5/2 > 2$, $g(4) = 11/4 < 4$, and g is increasing, so $g([2, 4]) \subset [2, 4]$. On this interval $g'(x) = 1/x^2 \leq 1/4 < 1$, so g is a contraction with a unique fixed point $\omega \in [2, 4]$. Indeed, $\omega = 3 + 1/\omega$, so $\omega = (3 + \sqrt{5})/2$.

Exercise 2.3.2 The second iterate f^2 is a nondecreasing map. Hence periods can only be 1 or 2. On the other hand, the map $f(x) = -x$ has period-2 points.

Exercise 2.5.3 $f(x) - x$ has a zero by the Intermediate-Value Theorem and is nonincreasing, hence its set of zeros is a point or an interval.

Exercise 2.6.2 If $x \in \bigcup_{\alpha \in A} O_\alpha$, then there is an $\alpha \in A$ such that $x \in O_\alpha$ and hence an $r > 0$ such that $B(x, r) \subset O_\alpha \subset \bigcup_{\alpha \in A} O_\alpha$. Now use that closed sets are the complements of open sets.

Exercise 2.6.3 $\{n \in \mathbb{Z} \mid d(n, 0) < 1\} = \{0\}$ and $\{n \in \mathbb{Z} \mid d(n, 0) \leq 1\} = \{-1, 0, 1\}$. Both are open and closed: $\{n \in \mathbb{Z} \mid d(n, 0) < 1\}$ is open by Exercise 2.6.1, and so is every point and hence every subset of $\mathbb{Z}$ by Exercise 2.6.2. Then every set is closed also because its complement is open.

Exercise 2.6.4 $\{n \in \mathbb{Z} \mid d(n, m) < 1\} = \{m\}$ is open by Exercise 2.6.1 for every m, and so is hence every subset of $\mathbb{Z}$ by Exercise 2.6.2.

Exercise 2.6.5 Interiors are open by Definition A.1.4 and Exercise 2.6.2. To verify that $\bar{A} \subset \bar{A}$ take $x \in \bar{A}$ and $r > 0$. Then there is a $y \in B(x, r/2) \cap \bar{A}$ and a $z \in B(y, r/2) \cap A$ by Definition A.1.4. Then $z \in A \cap B(x, r)$ and $x \in \bar{A}$ by Definition A.1.4.

Exercise 2.6.6 Show from Definition A.1.4 that ∂S is $\bar{S} \setminus \text{Int } S$, which is closed. If S is open, $x \in \partial S$, and $r > 0$, then $B(x, r) \cap S \neq \varnothing$ and $x \notin \text{Int } \partial S$ because $S \cap \partial S = \varnothing$ by Definition A.1.4. Now combine these: A boundary is closed and its boundary is the boundary of its complement, which is open.

Exercise 2.6.7 $\mathbb{R}$ is complete (one of the standard properties), $\mathbb{Q}$ is not because the sequence in Exercise 1.1.8 is a rational Cauchy sequence that does not converge (in $\mathbb{Q}$). $\mathbb{Z}$ and $[0, 1]$ are complete by Exercise 2.6.8. (For $\mathbb{Z}$ this also follows from the easy observation that a Cauchy sequence is eventually constant.)

Exercise 2.7.1 Use Exercise 2.6.8.

Exercise 3.1.2 $0 = \langle Av, w \rangle - \langle v, Aw \rangle = (\lambda - \mu)\langle v, w \rangle$.

Exercise 3.2.1 (a) Expanding node, (b) saddle, (c) expanding focus, and (d) expanding degenerate node.

Exercise 3.3.1 Here are the characteristic polynomials: (a) $-\lambda((3 - \lambda)(-3 - \lambda) + 8) + (2(-3 - \lambda) + 8) + 2(4 - 2(3 - \lambda)) = -\lambda(\lambda^2 - 1) + 2\lambda - 2 = (\lambda - 1)[-\lambda(\lambda + 1) + 2] = (1 - \lambda)[\lambda^2 + \lambda - 2] = (1 - \lambda)(\lambda - 1)(\lambda + 2)$; (b) $(-1 - \lambda)[(-1 - \lambda)^2 - 1] + 2(-1 - \lambda) = -(1 + \lambda)[(1 + \lambda)^2 + 1]$; (c) $(2 - \lambda)[(1 + \lambda)^2(2 - \lambda) + (2 - \lambda)] = (2 - \lambda)^2[(1 - \lambda)^2 + 1]$.

Exercise 4.1.8 The intersection of any two such sets is an invariant closed set that is a subset of each of them.

Exercise 4.1.10 Let $X = \{z \in \mathbb{C} \mid |z| \in \{1, 2\}\}$ and $f(z) = e^{2\pi i \alpha}$ for some $\alpha \in \mathbb{R} \setminus \mathbb{Q}$.

Problem 4.1.15 For transitivity it suffices to show that 1 is in the orbit closure. Every $g \in \mathbb{Z}_2 \setminus \mathbb{Z}_2^+$ is a limit of odd integers. For m odd and $n \in \mathbb{N}$ there exists $k \in \mathbb{N}$ such that $mk = 1 \pmod{2^n}$.

Exercise 4.2.8 If the lower bound of positive elements of Γ is a positive number a, then any element of the group is a multiple of a. Otherwise, there are both positive and negative numbers in Γ arbitrarily close to zero, and hence Γ is dense.

Exercise 4.3.3 $F(x + 1) - F(x) = 1/2(\sin(x + 1) - \sin x)$. This is a nonconstant continuous function, hence F is not the lift of any circle map.

Problem 4.4.4 The intersection is an interval by monotonicity. To show that it is nonempty we observe that $\rho_{0,b} = 0$ and $\rho_{1,b} = 1$ and use continuity. To obtain positive length we notice that Proposition 4.4.10 applies because $f_{a,b}$ is an entire function.

Problem 4.4.6 Use Propositions Proposition 4.4.9 and Proposition 4.4.10 as in the proof of Proposition 4.4.13.

Exercise 5.1.1 Otherwise $\sqrt{3}$ is a rational combination of 1 and $\sqrt{5}$. Squaring both sides of this equation implies that 3 is irrational.

Exercise 6.1.1 Preservation of orientation and length implies that $f(x) - f(0) =$ length$([f(0) f(x)]) = $ length$([0, x]) = x - 0$, so $f(x) = x + f(0)$.

Exercise 6.1.8 Consider $\mathbb{Q}$ and $\{O_q := \mathbb{Q} \setminus \{q\} \mid q \in \mathbb{Q}\}$.

Problem 6.1.10 The set of points whose orbits intersect any given open set is open by definition and dense because this set contains any dense orbit. Taking intersection over a countable set of balls of rational radius centered at the points of a dense orbit, one sees that the set of points whose orbits intersects all such balls (and hence are dense) is the intersection of countably many open dense sets. The statement follows by passing to the complements.

Exercise 6.2.5 The Lagrange equations are invariant with respect to reflection in a great circle. Hence their solutions (that is, geodesics) are invariant in the sense that the image of a geodesic under a reflection is again a geodesic. If the initial condition is tangent to a great circle, then uniqueness of solutions of the Lagrange equation forces the great circle to be a geodesic. Since every tangent vector is tangent to a great circle, those are the only geodesics. Thus the geodesic flow is periodic.

Exercise 6.4.1 Connect the points of intersection of the billiard table with the symmetry axes in the counterclockwise direction. This figure is invariant under both symmetries, hence at each point of intersection with the table the angle of incidence is equal to the angle of reflection.

Exercise 7.1.4 Every periodic point x is uniquely defined (coded) by the sequence of its visits to Δ_0 or Δ_1. Conversely, every sequence of zeroes and ones of length n produces exactly one period-n orbit. If, in particular, such a sequence is not periodic with a smaller period, then the corresponding periodic orbit has period exactly n.

Exercise 7.1.10 Describe the recursion relation as in Section 3.1.9 and project it to the torus $\mathbb{R}^2/(10\mathbb{Z})^2$, that is, mod 10. Alternatively, consider the orbit of $(1/10, 1/10)$ on the standard torus. What is the period? This is a careful way of using the fact that the last digit of a Fibonacci number is uniquely determined by the last digits of the two preceding ones.

Exercise 7.2.6 Let $p \in U, q \in V$ be periodic points (we use Proposition 7.1.10) and n their common period. The line from the first family passing through p is dense, as is the line from the second family passing through the point q. Hence each line intersects the other in a dense set of points that are thus heteroclinic.

Exercise 7.3.2 Show that the set of a binary expansion $0, a_1, a_2, \ldots$, for which $a_i = 1$ when i is not a multiple of 20, is uncountable.

Exercise 7.3.12 Draw the Markov graph for

$$\begin{pmatrix} 1 & 1 \\ 1 & 0 \end{pmatrix}.$$

Then consider its arrows as points and draw the Markov graph that shows which of the arrows in the first graph can be followed by which others. Compare with the Markov graph for

$$\begin{pmatrix} 1 & 1 & 0 \\ 0 & 0 & 1 \\ 1 & 1 & 0 \end{pmatrix}.$$

Exercise 7.4.1 The second iterate of every point outside $[0, 1]$ is negative. Any negative point goes to $-\infty$.

Exercise 7.4.3 Verify that

$$\begin{pmatrix} 2 & 1 \\ 1 & 1 \end{pmatrix} = \begin{pmatrix} 1 & 1 \\ 1 & 0 \end{pmatrix}^2$$

and check that the partition in Figure 7.4.4 is a Markov partition for

$$\begin{pmatrix} 1 & 1 \\ 1 & 0 \end{pmatrix}.$$

Problem 7.4.7 Assuming that the semiconjugacy is not bijective, we conclude that the image of an interval is a single point. But then the image of any iterate of that interval is also a single point. Using the Mean-Value Theorem one shows that the lift of f increases the length of any interval. By compactness there is a longest interval that is mapped to a point, a contradiction.

Exercise 8.1.1 The interval $[0, 3]$ can be covered by two balls of radius 1, for example, balls centered at $2/3$ and $8/3$, whereas the cover by 1-balls centered at $1/2$, $3/2$, and $5/2$ is minimal.

Exercise 8.1.3 $\log \mu < \log 4 + \log \mu < \log 4 - \log \lambda < \log 4 - \log 2 = \log 2 < - \log \mu$.

Exercise 8.3.2 Consider the direct product of any map f with positive topological entropy and an irrational rotation R_α. By Proposition 8.2.9(5), $h_{\text{top}}(f \times R_\alpha) = h_{\text{top}}(f) + h_{\text{top}}(R_\alpha) = h_{\text{top}}(f) > 0$.

Index

Printed in the United States
By Bookmasters